HEAT AND MASS TRANSFER IN THE BIOSPHERE

PART 1　　TRANSFER PROCESSES IN THE PLANT ENVIRONMENT

ADVANCES IN THERMAL ENGINEERING

Editors:
JAMES P. HARTNETT
THOMAS F. IRVINE, JR.

I	Blackshear	●	Heat Transfer in Fires: thermophysics, social aspects, economic impact
II	Afgan and Beer	●	Heat Transfer in Flames
III	deVries and Afgan	●	Heat and Mass Transfer in the Biosphere — Part 1 Transfer Processes in the Plant Environment
IV	Eckert and Goldstein	●	Measurements Techniques in Heat Transfer 2nd revised and augmented edition (in press)
V	Gutfinger	●	Current Topics in Thermal Sciences (in press)
VI	Hsu and Graham	●	Boiling Heat Transfer in Two Phase Flows (in press)
VII	Ginoux	●	Two Phase Flows in Power Machinery (in press)
VIII	Yovanovich	●	Advanced Heat Conduction (in preparation)
IX	Pfender	●	Phenomena in Electrical High-Intensity Discharges (in preparation)

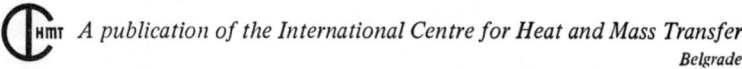 *A publication of the International Centre for Heat and Mass Transfer*
Belgrade

Institutional Members:

American Geophysical Union
American Institute of Chemical Engineers
American Society of Mechanical Engineers
Associazione Termotecnica Italiana
Canadian Society for Chemical Engineering
Canadian Society for Mechanical Engineering
Egyptian Society of Engineers
Indian National Committee for Heat and Mass Transfer
Institution of Chemical Engineers, London
Institution of Engineers of Australia
Institution of Mechanical Engineers, London
Israel Institute of Chemical Engineers
Koninklijk Instituut van Ingenieurs, Netherlands
National Committee for Heat and Mass Transfer of the
    Academy of Sciences of the USSR
Society of Chemical Engineers of Japan
Societé Française des Thermiciens
Verein Deutscher Ingenieure
Yugoslav Society of Heat Engineers

HEAT AND MASS TRANSFER IN THE BIOSPHERE

PART 1   TRANSFER PROCESSES IN THE PLANT ENVIRONMENT

EDITORS

D. A. deVries
Eindhoven University of Technology

N. H. Afgan
University of Belgrade

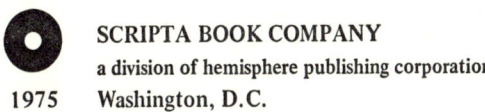

SCRIPTA BOOK COMPANY
a division of hemisphere publishing corporation
1975   Washington, D.C.

A HALSTED PRESS BOOK
JOHN WILEY & SONS
New York   London   Sydney   Toronto

Library of Congress Cataloging in Publication Data

Main entry under title:

Heat and mass transfer in the biosphere
Part 1   Transfer processes
          in the plant environment.

   (Heat and mass transfer in the biosphere; pt. 1)
(Advances in thermal engineering; v. 3)
   "Lectures and papers presented at the seminar on
'heat and mass transfer in the environment of vegetation'
... organized by the International Centre for Heat and
Mass Transfer and held at Dubrovnik, August 26-30, 1974".
   Includes index.
   1. Plant physiology—Congresses. 2. Botany—
Ecology—Congresses. 3. Bioenergetics—Congresses.
4. Heat—Transmission—Congresses. 5. Mass transfer—
Congresses. 6. Soils—Thermal properties—Congresses.
I. DeVries, Daniel A., ed. II. Afgan, Naim, ed.
III. International Centre for Heat and Mass Transfer.
IV. Series.
QH510.H4   pt.1   [QK754]   574.1'9121s   [581.1'9121]
ISBN 0-470-20985-2   74-28066

HEAT AND MASS TRANSFER IN THE BIOSPHERE

**Part 1   TRANSFER PROCESSES
           IN THE PLANT ENVIRONMENT**

Copyright © 1975 by Scripta Book Company. All rights reserved.
No part of this publication may be reproduced, stored in a
retrieval system, or transmitted, in any form or by any
means, electronic, mechanical, photocopying, recording, or
otherwise, without the prior written permission of the
publisher.

Scripta Book Company
a division of hemisphere publishing corporation
1025 Vermont Avenue, N.W.
Washington, D.C. 20005

Printed in the United States of America

CONTENTS

FOREWORD ix

PART I BASIC PROCESSES AND METHODS OF OBSERVATION

1 SOIL

1. Heat Transfer in Soils
   D. A. de Vries 5
2. Water Movement in Soil
   J. R. Philip 29
3. Thermodynamic and Rheological Peculiarities of Soil Water and Their Role in Energy- and Mass Transfer
   S. V. Nerpin 49
4. Heat and Water Transfer in a Natural Soil Environment
   R. D. Jackson, B. A. Kimball, R. J. Reginato, S. B. Idso and F. S. Nakayama . . . 67
5. The Early Stages of Infiltration into a Swelling Soil
   D. E. Smiles and P. M. Colombera 77
6. Simultaneous Heat and Mass Transfer in Soils with Subsurface Heated Porous Pipes
   D. L. Slegel, L. R. Davis and L. Boersma 87
7. Comments on Computer Modeling of a Moist Soil
   G. S. Vansteenkiste and F. de Schutter 97
8. Simulation of the Thermal Behaviour of Bare Soils for Remote Sensing Purposes
   A. Rosema 109
9. Surface Phenomena Connected with Evaporating Water and Condensing Water Vapour in Thin Capillaries
   N. V. Churaev 125

2 LOWER ATMOSPHERE

10. Aerodynamics of Vegetated Surfaces
    J. A. Businger 139
11. Heat and Mass Transfer Within Plant Canopies
    B. Legg and J. L. Monteith 167
12. Radiative Transfer in Vegetation
    J. M. Norman 187
13. General Principles of Natural Evaporation
    F. Kreith and W. D. Sellers 207
14. Methods of Observation of Heat and Mass Transfer in the Lower Atmosphere and in Plant Canopies
    A. Perrier 229
15. Simulation of Flow Above Forest Canopies
    W. Z. Sadeh 251
16. Energy and Mass Transfer in Vegetation by Electrochemical Analog
    P. H. Schuepp and K. D. White 265

17	Microclimatic Modeling of the Desert	
	J. W. Mitchell, W. A. Beckman, R. T. Bailey and W. P. Porter	275
18	An Approximate Analysis of the Momentum Balance for the Air Flow in a Pine Stand	
	J. D. Bergen	287
19	A Field Study of Atmospheric Exchange Processes Within a Vegetative Canopy	
	G. den Hartog and R. H. Shaw	299
20	Energy and Mass Exchange of a Native Grassland in Saskatchewan	
	E. Ripley and B. Saugier	311
21	Radiation Exchange in Plant Canopies	
	J. Ross and T. Nilson	327
22	An Eddy Correlation Method for the Determination of Momentum, Heat and Mass Transfer, Using Hot-Wire Anenometry	
	A. Baille and J. P. Chiapale	337
23	Measurement of Atmospheric Infrared Radiant Flux and Testing of Some Empirical Formulae for Estimating This Flux	
	C. L. Palland	345
24	Laser-Doppler Anemometry and Its Application to Flow Investigations in the Environment of Vegetation	
	F. Durst, G. Wigley and M. Zare	353

3 PLANTS

25	Water Transfer in Plants	
	P. G. Jarvis	369
26	Water Transport in Wheat	
	O. T. Denmead and B. S. Millar	395
27	Water Vapour Diffusion Porometry for Leaf Epidermal Resistance Measurements in the Field	
	C. J. Stigter	403
28	Heat and Mass Transfer From Real and Model Leaves	
	J. A. Clark and G. Wigley	413

PART II   APPLICATIONS

1 PHYTO-ENGINEERING

29	Energy and Mass Transfer in Plant Communities	
	A. A. Nichiporovich	427
30	Water Uptake by Vegetation	
	W. R. Gardner, W. A. Jury and J. Knight	443
31	A Numerical Method for Estimating the Modification of Heat Budget Introduced by Hedges	
	J. P. Chiapale	457
32	Modification of Land Roughness and Resulting Microclimatic Effects: A Field Study in Brittany	
	G. Guyot and B. Seguin	467
33	Reflectant Induced Modification of the Radiation Balance for Increased Crop Water Use Efficiency	
	R. Lemeur and N. J. Rosenberg	479
34	The Use of Anti-Transpirants to Control Water Consumption in Eco-Systems; An Experimental Study of Short- and Long- Term Effectiveness of Various Transpiration-Reducing Chemicals	
	F. Kreith and A. Taori	489

Contents

35 Water Transfer to Germinating Seeds as Affected by Soil Hydraulic Properties and Seed–Water Contact Impedance
    A. Hadas 501
36 Energy and Agriculture: A National Case Study
    G. Stanhill 513

2 POLLUTION IN THE PLANT ENVIRONMENT

37 Problems of Chemical Reaction and Biological Processes in Soils
    D. E. Elrick, P. H. Groenevelt and T. J. M. Blom 537
38 Prediction of Soil- and Ground-Water Pollution
    L. Wartena 549
39 Pollution in Plant Canopies
    A. C. Chamberlain 561
40 Transport of Micronic Particles from Atmosphere to Foliar Surfaces
    Y. Belot and D. Gauthier 583

INDEX 593

FOREWORD

This volume is the first in a series dealing with heat and mass transfer in the biosphere, to be published by Scripta Book Company. It contains the lectures and papers presented at the Seminar on "Heat and Mass Transfer in the Environment of Vegetation," which was organized by the International Centre for Heat and Mass Transfer.*

Part I of the present volume deals with transfer processes occurring in the soil, in the lower atmosphere, and in plants themselves. Observational methods and measuring techniques are also treated in this part.

Part II deals first with methods aiming at an increase of crop productivity, for which the term *phyto-engineering* is introduced. Second, it deals with pollution of soil water and the plant canopy in relation to plant growth and productivity.

The aim of this book as well as of the Seminar is to stimulate interest in this highly important field of research and engineering, a field that is of importance to agricultural scientists and engineers, micrometeorologists, hydrologists, crop physiologists, and physical ecologists.

The past decade has shown an increasing interest in environmental problems among scientists and engineers of various disciplines and among the public in general. This volume can help in presenting the present state of knowledge on transfer processes in the plant environment and in pointing out unsolved problems and avenues of further research.

Each part in this volume starts with the lectures presented at the Seminar, which also contain extensive lists of references. These are followed by short communications, many of which report on advanced methods of observation and computation, in addition to presenting new observational material.

It is my pleasure to acknowledge the important role played by my colleagues in the Seminar Committee in designing the seminar program, in suggesting names of lecturers, and in making a final selection of papers for this volume. They are: J. A. Businger, D. E. Elrick, W. R. Gardner, F. Kreith, J. L. Monteith, S. V. Nerpin, A. Perrier, and J. R. Philip. Each of them is an author or coauthor of one of the lectures. In particular Dr. J. R. Philip, Fellow of the Royal Society, had an important share in the design of the program during the early stages of preparation.

Also I acknowledge the cooperation of the Scientific Secretary of the International Centre for Heat and Mass Transfer, Professor N. H. Afgan, of the Staff of the Centre at Belgrade, and of Mr. W. Begell, Publisher of the Scripta Book Company.

Daniel A. deVries

*Held at Dubrovnik, August 26–30, 1974.

HEAT AND MASS TRANSFER IN THE BIOSPHERE

PART 1 TRANSFER PROCESSES IN THE PLANT ENVIRONMENT

PART I
BASIC PROCESSES AND METHODS OF OBSERVATION

SECTION 1
SOILS

HEAT TRANSFER IN SOILS

DANIEL A. de VRIES

Eindhoven University of Technology,
The Netherlands

ABSTRACT

Heat is transferred in soils mainly by conduction. Transport of latent heat by water vapour diffusion in the air-filled pores must also be taken into account.

The calculation of the overall thermal conductivity and the volumetric heat capacity of a soil from its composition is treated.

Some general features of the storage and release of heat during the annual and diurnal temperature cycles and their influence on the energy balance of the earth's surface are discussed. The penetration of frost into a soil is touched upon.

The main unsolved problems in the field of soil heat transfer are connected with the combined transfer of heat and moisture in soils. Present theories and their limitations are briefly discussed and suggestions for further work are given.

LIST OF SYMBOLS OF QUANTITIES AND THEIR SI—UNITS

a	thermal diffusivity	$m^2 \cdot s^{-1}$
a_j	axis of ellipsoid	m
c	specific heat	$J \cdot kg^{-1} \cdot K^{-1}$
C	volumetric heat capacity	$J \cdot m^{-3} \cdot K^{-1}$
d	damping depth	m
D	molecular diffusion coefficient	$m^2 \cdot s^{-1}$
D_T	macroscopic thermal diffusion coefficient	$m^2 \cdot s^{-1} \cdot K^{-1}$
D_θ	macroscopic moisture diffusion coefficient	$m^2 \cdot s^{-1}$
E	evaporation rate	s^{-1}
g	acceleration of free fall	$m \cdot s^{-2}$

g_j	shape factor	1
h	relative humidity	1
H	specific enthalpy	$J \cdot kg^{-1}$
k	weight factor	1
K	hydraulic conductivity	$m \cdot s^{-1}$
L	heat of vaporization	$J \cdot kg^{-1}$
M	molar mass	$kg \cdot mol^{-1}$
p	pressure	Pa
q	heat flux density	$W \cdot m^{-2}$
R	gas constant	$J \cdot K^{-1} \cdot mol^{-1}$
t	time	s
T	Temperature	$K, {}^\circ C$
x	volume fraction	1
z	vertical co-ordinate (positive downward)	m
θ	volumetric moisture content	1
Θ	temperature amplitude	${}^\circ C$
λ	thermal conductivity	$W \cdot m^{-1} \cdot K^{-1}$
ρ	density	$kg \cdot m^{-3}$
ϕ	mass flow density	$kg \cdot m^{-2} \cdot s^{-1}$
Φ	total moisture potential	m
Ψ	matrix potential	m
ω	circular frequency	s^{-1}

Subscripts

a	−	air	o	−	organic
i	−	number (1,2,...)	s	−	solid
j	−	number (1,2,3)	v	−	vapour
l	−	liquid	vs	−	vapour, saturated
m	−	mineral	w	−	water

1. INTRODUCTION. MECHANISMS OF THE HEAT TRANSFER IN SOILS

Heat transfer in soils is of importance in a description of plant environment in a number of ways. First of all it is one of the important terms in the energy balance of the earth's surface and consequently it directly influences the temperature regimes near the surface, both in the upper soil and lower air layers. Thus the entire biosphere is affected.

Secondly in many instances there exists a strong coupling between the heat transfer and the moisture transfer in the soil. The processes of evaporation and dew formation are therefore influenced by the thermal behaviour of the soil.

Thirdly the heat flow in the soil can be influenced, be it not controlled, by such measures as mulching, tillage, irrigation and drainage. Although generally these measures are not taken specifically for manipulating the soil heat transfer, this aspect can play an important role.

Some other fields of application of the study of heat transfer in soils, which are more of an engineering nature, are: heat transfer from underground electrical cables and heat transfer underneath buildings and roads, especially when accompanied by frost and thaw.

The mechanisms of heat transfer in soil are in order of importance: conduction, convection and radiation. Conduction occurs throughout the soil but the main flow of heat is through the solid and liquid parts. Convection in the usual sense is most times negligible, with the exception of rapid infiltration of water after irrigation or heavy rain. However, this does not hold for the transport of latent heat by water vapour, which contributes greatly to the heat transfer in the gasfilled pores. Radiation heat transfer is only of importance in dry soils at high temperatures and within large pores.

Most of the simple theory of heat transfer in soils deals with solving the heat conduction equation under given boundary and initial conditions or for periodic variations of temperature. In this case the main complications are caused by the inhomogeneity of the soil and the fact that the boundary condition at the surface is difficult to specify.

When the combined transport of heat and moisture must be considered, two coupled partial differential equations must be solved, which are essentially non-linear. The two dependent variables are temperature and moisture content. The situation is further complicated by the occurrence of hysteresis in the moisture transfer.

The thermal properties of soil will be dealt with in section 2 of this paper. Some simplified cases of soil heat transfer are treated in section 3 and the problems associated with the combined heat and moisture transport in section 4.

2. THERMAL PROPERTIES OF SOILS

2.1. Microscopic and macroscopic approaches

The differential equation for heat conduction in soils is written as

$$C \, \partial T/\partial t = \nabla(\lambda \nabla T) \tag{2.1}$$

Symbols of the quantities occurring in this equation are explained in the list at the beginning of this paper. Since the soil usually is not homogeneous, λ cannot be considered as a constant with respect to the spatial co-ordinates.

The application of (2.1) to a composite porous medium such as a soil requires some explanation. For each homogeneous part of the medium one should in principle solve a heat conduction equation with the proper initial condition and boundary conditions at the interfaces between such parts. This would already be an impossible task for a moist granular medium with a well defined geometric structure, e.g. a close packing of spheres, even in the absence of the complication of moisture transfer. Such a "microscopic" approach must be discarded altogether in the case of soils.

Instead, for both heat and moisture transfer one applies a "macroscopic" approach in which one envisions the existence of a unit cell of soil with linear dimensions that are large as compared with the grain and pore sizes and also large enough to contain a representative sample of the soil constituents. On the other hand the linear dimensions of a cell must be small in comparison with the length scales that can be attached to the variations of temperature and composition, including the moisture content.

Next "overall" values of the physical properties of a unit cell are introduced. These depend, of course, on the composition, the pore structure and the transfer phenomena found in the unit cell. A physical theory will aim at constructing a model from which the values of the overall physical properties can be calculated. If such a model does not exist one must have recourse to experiment for their determination. The validity of a model must, of course, always be tested against experiments.

In the case of heat conduction the macroscopic physical properties involved are the volumetric heat capacity and the thermal conductivity. Models for calculating these quantities are discussed in the remaining part of section 2.

2.2. The volumetric heat capacity of soils

The heat capacity of a unit cell can be found by addition of the heat capacities of the various phases. We distinguish between a number (subscript i) of solid phases (subscript s), liquid water (subscript w) and air (subscript a), and write:

$$C = \sum_i x_{si} C_{si} + x_w C_w + x_a C_a \qquad (2.2)$$

where x denotes the volumetric fraction of a phase. The summation extends over the various constituents of the solid phase, i.e. quartz, other minerals and organic matter. Each C-value is found as the product of the density, ρ, and the specific heat, c.

Most soil minerals have about the same densities and specific heats (see (1) for detailed information). For simplicity the various kinds of organic matter are considered to form a single phase. Representative values of ρ, C and λ are given in table 2.1. From these it will be clear that the contribution of air can be neglected. The values given are those for dry air, whereas the gas-filled pores also contain water vapour. The influence of this water vapour on the thermal properties of the air is neglected here.

Distinguishing only between mineral (subscript m) and organic (subscript o) matter, we have:

$$C = x_{sm} C_{sm} + x_{so} C_{so} + x_w C_w \qquad (2.3)$$

Table 2.1
Thermal properties and densities of soil materials, water and air at $10^{\circ}C$ and of ice at $0^{\circ}C$.

Substance	$\rho (kg/m^3)$	$C(J/m^3 \cdot K)$	$\lambda (W/m \cdot K)$
Quartz	$2.66 \cdot 10^3$	$2.0 \cdot 10^6$	8.8
Other minerals	$2.65 \cdot 10^3$	$2.0 \cdot 10^6$	2.9
Organic matter	$1.3 \cdot 10^3$	$2.5 \cdot 10^6$	0.25
Water	$1.0 \cdot 10^3$	$4.2 \cdot 10^6$	0.57
Ice	$0.92 \cdot 10^3$	$1.9 \cdot 10^6$	2.2
Air	1.25	$1.25 \cdot 10^3$	0.025

with

$$x_{sm} + x_{so} + x_w = 1 - x_a \qquad (2.4)$$

For a typical mineral soil with $x_{sm} = 0.55$ and $x_{so} = 0$, C-values range from 1.1 MJ/(m^3·K) at complete dryness ($x_w = 0$) to 3.0 MJ/(m^3·K) at saturation ($x_w = 0.45$). For a typical peat soil, with $x_{sm} = 0$ and $x_{so} = 0.20$, C-values range from 0.5 MJ/(m^3·K) at complete dryness to 3.4 MJ/(m^3·K) at saturation ($x_w = 0.80$).

2.3. Thermal conductivity of soils

Table 2.1 shows large differences in the thermal conductivity values of the various soil constituents. In general the thermal conductivity of a soil depends in a rather complicated way on its composition, in particular its water content, x_w. Typical curves for a quartz sand, a loam and a peat soil are given in fig. 2.1.

The general trend of the curves for mineral soils can be understood qualitatively as follows. At complete dryness the heat flow passes mainly through the grains, but has to bridge the air-filled gaps between the grains around their contact points (fig. 2.2a). At very low water contents the soil particles are covered by thin adsorbed water layers (not shown in fig. 2.2). The thickness of these layers increases with increasing water content. At a certain x_w liquid water rings start to form around the contact points between grains; they show a curved air-water interface (fig. 2.2b). From this point on the thermal conductivity increases rapidly with increasing x_w, until the rings almost completely fill the original gap (fig. 2.2c). When x_w increases still further complete pores are filled with water, up to saturation. This is reflected by the slower increase of λ with x_w.

A physical model for calculating the thermal conductivity of soils in dependence of their water content was given by the author (1,2,3). The model was based on ideas put forward by Burgers (4) and others (see 2 for full references) The soil is considered to consist of water as a continuous medium in which soil grains and small air pockets are distributed. The overall thermal conductivity of the soil can be expressed as:

$$\lambda = \frac{x_w \lambda_w + \sum_i k_i x_i \lambda_i + k_a x_a \lambda_a}{x_w + \sum_i k_i x_i + k_a x_a} \qquad (2.5)$$

Heat Transfer in Soils

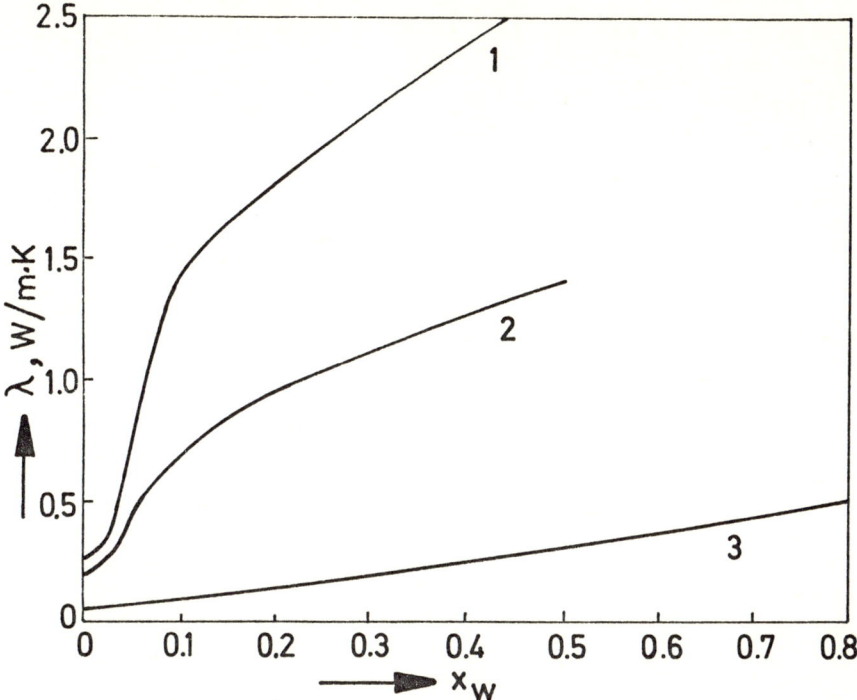

Fig. 2.1. Thermal conductivity, λ, in relation to volumetric soil water content x_w, at $10°C$. Curve 1: quartz sand with $x_s = 0.55$. Curve 2: loam with $x_s = 0.50$. Curve 3: peat soil with $x_s = 0.20$.

Here the summation extends over the different solid soil constituents, which are now characterized by their thermal conductivity and shape. The multiplication factors k_i are easily seen to represent the ratio of the space average of the temperature gradient in the soil grains of kind i and the space average of the temperature gradient in the water. k_a similarly represents this ratio for

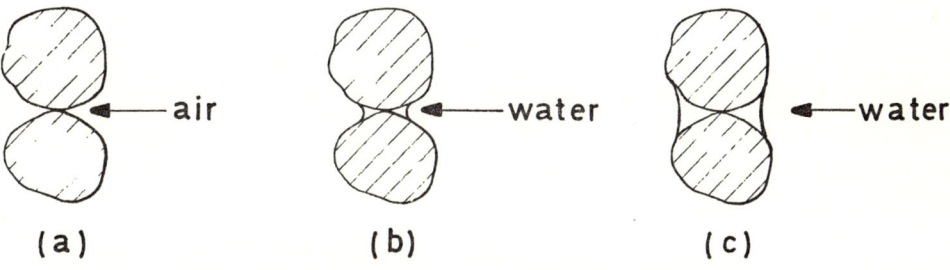

Fig. 2.2 Formation of water rings (see text).

the gradients in the air and the water in the soil. Of course, $k_i < 1$ if $\lambda_i/\lambda_w > 1$ and reversely.

For a grain in the form of an ellipsoid (with principal axes a_1, a_2, a_3) one has for a temperature gradient in the direction of a_j:

$$k_{ij} = \left[1 + \left(\frac{\lambda_i}{\lambda_w} - 1\right) g_1\right]^{-1} \tag{2.6}$$

with

$$g_j = \tfrac{1}{2} a_1 a_2 a_3 \int_0^\infty \frac{du}{(a_1^2+u)^{\frac{3}{2}} (a_2^2+u)^{\frac{1}{2}} (a_3^2+u)^{\frac{1}{2}}} \tag{2.7}$$

and $\qquad g_1 + g_2 + g_3 = 1 \tag{2.8}$

For a random distribution of the axial directions in space one has

$$k_i = \tfrac{1}{3} (k_{i1} + k_{i2} + k_{i3}) \tag{2.9}$$

Hence, k is a weight factor depending on the thermal conductivity and the shape of the enclosure.

In most cases a good result is obtained by considering the soil grains to be spheroids with axes $a_1 = a_2 = na_3$. Then they are characterized by a single shape factor $g_1(n)$.

For the air enclosures this shape factor is deduced by interpolation between $\tfrac{1}{3}$ (spherical enclosures) near water saturation, to a value for an oblate spheroid with n of the order of 10 at low water contents. Space does not permit to elaborate this point and the reader is referred to (1) for a detailed description.

The influence of latent heat transfer in the air-filled pores is proportional to the temperature gradient in these pores. It can therefore be taken into account by adding to the thermal conductivity of air an apparent conductivity due to the evaporation and condensation of water vapour. This apparent conductivity rises rapidly with increasing temperature. Its value is treated in section 4.

The model breaks down at low moisture contents, when water no longer can be considered as a continuous medium that envelops the other consituents. This usually occurs at x_w-values somewhere in the lower half of the steepest part of the λ versus x_w curve.

At complete dryness and at very low water contents the same model applies with air as the continuous medium. However, the theory needs correction when the ratio of the conductivity of the enclosures to that of the continuous medium becomes of the order of 10^2. This applies to λ_s/λ_a for mineral soils. The corrected values are found by introducing a semi-empirical multiplication factor 1.25 on the right hand side of eq. (2.5).

Values of λ for x_w-values between dryness and the value where water can be considered as the continuous medium must be found by interpolation.

The model allows the λ-values to be found with an accuracy of usually better than 5%, except in the interpolation range, where the error becomes of the order of 10%. It has been extensively tested by the author and by other investigators, both in the laboratory and in the field. Since the model rests on a clear physical basis, it can only go wrong when the configuration is different from the one assumed. Its application is clearly restricted to granular media and it does not apply, for instance, to porous media in which the solid matrix and the pore space each have a continuous structure.

The model can also be applied in two steps to calculate the thermal conductivity of a soil with aggregated particles (5). In the first step the conductivity of the aggregates is calculated, in the second step that of the soil as a whole.

3. TEMPERATURE VARIATION AND HEAT FLUX IN SOILS

3.1. General considerations

Equation (2.1) can be solved by known methods when λ and C are constant or known functions of the space co-ordinates and when appropriate initial and boundary conditions are given (see (6)). In practice these requirements can only be met for soils under exceptional circumstances. For a real soil under field conditions the soil composition changes in all directions and with time. Hence, this also applies to the thermal properties of the soil. In addition they are temperature dependent, so that the differential equation for temperature is non-linear. Modern numerical methods in combination with computer use are able to cope with these difficulties, at least in principle, but in practice their application is usually limited to cases where the temperature varies in one dimension only.

The question of defining proper boundary conditions is even more complicated. At the soil surface the heat flux has to comply with the energy balance equation. This means that one has to deal with a complex boundary condition, governed by radiation and heat and mass transfer in the atmosphere. At a certain depth in the soil one may set the temperature at a constant value, which has to be measured or estimated.

The corresponding question for heat transfer from the surface to the air is even more complicated, because one has to account for free and forced convection and evaporation into the atmospheric boundary layer. An appropriate upper boundary condition is also difficult to specify. Finally, in the case of a vegetated surface a plant-air layer separates the soil surface from the atmosphere. An attempt to find solutions under simplified boundary conditions for a bare soil is presented by Rosema (7) at this Seminar.

The specification of initial conditions meets with problems of a similar kind. Stationary solutions have no meaning for field application. Fortunately periodic solutions do and these will be discussed below. They allow us to draw a number of useful conclusions even by considering a greatly simplified case.

For many applications no great accuracy in estimating the soil heat flux is required. This holds, for instance, for most energy balance considerations.

3.2. Periodic variations of temperature

Variations of soil composition, including those of water content, are usually much more pronounced in the vertical than in a horizontal direction. Therefore, equation (2.1) is simplified to the one-dimensional case:

$$C \frac{\partial T}{\partial t} = \frac{\partial}{\partial z}\left(\lambda \frac{\partial T}{\partial z}\right) \tag{3.1}$$

When C and λ are uniform in depth and constant in time we arrive at the simple diffusion equation

$$\partial T/\partial t = a\, \partial^2 T/\partial z^2 \tag{3.2}$$

where $a = \lambda/C$ is the thermal diffusivity of the soil. Values of this quantity for the three soils of fig. 2.1 as a function of the water content are given in fig. 3.1. For the mineral soils the thermal diffusivity shows a maximum value at a relatively low value of x_w. For the organic soil it changes little over the whole range of water content.

For a sinusoidal variation of surface temperature we obtain, with the boundary conditions

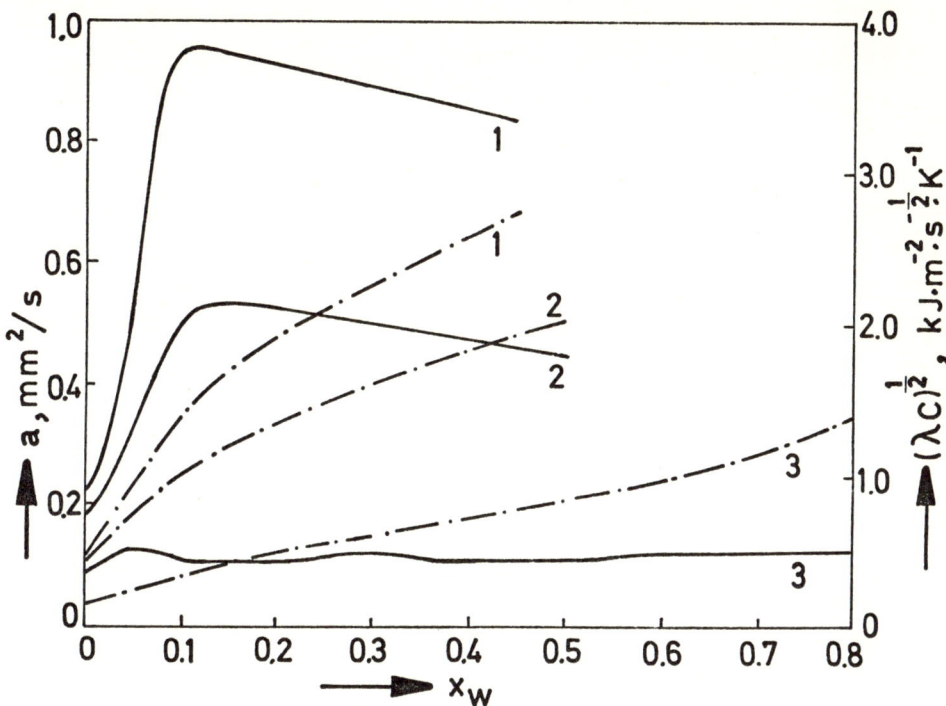

Fig. 3.1 Values of thermal diffusivity, a (solid lines), and $(\lambda C)^{\frac{1}{2}}$ (broken lines), in relation to x_w for the soils of fig. 2.1.

$$T(t,0) = T_a + \Theta_o \cos\omega t$$
$$T(t,\infty) = T_a = \text{constant} \quad (3.3)$$

the well-known solution:

$$T(t,z) = T_a + \Theta_o e^{-z/d} \cos(\omega t - z/d) \quad (3.4)$$

with $\quad d = (2a/\omega)^{\frac{1}{2}} \quad (3.5)$

d is called the damping depth for obvious reasons. The heat flow density is found as:

$$q(t,z) = -\lambda \partial T/\partial z = \Theta_o (\lambda C \omega)^{\frac{1}{2}} e^{-z/d} \cos(\omega t - z/d + \pi/4) \quad (3.6)$$

For z = 0 we have

$$q(t,0) = \Theta_o (\lambda C \omega)^{\frac{1}{2}} \cos(\omega t + \pi/4) \quad (3.7)$$

One of the important effects of heat transfer in the soil is the storage of heat during day time and its release at night. The same applies to storage during spring and summer and release during autumn and winter. The greatly simplified solution given above allows us to make an estimate of the order of magnitude of the heat fluxes involved.

The penetration depth of temperature "waves" in soil is governed by the quantity d. At a depth of 5d the temperature variations are almost completely damped out. For the diurnal variation $\omega = 2\pi/86400 \text{ s}^{-1} = 0.727 \cdot 10^{-4} \text{ s}^{-1}$ and d ranges from 0.080 m till 0.162 m for the sand of fig. 3.1; from 0.072 m till 0.119 m for the loam and it varies around 0.055 m for the peat soil. Hence, the penetration depth of the diurnal variation is of the order of 0.3 m till 0.8 m. For the annual variation these values are $(365)^{\frac{1}{2}}$ or roughly 20 times larger. Rapid temperature variations, caused for instance by intermittent shading by small clouds, have a smaller damping depth.

Values of $(\lambda C)^{\frac{1}{2}}$ are also given in fig. 3.1. They must be multiplied by $\Theta_o \omega^{\frac{1}{2}}$ to obtain the amplitude of the heat flux density at the soil surface. This enables us to estimate the order of magnitude of this heat flux density for energy balance considerations.

The energy balance of the earth's surface in its simplest form can be written as (8,9):

$$q_{rad} + q_{soil} + q_{air} + q_{ev} = 0 \qquad (3.8)$$

where q_{rad} denotes the net radiation flux density, q_{soil} the heat flux density in the soil, q_{air} that in the air, and q_{ev} the latent heat flux density due to evaporation. All terms are counted positive when the flow is directed towards the surface, hence, $q_{soil} = -q(t,0)$. We shall focus our attention on the diurnal and the annual variation of q_{soil} in relation to the other terms in eq. (3.8).

For the diurnal temperature variation $\omega^{\frac{1}{2}} = 8.53 \cdot 10^{-3} \text{ s}^{-\frac{1}{2}}$, whilst Θ_o usually lies between 5 °C and 15 °C and may attain a value of 25 °C under exceptional conditions. The highest values of Θ_o are usually found with clear skies, light winds and dry soils. Under dry conditions little or no evaporation occurs, so that q_{ev} is a small term in the energy balance, but since $(\lambda C)^{\frac{1}{2}}$ is small too for a dry soil most of the energy is transferred to the atmosphere as sensible heat (q_{air}). When the soil is moist most of the radiative energy is used up in evaporation and the temperature amplitude Θ_o usually is much smaller. Therefore, the amplitude of q_{soil} does not vary greatly with soil moisture content. From the numerical values given above it can be seen that on the average it is about 70 W/m^2. This value holds for bare mineral soils. If a dense and tall vegetation exists, the value will be much lower under the same climatic conditions, due to the shielding effect of the plant-air layer.

For the annual variation $\omega^{\frac{1}{2}} = 0.446 \cdot 10^{-3}$ s$^{-\frac{1}{2}}$, whilst the temperature amplitudes are of the same order as the diurnal ones, except near the equator, where they become even smaller. In this case also the temperature amplitudes are highest under arid and lowest under humid conditions. The average value of the amplitude of q_{soil} will therefore lie between 1 and 5 W/m^2. Usually it is small in comparison with the amplitude of q_{rad}. This is the reason that, for energy balance considerations, a rough estimate of q_{soil} is sufficient when periods of several days or longer are considered.

It will be clear that more precise estimates can be made by taking into account the variation of the soil thermal properties with depth and time. Analytical solutions can be obtained for a soil consisting of two or more homogeneous layers (cf. (1)) with constant thermal properties. In this way the influence of a mulch can be described.

Also the mathematical treatment can be extended to other boundary conditions and a non-periodic variation of temperature. For agricultural or horticultural purposes a prediction of night minimum temperatures is of great importance in connexion with the prevention of night frost damage. From fig. 3.1 it follows that peat soils are expecially dangerous in this respect.

3.3. Frost penetration

The problem of the penetration of a frost front into a soil is also of great practical interest. Mathematically it is often treated as a problem of heat conduction in a two-layer system, consisting of a frozen and a non-frozen part, both with uniform thermal properties, and with latent heat production at the moving interface between the layers (6). The penetration depth is then calculated as a function of time. Lefur et al. (10) have given a general analytic solution for this problem with a surface temperature that is a prescribed function of time.

In some soils the frost front can become stagnant or slowly moving and then water moves towards the front, where it freezes and gives rise to the formation of ice lenses. The physics of this phenomenon is only partly understood; the water movement is caused by the difference in chemical potential of the water at the ice front and at lower depths in the soil. A quantitative theory explaining the cause and the rate of water movement towards the ice front has not yet been given. Space does not permit to elaborate this point and the reader is referred to the literature on the subject (11,12,13,14).

4. COMBINED HEAT AND MOISTURE TRANSFER

4.1. General considerations

The theory of combined heat and moisture transfer in soils has to describe the moisture transfer under the combined influence of gradients of temperature and moisture content on the one side and the heat transfer under the influence of temperature gradients and mass flow of moisture on the other side. The theoretical treatment is greatly complicated by the interaction of soil water and the solid matrix, and the fact that soil water and soil air are multi-component systems.

Fortunately, the influence of moisture movement on the transfer of heat is under most circumstances limited to the air-filled pore space and can be accounted for rather simply. Also, the influence of temperature gradients on the moisture transfer in the liquid phase is often small in comparison with the influence of gradients of moisture content. The transfer of liquid water can then be described without taking coupling between heat and moisture transfer into account. This applies to many conditions for soils that are sufficiently wet, so that moisture transfer in the liquid phase is large in comparison with transfer in the vapour phase.

In completely dry soils the problem does not arise, of course. Therefore, the theory of combined heat and moisture transfer has to deal with situations where: (i) transport in the vapour and the liquid phases are of a comparable order of magnitude; (ii) the moisture transfer is caused by temperature differences only. The first situation is often encountered in the upper soil layers.

The main mechanism of vapour transfer in soils is by diffusion of water vapour in soil air. When the pore diameter is not large in comparison with the mean free path in the gas slip phenomena must also be considered. Liquid water movement is often due to the influence of gravity and of capillary forces when the soil is wet, and to the influence of adsorptive forces when the soil contains only a small amount of water.

Heat Transfer in Soils

Complicating factors, that are often disregarded, are: the influence of solutes in the soil water, electric forces and shrinkage or swelling in colloidal soils, the influence of hysteresis, surface phenomena at the liquid solid interface. Some of these are treated in the lectures of J.R. Philip (15) and S.V. Nerphin (16). A textbook on soil water phenomena was written by E.C. Childs (17), a comprehensive account of the theory of heat and mass transfer in porous media with application to building materials has been given by Luikov (18). Recently W.A. Jury (19) has presented a critical review of soil science literature on the subject as part of his doctor's thesis.

4.2. The influence of vapour movement

The way to incorporate the influence of vapour distillation on heat transfer in porous media was first suggested by O. Krischer and H. Rohnalter (20). Under conditions of uniform and constant total pressure, p, the vapour flux density, ϕ_v, in an air-filled pore (neglecting thermal diffusion) can be written as:

$$\vec{\phi}_v = -D \frac{p}{p-p_v} \frac{M}{RT} \nabla p_v \tag{4.1}$$

where D is the diffusion coefficient of water vapour in air, p_v its partial pressure, M its molar mass and R the universal gas constant.

The value of D can be found from

$$D = c \, (p_o/p) \, (T/T_o)^n \tag{4.2}$$

with p_o = 1 atm = $1.01325 \cdot 10^5$ Pa, T_o = 273.15 K, c = 21.7 mm^2/s and n = 1.88 (see references 21,22,23).

The partial vapour pressure can be expressed as the product of the saturation vapour pressure, p_{vs}, and the relative humidity, h:

$$p_v = h p_{vs} \tag{4.3}$$

Except at low moisture contents h differs little from 1 and, since p_{vs} depends on T only, we have:

$$\nabla p_v = h(dp_{vs}/dT)(\nabla T)_a \approx (dp_{vs}/dT)(\nabla T)_a \tag{4.4}$$

where $(\nabla T)_a$ is the temperature gradient in the air-filled pore. Hence, water will evaporate at the warm end of a pore and condense at its cold end, thereby transferring latent and sensible heat. The latter transfer is negligible, however, because of the small vapour density.

The latent heat flux density by vapour diffusion can, with the heat of vaporization, L, be written as:

$$L\vec{\phi}_v = -LD \frac{p}{p-p_v} \frac{M}{RT} \frac{dp_{vs}}{dT} (\nabla T)_a \tag{4.5}$$

and the total heat flux density inside an air-filled pore as:

$$\vec{q}_a = -\lambda_a (\nabla T)_a + L\vec{\phi}_v = -(\lambda_a + \lambda_{vs})(\nabla T)_a \tag{4.6}$$

It is seen that the thermal conductivity of air is increased by λ_{vs}, the apparent contribution due to vapour diffusion, given by:

$$\lambda_{vs} = LD \frac{p}{p-p_v} \frac{M}{RT} \frac{dp_{vs}}{dT} \tag{4.7}$$

Values of this quantity following from (4.2) and (4.7) for $p = p_o = 1$ atm are given in fig. 4.1.

From figure 4.1 and the argument given in section 2.3 it follows that the latent heat transfer is very effective in increasing the thermal conductivity of soils, since it multiplies the conductivity of the air-filled pores by a factor ranging from 2 at 0 °C to 20 near 60 °C. The value of $\lambda_a + \lambda_{vs}$ becomes equal to that of λ_w near 62 °C. At this temperature the thermal conductivity of a soil is therefore independent of its moisture content as long as h differs little from 1. This was confirmed experimentally (2,20).

The overall thermal conductivity of the soil can be calculated as described in section 2.3 by substituting $\lambda_a + \lambda_{vs}$ for λ_a. The merit of this procedure lies in the fact that it incorporates the microscopic treatment of vapour movement into a theory for predicting the macroscopic thermal conductivity.

At low moisture contents, when the liquid is held by adsorption forces, the relative humidity is less than 1 and depends on both temperature and liquid moisture content, $\theta_1 = x_w$; θ_1 is used in this section in order to bring the notation in accordance with previous literature. Now the vapour pressure gradient in (4.1) can be expressed as:

$$\nabla p_v = h \frac{dp_{vs}}{dT} (\nabla T)_a + p_{vs} \frac{\partial h}{\partial T} (\nabla T)_a + p_{vs} \frac{\partial h}{\partial \theta_1} (\nabla \theta_1)_a \tag{4.8}$$

Usually the second term on the right hand side of (4.8) is small in comparison with the first one, so that the apparent conductivity by vapour diffusion becomes $h\lambda_{vs}$. However, the third term on the right hand side need not be

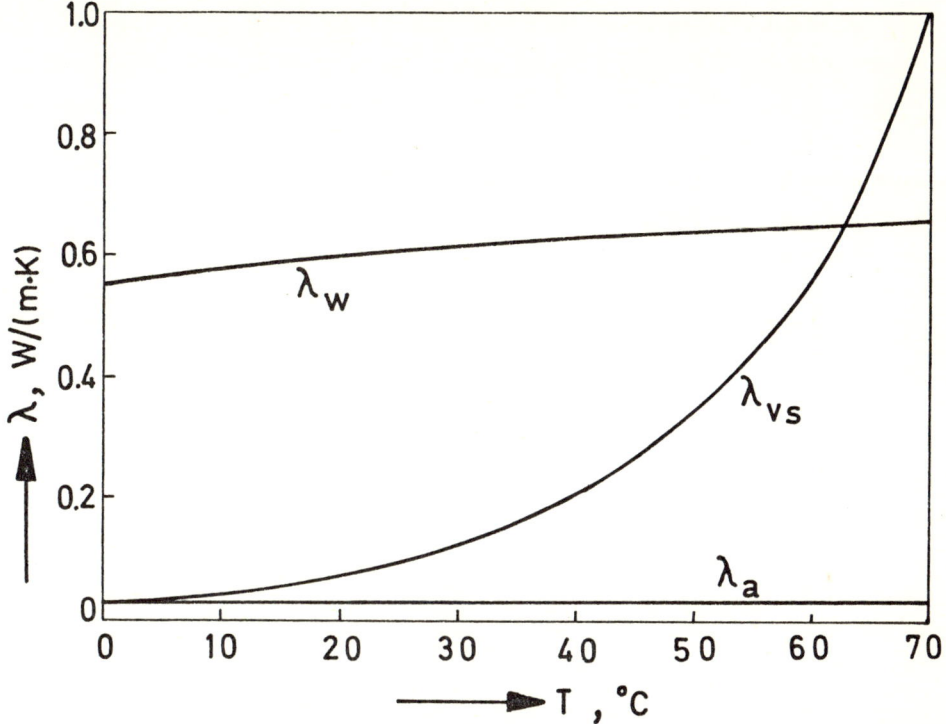

Fig. 4.1 Values of saturated apparent thermal conductivity due to vapour distillation, λ_{vs}, at p = 1 atm, and conductivities for water and air.

negligible, which implies that the heat transfer will be influenced by moisture gradients.

4.3. Combined transfer of heat and moisture

Macroscopic theories of combined heat and moisture transfer in soils have been developed along two different lines. In the first approach one attempts to identify the separate transfer phenomena occurring in the soil and to develop equations for the macroscopic fluxes of heat and moisture on the basis of a physical model of the soil system. In the second approach one applies the formalism of irreversible thermodynamics to the coupled phenomena of heat and moisture transfer and attempts to identify the relevant fluxes and forces. The fluxes are then expressed as linear functions of the forces through the so-called phenomenological relations. Neither approach is able to do full justice to the complexity of the system that it tries to describe.

A theory was developed along the first line by J.R. Philip and the author (24,25,26). The model is based on the concept of viscous flow of liquid water under the influence of gravity and of capillary and adsorption forces, and on the concept of vapour movement by diffusion. Further, local thermodynamic equilibrium between liquid and vapour is assumed.

The theory of liquid flow is described in J.R. Philip's contribution to this Seminar (15). It is based on Darcy's law, which is written as:

$$\vec{\phi}_1 = -\rho_1 K \nabla \Phi \qquad (4.9)$$

where Φ is the total moisture potential. In hydrology this quantity is usually expressed as energy per unit weight, so that its SI-unit becomes $kg \cdot m^2 \cdot s^{-2}/kg \cdot m \cdot s^{-2} = m$. K is the hydraulic conductivity; its SI-unit is m/s in the present notation. K depends on the liquid moisture content, θ_1, and on the distribution of the liquid water inside the pore system. It is inversely proportional to the viscosity and therefore also dependent on temperature. The liquid density, ρ_1, is considered to be constant throughout.

The moisture potential can be written as:

$$\Phi = \Psi - z \qquad (4.10)$$

where Ψ is the so-called moisture or matrix potential, due to the interaction with the solid matrix, whilst z is the gravity potential (note that z is counted positive downward). Ψ is negative in unsaturated soils; it also depends on θ_1 and on the liquid distribution inside the pore system. Even with a rigid matrix this distribution is not a unique one for a given value of θ_1. It depends on the past history of wetting and drying, which gives rise to hysteresis effects. Hence, Ψ and also K are multivalued functions of θ_1. These functions can only be considered as unique when θ_1 decreases or increases monotonously with time. Ψ also depends on temperature. When capillary forces are considered Ψ is proportional to the surface tension of the soil water and the cosine of the contact angle for the solid - water - air system.

Considering θ_1 and T as the two independent variables in which the other quantities are expressed, one obtains:

$$\vec{\phi}_1/\rho_1 = -D_{\theta 1} \nabla \theta_1 - D_{T1} \nabla T + K\vec{k} \qquad (4.11)$$

with

$$D_{\theta 1} = K(\partial \Psi / \partial \theta_1)_T \qquad (4.12)$$

Heat Transfer in Soils

$$D_{T1} = K(\partial \Psi / \partial T)_{\theta_1} \tag{4.13}$$

$\vec{k}$ is the unit vector in the positive z-direction.

The macroscopic vapour flow density is derived from eq. (4.1), which holds for a single pore, by introduction of a factor $f(\theta_1)$ on the right hand side, which describes the diffusion inside the pore system:

$$\vec{\phi}_v = -f(\theta_1) D \frac{p}{p-p_v} \frac{M}{RT} \nabla p_v \tag{4.14}$$

The assumption of local thermodynamic equilibrium links p_v to Ψ via the relation:

$$p_v = p_{vs} h = p_{vs} \exp(Mg\Psi/RT) \tag{4.15}$$

The main problem to be solved is finding a suitable expression for $f(\theta_1)$. The usual way to describe the influence of the pore system, i.e. by a diffusion cross section (equal to x_a) and a tortuosity factor (<1), does not apply to partially wetted porous media. Philip and De Vries (24) pointed out that a series-parallel process of transfer of vapour and liquid under influence of a temperature gradient occurs in part of the pore system. By this process both the diffusion cross section and the tortuosity factor are increased. They suggested a simple form for $f(\theta_1)$; a more refined treatment was given by J. van der Kooi (27). Space does not permit to elaborate this point here.

Combining (4.14) and (4.15) one obtains:

$$\vec{\phi}_v/\rho_1 = -D_{\theta v} \nabla \theta_1 - D_{Tv} \nabla T \tag{4.16}$$

with

$$D_{\theta v} = f(\theta_1) D \frac{p}{p-p_v} \frac{M}{RT} \frac{1}{\rho_1} \left(\frac{\partial p_v}{\partial \theta_1}\right)_T \tag{4.17}$$

$$D_{Tv} = f(\theta_1) D \frac{p}{p-p_v} \frac{M}{RT} \frac{1}{\rho_1} \left(\frac{\partial p_v}{\partial T}\right)_{\theta_1} \frac{(\nabla T)_a}{\nabla T} \tag{4.18}$$

$(\nabla T)_a$ is now the average temperature gradient in an air-filled pore. From the model presented in section 2.3 it follows that

$$(\nabla T)_a / \nabla T = k_a/(x_w + \sum_i k_i x_i + k_a x_a) \tag{4.19}$$

The total moisture flow density can be found by addition of (4.11) and (4.16):

$$\vec{\phi}/\rho_1 = -D_\theta \nabla \theta_1 - D_T \nabla T + K\vec{k} \tag{4.20}$$

It must be noted that the two types of flow do interact so that strictly speaking they are not additive.

Application of the principle of mass conservation leads to:

$$\partial \theta_1 / \partial t = - \nabla \cdot \vec{\phi}_1 / \rho_1 - E \qquad (4.21)$$

where E is the evaporation rate inside the pore system. Its value follows from the vapour-liquid equilibrium condition applied to a unit cell. The final form of (4.22) becomes:

$$[1+F_1(\theta_1,T)] \frac{\partial \theta_1}{\partial t} + F_2(\theta_1,T) \frac{\partial T}{\partial t} = \nabla \cdot (D_\theta \nabla \theta_1) + \nabla \cdot (D_T \nabla T) - \partial K / \partial z \qquad (4.22)$$

Expressions for the functions F_1 and F_2 are given in the original papers (25, 26). For many applications they can be set to zero (19).

The macroscopic flow density of enthalpy, q_H, associated with the mass flow, is:

$$\vec{q}_H = H_1 \vec{\phi}_1 + H_v \vec{\phi}_v \qquad (4.23)$$

where H is the specific enthalpy. The principle of conservation of heat can in our system, where the pressure is constant, be expressed as:

$$\rho_1 \partial H / \partial t = -\nabla \cdot \vec{q} - \nabla \cdot \vec{q}_H \qquad (4.24)$$

where H is the total specific enthalpy for a unit cell. With proper substitutions for H, $\vec{q}$ and $\vec{q}_H$ the following equation for $\partial T / \partial t$ is obtained:

$$\left[[C+F_3(\theta_1,T)] \right] \frac{\partial T}{\partial t} + F_4(\theta_1,T) \frac{\partial \theta_1}{\partial t} = \nabla \cdot (\lambda_* \nabla T) + L\rho_1 \nabla \cdot (D_{Tv} \nabla T) +$$
$$L\rho_1 \nabla \cdot (D_{\theta v} \nabla \theta_1) - \rho_1 c_1 (\vec{\phi}_1 \cdot \nabla T) - \rho_v c_{pv} (\vec{\phi}_v \cdot \nabla T) \qquad (4.25)$$

λ_* is a hypothetical thermal conductivity for a soil without mass flow of moisture. The influence of vapour diffusion caused by a temperature gradient is now given by the second term on the right hand side of (4.25). Applying the argument of section 4.2 the first two terms on the right hand side can be replaced by $\nabla \cdot (\lambda \nabla T)$, where λ has the same meaning as in section 4.2. This ambiguity arises from the limitations inherent in the macroscopic approach.

Often one can neglect the second term on the left hand side and the last three terms on the right hand side of (4.25). Also F_3 is usually small compared with C, so that the original equation (2.1) can be applied.

Equations (4.22) and (4.25) together form the differential equations from which θ_1 and T should be solved with proper initial and boundary conditions.

It must be remembered that all parameters in these equations, like K, D_θ, D_T, are intricate functions of T and especially of θ_1.

The thermodynamic theory starts from an expression for the production of entropy as caused by heat and moisture transfer (28,29). The forces are the gradients of T^{-1} and of the chemical potential of soil water at constant temperature. The latter is proportional to $(\nabla\Psi)_T$ or $(\nabla\theta)_T$. Next, phenomenological equations are obtained expressing the fluxes of moisture and of heat as linear functions of $\nabla\theta$ and ∇T. The four phenomenological coefficients occurring in these equations are unknown functions of θ and T. According to Onsager's theorem the cross-coupling coefficients are equal when fluxes and forces are properly chosen. It should be noted that in the theory of Philip and De Vries the soil is also characterized by three functions of θ_1 and T, viz. Ψ, K and λ. Jury (19) has shown how the thermodynamic flux equations can be related to the corresponding equations in the approach of Philip and De Vries.

At first sight the thermodynamic theory has the advantages of being quite general and of reducing the number of unknown parameters by one. By its nature it does not lead to expressions for these parameters, so that they must be determined experimentally or be deduced from a physical model.

A more serious disadvantage of the thermodynamic theory lies in the gross simplification of the soil system as a continuous system in which heat and moisture transfer are the only coupled phenomena under consideration. In fact this is a much cruder macroscopic approach than the one sketched in the previous pages. It does not deal with vapour diffusion and with the interaction between heat transfer in the matrix and in the pore system, and it treats the phase change of water only in an implicit way.

The theory in this form has been applied to calculate the thermomolecular pressure effect and the mechanocaloric effect in discontinuous systems (30). However, it is applied here to the multiphase and multicomponent soil system. The present author therefore believes that its applicability to soils is much more limited than the model approach, even although the latter does not necessarily comply with Onsager's theorem.

Various attempts have been made to test the theories described in this section. They all met with the difficulty of determining the unknown parameters as functions of T and expecially of θ_1. In addition, the effects to be measured are often very small and too little is known about the influence of the simplifications that have been introduced. In a number of cases the theory of Philip and De Vries has been able to predict or explain the rate of moisture transfer under the influence of a temperature gradient. The theory has also been applied with reasonable success to the combined heat and moisture transfer in cellular concrete roofs (27).

There are instances in which no agreement between theory and experiment was obtained (19). Especially the value of D_{T1} is uncertain because it is very difficult to measure the temperature dependence of Ψ (see eq. (4.13)).

CONCLUDING REMARKS

From the preceding paragraphs it follows that basically much is known about thermal properties of soils and heat transfer in soils. Further studies are required on the combined heat and moisture transfer in the moisture range where the liquid and vapour flow rates are of comparable magnitude, and on liquid transfer under the influence of a temperature gradient. Separate studies are also required on the physics of moisture transfer near a frost front in the soil.

Complicating factors, not included in the present discussion, are the influence of solutes and of hysteresis. Further study of these is also required.

In application to field situations the effects of the diurnal, seasonal and annual temperature variations on moisture transfer should be studied in more detail.

Note Added in Proof

During a discussion on the application of irreversible thermodynamics to heat and moisture transfer in soils, held during the Seminar, Dr. P.H. Groenevelt has demonstrated how the various physical processes occurring in the soil can be incorporated in the thermodynamic theory in such a way that the Onsager relations are satisfied (cf. also (19)).

REFERENCES

(1) Van Wijk, W.R. (Ed.), Physics of plant environment, North-Holland Publ. Cy., Amsterdam, 1963.
(2) De Vries, D.A., Mededelingen Landbouwhogeschool, Wageningen 52, 1-73, 1952.
(3) De Vries, D.A., Bulletin Inst.Int. du Froid, Annexe 1952-1, 115-131, 1952.
(4) Burger, H.C., Phys. Zs. 20, 73-76, 1915.
(5) De Vries, D.A., Landbouwkundig Tijdschrift 65, 676-683, 1953.
(6) Carslaw, H.S. and Jaeger, J.C., Conduction of heat in solids, Clarendon Press, Oxford, 1959.
(7) Rosema, A., this Seminar, Section I, Soil, 1974.
(8) Geiger, R., Das Klima der bodennahen Luftschicht, F. Vieweg und Sohn, Braunschweig, 1961.
(9) De Vries, D.A., in: Environmental control of plant growth (Ed.L.T. Evans), Academic Press, London, 1963.
(10) Lefur, B., Bataille, J. and Aguirre-Puente, J., C.R. Acad. Sc. B 259, 1483-1485, 1964.
(11) Koopmans, R.W.R. and Miller, R.D., Soil Sci.Soc.Amer.Proc. 30, 680, 1966.
(12) Hoekstra, P., Soil Sci.Soc.Amer.Proc. 33, 512-518, 1969.
(13) Aguirre-Puente, J. and Philippe, A., Revue Générale de Thermique no. 96, 1123-1141, 1969.
(14) Gupta, J.P., Moisture migration and heat transfer in wet sand during freezing, Ph.D.Thesis, Univ. of Pennsylvania, 1972.
(15) Philip, J.R., this Seminar, Section I, Soil, 1974.
(16) Nerpin, S.V., this Seminar, Section I, Soil, 1974.
(17) Childs, E.C., An introduction to the physical basis of soil water phenomena, John Wiley and Sons, London, 1969.
(18) Luikov, A.V., Heat and mass transfer in capillary-porous bodies, Pergamon Press, Oxford, 1966.
(19) Jury, W.A., Simultaneous transport of heat and moisture through a medium sand, Ph.D.Thesis, Univ. of Wisconsin, 1973.
(20) Krischer, O. and Rohnalter, H., V.D.I. Forschungsheft 402, 1940.
(21) De Vries, D.A. and Kruger, A.J., Proc. C.N.R.S. Symp. "Phénomènes de transport avec changement de phase dans les milieux poreux ou colloidaux", No. 160, Paris, 1967.

(22) Dijkema, K.M., Stouthart, J.C. and De Vries, D.A., Proc.4th All-Union Heat and Mass Transfer Conference, Minsk, Vol. 7, 577-583, 1972.
(23) Kruger, A.J., Daey Ouwens, C.C.H.T., Te Velde, K., and De Vries, D.A., Appl.Sci. Res. 22, 390-399, 1970.
(24) Philip, J.R. and De Vries, D.A., Trans.Amer.Geophys.Union 38,222-232,594, 1957.
(25) De Vries, D.A., Trans.Amer.Geophys.Union 39, 909-916, 1958.
(26) De Vries, D.A., De Ingenieur 74, O 45-53, 1962.
(27) Van der Kooi, J., Moisture transport in cellular concrete roofs, Ph.D. Thesis Eindhoven Univ.of Technology, Waltman, Delft, 1971.
(28) Taylor, S.A. and Cary, J.W., Soil Sci. Soc.Amer.Proc. 28, 167-172,1964.
(29) Cary, J.W., Soil Sci.Soc.Amer.Proc. 30, 428-433, 1966.
(30) De Groot, S.R. and Mazur, P., Non-equilibrium thermodynamics, North-Holland Publ. Cy., Amsterdam, 1962.

WATER MOVEMENT IN SOIL

JOHN PHILIP

*CSIRO Division of Environmental Mechanics,
Canberra City, Australia.*

ABSTRACT

This review outlines the mathematical-physical analysis of water movement in unsaturated nonswelling soils. The characterization of the soil by the dependence of <u>hydraulic conductivity</u> and <u>moisture potential</u> on moisture content is the basis for a quantitative predictive system for nonhysteretic phenomena. The strongly nonlinear Fokker-Planck and diffusion equations involved are discussed. A limited extension to swelling soils is treated briefly.

LIST OF SYMBOLS

D	moisture diffusivity	$m^2 \, sec^{-1}$
D*	moisture diffusivity (horizontal swelling system)	$m^2 \, sec^{-1}$
D**	moisture diffusivity (vertical swelling system)	$m^2 \, sec^{-1}$
e	void ratio (ratio of volume of soil to volume of soil particles)	
g	acceleration due to gravity (≈ 9.8)	$m \, sec^{-2}$
H	relative humidity	
K	hydraulic conductivity	$m \, sec^{-1}$
m	material coordinate	m
P	vertical stress	m
p	pressure	Nm^{-2}
R	gas constant for water vapour (≈ 461.5)	$J \, kg^{-1} \, °K^{-1}$
r	radial space coordinate	m
r_o	radius of cavity surface	m
T	absolute temperature	$°K$
t	time	sec

V	vector flow velocity	m sec^{-1}
V_r	vector flow velocity in rest frame of soil particles	m sec^{-1}
v	asymptotic velocity of moisture profile	m sec^{-1}
x	horizontal space coordinate	m
Z	depth of water table	m
z	vertical space coordinate, positive downward	m
α	coefficient in Eq. (16)	
γ	apparent wet specific gravity = $(\vartheta + \gamma_s)/(1 + e)$	
γ_s	specific gravity of soil particles	
θ	volumetric moisture content (fraction of bulk volume occupied by water)	
θ_o	initial value of θ	
θ_1	final value of θ	
ϑ	moisture ratio (ratio of volume of water to volume of soil particles) = $\theta(1 + e)$	
Ξ	potential of external forces	m
ρ	density of water	kg m^{-1}
Φ	total potential	m
ϕ_1	similarity variable for one-dimensional absorption, defined by Eq. (8)	m sec$^{-\frac{1}{2}}$
ϕ_n	coefficient of nth term of series in Eq. (10)	m sec$^{-n/2}$
ϕ_{*n}	coefficient of nth term of series in Eq. (9)	m^n sec$^{-n/2}$
Ψ	moisture potential	m
Ω	overburden potential	m
∇	gradient of a scalar	
$\nabla \cdot$	divergence of a vector	

1. INTRODUCTION

Dubrovnik is indeed a happy choice of venue for this Seminar: and not only because of the sun, the sea, and the splendours of the city. The ancient city-state of Ragusa was, for centuries, the principal avenue of exchange and cross-fertilization between two great cultures, the Latin and the Slavonic: and it is our express hope that this Seminar will, like Ragusa, provide a stimulating and productive bridge between two worlds: on the one hand, the world of engineering specialists in heat and mass transfer; and, on the other, that of environmental scientists studying flow and transport processes in soils, plants, and the lower atmosphere.

My task in all this is to offer a review of the processes of water movement in soil. It is appropriate to recall in this context that Dubrovnik's most illustrious son, Ruđer Josip Bošković, F.R.S., turned his hand, when required, to the problems of soil water and agricultural hydrology. His report to Pope Clement XIII on the drainage of the Pontine marshes in 1764 served as the basis for all subsequent work on the marshes. (Bošković, born in Ragusa in 1721, was one of the last great polymaths: he excelled as poet, diplomat, geodesist, applied mathematician, and natural philosopher. His original and perceptive writings include the precursors of modern atomic theory, of the calculus of observations, and of non-Euclidean geometry.)

2. SOIL WATER AND THE HYDROLOGIC CYCLE

In its natural state the soil is normally unsaturated: that is, it contains both water and air. Most of the water involved in the hydrologic cycle is located in unsaturated soil between the time of its arrival as rain at the soil surface and that of its return to the atmosphere. A small fraction of precipitation does not enter the soil, but moves overland directly into streams or lakes; and a second small fraction percolates downward through the unsaturated 'zone of aeration' and joins the groundwater, i.e. the water in the deep, habitually-saturated strata of soil, alluvia, or rocks. In a dry country such as Australia, as much as 93% of the precipitation enters the soil; and, of this, 92% returns directly to the atmosphere, only about 1% reaching the groundwater [1,2].

The processes of water movement in unsaturated soil thus play a central part in the scientific study of the terrestrial sector of the hydrologic cycle and in the related problems of irrigated and dry-land agriculture, of plant ecology, and of the biology of soil flora and fauna. They are, in addition, of great significance in connection with the transport through the soil of materials in solution, such as natural salts, fertilizers, and urban and industrial wastes and pollutants. Specific phenomena of great interest and importance include: infiltration (the entry into the soil of water made available at its surface); drainage and retention of water in the soil strata; extraction of soil water by plant roots; and evaporation of water from the soil.

It will be understood that the soil, the plant, and the atmosphere form a thermodynamic continuum for water transfer [3,4], so that it is somewhat artificial to discuss the soil-path of the water in isolation, as I am obliged to do here. Lecturers to other sessions of this Seminar will treat the transfer of water in plants and in the lower atmosphere, and their reviews will form essential supplements to the present paper.

The major part of this review is devoted to an outline of the mathematical-physical approach to the analysis of water movement in unsaturated nonswelling soils which has been developed over the past 20 years or so, principally in North America, England, Western Europe, and Australia. The penultimate Section 8 provides a brief account of recent work in which the approach is extended to apply to the small, but important class of swelling soils (i.e. soils of high colloid content for which the property of volume change must be taken into account in the analysis).

3. HYDRAULIC CONDUCTIVITY OF UNSATURATED NONSWELLING SOILS

We may write Darcy's law for <u>saturated</u> media, specialized to water flow, as

$$V = -K\nabla\Phi. \qquad (1)$$

V is the vector flow velocity, Φ is the total potential, and K is the <u>hydraulic conductivity</u>. The engineering device of expressing potentials per <u>unit weight</u> simplifies our equations and units: K then has the dimensions $[\text{length}][\text{time}]^{-1}$. We note that

$$\Phi = p/\rho g + \Xi.$$

V and Φ are averages over regions with dimensions large compared with those of the individual pore. Much effort has gone into alleged 'proofs' of Darcy's law: but it suffices to observe that there are both hydrodynamic and statistical elements to the law. The <u>hydrodynamic</u> element is that the Navier-Stokes equation (combined with the incompressibility and no-slip conditions) is linear in the limit as the Reynolds number approaches zero; the <u>statistical</u> element is that, although it is not feasible to know details of the internal geometry or of the distribution of velocity and potential on the microscopic scale, the medium is sufficiently homogeneous that local mean quantities such as V and Φ can be defined and exist [5,6].

K is a scalar for isotropic soils and a symmetrical second-order tensor for anisotropic soils. We limit the discussion specifically to isotropic soils, but extension to anisotropic soils is straightforward.

The basic concepts of flow in unsaturated nonswelling soils are due primarily to Buckingham [7]. He suggested that Darcy's law should hold for them in a modified form with K a function of θ, the volumetric moisture content. Richards [8] and others [9,10] provided experimental confirmation and established the general character of $K(\theta)$. For obvious physical reasons [2, 3] K decreases rapidly, through as much as six or more decades [11], as θ decreases from its saturation value through the range of interest.

Fig. 1 shows a typical $K(\theta)$ relationship. We therefore rewrite (1) in the form appropriate to unsaturated nonswelling soils:

$$V = -K(\theta)\nabla\Phi. \qquad (2)$$

4. TOTAL POTENTIAL AND MOISTURE POTENTIAL OF WATER IN NONSWELLING SOILS

In unsaturated soils the water is not free in the thermodynamic sense because of capillarity, adsorption, and electrical double layers [12,13]. Capillarity is dominant in wet, coarse-textured media, and adsorption assumes its greatest importance in dry media. Double-layer effects may be significant in fine-textured media exhibiting colloidal properties. Buckingham [7] was the first to appreciate that the conservative forces governing the equilibrium and movement of soil-water are amenable to treatment through their associated scalar potentials.

We define such potentials relative to the reference state of water (of composition identical to the soil solution) at atmospheric pressure and

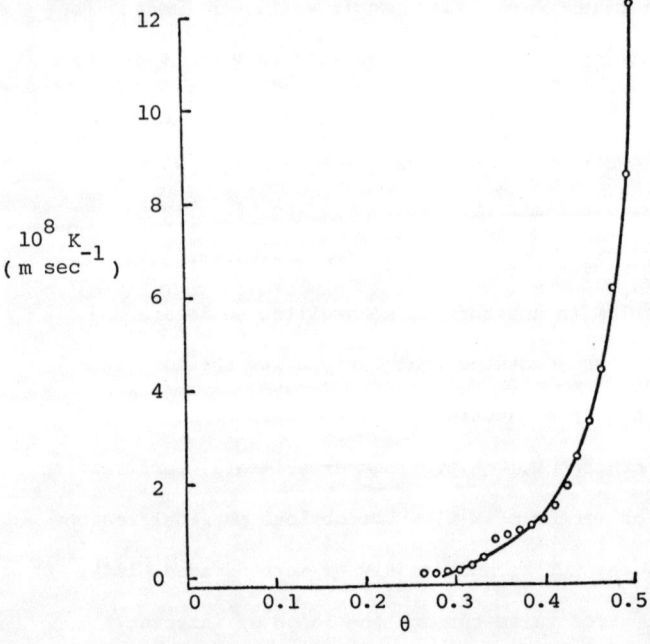

Fig. 1. Dependence of hydraulic conductivity, K, on moisture content, θ, for Yolo light clay [9].

datum elevation z = 0. We then have

$$\Phi = \Psi - z. \qquad (3)$$

Ψ, <u>the moisture potential</u>, is the potential of the forces arising from local interactions between soil and water [14]. It is not essential either to know or to specify these forces in detail: it suffices that Ψ can be measured by well-established techniques [15, 16, 17]. In water-wet nonswelling soils $\Psi = 0$ at saturation and decreases with θ to very large negative values (typically -10^4 m) at the dry end of the moisture range of interest. Fig. 2 depicts the $\Psi(\theta)$ relation for the soil for which $K(\theta)$ is given in Fig. 1.

The partial volumetric Gibbs free energy associated with the local soil-water interaction is $\rho g \Psi$ and it follows that (in the absence of solutes) the liquid and vapour systems are connected at equilibrium by the relation

$$H = \exp g\Psi/RT. \qquad (4)$$

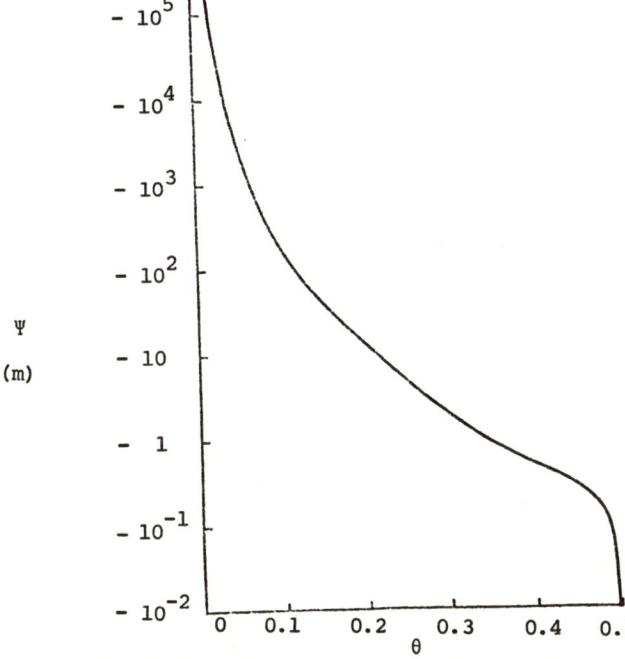

Fig. 2. Dependence of moisture potential, Ψ, on moisture content, θ, for Yolo light clay [9].

5. GENERAL PARTIAL DIFFERENTIAL EQUATION OF FLOW IN UNSATURATED NONSWELLING SOILS

Combining (2) and (3) with the continuity requirement yields

$$\partial \theta / \partial t = \nabla \cdot (K \nabla \Psi) - \partial K / \partial z. \qquad (5)$$

When the relations between K, Ψ, and θ are single-valued, (5) may be rewritten in terms of a single dependent variable. In terms of θ, the equation is

$$\frac{\partial \theta}{\partial t} = \nabla \cdot (D \nabla \theta) - \frac{dK}{d\theta} \cdot \frac{\partial \theta}{\partial z}. \qquad (6)$$

Both the <u>moisture diffusivity</u> D, defined by

$$D = K d\Psi/d\theta \; ,$$

and the coefficient $dK/d\theta$ are, in general, strongly-varying functions of θ. $D(\theta)$ for the soil of Figs. 1 and 2 is shown in Fig. 3. In terms of Ψ, (5) becomes

$$\frac{d\theta}{d\Psi} \cdot \frac{\partial \Psi}{\partial t} = \nabla \cdot (K \nabla \Psi) - \frac{dK}{d\Psi} \cdot \frac{\partial \Psi}{\partial z} \; , \qquad (7)$$

with the various coefficients functions of Ψ.

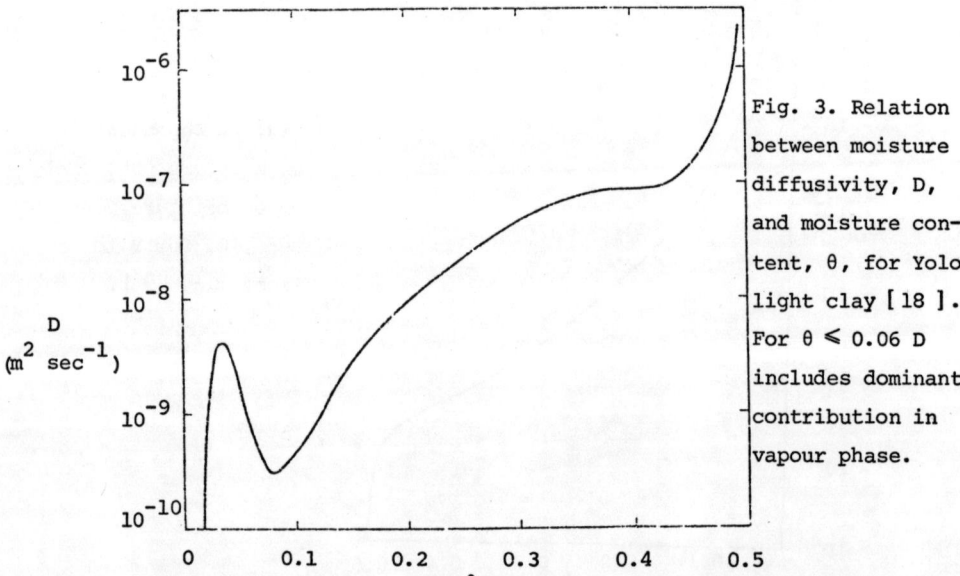

Fig. 3. Relation between moisture diffusivity, D, and moisture content, θ, for Yolo light clay [18]. For $\theta \leqslant 0.06$ D includes dominant contribution in vapour phase.

Richards [8] developed (5) and (7). Childs and George [19] recognized the diffusion character of (6) for a horizontal one-dimensional system. Klute [20] explicitly derived (6). The approach was extended [2,18,21] to include water transfer in vapour and adsorbed phases in the same formalism.

The strong nonlinearity of Fokker-Planck equations (6) and (7) cannot be ignored, and progress in unsaturated flow studies depends centrally on their solution. We return to a discussion of nonlinear Fokker-Planck and diffusion equations in Section 7 below.

6. APPLICATIONS AND EXTENSIONS

Appropriate solutions of (6) and (7) accord with experiment and offer insight into the physics of such diverse phenomena as infiltration [2,5,21-27], capillary rise [28], evaporation from soil [2,29], drainage and retention of soil-water [3,30,31,32], and extraction of soil-water by plant roots [3,11,33].

Limitations to this approach have been reviewed elsewhere [5]. The analysis has been extended to nonisothermal systems by Philip and de Vries [34,35,36]. This involves simultaneous moisture and heat transfer, and provides, *inter alia*, the quantitative theory of drying of porous bodies [29]. Jury [37] has provided a useful and incisive commentary on this work by reinterpreting it in terms of the formulation of the thermodynamics of irreversible processes.

Some progress has been made with the extension to hysteretic systems [primarily those with hysteresis in the $\Psi(\theta)$ relation], including the problem of the mathematical representation of hysteretic properties. Miller and Miller [38] gave a penetrating but qualitative discussion. Poulovassilis [39] gave the first account in terms of the independent domain model. Childs [40] extended the flow equation to hysteretic systems. Similarity hypotheses

[41,42] simplify the problem, though there remains some question as to the adequacy of the independent domain model [43,44,45].

Extensions to take account of the complication of aggregation (or cracking) [46,47] and the effects of soil air [48-51] and of hydrodynamic stability [52,53,54] are also under way. We leave discussion of the extension to swelling soils to Section 8.

7. SOLUTION OF NONLINEAR DIFFUSION AND FOKKER-PLANCK EQUATIONS

Equations (6) and (7) may be solved by the brute-force use of high-speed computers, so long as the problem is well-posed and the computation scheme stable. Much work of this type has, in fact, been done in the last 15 years or so. Here, however, we discuss quasi-analytical and analytical methods, since they tell us more about the fundamental structure of the solutions and the general character of the phenomena which they describe. 'Quasi-analytical' solutions depend on the methods of mathematical analysis for the establishment of their basic form, even though some 'coefficients' require to be determined numerically. When the solution may be found completely by mathematical analysis, we call it 'analytical'. Both analytical and quasi-analytical solutions are 'exact' in the (somewhat optimistic) usage of fluid mechanics [55].

The principal exact solutions and the relevant techniques have been reviewed in detail in [5], [56], and [57], so the treatment here is very condensed. We discuss primarily the transient problems which arise when water is suddenly supplied or removed in consequence of a step-function change of θ or Ψ on (at least part of) the boundary of a mass of a homogeneous soil at some initial uniform moisture content.

A fundamental solution is that for horizontal one-dimensional systems. This is a similarity solution of the one-dimensional nonlinear diffusion equation [58], namely

$$x(\theta,t) = \phi_1(\theta)t^{\frac{1}{2}} \tag{8}$$

The function ϕ_1 is the solution of a nonlinear ordinary equation which may be found by a simple and accurate numerical technique [21,59]. ϕ_1 may be obtained analytically for an indefinitely large class of functional forms of $D(\theta)$ [60]. The importance of this solution is that it provides the leading term of the perturbation solution of related problems involving the complications of (arbitrary) two-and three-dimensional geometry and/or gravity. The latter is embodied in the first-order term on the right of (6).

The solutions for unsteady flow from (or to) a cylindrical or spherical cavity (no gravity) is then [61]

$$r(\theta,t) - r_0 = \phi_1(\theta)t^{\frac{1}{2}} + \phi_{*2}(\theta)r_0^{-1}t + \phi_{*3}(\theta)r_0^{-2}t^{3/2} + \ldots \tag{9}$$

The functions ϕ_{*2}, ϕ_{*3}, etc. are solutions of linear ordinary equations which are readily solved numerically [62]. For three dimensions the function ϕ_{*2} is just twice that for two dimensions. The radius of practical convergence of expansion (9) is limited. The three-dimensional solution is supplemented by a steady exact solution asymptotically valid for large t.

The solution for one-dimensional infiltration (vertically downward flow) is of the form [2,21,62]

$$z(\theta,t) = \phi_1(\theta)t^{\frac{1}{2}} + \phi_2(\theta)t + \phi_3(\theta)t^{3/2} + \ldots \tag{10}$$

Here again ϕ_2, ϕ_3, etc. are the easily-obtained solutions of linear ordinary equations. For the example depicted in Fig. 4, the first four terms of (10) suffice to yield an accurate solution (error $\leq 0.5\%$) for t as large as 10^6 sec. The same solution applies for capillary rise (vertical upward flow) [28] with the sign of the odd terms on the right of (10) reversed. The

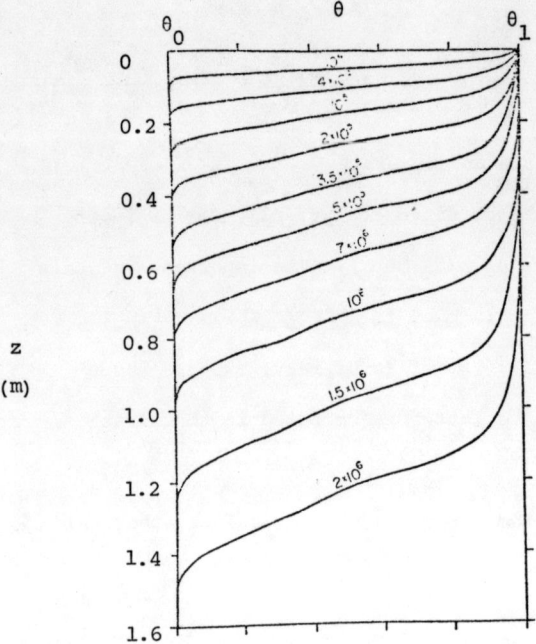

Fig. 4. Computed moisture profiles for one-dimensional infiltration in Yolo light clay [2,21,22]. Numerals on each profile represent value of t (sec) at which profile is realized. Profiles for $t \leq 10^6$ sec calculated from the first four terms of series (10); those for $t > 10^6$ sec based on (11).

radius of practical convergence is, however, less; and the solution is supplemented by the exact solution in the limit as $t \to \infty$.

A double perturbation of (8) has been used in a study of the early stages of unsteady flow from cylindrical and spherical cavities (and semi-circular furrows and hemispherical basins) with gravity included [63].

These various solutions based on perturbation of a similarity solution have close analogies which will be familiar to heat and mass transfer engineers - for example the perturbation by Howarth [64] of the Blasius [65] flat plate boundary layer.

Solution (10) is supplemented by <u>asymptotic solution</u>, valid for large t,

$$z(\theta,t) = (t - t_0)v + \zeta(\theta). \tag{11}$$

The velocity of the moisture profile, v, is given by

$$v = [K(\theta_1) - K(\theta_0)]/(\theta_1 - \theta_0).$$

ζ follows from a simple quadrature and t_0 from a matching procedure [22].

The solution for large t thus consists of the uniform downward movement of a wave of constant shape. For the example of Fig. 4, this solution is used for $t > 10^6$ sec.

The one-dimensional vertical form of (6) is recognizable as a 'more non-linear' variant of the equation of Burgers [66], well-known to heat and mass transfer engineers in the contexts of turbulence and shock waves. As Lighthill [67] showed, the quasi-equilibrium between convection and diffusion implicit in appropriate asymptotic solutions of Burgers' equation yields a plane shock-wave of constant shape and uniform velocity - of the same general character as (11).

Knight [57] is currently studying exact solutions of Burgers' equation relevant to soil water problems. Unsurprisingly, he finds a Burgers model more informative and more accurate than simple linearization [28,61,68].

Various other <u>exact solutions</u> are available. Steady two- and three-dimensional solutions may be found through quasilinearization [5,61,69-75] based on a transformation due to Kirchoff [76]. Limited ranges of exact solutions are available for flux boundary conditions [5,77] and for heterogeneous media [57,78,79].

Where similarity methods apply they have the great virtue of reducing the number of independent variables by one. <u>Integral methods</u>, on the other hand, offer the prospect of effectively achieving the same simplification for the many real-world problems which stubbornly refuse to fit the Procrustean bed of similarity. Integral methods have been much used in heat and mass transfer engineering, and it is of interest that they also have a place in soil water studies.

Green and Ampt [80] unconsciously used a primitive integral method by simply supposing the advancing moisture profiles to be step functions. This assumption is exact for D proportional to a Dirac delta-function at the wet end of the moisture range [24,81].

Parlange [82,83,84] proposed an integral method involving iteration. Parlange's first approximation is that of Macy [85]: the higher approximations, however, abandon integral continuity, with the consequence that the procedure is nonconvergent and oscillatory even in favourable cases [86]. A related technique [87] enables improved initial estimates of 'shape' and preserves integral continuity in all approximations. The iterative procedure is consequently stable and is rapidly convergent in favourable cases.

8. ONE-DIMENSIONAL WATER MOVEMENT AND VOLUME CHANGE IN SWELLING SOILS

Three new basic elements enter the extension of the flow theory to swelling soils:

A. In unsteady swelling systems, the soil particles are, in general, in motion, so that it must be recognized [88] that Darcy's law applies to flow relative to the soil particles. We therefore replace (2) by

$$V_r = -K(\vartheta)\nabla\Phi . \quad (12)$$

B. For one-dimensional systems involving self-weight and/or surface loading, equation (3) must be generalized to include the <u>overburden potential</u> Ω [89], so that it becomes

$$\Phi = \Psi + \Omega - z. \quad (13)$$

It is convenient to take Ψ as the 'unloaded' moisture potential, and Ω is then the contribution to Φ due to normal stress, P.

C. Whereas $K(\theta)$ and $\Psi(\theta)$ provide a sufficient hydrodynamic characterization of nonswelling soils (for nonhysteretic processes), we now require $K(\vartheta)$, $\Psi(\vartheta)$, and $e(\vartheta,P)$, as well as the particle specific gravity γ_s. Fig. 5 shows a typical $e(\vartheta,P)$ relation for a swelling soil. For mineral soils $\gamma_s \approx 2.7$.

Unsteady flow problems are most conveniently analyzed through use of the Lagrangian or material coordinate [91,92,93] such that

$$\nabla m = (1 + e)^{-1}. \quad (14)$$

For two-component (soil, water) and (unloaded) three-component (soil, water, air) horizontal systems, the resulting flow equation is [93,94,95]

$$\frac{\partial \vartheta}{\partial t} = \frac{\partial}{\partial m}\left[D^* \frac{\partial \vartheta}{\partial m}\right], \quad (15)$$

with

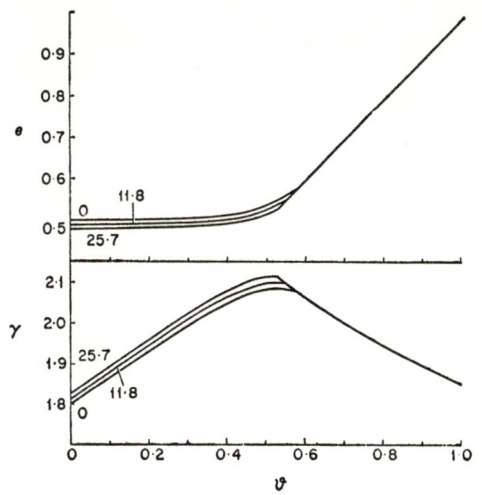

Fig. 5. The function $e(\vartheta,P)$ and $\gamma(\vartheta,P)$ for the illustrative swelling soil [90]. e is the void ratio, γ the apparent wet specific gravity. The numerals on the curves denote values of P (m).

$$D^* = (1 + e)^{-1} K d\Psi/d\vartheta.$$

This equation is consonant with experiment [93] and avoids limitations of the classical soil-mechanical analysis of Terzaghi [96].

The overburden potential Ω enters the analysis of vertical and loaded systems. It has been shown by Bolt [97,98] that

$$\Omega = \alpha P \text{ with } \alpha(\vartheta,P) = P^{-1} \int_0^P (\partial e/\partial \vartheta)_{\vartheta=\vartheta,P} dP. \qquad (16)$$

For a vertical column at equilibrium, then,

$$\Phi = \Psi - z + \alpha[P(0) + \int_0^z \gamma \, dz] = \text{constant} = -Z. \qquad (17)$$

Here P(0) is the vertical stress due to loading at the upper surface z = 0. Fig.5 shows γ for our illustrative soil. In the valid approximation that the P-dependence of α and γ is weak, the solution of (17) is available in closed form. There are three classes of solution representing three distinct types of equilibrium profile, defined through ϑ_P, the moisture ratio at which γ assumes its maximum value [89]. We have: (A) <u>Hydric</u> profiles for which the surface moisture ratio $\vartheta_0 > \vartheta_P$ and $\partial \vartheta/\partial z < 0$; (B) <u>Pycnotatic</u> profiles with $\vartheta_0 = \text{constant} = \vartheta_P$; (C) <u>Xeric</u> profiles for which $\vartheta_0 < \vartheta_P$ and $\partial \vartheta/\partial z > 0$. See Fig. 6.

Continuation of the analysis yields the steady state vertical flow equation. This may also be solved, and the solutions indicate that steady vertical flows are possible only for certain combinations of values of ϑ at the surface

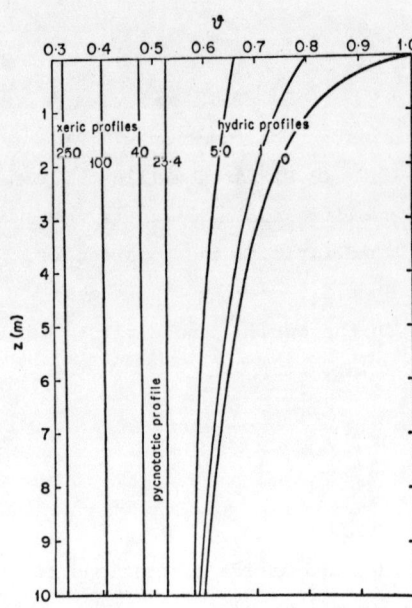

Fig. 6. Equilibrium moisture profiles for a soil characterized (inter alia) by Fig. 5. The numerals on the profiles denote values of Z, the depth to the water table (m).

and in depth [99]. The further step is the integrodifferential equation for unsteady flow and volume change in vertical swelling systems [99]. It suffices here to observe that this is of the form

$$\frac{\partial \vartheta}{\partial t} = \frac{\partial}{\partial m}\left[D^{**}\frac{\partial \vartheta}{\partial m}\right] + \text{second term}, \qquad (18)$$

with

$$D^{**} = (1 + e)^{-1} K\{d\Psi/d\vartheta + P(0)d\alpha/d\vartheta\}.$$

The second term on the right of (18) is negligibly small at small t for many unsteady phenomena of interest. Here also, then, we have the possibility of perturbing similarity solutions of the nonlinear diffusion equation.

The theory outlined in this Section differs profoundly from the classical hydrologic theory which takes no account of swelling and neglects the contribution of Ω to Φ.

The simplest general statement one can offer on the influence of swelling is this: the net effect of gravity on the equilibrium and flow of water in swelling soils is approximately $(1 - \gamma\alpha)$ times that in nonswelling ones. For a mineral soil this factor is about -1 at saturation, decreasing to 0 at $\vartheta = \vartheta_p$, and approaching $+1$ as $\alpha \to 0$ at small values of ϑ. Various 'intuitions' of the hydrologist are therefore invalidated.

Equilibrium moisture distributions for a swelling soil are thus totally different in character from those for a nonswelling soil; and an attempt to interpret equilibria in a swelling soil through classical concepts may be quite misleading [100]. The differences carry over to flow processes: the course of infiltration in a swelling soil evidently has analogies with that of capillary rise in a nonswelling one: and evaporation from an initially wet swelling soil will not exhibit the sharp transition between constant-rate and falling-rate phases which is characteristic of nonswelling soils [2,29]. These developments have evident practical consequences for groundwater hydrology [90] and irrigation technology [101].

CONCLUSION

One final word. There is little in this review which will be novel to the soils specialists here, and some will complain that my treatment has been oversimplified. I feel, however, that my real task has been to convey to the engineers at this Seminar something of the content and the flavour of modern physical studies of soil water. They may find some of this work of direct interest in their own studies of other porous media or other diffusion systems. Equally, I hope that they will see ways in which their engineering experience and expertise can help along our attack on the many problems of soil water which remain to be solved.

REFERENCES

[1] Nimmo, W.H.R. 1949: Jour. Inst. Engrs. Australia 21: 29-34.
[2] Philip, J.R. 1954: Jour. Inst. Engrs. Australia 26: 255-259.
[3] Philip, J.R. 1957: Proc. 3rd Int. Congr. Irrig. Drainage 8. 125-8.154.
[4] Philip, J.R. 1966: Ann. Rev. Plant Physiol. 17: 245-268.
[5] Philip, J.R. 1969: Adv. Hydroscience 5: 214-296.
[6] Philip, J.R. 1970: Ann. Rev. Fluid Mechs. 2: 177-204.
[7] Buckingham, E. 1907: U.S.D.A. Bur. Soil Bull. 38.
[8] Richards, L.A. 1931: Physics 1: 318-333.
[9] Moore, R.E. 1939: Hilgardia 12: 383-426.
[10] Childs, E.C., and Collis-George, N. 1951: Proc. Roy. Soc. A201: 392-405.
[11] Gardner, W.R. 1960: Soil Sci. 89: 63-73.
[12] Edlefsen, N.E., and Anderson, A.B.C. 1943: Hilgardia 16: 31-299.
[13] Schofield, R.K. 1935: Trans. 3rd. Int. Congr. Soil Sci. 1: 30-33.
[14] Philip, J.R. 1970: Soil Sci. 109: 294-298.
[15] Richards, L.A. 1949: Soil Sci. 68: 95-112.
[16] Croney, D., Coleman, J.D., and Bridge, P.M. 1952: Road Res. Tech. Paper 24.
[17] Holmes, J.W., Taylor, S.A., and Richards, S.J. 1967: pp 275-303 of "Irrigation of Agricultural Lands" (Eds. Hagen, R.M., Haise, H.R., and Edminster, T.W.) Amer. Soc. Agronomy: Madison.
[18] Philip, J.R. 1955: Proc. Nat. Acad. Sci. India (Allahabad) 24A: 93-104.
[19] Childs, E.C., and George, N.C. 1948: Disc. Faraday Soc. 3: 78-85.
[20] Klute, A. 1952: Soil Sci. 73: 105-116.
[21] Philip, J.R. 1957: Soil Sci. 83: 345-357.
[22] Philip, J.R. 1957: Soil Sci. 83: 435-448.
[23] Philip, J.R. 1957: Soil Sci. 84: 163-178.
[24] Philip, J.R. 1957: Soil Sci. 84: 257-264.
[25] Philip, J.R. 1957: Soil Sci. 84: 329-339.
[26] Philip, J.R. 1958: Soil Sci. 85: 278-286.
[27] Philip, J.R. 1958: Soil Sci. 85: 333-337.
[28] Philip, J.R. 1966: "Water in the Unsaturated Zone" Symp. Wageningen UNESCO 1: 471-478.
[29] Philip, J.R. 1957: J. Meteorology 14: 354-366.
[30] Staple, W.J., and Lehane, J.J. 1954: Canad. J. Agr. Sci. 34: 329-342.
[31] Day, P.R., and Luthin, J.N. 1956: Soil Sci. Soc. Amer. Proc. 20: 443-447.

[32] Whisler, F.D., and Watson, K.K. 1968: J. Hydrol. 6: 277-296.
[33] Cowan, I.R. 1965: J. Appl. Ecol. 2: 221-239.
[34] Philip, J.R., and Vries, D.A. de 1957: Trans. Amer. Geophys. Un. 38: 222-232.
[35] Vries, D.A. de 1958: Trans. Amer. Geophys Un. 39: 909-916.
[36] Vries, D.A. de, and Philip, J.R. 1957: J. Geophys. Res. 64: 386-388.
[37] Jury, W.A. 1973: "Simultaneous transport of heat and moisture through a medium sand." Ph.D. Thesis, Univ. of Wisconsin.
[38] Miller, E.E., and Miller, R.D. 1956: J. Appl. Phys. 27: 324-332.
[39] Poulovassilis, A. 1962: Soil Sci. 93: 405-412.
[40] Childs, E.C. 1964: Soil Sci. 97: 173-178.
[41] Philip, J.R. 1964: J. Geophys. Res. 69: 1553-1562.
[42] Mualem, Y. 1973: Water Resources Res. 9: 1324-1331.
[43] Topp, G.C., and Miller, E.E. 1966: Soil Sci. Soc. Amer. Proc. 30: 156-162.
[44] Topp, G.C. 1971: Soil Sci. Soc. Amer. Proc. 35: 219-225.
[45] Poulovassilis, A., and Childs, E.C. 1971: Soil Sci. 112: 301-312.
[46] Philip, J.R. 1968: Australian J. Soil Res. 6: 1-19.
[47] Philip, J.R. 1968: Australian J. Soil Res. 6: 21-30.
[48] Youngs, E.G., and Peck, A.J. 1964: Soil Sci. 98: 290-294.
[49] Peck, A.J. 1965: Soil Sci. 99: 327-334.
[50] Peck, A.J. 1965: Soil Sci. 100: 44-51.
[51] Morel-Seytoux, H.J. 1973: Adv. Hydrosci. 9: 119-202.
[52] Hill, D.E., and Parlange, J.-Y. 1972: Soil Sci. Soc. Amer. Proc. 36: 697-702.
[53] Philip, J.R. 1972: Soil Sci. 113: 294-300.
[54] Raats, P.A.C. 1973: Soil Sci. Soc. Amer. Proc. 37: 681-685.
[55] Van Dyke, M. 1954: "Perturbation Methods in Fluid Mechanics." Academic Press: New York.
[56] Braester, C., Dagan, G., Neuman, S., and Zaslavsky, D. 1971: Rep. Hydrodynamics Hydr. Eng. Lab., Technion.
[57] Philip, J.R. 1974: Soil Sci. 117 (In press).
[58] Boltzmann, L. 1894: Ann. Phys. (Lpz.) 53: 959-964.
[59] Philip, J.R. 1955: Trans. Faraday Soc. 51: 885-892.
[60] Philip, J.R. 1960: Australian J. Phys. 13: 1-12.
[61] Philip, J.R. 1966: "Water in the Unsaturated Zone". Symposium Wageningen UNESCO. 1: 503-525.
[62] Philip, J.R. 1957: Australian J. Phys. 10: 43-53.
[63] Philip, J.R. 1969: Australian J. Soil Res. 7: 213-221.
[64] Howarth, L. 1938: Proc. Roy. Soc. A164: 547-579.
[65] Blasius, H. 1908: Z. Math.u. Phys. 56: 1-37.
[66] Burgers, J.M. 1948: Adv. Appl Mech. 1: 171-199.
[67] Lighthill, M.J. 1956: pp. 250-351 of "Surveys in Mechanics" (Eds. Batchelor, G.K. and Davies, R.M.). Cambridge U.P.
[68] Philip, J.R. 1966: "Water in the Unsaturated Zone". Symposium Wageningen UNESCO 1: 471-478.
[69] Philip, J.R. 1968: Water Resources Res. 4: 1039-1047.
[70] Philip, J.R. 1971: Soil Sci. Soc. Amer. Proc.: 35: 867-871.
[71] Wooding, R.A. 1968: Water Resources Res. 4: 1259-1273.
[72] Raats, P.A.C. 1970: Soil Sci. Soc. Amer. Proc. 34: 709-714.
[73] Raats, P.A.C. 1971: Soil Sci. Soc. Amer. Proc. 35: 689-694.
[74] Raats, P.A.C. 1972: Soil Sci. Soc. Amer. Proc. 36:397-401.
[75] Zachmann, D.W., and Thomas, A.W. 1973: Soil Sci Soc. Amer.Proc.37:495-500.
[76] Kirchhoff, G. 1894: Vorlesungen über die Theorie der Warme. Barth: Leipzig.

[77] Knight, J.H. and Philip, J.R. 1974: J. Eng. Maths. 6 (in press).
[78] Philip, J.R. 1967: Australian J. Soil Res. 5: 1-10.
[79] Philip, J.R. 1972: Soil Sci. Soc. Amer. Proc. 36: 268-273.
[80] Green, W.H., and Ampt, G.A. 1911: J. Agr. Sci. 4: 1-24.
[81] Philip, J.R. 1973: Soil Sci. 116: 328-335.
[82] Parlange, J.-Y. 1971: Soil Sci. 111: 134-137.
[83] Parlange, J.-Y. 1971: Soil Sci. 111: 170-174.
[84] Parlange, J.-Y. 1971: Soil Sci. 112: 313-317.
[85] Macey, R.I. 1959: Bull Math. Biophys. 21: 19-32.
[86] Knight, J.H., and Philip, J.R. 1973: Soil Sci. 116: 407-416.
[87] Philip, J.R., and Knight, J.H. 1974: Soil Sci. 117: (in press).
[88] Gersevanov, N.M. 1937: "The Foundations of Dynamics of Soils". Stroiizdat: Moscow-Lenigrad, 3rd Ed.
[89] Philip, J.R. 1969: Australian J. Soil Res. 7: 99-120.
[90] Philip, J.R. 1971: pp. 95-107 of "Salinity and Water Use" (Eds. Talsma, T., and Philip, J.R.). Macmillan: London.
[91] Hartley, G.S., and Crank, J. 1949: Trans. Faraday Soc. 45: 801-818.
[92] McNabb, A. 1960: Q. Appl. Math. 17: 337-347.
[93] Smiles, D.E., and Rosenthal, M.J. 1968: Australian J. Soil Res. 6: 237-248.
[94] Philip, J.R. 1968: Australian J. Soil Res. 6: 249-267.
[95] Philip, J.R., and Smiles, D.E. 1969: Australian J. Soil Res. 7: 1-19.
[96] Terzaghi, K. von 1923: Sitzb. Akad. Wiss (Wien). Abt. 2a, 132: 125-138.
[97] Philip, J.R. 1970: Water Resources Res. 6:1248-1251.
[98] Groenevelt, P.H., and Bolt, G.H. 1972: Soil Sci. 113: 238-245.
[99] Philip, J.R. 1969: Water Resources Res. 5: 1070-1077.
[100] Philip, J.R. 1969: Australian J. Soil Res. 7: 121-141.
[101] Philip, J.R. 1972: Proc. 8th Int. Congr. Irrig. Drainage C.13-C.28.

3

THERMODYNAMIC AND RHEOLOGICAL PECULIARITIES OF SOIL WATER AND THEIR ROLE IN ENERGY—AND MASS TRANSFER

S. V. NERPIN

Thermodynamic and rheological peculiarities of soil water have been considered, which is related to surface phenomena and structural properties of water. Possible scope for applying linear thermodynamic equations has been treated on that basis.

NOMENCLATURE

- μ_i : chemical potential of i-component in the solution
- μ_0 : chemical potential of the pure solvent
- T : temperature
- κ : Boltzmann constant
- c : ratio of the number of dissolved ions to the number of the solvent molecules
- p : pressure
- V : flow velocity
- η_0, η_∞ : liquid viscosity with disturbed and undisturbed structure, respectively
- Δp : interphase pressure difference
- ζ : function depending only on concentration c
- χ : function depending only on pressure p and temperature T
- U : function depending only on coordinates characterizing the position of the molecule (ion) in the space
- V_c : partial volume of the solvent molecules
- A : the constant in the expression for forces of

molecular interaction between the condensed bodies

P : osmotic pressure

P_m, P_i, P_s : molecular, ion-electrostatic and structural constituents of disjoining pressure, respectively

τ, τ_0 : shear stress and yield stress, respectively

INTRODUCTION

One may indicate some thermodynamic and rheological peculiarities of soil water which affect the nature and intensity of energy- and masstransfer. These features are due to the action of surface forces at the interfaces between the solid, liquid and gaseous components of the system and structural properties of water.

Deryaguin's concept of disjoining pressure in thin layers of the liquid separating solid particles or covering their surface in the form of thin films may be referred to as one of the most important results of investigations for the last decades. The notion of disjoining pressure of thin layers which is phenomenologically similar to that of osmotic pressure takes into account the forces of molecular interaction in the system, ion-electrostatic forces and structural changes in boundary layers.

The next most significant result of the latest investigations is the refined conception of polymorphism in wall-boundary layers of polar liquids. The structure of boundary layers determines nonuniformity of the liquid across a thin layer relating to such values as enthalpy, structural velocity, dielectric permeability, etc.

Another factor influencing modern development of the theory of energy- and masstransfer in soils and grounds is comprehension of the role of hydrogen bonds in rheological properties of liquids.

Thus, there appeared a basis for developing the theory of non-linear infiltration which takes into consideration the dependence of kinetic coefficients on tranfer and field potentials, the value of thermodynamic forces and structure relaxation of the liquid when flowing through nonuniform pore space and in the case of nonstationary nature of external fields.

OSMOTIC AND DISJOINING PRESSURES

The movement of soil water as any mechanical movement means the absence of thermodynamic equilibrium in the system, the condition for which is constant chemical potential and temperature. Chemical potential of the solvent molecules (ions) and that of the dissolved substance may be written as follows:

$$\mu = \kappa T \zeta(c) + \chi(p,T) + U(x,y,z) \tag{I}$$

From this expression it follows that mechanical equilibrium may be disturbed due to space inconstancy in temperature, pressure, solution concentration, and field potentials, such as gravitational and electric external fields, molecular interactions. It is important to note that in liquid mixtures when viscosity forces act, mechanical equilibrium of the whole solution is disturbed if the condition of chemical potential constancy is not observed for any of solution components. For example, if the external field by itself does not change the chemical potential of the solvent molecules and no liquid movement would occur in the absence of ions in the solution, with ions in the solution there occurs not only ion diffusion within the solution in the direction of the electric field, but the movement of solvent molecules under the action of viscous forces too. Only in the case of electroneutrality of the solution in every point of the space viscous forces arising due to relative displacement of ions would be compensated for. When there is no compensation mentioned, there appears something like an external force field $u(x,y,z)$

affecting the solvent molecules, the constancy of chemical potential of these latter being disturbed too.

From Eq.(I) it follows that inconstancy of only one or two constituents of chemical potential is not a sufficient condition for arising of the movement. Only if the derivative of the sum of all the three constituents differs from zero, the liquid movement arises. Hence, the action of the force $X_u = \partial U(x,y,z)/\partial x$, for example, may be fully compensated for by the action of the force $X_\chi = \partial \chi(p,T)/\partial x$, if $X_u = X_\chi$, or by the action of the force $X_\zeta = \partial[\kappa T \zeta(c)]/\partial x$ when $X_u = X_\zeta$.

Thus, we arrive at a generalized concept of partial forces as space derivatives of chemical potential components of molecules (ions). The origin of those forces will be of no importance when formalizing the expressions for hydraulic flows. It is all the same whether they are connected with changes in temperature, concentration of dissolved substances, pressure, or with the presence of any external gravitational fields.

When considering both equilibrium state of soil water and its movement it appeared to be convenient to substitute the components related to osmotic and disjoining pressures for those of chemical potential of the solvent molecules. The convenience is likely due to the fact that the concept of osmotic pressure of the solution and that of disjoining pressure of thin layers are very visual and though the above substitution is formal, it does not change essentially either the equilibrium conditions of the system or kinetic transfer equations.

Let us consider an osmotic cell consisting of two parts separated with a partition permeable for the solvent molecules and impermeable for the molecules of the dissolved substance. Assume that the left part of the cell contains a pure solvent and the right one - a solution of c concentration (Fig.I).

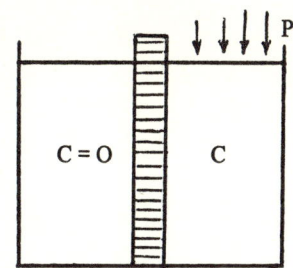

Fig.I Sketch illustrating
an osmotic cell

Since in the left part the chemical potential of the solvent molecules is μ_0 and in the right part it is $\mu_0 - c\kappa T$, the potential difference in both parts will be $\Delta\mu = -c\kappa T$, which is the evidence of the solvent flowing from the left part into the right one. The flow may be arrested if chemical potential of the solvent molecules in the right part of the cell is raised by $\Delta\mu$. That may be realised, for example, by additional pressure. As at the pressure change by P the chemical potential of the solvent molecules changes by $\Delta\mu = V_c P$, one should increase the pressure in the right part by $P = c\kappa T/V_c$ to stop the flow through the partition. This additional pressure is called osmotic.

From the above it follows that the effect of the solvent flowing from one part of the cell into the other will be the same when potential difference is caused by concentration difference and when potential difference is connecσted with different pressures in the right and the left parts. Therefore, when describing the flow through the partition we may reduce the chemical potential components depending on concentration to those depending on pressure. To fully take into account the flow effects in the osmotic cell considered, which as a matter of fact are connected with the difference in the concentration of the dissolved substances in the left and the right parts of the cell, it is sufficient to assume, for example, that as the solution-air interface there exists a negative pressure difference $\Delta p = -c\kappa T/V_c$

Let us show that potentials of the solvent molecules may be too formally reduced to pressure effects. Such potentials are actually connected with the existence of the molecular force field or ion-electrostatic interaction in the presence of a charge on the surface of the particles immersed into the electrolyte solution.

Let us consider two particles separated with a clearance h and immersed into the liquid (Fig.2). From the theory of molecular forces it is known that chemical potential of the molecules within the layer μ_h and in the volume beyond it μ_∞ is not the same. The value of $\Delta\mu = \mu_h - \mu_\infty$ may be both negative and positive depending on the nature of particles and the liquid surrounding them. In the first case the liquid will flow out of the clearance and the particles will draw closer together; in the second case the liquid will flow into the clearance and the particles will move away from each other. In the simplest case when the layer is uniform by its structure, the potential difference is expressed by the relationship $\mu_h - \mu_\infty = AV_c/h^3$
To reduce difference potentials of μ_h and μ_∞, connected

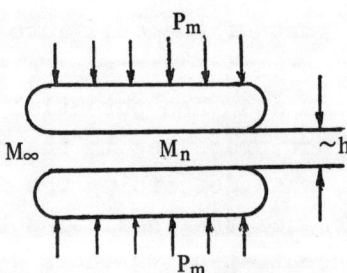

Fig.2 Schematic representation of the concept 'disjoining pressure'

with the action of the molecular forces field, to pressure difference one should adhere to the same consideration as in the case of osmotic pressure. To bring the system into equilibrium it is sufficient to increase or decrease the pressure in the layer by the value of $P_m = \Delta\mu/V_c$ depending on the sign A . That additional pressure was called by Deryaguin the disjoining pressure. Hence, the flow of the liquid from the layer to the bulk and vice versa will be the same both in the case of potential difference caused by the field of molecular forces and in the case of potential difference connected with the difference in pressure of the layer and the bulk. Consequently describing a liquid flow it is possible to reduce the chemical potential constituents depending on the field of molecular forces to the chemical potential constituents depending on pressure. For this it is enough to suppose that on the thin layer boundary there exists interphase pressure difference $\Delta p = -P_m$ thanks to which we may fully take into account the effects actually connected with the action of the molecular field.

The field of molecular forces acting between the condensed bodies is not the only cause of changes in chemical potential of the molecules within the thin layer. There are two more causes for that. One of them is overlapping of diffuse ionic atmospheres in the thin layer, the other is overlapping of bound layers of water posessing another structure than that of the liquid in the bulk.

Considerations similar to the case of osmotic pressure and the molecular constituent of disjoining pressure allow us to draw a conclusion about the possible formal substitution of interphase pressure differences Δp_i and Δp_s related to the corresponding constituents of disjoining pressure by $\Delta p_i = -P_i; \Delta p_s = -P_s$ for the chemical potential constituents connected with the existence of the ion-electrostatic field in the thin layer and with modification of water structure in the wall-boundary layers.

The form of the functions $P_m(h)$ and $P_i(h)$ can be found in many works, such as /1, 2/.

As to the functions $P_s(h)$ its form may be deduced from experimental data /3/.

RHEOLOGICAL PROPERTIES OF SOIL WATER

Soil moisture is a solution containing different ions and in some cases colloid particles of mineral or organic origin. Water may enter the soil and ground from aquifers when moving to the evaporation front as a result of infiltration from reservoirs, rainfalls and irrigation, condensation and thawing of snow. In all these cases rheological properties of soil water manifesting themselves as inner forces arising during its movement in the soil pores must be different. These differences are determined by distinguishing features of water structure exhibiting both at the transition from one phase state into another and depending on the conditions in which the phase exists, i.e. pressure, temperature, nature and intensity of external fields. Modifications of ice and conditions of their originating have been studied best. As for the liquid state, only during the two last decades there appeared some concepts of various modifications of liquid water. Investigations by B.V.Deryaguin and his collaborators /4, 5/ showed that polar liquids unlike nonpolar ones could form polymolecular layers on the solid surfaces, the layers possessing both thermodynamic and mechanical properties different from the rest of the liquid. Based on such ideas there was developed a theory of equilibrium, stability and movement of the liquid in thin films with regard for boundary layers as a peculiar phase /6, 2/; this theory has lately found an ever increasing experimental confirmation /7, 8/. Origination of the modified state in the liquid just at the boundary surfaces and the ability of forming boundary phases only for polar liquids indicates the role surface forces and dipole moments of liquid molecules in this phenomenon.

The studies by V.D.Sobolev and N.V.Churaev on water movement in ultrathin capillaries /9/ give some grounds for inferring that thickness of boundary layers decreases with temperature rise and they fully 'melt' at the temperature of about 60°C.

To account for many experimental data it appears insufficient to know that water structure in the pores is of two forms: one form refers to the water in boundary layers covering the charged surfaces of the particles, and the other refers to normal water beyond these layers. For example, when studying mechanocaloric effect /10/ to explain inversion of temperature field at the increase of hydraulic gradient it appeared necessary to assume the existence of a layer separating hydrated shells and the rest of water contained in the pores. This layer should have the least ordered structure and hence enthalpy of higher value not only as compared to the wall-boundary hydrated shells but also to the water in its normal state.

At present a great number of experimental data has been obtained favouring such concepts of the state of water in pores of disperse materials. The most complete review mainly of western studies may be found in the work /11/. At about the same period the idea of the three-layer model of near-interface water was developed in the papers /10, 12/.

The share of modified water in the bulk of soil water depending on dispersity and density of the system, particle nature, constitution of exchange complex and temperature may be commonsurable with the amount of remaining water unaffected by the action of surface forces and being in its normal state. Therefore rheological properties of soil water should be considered with regard for the possible existence of several water modifiactions.

As stated above, the peculiarities of water structure

manifest themselves not only in the existence of its various modifications but in the change of properties within one modification depending on the origin of the liquid phase and time elapsed since the moment of phase transformation.

A most peculiar feature of water in the existence of hydrogen bonds between molecules, which forms a framework that is retaining in the liquid state too at simultaneous existence of molecules beyond the frame. Such structure of water brings its solid and liquid states even closer together as compared to the liquids without hydrogen bonds.

In recent years it has been stated by several investigations that after phase transitions of water its structure acquires its steady state corresponding to the given temperature and pressure not at once; it retains the traces of the preceding state exhibiting "memory" of it for some minutes or even hours. The developed theory based on the two-structural models allows interpreting most thermodynamic and physicochemical peculiarities of water and other liquids with hydrogen bonds. However, up to now the state of the theory does not enable us to estimate quantitatively such rheological characteristics of a liquid as relaxation periods of individual molecules and their structural associates, viscosity of the liquid, its yield stress and strain moduli at different frequencies of external influence. Therefore, experimental data are the main sources of comprehension of these phenomena.

As early as in the fiftieth /13/ it was shown experimentally that at the room temperature water and water solutions of salts can exhibit static shear stress which retains for an indefinitely long period of time. Wishing to emphasise the small value of the yield stress ($\tau_0 \approx 3 \cdot 10^{-4}$ Ns/m²) compared to the ultimate strength of water under high frequencies of loading, we denoted such values as traces of shear strength. Later detailed investigations by N.Ph.Bondarenko showed that not only water possesses the ability to exhibit 'static' shear strength but all the liquids with hydrogen bonds as well /12, 14, 15/. Besides, he revealed that ultimate shear strength decreases with temperature increase and it approaches zero at 60°, which is in full agreement with temperature change of the hydrogen bonds energy /14, 15/.

The idea of water ability to exhibit 'static' shear strength has been used when choosing a rheological model for water to generalize Darcy's law. A good approximation was shown to be attained when applying Shwedov-Bingham's equation $\tau = \tau_0 + \eta dV/dy$ instead of Newton's law of liquid friction $\tau = \eta dV/dy$.

Darcy's law generalized on that basis has been used to solve a number of nonlinear infiltration problems /13, 16, 17, 18/. However, that approximation based on Shwedov-Bingham's equation embraces only those problems where the condition $\tau > \tau_0$ holds true for the whole field of filtration. At the same time it is obvious that relaxation processes in the liquid proceed if $\tau < \tau_0$ too; they should proceed as a very slow filtration being manifestation of the liquid "creeping" when shear stress does not exceed ultimate shear strength at which the structure of the liquid is still not broken.

Modern concepts of water structure based on vast experimental data allowed us to suggest the idea of applying in the case of water the rheological model for structural liquids widely used in the investigations by N.A. Rebinder, his disciples and collaborators /19/.

Such a model is shown in Fig.3. Mathematical description of the model based on the linear approximation (Fig. 4) was suggested by /20/ in the form of

$$\frac{dV}{dy}(\tau) = \begin{cases} \tau \frac{1}{\eta_\infty} - \tau_0 \frac{1}{\eta_\infty} - \frac{1}{\eta_0} &, \tau \geqslant \tau_0, \\ \tau \frac{1}{\eta_0} & \tau \leqslant \tau_0 \end{cases}$$

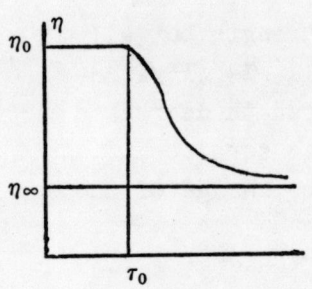 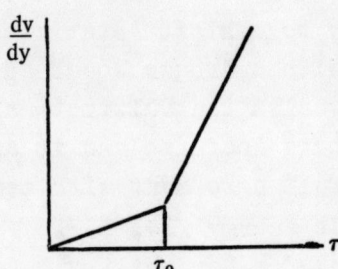

Fig.3 A rheological curve for thixotropic structures

Fig.4 Piecewise-linear approximation of the function

The essence of the model is seen to be as follows: when the rate of relative deformations is lower than some ultimate value, the viscosity is constant η_0. Thereby the structure of the liquid is still not broken and therefore the appropriate shear stress may be referred to as ultimate shear strength τ_0. At the increase of stress and consequently of strain rate the structure begins to be disturbed and the value of viscosity being now a function of shearing stresses commences to decrease.

At the values of shear stresses and strain rates resulted in a complete breaking of the structure the value of viscosity becomes again constant η_∞, but of another lower order than η_0. Thus, speaking of rheological properties of water contained in the pores of soils and grounds one should bear in mind its possible existence in different modifications connected both with the properties of water itself (polarity of molecules, the presence of hydrogen bonds) and the action of surface forces.

Peculiarities in the processes of water transfer under the action of forces of various nature can be interpreted best assuming the reality of a three-layer model for water. The first layer covering the charged surface of particles has the most ordered structure and hence anomalous properties compared to the normal ones. These hydrated shells are covered with an intermediate layer of a less ordered structure; that layer separates the shells from the rest of the liquid with normal structure and properties.

To estimate the role of rheological properties of water contained in the pores of soils and grounds it is necessary to know the parameters of rheological curves for all the supposed structural states of water when considering various processes of its transfer.

For the present state of investigations the following approximate values may be recommended for estimated calculations:

- for normal water structure

$$\eta_\infty = 10^{-3} \text{ N}\cdot\text{s/m}^2; \quad \eta_0 = 10\,\eta_\infty; \quad \tau_0 = 5\cdot 10^{-4} \text{ N/m}^2$$

- for hydrated boundary layers

$$\eta_\infty = 10^{-3} \text{ N}\cdot\text{s/m}^2; \quad \eta_0 = 10^2\,\eta_\infty; \quad \tau_0 = 10 \text{ N/m}^2$$

- for the intermediate disordered layer

$$\eta_\infty = 10^{-3} \text{ N}\cdot\text{s/m}^2; \quad \eta_0 = \eta_\infty; \quad \tau_0 = 0$$

The data obtained for such evaluation are given, for example, in /21/.

ON APPLICABILITY OF THE PHENOMENOLOGICAL EQUATIONS OF IRREVERSIBLE TRANSFER TO THE TRANSFER IN SOILS AND GROUNDS

All stated above indicates the limitations in using linear equations of irreversible thermodynamics in the problems of energy- and masstransfer occuring in the soil. Since all the kinetic coefficients L_{ik} at $i \neq k$ in the equations for the flow $J_i = \Sigma L_{ik} X_k$ contain viscosity, and as viscosity of a thixotropic liquid bodies, in a general case, depends on thermodynamic force X_k, the applicability of linear equations in the form of $J_i = \Sigma L_{ik} X_k$ to soils and grounds should be examined specially. One may indicate only extreme cases when nonlinear effects practically disappear and phenomenological equations of irreversible transfer may be applied without reserve.

Such is the case of the liquid transfer at temperature of about 60°C when water in the bulk of pores becomes a Newtonian liquid. The other cases are connected with the value of shear stress in the flow. Nonlinear effects lack both at the condition $\tau < \tau_0$ and when $\tau \gg \tau_0$. In these cases viscosity also does not depend on shear stresses and hence on the forces X_K as well; its values are constant, namely η_0 at $\tau < \tau_0$ and η_∞ at $\tau \gg \tau_0$.

One more extreme case may be mentioned relating to relaxation of the liquid structure. If in the considered system there exists a periodical cycle of transient states related to the space nonuniformity of pores or to nonstationarity of the process, in the case when relaxation period of the structure is much longer than the period of the cycle, viscosity will also remain constant retaining the intermediate value between η_0 and η_∞, which corresponds to the maximum shear stresses arising at periodical variation of the force X_K.

In all the other cases neglecting the existing relationship between the values of kinetic coefficients and these of thermodynamic forces should be determined by a degree of approximation of the model depending on its application and accuracy of calculations performed on its basis.

CONCLUSION

The lecture deals only with main thermodynamic and rheological peculiarities of soil water that define the equations for the flow through the section of individual pores. Such an approach may also be sufficient for considering total flows through the pore system. However, it holds true only in the case of the simplest models of porous bodies, for exmaple, for a filter with the system of nonintersecting pores. In the case of real porous bodies there arise new problems related, in particular, to kinematic structure of the flows and developing macrononuniformity of the fields of forces at unsteady processes of water transfer in unsaturated soils and grounds. In general, simulation of flows in unsaturated systems gives rise to many problems connected with thermodynamics, which is beyond the scope of the present lecture. For example, the problem of the relation between the potential of the solution and its quantity in unit volume of the system, the nature of the relation between the potential and the rate of drying or wetting of the system, etc. May be some of these problems will be considered in other communications.

REFERENCES

/1/ Deryaguin, B.V. (1955): On the concept of 'disjoining pressure', its value and role in statics and kinetics of thin layers of liquids. Koloidny zhurnal, tom 17, vypusk 3, 207-214.

/2/ Nerpin, S.V. and A.Ph.Chudnovsky: (1967). Soil Physics (a book, 583 pages, Moscow).

/3/ Churaev, N.V. and I.G.Ershova: (1971). Anomalous behaviour of liquids at evaporation from quartz capillaries. Koloidny zhurnal, tom 33, N 5, 913-918.

/4/ Deryaguin, B.V.: (1952). Solvate layers as special interfacial phases studied by direct methods. Trudy vsesoyusnoi konferenzii po koloidnoi khimii. Kiev, Izdanie Ukrainskoi Akademii nauk, 26-51.

/5/ Deryaguin, B.V., V.V.Karasev and Z.M.Zorin: (1954). On peculiar aggregate states of liquids in the layers interfacial with the solid. In the book: Structure and Physical Properties of a Substance in the Liquid State, 141-159. Izdanie Kievskogo universiteta.

/6/ Nerpin, S.V. and Deryaguin, B.V.: (1956). Flow kinetics and steadiness of thin liquid layers on hard substrate with regard for a solvate shell as a special phase. Doklady Akademii nauk SSSR, tom 100, N 1, 17-20.

/7/ Volkova, V.J., S.V.Nerpin and O.G.Usiarov: (1971). Effect of electrolytes on thickness and steadiness of water wetting films. Sbornik trudov po agronomicheskoi fizike, vypusk 32, 184-191.

/8/ Viktorina, M.M., B.V.Deryaguin, I.G.Ershova and N.V. Churaev: (1971). Effect of the area of wetting films on their steadiness. Doklady Akademii nauk SSSR, tom 200, N6, 1306-1309.

/9/ Sobolev, V.D.: (1971). Study of water movement in microcapillaries. Abstract of Cand.thesis. Leningrad.

/10/ Deryaguin, B.V., J.Shutor, S.V.Nerpin and M.A.Arutjunian: (1965). Study of thermoosmotic effect for water in glass capillaries. Doklady Akademii nauk SSSR, tom 161, N 1, 147-150.

/11/ Drost Hansen W.: (1969). Structure of water near solid interfaces. Industrial Eng.CheM.,vol.61, p.10-47.

/12/ Bondarenko,N.Ph.: (1966). Study of nature of filtration anomalies. Abstract of Doc.thesis. Leningrad.

/13/ Nerpin, S.V. and N.Ph.Bondarenko: (1957). Study of mechanical properties of thin liquid layers by filtration method. Doklady Akademii nauk SSSR, tom 114, N 4, 833-836.

/14/ Bondarenko, N.Ph.: (1967). On nature of filtration anomalies of liquids. Doklady Akademii nauk SSSR, tom 177, No 2, 383-386.

/15/ Bondarenko, N.Ph.: (1968). Effect of intermolecular hydrogen bonds on liquid flow in capillaries. Zhurnal fizicheskoi khimii, tom 42, vypusk 1, 225-226.

/16/ Jonat, V.A.: (1960). Calculation of systematic draining in two-layer soils. Sbornik: Isushenie bolotnykh i zabolochenych pochv, 246-256. Minsk, Belorussian Academy of Agricultural Sciences.

/17/ Churaev, N.V.: (1964). Infiltration of structurized liquids through heteroporous bodies. Izvestia Akademii nauk SSSR (mechanics and machinebuilding), N 1, 136-140.

/18/ Bondarenko, N.Ph.: (1966) Calculation of vertical drainage with regard for shear strength traces. Doklady VASKHNIL, N 9, 41-44.

/19/ Mikhailov, N.V. and P.A.Rebinder: (1955). On structural-mechanical properties of diperse and high-molecular systems. Koloidny zhurnal, tom 17, N 2, 107-119.

/20/ Nerpina, N.S.: (1970). Two-dimensional filtration problems of rheologically complex fluids. Intern. seminar on Heat- and Mass Transfer in Rheologically complex fluids. Jugoslavia, Herceg-Novi.

/21/ Nerpina, N.S. and V.A.Yangarber: (1972). Generalization of Darcy law for rheologically complex liquids and error estimation of calculations based on Darcy linear approximation. Symposium on fundamentals of transport phenomena in porous media. Canada.

HEAT AND WATER TRANSFER IN A NATURAL SOIL ENVIRONMENT

R. D. JACKSON, B. A. KIMBALL, R. J. REGINATO,
S. B. IDSO, and F. S. NAKAYAMA

*Agricultural Research Service, United States Department of Agriculture,
United States Water Conservation Laboratory,
Phoenix, Arizona, USA*

ABSTRACT

Experiments conducted in a natural field soil yielded heat and water fluxes as functions of depth and time which were compared with theoretical values. Calculated and measured water fluxes agreed well at night but diverged during the day. Calculated and measured heat fluxes agreed well at night but diverged particularly at the shallow depths at other times.

NOTATION

D_a	= diffusion coefficient for water vapor in air	$cm^2\ sec^{-1}$
$D_{\theta\ell}$	= isothermal liquid diffusivity	$cm^2\ sec^{-1}$
$D_{\theta v}$	= isothermal vapor diffusivity	$cm^2\ sec^{-1}$
$D_{T\ell}$	= thermal liquid diffusivity	$cm^2\ sec^{-1}\ °C^{-1}$
D_{Tv}	= thermal vapor diffusivity	$cm^2\ sec^{-1}\ °C^{-1}$
h	= relative vapor pressure	
$J_{\theta m}$	= measured soil-water flux	$cm\ sec^{-1}$
$J_{\theta c}$	= calculated soil-water flux	$cm\ sec^{-1}$
$J_{\theta\ell}$	= $-D_{\theta\ell}\nabla\theta$ = water flux due to liquid flow	$cm\ sec^{-1}$
$J_{\theta v}$	= $-D_{\theta v}\nabla\theta$ = water flux due to vapor diffusion	$cm\ sec^{-1}$
$J_{\theta Tv}$	= $-D_{Tv}\nabla T$ = water flux due to thermal vapor diffusion	$cm\ sec^{-1}$
$J_{\theta p}$	= water flux assuming all terms multiplied by the temperature gradient are zero	$cm\ sec^{-1}$
J_{hm}	= measured heat flux	$cal\ cm^{-2}\ sec^{-1}$
J_{hc}	= calculated heat flux	$cal\ cm^{-2}\ sec^{-1}$
J_{hp}	= heat flux due to conduction only	$cal\ cm^{-2}\ sec^{-1}$

J_{hTv} = heat flux due to vapor distillation on a temperature gradient — cal cm^{-2} sec^{-1}

$J_{h\theta v}$ = heat flux attributable to vapor movement due to a water content gradient — cal cm^{-2} sec^{-1}

K = hydraulic conductivity — cm sec^{-1}

k = unit vector in vertical direction

L = heat of vaporization — cal g^{-1}

T = temperature — °C

t = time — sec

z = vertical distance, positive upwards — cm

θ = volumetric water content — cm^3 cm^{-3}

λ = thermal conductivity — cal cm^{-1} sec^{-1} °C^{-1}

λ_a = thermal conductivity of dry air — cal cm^{-1} sec^{-1} °C^{-1}

λ_{av} = thermal conductivity of air plus apparent conductivity due to water vapor distillation — cal cm^{-1} sec^{-1} °C^{-1}

ρ_o = density of saturated water vapor — g cm^{-3}

ρ_ℓ = density of liquid water — g cm^{-3}

INTRODUCTION

Water movement in soil is influenced by temperature and temperature gradients, and heat flow is simultaneously influenced by the movement of soil water. Theories to describe this coupled system have been developed by Philip and de Vries [1], and de Vries [2]. Most tests of these theories have been restricted to laboratory experiments. Notable exceptions are the work of Rose [3,4] on water flux and of Wierenga and de Wit [5] on heat flux. In this paper, we consider both heat and water flux within the top 10 centimeters of a field soil under natural, diurnally varying environmental conditions

EXPERIMENTAL

Two weighing lysimeters and a surrounding 91- by 73-m plot were irrigated with about 10 cm of water on 17 September 1973. Soil-water contents, soil temperatures, and evaporation were measured at 20-minute intervals on 2 days, 21 September and 2 October. Water contents were measured at depth intervals of 0 to 0.2, 0 to 0.5, 0 to 1, 1 to 2, 2 to 4, 4 to 6, 6 to 8, and 8 to 10 cm in a 20- by 20-m subplot near the lysimeters. These data were smoothed with time using a Fourier transform technique [6]. Then they were smoothed with depth by statistically fitting log-polynomial curves to the water contents as a function of depth for each time period. Smoothed water contents and water content gradients were obtained from these curves.

Water fluxes at the surface ($J_{\theta m 0}$) were obtained directly from the lysimeters, and water fluxes at a particular depth, z, were calculated from

$$J_{\theta m z_2} = J_{\theta m z_1} - \int_{z_1}^{z_2} (\partial\theta/\partial t) dz. \qquad (1)$$

Water fluxes obtained using equation (1) will be referred to as measured fluxes.

Soil temperatures were measured using four thermocouples at about 1-mm depth (just covered with a thin layer of soil), two at about 5 mm, one at each 1/2-cm interval to 10 cm, one at each 2-cm interval to 30 cm, and one each at 48, 64, 96, and 128 cm. The temperature data were smoothed and gradients were obtained in a manner similar to the water content data. Soil-heat flux (J_{hm}) profiles were obtained for each time interval using a recently developed null-alignment method [7].

CALCULATION OF HEAT AND WATER FLUX

<u>Water flux.</u> Soil-water flux ($J_{\theta c}$) was calculated using the relation [1,2]

$$J_{\theta c} = -D_{\theta \ell} \nabla \theta - D_{\theta v} \nabla \theta - D_{T\ell} \nabla T - D_{Tv} \nabla T - Kk. \qquad (2)$$

The diffusion coefficients $D_{\theta \ell}$ and $D_{\theta v}$ (figure 1) were based on experimental data [8,9,10,11] for T = 25°C. The temperature dependence of $D_{\theta \ell}$ was taken as the ratio of surface tension to viscosity of water, and temperature dependencies of D_a and ρ_o were used to adjust $D_{\theta v}$ to the various temperatures encountered in the experiment. Temperature-dependent coefficients obtained in this manner agreed well with previously measured values [8,11]. The temperature dependence of D_{Tv} and $D_{T\ell}$ is inherent in their calculation. The hydraulic conductivity, K, was calculated but was negligible for the water contents in this experiment. The term $D_{T\ell}$ is also small. For comparison, an "isothermal" soil-water flux ($J_{\theta p}$) was calculated, by assuming that the third and fourth terms on the right-hand side of equation (2) were zero.

<u>Heat flux.</u> Soil-heat flux (J_{hc}) was calculated using the relation

$$J_{hc} = -\lambda \nabla T - \rho L D_{\theta v} \nabla \theta. \qquad (3)$$

The thermal conductivity, λ, was obtained via the method of de Vries [12,13], which accounts for heat conduction in the solid, liquid, and gaseous phases.

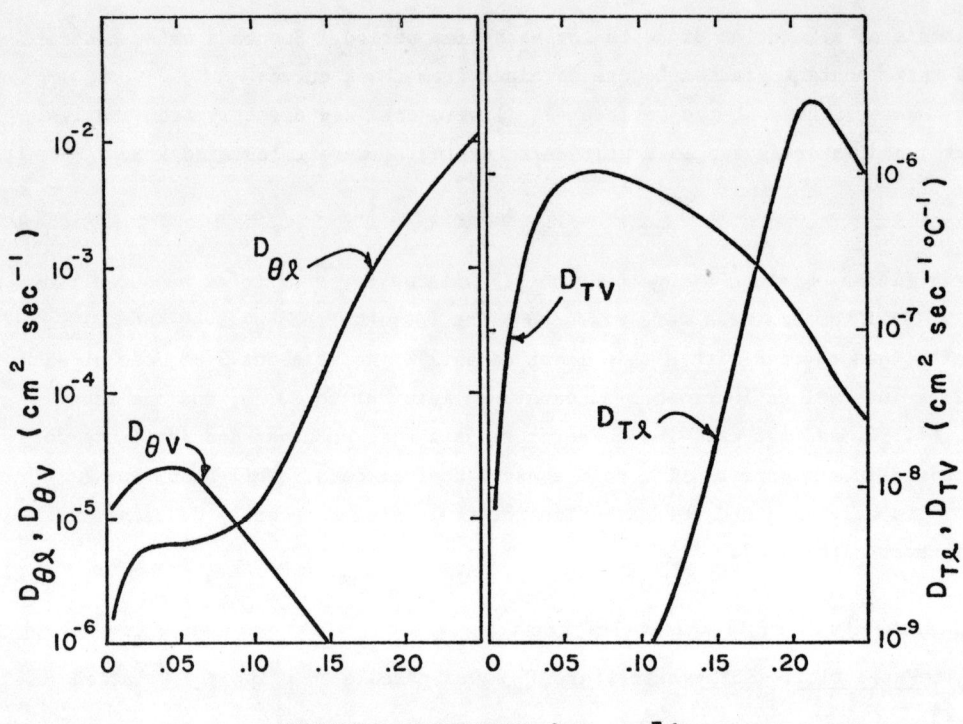

Figure 1. Isothermal liquid diffusivity ($D_{\theta\ell}$), isothermal vapor diffusivity ($D_{\theta v}$), thermal liquid diffusivity ($D_{T\ell}$), and thermal vapor diffusivity (D_{Tv}) as a function of water content at 25°C.

In the gas phase, it includes the conductivity of air and the latent heat carried by vapor distillation, i.e.,

$$\lambda_{av} = \lambda_a + hLD_a d\rho_o/dT. \qquad (4)$$

For comparison, the soil-heat flux due only to conduction (J_{hp}) was calculated by assuming that the second term on the right of equation (4) was zero. Equation (4) and the method of calculating λ enter into the calculation of D_{Tv}.

The heat flux attributable to vapor distillation along a temperature gradient (J_{hTv}) was obtained as the difference $J_{hTv} = -\lambda\nabla T - J_{hp}$, and the heat flux attributable to vapor movement due to a water content gradient was computed as $J_{h\theta v} = -\rho_\ell LD_{\theta v}\nabla\theta$.

RESULTS AND DISCUSSION

Soil-heat and soil-water fluxes were measured and calculated each 20 minutes for 2 days, 21 September (high water content) and 2 October (low water content). The 1200- and 2200-hour time periods, midday and near midnight, were selected to represent this large volume of data since these time periods depict the daily extremes in temperatures and water contents.

The fluxes for the high water content day at 1200 hours are shown in figure 2. The measured ($J_{\theta m}$) and calculated ($J_{\theta c}$) water fluxes agree well within the first centimeter (figure 2A). Below 1 cm they widely diverge since the measured flux becomes negative and the calculated flux remains positive. Most of the slight difference between $J_{\theta c}$ and $J_{\theta p}$ is due to $J_{\theta Tv}$. The term $J_{\theta \ell}$ is the predominate component of the various water fluxes for this data and time. The calculated heat flux agrees better with the measured flux when only conduction mechanisms are assumed to be operative (Figure 2D). When the complete vapor term is used, the calculated flux is more negative than the measured flux.

Data for 2200 hours (21 September) show that temperatures are much lower than at 1200 hours and that the temperature gradients are all small and positive (figure 3C). Both the water and heat fluxes were predicted reasonably well by the theory (figure 3A, 3D).

Data for 1200 hours (2 October) at low water contents show the measured and calculated water fluxes do not agree within the first centimeter of soil, but agree reasonably well below that depth (figure 4A). However, at the 0.2-cm depth, $J_{\theta m}$ and $J_{\theta p}$ closely agree. Only at this shallow depth does neglecting the thermal diffusion components improve the agreement.

Agreement between calculated and measured values was reasonable for heat flux below 5 cm (figure 4D). Above 5 cm, calculated values were more negative than measured values. The divergence appears to stem from the term J_{hTv} for heat flux (figure 4E) and $J_{\theta Tv}$ for water flux (figure 4B). Both terms are too large in absolute magnitude. The J_{hTv} term is apparently too large in absolute magnitude by a factor of 2 for all four time periods.

Figure 5 presents data for 2200 hours on 2 October. As in figure 3, with low temperatures and low temperature gradients, measured and calculated values agreed well. However, at other nighttime periods agreement was not always as good as is shown in figures 3 and 5.

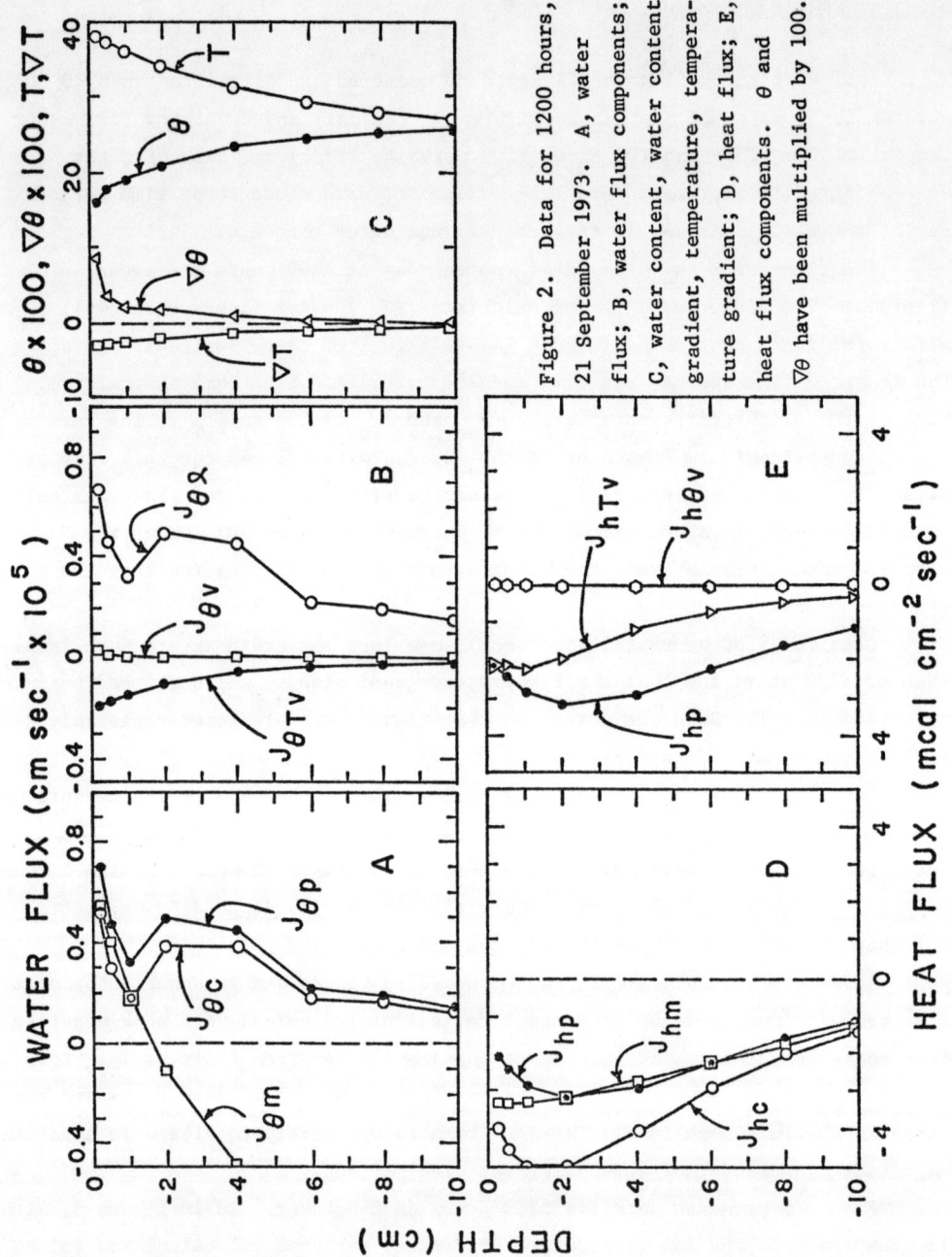

Figure 2. Data for 1200 hours, 21 September 1973. A, water flux; B, water flux components; C, water content, water content gradient, temperature, temperature gradient; D, heat flux; E, heat flux components. θ and $\nabla\theta$ have been multiplied by 100.

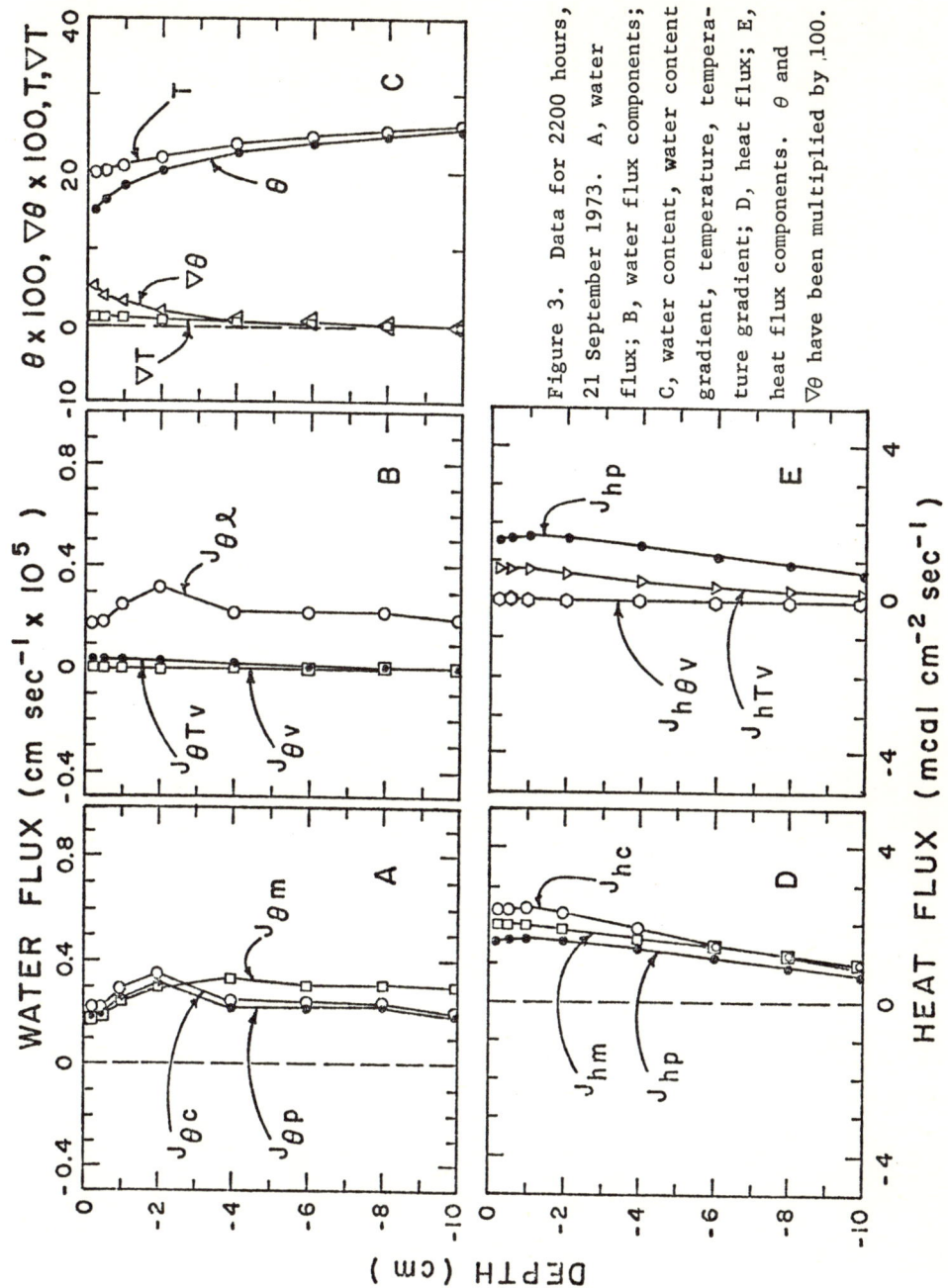

Figure 3. Data for 2200 hours, 21 September 1973. A, water flux; B, water flux components; C, water content, water content gradient, temperature, temperature gradient; D, heat flux; E, heat flux components. θ and $\nabla\theta$ have been multiplied by .100.

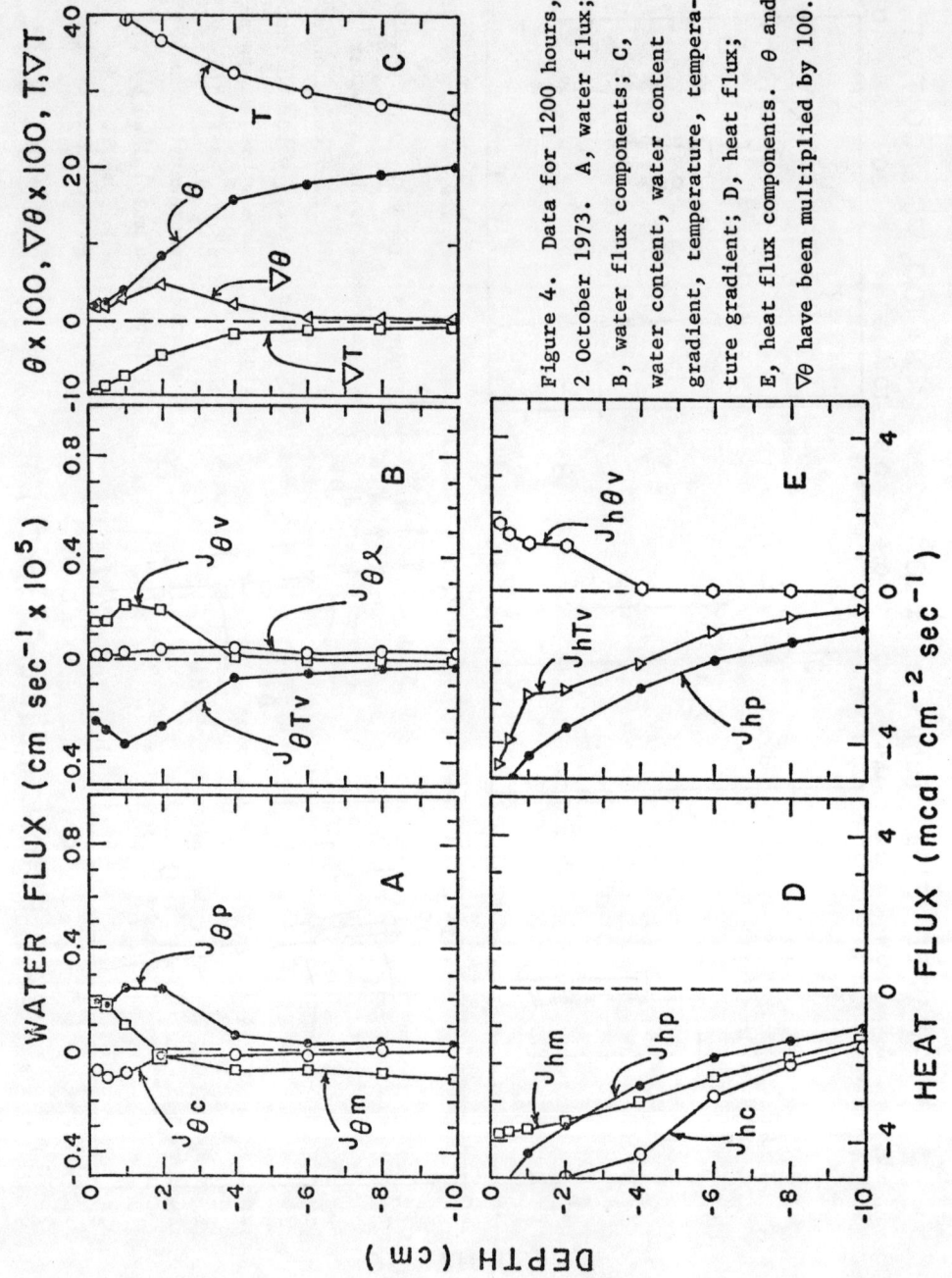

Figure 4. Data for 1200 hours, 2 October 1973. A, water flux; B, water flux components; C, water content, water content gradient, temperature, temperature gradient; D, heat flux; E, heat flux components. θ and $\nabla\theta$ have been multiplied by 100.

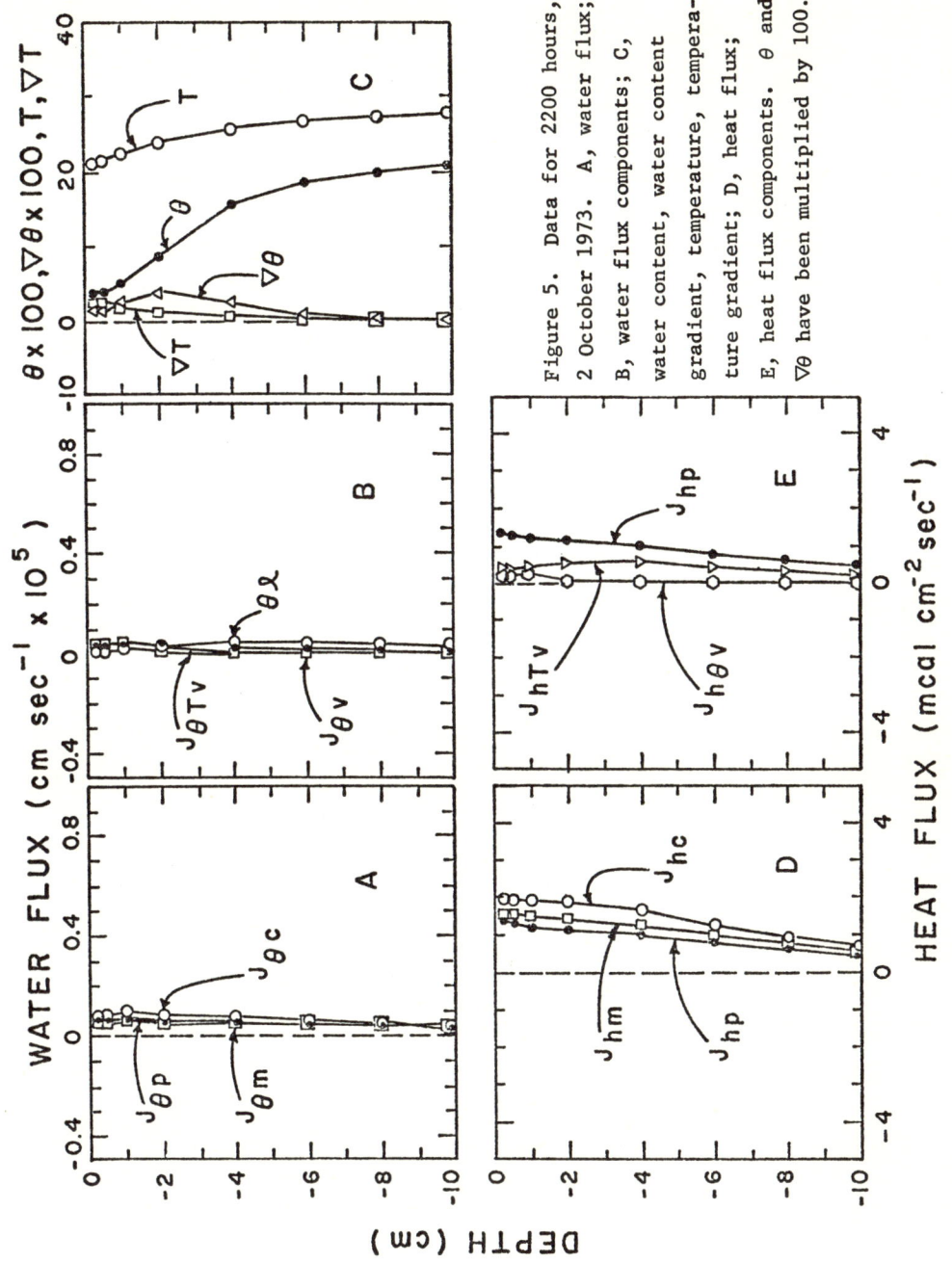

Figure 5. Data for 2200 hours, 2 October 1973. A, water flux; B, water flux components; C, water content, water content gradient, temperature, temperature gradient; D, heat flux; E, heat flux components. θ and $\nabla\theta$ have been multiplied by 100.

CONCLUDING REMARKS

Experimental data of the accuracy required for stringent tests of simultaneous heat and water transfer theories are extremely difficult to obtain. Several sources of error may have affected the measured data used here. For instance, the $D_{\theta\ell}$ values may be too large at the higher water contents. Refinements in the theory underlying the various coefficients may improve the prediction of both heat and water flux. Such refinements appear particularly necessary for prediction of vapor fluxes due to thermal gradients.

LITERATURE CITED

[1] Philip, J. R., and D. A. de Vries. 1957. Moisture movement in porous materials under temperature gradients. Trans. Amer. Geophys. Union 38:222-232.
[2] de Vries, D. A. 1958. Simultaneous transfer of heat and moisture in porous media. Trans. Amer. Geophys. Union 39:909-916.
[3] Rose, C. W. 1968. Water transport in soil with a daily temperature wave. I. Theory and experiment. Aust. J. Soil Res. 6:31-44.
[4] Rose, C. W. 1968. Water transport in soil with a daily temperature wave. II. Analysis. Aust. J. Soil Res. 6:45-57.
[5] Wierenga, P. J., and C. T. de Wit. 1970. Simulation of heat transfer in soils. Soil Sci. Soc. Amer. Proc. 34:845-848.
[6] Kimball, B. A. Smoothing data with Fourier transformations. Agron. J. (in press)
[7] Kimball, B. A., and R. D. Jackson. Soil heat flux determination: A null-alignment method. (in preparation)
[8] Jackson, R. D. 1963. Temperature and soil-water diffusivity relations. Soil Sci. Soc. Amer. Proc. 27:363-366.
[9] Jackson, R. D. 1964. Water vapor diffusion in relatively dry soil: I. Theoretical considerations and sorption experiments. Soil Sci. Soc. Amer. Proc. 28:172-176.
[10] Jackson, R. D. 1964. Water vapor diffusion in relatively dry soil: II. Desorption experiments. Soil Sci. Soc. Amer. Proc. 28:464-466.
[11] Jackson, R. D. 1965. Water vapor diffusion in relatively dry soil: IV. Temperature and pressure effects on sorption diffusion coefficients. Soil Sci. Soc. Amer. Proc. 29:144-148.
[12] de Vries, D. A. 1952. The thermal conductivity of granular materials. Annexe 1952-1 Bul. Institute Interm. du Froid. pp. 115-131.
[13] de Vries, D. A. 1963. Thermal properties of soils. In Van Wijk, W. R. Physics of the plant environment. John Wiley & Sons, Inc., New York, pp. 210-235.

5

THE EARLY STAGES OF INFILTRATION INTO A SWELLING SOIL

DAVID E. SMILES and PETER M. COLOMBERA

CSIRO Division of Environmental Mechanics, Canberra, Australia

ABSTRACT

A theory of water movement in swelling soils exists. Some aspects of this theory have been examined for two-phase systems, but systems containing air present experimental as well as conceptual difficulties.

A simple experiment is described in which infiltration occurs into an initially dry swelling soil that is restrained laterally but may swell vertically. Some physical aspects of the process are demonstrated, as well as the Lagrangian formalism necessary for its mathematical description.

SYMBOLS USED:

Φ = total potential of soil water (cm)

Ψ = moisture potential (cm)

Z = gravitational potential of soil water (cm)

Ω = overburden potential of soil water (cm)

v = volume flux of water (cm s^{-1})

k = hydraulic conductivity of soil (cm s^{-1})

z = vertical coordinate (cm)

ϑ = moisture ratio

e = void ratio

t = time (s)

m = material (Lagrangian) coordinate (cm)

D_m = moisture diffusivity in Lagrangian system (cm^2 s^{-1})

θ = volumetric moisture content

i = cumulative volume of infiltrated water (cm)
S = sorptivity (cm s$^{-\frac{1}{2}}$)
$\lambda = mt^{-\frac{1}{2}}$ (cm s$^{-\frac{1}{2}}$)
D = moisture diffusivity in Eulerian space (cm^2 s^{-1})
$\Lambda = zt^{-\frac{1}{2}}$ (cm s$^{-\frac{1}{2}}$)
g = gravitational constant (cm s^{-2})
ΔPE = change in gravitational potential energy (g s^{-2})
ρ_w = density of water (g cm^{-3})
ρ_s = soil solid density (g cm^{-3})

INTRODUCTION

Although a number of practical and theoretical problems of water flow in two-phase swelling systems have been studied [1,2,3,4,5], there appears to have been little work on corresponding three-phase situations despite a preliminary theoretical approach by Philip and Smiles [6].

The work that has been performed appears, in many cases, to examine situations which, though they give insights into the phenomena affecting water movement in swelling soils, fail to realise the conditions necessary to permit test of the theory proposed by Philip and Smiles. For example, the work of Collis-George and his school [7,8,9,10] has involved study of the process of infiltration into columns of aggregated swelling soils. Rigid columns of various diameters have been used, elastic columns have been used, thermal effects and structural breakdown have been observed, and manometric pressure changes measured. These experiments have identified aspects of the experimental method and of the interaction of water with swelling soil that should be important, but no test of the theory has been attempted. Similarly, Warkentin and his co-workers [11,12] have examined infiltration in confined and 'unconfined' swelling samples, but again these experiments were not examined in terms of the proposed theory; indeed, Gumbs and Warkentin [12] appear to have used conventional non-swelling soil theory to calculate diffusivity functions.

Since ad-hoc experimentation is inevitably less productive in the real world than established predictive theory, it appears important that the theoretical approach of Philip and Smiles [6] should be examined in appropriately designed experiments that either show it to be false, or provide the necessary soil characteristics to permit its further predictive use. In the first instance, it appears sensible to make these experiments as simple as is consistent with the demands of the theory and yet still demonstrate the central phenomenon of macroscopic one-dimensional volume change accompanying water movement. It is perhaps unwise to generalize on the nature of restraints experienced by field soils, but it seems reasonable [13] to examine first the case of a soil constrained to swell vertically but not horizontally. This communication describes some experiments in which an attempt is made to attain these conditions.

THEORY

The vertical flow of water in a swelling soil may be described in terms of theory based on (i) the total potential, Φ, of the water given by

$$\Phi = \Psi + Z + \Omega \tag{1}$$

where Ψ is the moisture potential, Z is the gravitational potential and Ω the overburden potential [13], (ii) Darcy's law,

$$v = - K \frac{\partial \Phi}{\partial z} \tag{2}$$

in which z is the vertical coordinate, K is the hydraulic conductivity, and v the volume flux relative to the rest frame of the particles, and (iii) the continuity statement in an appropriate coordinate system.

It is necessary that both K and Ψ, and in addition the void ratio, e, be known functions of the moisture ratio, ϑ. See [13] for a full discussion.

Experience with saturated systems [14], as well as data of Gumb and Warkentin [12], suggest that for these situations the vertical movement of water following a sustained step change in ϑ at a boundary may be considered as a sorption phenomenon for quite long periods of time. We thus examine here the movement of water into a vertical column of swelling soil whose initial moisture ratio is constant (ϑ_n) and where the ϑ at the surface, at time t = 0, is set and maintained at ϑ_o. The 'early' stages of flow are examined in terms of the 'gravity free' equation

$$\frac{\partial \vartheta}{\partial t} = \frac{\partial}{\partial m} \left(D_m \frac{\partial \vartheta}{\partial m} \right) \tag{3}$$

in which

$$D_m = \frac{K}{(1 + e)} \frac{d\Psi}{d\vartheta} \tag{4}$$

and m is a material coordinate defined by

$$m = \int_0^z (1 + e)^{-1} \, dz. \tag{5}$$

z is defined positive downwards. In terms of m-space, the experimental conditions to be realised are

$$\vartheta = \vartheta; \; m = 0; \; t \geqslant 0$$
$$\vartheta = \vartheta; \; m > 0; \; t = 0 \tag{6}$$

EXPERIMENTAL

A vertical column approximately 30 cm long was constructed of perspex sections 1 cm long and 2.25 cm internal diameter. The composite column was mounted in a glass tube of internal diameter sufficient to permit ready vertical movement of the perspex sections.

The column was packed with a fine soil sample that had passed a 0.25 mm sieve. A sample of topsoil from a chermozemic type black soil containing approximately 65% clay size material was used. The soil in the column was restrained laterally but restrained vertically only by self-weight.

Water was permitted to infiltrate into this column. The water was applied to the surface of the column through a porosity 1 sintered glass plate at the surface of which the water pressure was initially -1 cm. The water source was mounted on a horizontal lever that maintained a small constant stress on the soil surface through the sintered plate, and at the same time maintained contact between the plate and the soil surface as the latter rose during the infiltration process.

The reservoir supplying water to the sintered plate was maintained at a constant height throughout the infiltration process. As a result, the suction at the surface increased by 2-3 cm during the experiment. Since the soil column was initially air dry, this small change was not considered to significantly affect the process.

During each experiment, the position of the soil surface relative to a fixed datum was recorded, as was the position of the wetting front, and the cumulative volume of infiltrated water, i.

At the conclusion of the experiment, the column was destructively sampled and the water content distribution determined. Four experiments were performed and terminated after 2.16×10^3, 4.2×10^3, 7.74×10^3, and 9.64×10^3 s.

Fig. 1 shows the cumulative volume of infiltrated water during each experiment. Figs. 2 show the moisture content profiles at the conclusion of the experiments. Fig. 2a presents $\vartheta(mt^{-\frac{1}{2}})$, and Fig. 2b $\theta = \vartheta/(1 + e)$ as a function of $zt^{-\frac{1}{2}}$. Fig. 3 shows the way the soil surface and the wetting front move during the course of the experiments.

DISCUSSION

Fig. 1 shows that the cumulative infiltration i is linear with respect to (time)$^{\frac{1}{2}}$ for at least 6×10^3 s. This is consistent with the assumption that Equation (3) is an appropriate flow equation for the time period involved, and that the conditions (6) are effectively realised. It should be noted that it is not common [16] for $i(\sqrt{t})$ to be linear for such extended periods of time in non-swelling soils. Fig. 1 also shows excellent agreement between four experiments and indicates for the conditions of the experiment a sorptivity, S, of 0.111 cm s$^{-\frac{1}{2}}$.

Fig. 2a, however, provides the most convincing evidence that the process of infiltration in this material may be considered in terms of Equations (3) and (6) for the time period involved. This arises because the substitution $mt^{-\frac{1}{2}} = \lambda$ eliminates m and t from conditions (6) and reduces (3) to a non-linear equation in ϑ and λ. By implication, therefore, if (3) and (6) are appropriate, $\vartheta(\lambda)$ should be unique, as the experimental data indicate.

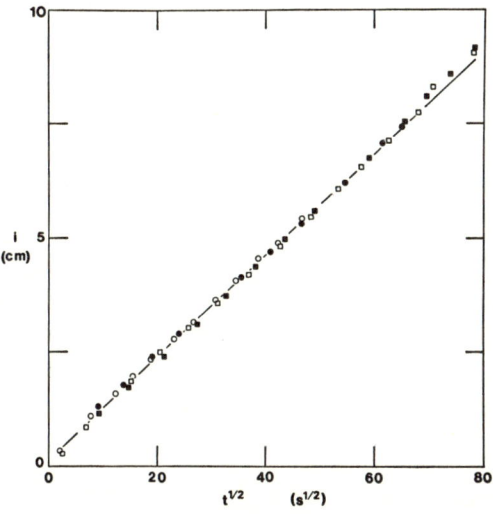

Fig. 1. Cumulative infiltration, i, presented with respect to the square root of time, $t^{\frac{1}{2}}$, for the four experiments described in the text. In this and subsequent figs. the experiments are identified by the following symbols.

○ = 2.16 × 10^3 s
● = 4.20 × 10^3 s
□ = 7.74 × 10^3 s
■ = 9.64 × 10^3 s

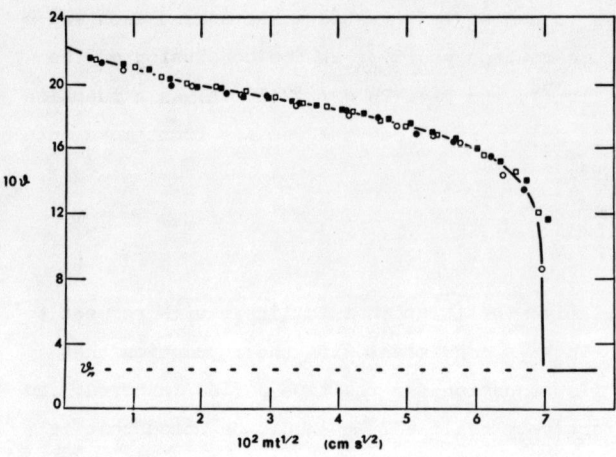

Fig. 2a. Moisture ratio, ϑ, graphed as a function of $mt^{-\frac{1}{2}}$ for the experiments described in the text.

Furthermore, $\int_{\vartheta n}^{\vartheta_0} \lambda \, d\vartheta = 0.113$ cm s$^{-\frac{1}{2}}$ agrees, as it should, with S calculated from Fig. 1. Dm(ϑ) calculated using the method of Matano [17] and Fig. 2a is shown in Fig. 4, thence, following Philip and Smiles [6], D may be calculated from the equation

$$D_m = \left[(1 + e)^{-2} - \vartheta(1 + e)^{-3} \, de/d\vartheta \right] D.$$

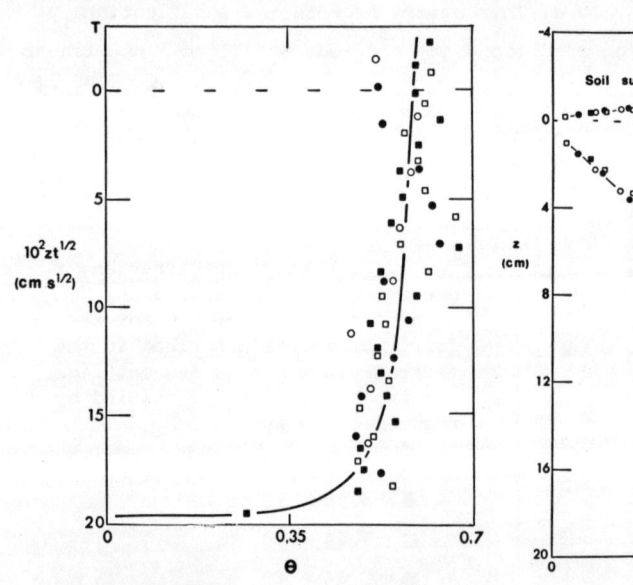

Fig. 2b. Volumetric moisture content, θ, graphed as a function of $zt^{-\frac{1}{2}}$ for the experiments described in the text.

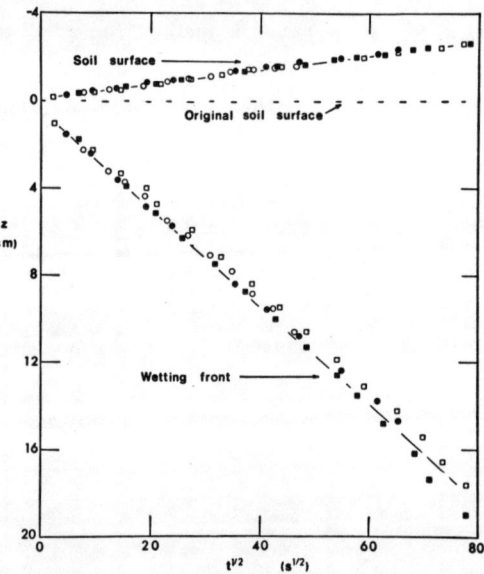

Fig. 3. Displacements of the soil surface and the wetting front as functions of the square root of time, $t^{\frac{1}{2}}$.

The Early Stages of Infiltration into a Swelling Soil

Only if $de/d\vartheta = 0$ will the calculation of D based on $\theta(zt^{-\frac{1}{2}})$ [12] be correct, even though $\theta(zt^{-\frac{1}{2}})$ preserves similarity, and of course, if $de/d\vartheta = 1$,

$$D_m = (1 + e - \vartheta)(1 + e)^{-3} D .$$

This situation appears to occur for $\vartheta > 1.4$ in the experiments described here. The phenomenon is not strictly normal shrinkage, however, because $\vartheta < e$.

Referring now to Fig. 2b, we note that $\theta(zt^{-\frac{1}{2}})$ appears to maintain similarity, but because of the problem of measuring bulk volume in the swelling column and sampling appropriately, the error in θ is very great, despite the fact that ϑ can be measured with great accuracy. The experimental data of Fig. 3 confirm that the variation in θ is a consequence of error of measurement rather than real variation in bulk density. Had variation in bulk density occurred we would not have observed such marked linearity in the relations between $t^{\frac{1}{2}}$ and the displacements of the soil surface and the wetting front. This figure demonstrates a basic difference between infiltration in non-swelling soils and swelling soils: viz. that in the former case the gravitational potential energy of the liquid decreases throughout the process and that of the solid remains constant; while in a swelling soil the potential energy of the solid increases as that of the water decreases.

If there is a net decrease in the gravitational potential energy of the soil system, the process retains aspects in common with infiltration in a non-swelling soil: if the net change is zero, Equation (3) exactly describes the process; a net increase in gravitational potential energy indicates that the infiltration process is analogous to capillary rise in a non-swelling soil.

If the change in gravitational potential energy of the water (ΔPE_w) is defined as the difference in potential energy of the spatial water distribution between times t and t = 0, then choosing for convenience $\Lambda = zt^{-\frac{1}{2}} = 0$ as datum

$$\Delta PE_w/gt = \rho_w [\int_0^T \theta \Lambda \, d\Lambda - \int_\infty^0 (\theta - \theta_n) \Lambda \, d\Lambda]$$

while the corresponding change for the solid, ΔPE_s, is given by

$$\Delta PE_s/gt = \rho_s [\int_0^T (\theta/\vartheta) \Lambda \, d\Lambda - \int_\infty^0 (\theta/\vartheta - \theta_n/\vartheta_n) \Lambda \, d\Lambda]$$

in which $\rho_s = 2.7$, θ_n corresponds to ϑ_n, and $\Lambda = T$ is the top of the swelling soil column. It should be noted that $\Lambda = zt^{-\frac{1}{2}} = 0$ gives the position of the top of the column at t = 0.

For Fig. 2b, $\Delta PE_w/gt \simeq -10.3 \times 10^{-3}$ g cm^{-1} s^{-1} while $\Delta PE_s/gt \simeq +4.4 \times$

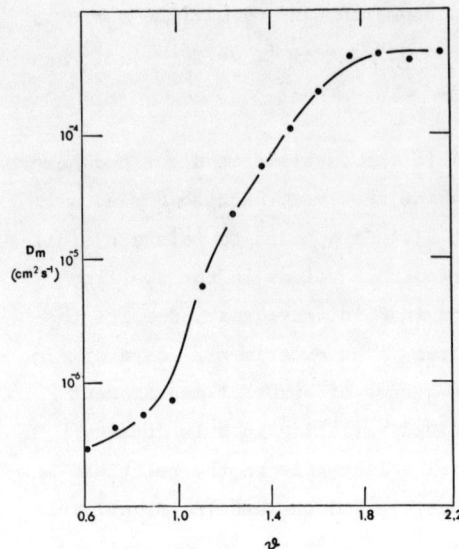

Fig. 4. Relationship between D_m and the moisture ratio, ϑ.

10^{-3} g cm^{-1} s^{-1}. For this material, therefore, the analogy is retained with infiltration in a non-swelling soil, although the apparent effect of gravity is reduced. This situation therefore differs from the two-phase infiltration described in [15] where water displaces only the more dense solid, and implies that some generalizations [13] concerning infiltration into swelling soils must be qualified for three-phase systems.

Finally, referring to Fig. 4, we note that the $D_m(\vartheta)$ relation is quite different to that observed for two-phase swelling systems where D_m is a decreasing function of ϑ. Further work is in progress to determine the range of expectation of $D_m(\vartheta)$ as well as the $\Psi(\vartheta)$ and $e(\vartheta)$ relations necessary to more fully characterise the material.

In conclusion, we note that the experimental method appears to impose conditions such that one can realistically examine the process of infiltration in a swelling soil, and test existing theory. The technique may be modified to measure stress, water pressure, and thermal effects. In addition, the experiments described here identify some of the physical processes that occur during infiltration into a swelling soil, and demonstrate the use of the Lagrangian mathematics necessary for its formal description.

REFERENCES

1. McNabb, A. 1960. Quarterly Appl. Maths. 17, 337-47.

2. Raats, P.A.C. 1965. Ph.D. Thesis University of Illinois, Urbana.
3. Gibson, R.E., England, G.L., and Hussey, M.J.L. 1967. Geotechnique 17, 261-73.
4. Shirato, M., Sambuichi, M., Kato, H., and Aragaki, T. 1969. Amer. Inst. Ch. Eng. Journal 15, 405-9.
5. Smiles, D.E., and Rosenthal, M.J. 1968. Aust. J. Soil Res. 6, 237-48.
6. Philip, J.R., and Smiles, D.E. 1969. Aust. J. Soil Res. 7, 1-19.
7. Collis-George, N., and Lal, R. 1970. Aust. J. Soil Res. 8, 195-207.
8. Collis-George, N., and Lal, R. 1971. Aust. J. Soil Res. 9, 107-16.
9. Lal, R., Bridge, B.J., and Collis-George, N. 1970. Aust. J. Soil Res. 8, 185-93.
10. Collis-George, N., and Lal, R. 1973. Aust. J. Soil Res. 11, 93-105.
11. Yong, R.N., and Warkentin, B.P. 1972. Proc. 2nd Symp. on Fundamentals of Transport Phenomena in Porous Media IAHR-ISSS. Vol. 1, pp. 306-19. (Guelph, Canada.)
12. Gumbs, F.A., and Warkentin, B.P. 1972. Soil Sci. Soc. Amer. Proc. 36, 720-4.
13. Philip, J.R. 1971. In 'Salinity and Water Use'. Proc. Symp. Aust. Acad. Sci., Canberra, 1971. pp. 95-107. (Macmillan: London.)
14. Smiles, D.E. 1974. Trans. Int. Congr. Soil Sci. 10th (Moscow) - in press.
15. Smiles, D.E. 1974. Soil Sci. - in press.
16. Talsma, T. 1969. Aust. J. Soil Res. 7, 269-76.
17. Matano, C. 1932. Jap. J. Phys. 8, 109-33.

SIMULTANEOUS HEAT AND MASS TRANSFER IN SOILS WITH SUBSURFACE HEATED POROUS PIPES

DAVID SLEGEL, LORIN DAVIS, and LARRY BOERSMA

Oregon State University, Corvallis, Oregon, U.S.A. 97331

ABSTRACT

The development of transient equations describing the heat and mass transfer in soils is presented. Both liquid and vapor phases are considered. The results of solving these equations numerically on a digital computer for temperature and moisture content are presented. Boundary conditions were chosen to represent a subsurface soil warming and irrigation system consisting of a series of submerged parallel porous, heated pipes. The spacing and depth of the pipes were varied to find their effect on temperature and moisture content distributions.

NOMENCLATURE

c_p	specific heat at constant pressure (cal/g K)
c_v	specific heat at constant volume (cal/g K)
D	vapor diffusivity (cm^2/sec)
E	evaporation rate (g/cm^3 sec)
g	acceleration of gravity (cm/sec^2)
G	Gibbs free energy (cal/g)
h	enthalpy (cal/g)
h_{fg}	heat of vaporization (cal/g)
i	vertical unit vector
k	hydraulic conductivity (cm/sec)
P	pressure (dynes/cm^2)
q	mass flux (g/cm^2 sec)
Q	heating rate (cal/cm^3 sec)

R	universal gas constant (cal/g K)
S	porosity (cm^3/cm^3)
t	time (sec)
T	temperature (K)
U	internal energy (cal/cm^3)
v	velocity (cm/sec)
z	gravitational potential or the vertical coordinate (cm)
∇	gradient operator (cm^{-1})
θ	water content (cm^3/cm^3)
λ	thermal conductivity (cal/cm sec K)
ρ	density (g/cm^3)
ϕ	total potential (cm)
ψ	thermodynamic potential of water in soil (cm)

Subscripts

ℓ	liquid
v	vapor
vo	vapor in air-filled pore
w	pure liquid water

INTRODUCTION

The projected increase in demand for electrical energy has created a greater awareness of the need to find beneficial uses for the "waste heat" from electrical power plants. The term "waste heat" describes the heat rejected from the plant in the form of warm, condenser cooling water. The search for beneficial uses has led to many warm water utilization schemes [1]. Among these is a proposed subsurface piping system carrying the condenser cooling water, which is utilized for heating and irrigating soil for agricultural purposes. This application might extend the growing season, accelerate crop growth, and provide an efficient method of irrigation. The method has been field tested on a limited scale [2]. This paper presents the development of a set of partial differential equations which describe the combined heat and mass transfer in soils. The results of expressing the equations in finite difference form and solving numerically on a digital computer are presented for a sandy soil containing a system of parallel, heated, porous pipes. The spacing and depth of the pipes were varied to present a method for optimization of the depth and spacing for a given soil, crop, and location. The parameters for optimization calculated were the increase in moisture content, the percent of the soil maintained above 24 C, and the average increase in temperature.

Much of the work in the field of combined heat and mass transfer in soils has dealt with the thermodynamics of irreversible processes. Cary and Taylor [3, 4] have developed equations from the theory of irreversible thermodynamics for the flux of heat and moisture in soils for saturated and unsaturated conditions, and have experimentally found the phenomenological coefficients for these equations for various soils. Cary [5] has extended these experiments and equations to liquid, vapor and heat transfer in soils. These equations, however, are not applied to continuity or energy equations.

Philip and de Vries [6] have adopted a more classical approach utilizing Darcy's law for liquid flux,

$$q_\ell = -\rho_w k \nabla \phi,$$

which they transformed to

$$q_\ell = -D_{T\ell} \nabla T - D_{\theta\ell} \nabla \theta - \rho_w k i.$$

A diffusion equation for vapor transfer was similarly transformed from dependence on the gradient of vapor density to dependence on the gradients of temperature and liquid water content. These equations were then substituted into the continuity equations. Subsequently, de Vries [7] utilized the equations for liquid and vapor flow, and the continuity equations for vapor and liquid phases to develop an equation for evaporation rates. An energy equation was also developed utilizing the moisture flux equations. These equations were then simplified to those corresponding to a steady-state one-dimensional case in which the liquid and temperature gradients are expressed in terms of the diffusion coefficients and the heat and mass fluxes.

Fritton et al. [8], using measured diffusion coefficients, compared results of the equations of Philip and de Vries to experimental results for soil exposed to several surface conditions. To obtain agreement between experimental and theoretical results they had to divide the measured water diffusivity by factors of 35 and 10. This discrepancy was attributed to hysteresis in the water tension curve and to the temperature dependence of the hydraulic conductivity.

ANALYSIS

In the present investigation Darcy's equation for liquid flow in porous media was used, which can be written as,

$$q_\ell = -\rho_w k \nabla \phi. \qquad (1)$$

Vapor flow was expressed by the diffusion equation as

$$q_v = -D\nabla \rho_{vo}. \tag{2}$$

The continuity equations for vapor and liquid phases are respectively

$$\frac{\partial \rho_v}{\partial t} + \nabla \cdot q_v = E, \tag{3}$$

and

$$\frac{\partial \rho_\ell}{\partial t} + \nabla \cdot q_\ell = -E. \tag{4}$$

Since $\rho_v = (S-\theta)\rho_{vo}$ and $\rho_\ell = \theta \rho_w$, the continuity equations may be written as

$$\frac{\partial}{\partial t}[(S-\theta)\rho_{vo}] - \nabla \cdot D\nabla \rho_{vo} = E,$$

and

$$\frac{\partial}{\partial t}(\theta \rho_w) - \nabla \cdot \rho_w k \nabla \phi = -E.$$

By comparison the term $-\rho_{vo}(\partial \theta / \partial t)$ can be eliminated from the vapor equation since $\rho_{vo} \ll \rho_w$. The vapor continuity equation was rewritten as

$$E = (S-\theta)\frac{\partial \rho_{vo}}{\partial t} - \nabla \cdot D\nabla \rho_{vo}. \tag{5}$$

Since ρ_w is constant, the liquid continuity equation becomes

$$\rho_w \frac{\partial \theta}{\partial t} - \nabla \cdot k\nabla \phi = -E. \tag{6}$$

The energy equation is written as

$$Q = \frac{\partial U}{\partial t} + \nabla \cdot \rho_j v_j h_j, \tag{7}$$

where the subscript j denotes summation for all species, and where

$$U = \rho_\ell U_\ell + \rho_v U_v + \rho_{soil} U_{soil} + \phi \quad \text{and} \quad Q = \nabla \cdot \lambda \nabla T.$$

After substituting the above, the energy equation becomes

$$\nabla \cdot \lambda \nabla T = \frac{\partial}{\partial t}(\rho_\ell U_\ell + \rho_v U_v + \rho_{soil} U_{soil} + \phi) + \nabla \cdot (\rho_\ell h_\ell v_\ell + \rho_v h_v v_v).$$

For the range of moisture contents considered, ϕ is small and neglected. After expansion the energy equation becomes

$$\nabla \cdot \lambda \nabla = \rho_\ell \frac{\partial U_\ell}{\partial t} + \rho_v \frac{\partial U_v}{\partial t} + \rho_{soil} \frac{\partial U_{soil}}{\partial t} + \rho_\ell v_\ell \cdot \nabla h_\ell + \rho_v v_v \cdot \nabla h_v$$

$$+ h_\ell (\nabla \cdot \rho_\ell v_\ell + \frac{\partial \rho_\ell}{\partial t}) + h_v (\nabla \cdot \rho_v v_v + \frac{\partial \rho_v}{\partial t})$$

$$- P_\ell / \rho_\ell \frac{\partial \rho_\ell}{\partial t} - P_v / \rho_v \frac{\partial \rho_v}{\partial t} .$$

After neglecting the mechanical work terms, $P/\rho(\partial \rho / \partial t)$, substitution of equations 5 and 6 and the relations, $dU = c_v dT$ and $h_{fg} = h_v - h_\ell$ yields

$$\nabla \cdot \lambda \nabla T = (\rho_\ell c_{v_\ell} + \rho_{soil} c_{v_{soil}}) \frac{\partial T}{\partial t} + \rho_\ell c_{p_\ell} v_\ell \cdot \nabla T + h_{fg} E + \rho_v c_{p_v} v_v \nabla T .$$

Since $\rho_w v_\ell = -\rho_w k \nabla \phi$, $\rho_\ell = \theta \rho_w$, and $\rho_v c_{p_v} \ll \rho_\ell c_{p_\ell}$, the energy equation becomes

$$\nabla \cdot \lambda \nabla T = (\theta \rho_w c_{v_\ell} + \rho_{soil} c_{v_{soil}}) \frac{\partial T}{\partial t} - \theta \rho_w c_{p_\ell} k \nabla \phi \cdot \nabla T + h_{fg} E . \qquad (8)$$

In order to find ρ_v as a function of temperature and moisture content, local thermodynamic equilibrium is assumed. The free energy of the water in the soil is

$$G_\ell = g \psi$$

assuming $G = 0$ for pure water. The free energy of the vapor is

$$G_{vap} = RT \ln P_v / P_{vo} ,$$

where P_v is the vapor pressure and P_{vo} is the saturated vapor pressure. For equilibrium $G_\ell = G_{vap}$ and therefore

$$RT \ln P_v / P_{vo} = g \psi ,$$

or

$$\frac{P_v}{P_{vo}} = e^{\psi g / RT} .$$

Using the ideal gas law for constant temperature at the liquid-vapor interface, $P/\rho =$ constant, then the relationship becomes

$$\frac{\rho_{v_o}}{\rho_{v_{sat}}} = e^{\psi g/RT}, \tag{9}$$

Equations 5, 6, 8 and 9 now provide four equations which yield solutions for ρ_v, E, T and θ, when data for ψ, k, D and λ is provided.

One difference between these equations and those of de Vries is that these equations contain gradients of vapor density and water tension whereas de Vries' transformed equations contain temperature and water content gradients only. Equations 5, 6, 8, and 9 have been expressed in finite difference form and solved numerically using an implicit scheme on a CDC 3300 digital computer.

RESULTS

Several test cases were run to check the results of the computer program against the results of laboratory experiments. Plots of the isotherms and lines of constant moisture content for one computer test case are presented in Figure 1. Also shown in Figure 1 for comparison are selected experimental results which are underlined. The experiment [9] modeled by Figure 1 was for a sandy soil in a container with insulated sides and bottom, atmospheric conditions 22.7 C and 70% relative humidity, and with a pipe on the wall at 32 cm depth. The pipe was maintained at 29 C and the moisture content of the adjacent soil was maintained at .27 cm^3/cm^3. The measured and calculated results of Figure 1 are in good agreement.

Having thus validated the procedure, calculations were made to obtain temperature and water content profiles for various spacings and depths for summer conditions (August) and winter conditions (January). The results shown were obtained by imposing heat and irrigation from the subsurface sources upon initial temperature and water content distributions, while assuming (i) average weather conditions for January or August, typical of the Willamette Valley in Oregon, (ii) a constant temperature of 10 C at a depth of 1200 cm, (iii) a saturated condition at a depth of 200 cm with the water content θ = 0.45 cm^3/cm^3. The computational program was run until the initial temperatures and water contents approached constant values. Heating and water application were then initiated and calculations proceeded for 31 days using one day time steps. The weather conditions used were obtained by averaging the past conditions for each day over a ten year period. Therefore, the ambient conditions used varied from day to day. The pipe temperatures were chosen to correspond to the condenser cooling water temperature for a power plant near Portland, Oregon using a cooling tower as the primary heat sink. These temperatures are 41 C for August and 29 C for January. The water content at the pipe was assumed to be 0.44 cm^3/cm^3. Data for ψ, k, and λ were for a sandy soil.

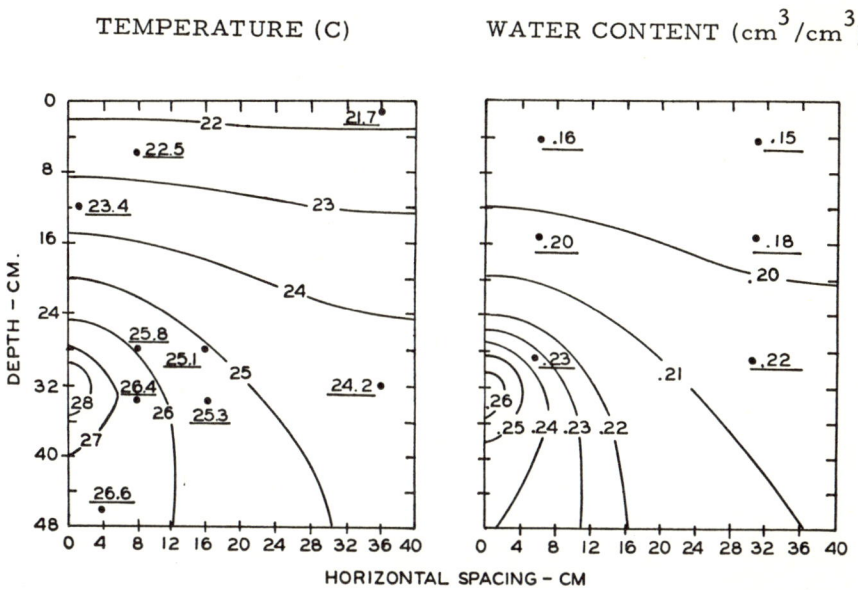

Figure 1. Test case isotherms (C) and lines of constant moisture content (cm^3/cm^3). Experimental values are underlined. Because of symmetry only one segment of the total profile is shown.

Calculations were made for pipe depths of 50 and 100 cm and for spacings of 140, 280, and 560 cm. Figure 2 presents typical lines of constant temperature and moisture content as functions of depth and horizontal distance from the pipe for August weather conditions. The initial temperatures and moisture contents are presented to the right of the plots. Table 1 presents the percent of the root zone maintained above 24 C, the average increase in temperature in the root zone, and the average increase in moisture content in the root zone for root zones of 100 and 200 cm depth.

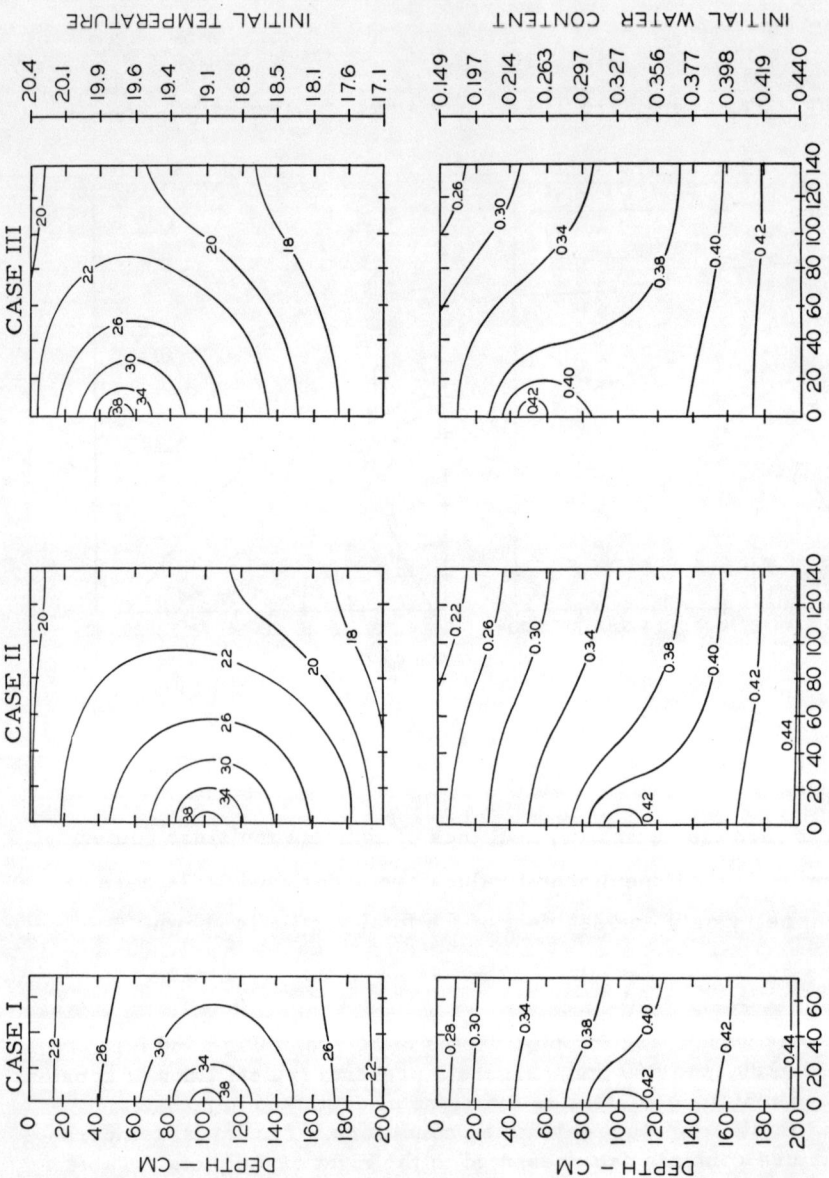

Figure 2. Predicted moisture content (cm^3/cm^3) in the bottom row of figures and temperatures (C) in the top row of figures in a sandy soil for three combinations of depth and spacing (cm) of a heated porous pipe system and summer weather conditions.

Table 1. Percent of root zone above 24 C, average temperature increase, and average increase in moisture content.

Root Zone Depth	Spacing	Pipe Depth (cm)		Pipe Depth (cm)		Pipe Depth (cm)	
		50	100	50	100	50	100
cm	cm	Percent of Soil Above 24 C		Ave. Temp. Increase (C)		Ave. Moisture Increase (cm^3/cm^3)	
January							
200	140	1.08	2.19	5.12	6.06	.0634	.0473
	280	0.57	0.86	2.44	2.83	.0386	.0264
	560	0.42	0.77	1.44	1.76	.0254	.0163
100	140	2.16	1.50	7.99	6.16	.104	.0673
	280	1.14	0.58	4.22	3.05	.0654	.0385
	560	0.84	0.46	2.42	1.91	.0423	.0237
August							
200	140	65.6	77.0	7.12	8.50	.0846	.0638
	280	20.7	28.2	2.31	3.76	.0569	.0397
	560	14.5	16.4	1.99	1.65	.0337	.0211
100	140	83.2	72.2	9.04	7.65	.144	.0997
	280	36.8	28.2	4.15	3.69	.101	.0640
	560	10.0	16.3	2.62	1.60	.0601	.0346

CONCLUSIONS

In Table 1, it is seen that in January a very small percent of the soil has been heated above 24 C. This is due to lower air temperatures and to lower pipe temperatures. The results for August indicate much higher heating effects. In all cases increased pipe spacing produced decreased moisture and heating effects, while the heating effects were greatest for pipe depths that coincided with the middle of the root zone. For a pipe depth of 50 cm the system was able to maintain high moisture content, except for the 560 cm spacing in August. Increasing the pipe depth from 50 to 100 cm reduced the effect on water content.

It is felt that the methods presented in this paper could be of great use for analyzing the effect of depth and spacing on temperature and water content distributions for field systems of subsurface warm porous pipes. From the information obtained from such an analysis, an indication as to the optimum spacing and depth of the piping system can be obtained. The results indicate that such systems might be very successful for subsurface irrigation and heating. The method presented in this work could be improved by extending the calculations over the entire growing season, and by adding a model for the presence of absorption of water by plant roots.

REFERENCES

1. Boersma, L. and Rykbost, K.A. 1973. Integrated Systems for Utilizing Waste Heat from Steam Electric Plants. Journ. Env. Qual. 2:179-188.

2. Sepaskhah, A.R., Boersma, L., Davis, L.R. and Slegel, D.L. 1973. Experimental Analysis of a Subsurface Soil Warming and Irrigation System Utilizing Waste Heat. Paper 73-WA/HT-11, presented at the Winter Annual Meeting ASME, Detroit, Michigan, November 11-15. United Engineering Center, 345 East 47th Street, New York, N.Y. 10017. 12 pp.

3. Cary, J.W. and Taylor, S.A. 1962. The Interaction of the Simultaneous Diffusion of Heat and Water Vapor. Soil Sci. Soc. Proc. 26:413-416.

4. Cary, J.W. and Taylor, S.A. 1962. Thermally Driven Liquid and Vapor Phase Transfer of Water and Energy in Soil. Soil Sci. Soc. Proc. 26:417-420.

5. Cary, J.W. 1965. Water Flux in Moist Soil: Thermal Versus Suction Gradients. Soil Sci. Soc. Amer. 100:168-175.

6. Philip, J.R. and de Vries, D.A. 1957. Moisture Movement in Porous Materials Under Temperature Gradients. Trans. Am. Geophys. Union 38:222-232.

7. de Vries, D.A. 1958. Simultaneous Transfer of Heat and Moisture in Porous Media. Trans. Am. Geophys. Union 39:909-916.

8. Fritton, D.D., Kirkham, D. and Shaw, R.H. 1970. Soil Water Evaporation, Isothermal Diffusion and Heat and Water Transfer. Soil Sci. Soc. Proc. 34:183-189.

9. Sepaskhah, A.R. 1974. Experimental Analysis of Subsurface Heating and Irrigation on the Temperature and Water Content of Soils, Ph.D. Thesis, Oregon State University.

COMMENTS ON COMPUTER MODELING OF A MOISTED SOIL

G.C. VANSTEENKISTE and F. DE SCHUTTER

University of Ghent - Ghent - Belgium

ABSTRACT

It has always been a tremendous task for numerical analyses to solve non-linear partial differential equations. The *modal technique*, presented in this paper, approaches the problem on his own specific way, transforming the *partial* differential equation into two infinite sets of *ordinary* differential equations, very well suited to be solved on a hybrid computer. Since this technique gives rise to only differential equations, it can be classified as a continuous time, continuous space method. Application of this technique to the moisture-flow equation shows that this method gives extremely good results, which are in *close agreement with experimental data*.

PROBLEM STATEMENT

The differential equation for the rate of change of water content in a moisted soil is :

$$\frac{\partial u}{\partial t} = \frac{\partial}{\partial x} (D\frac{\partial u}{\partial x} + K)$$

with u = water content
D = diffusivity
K = hydraulic conductivity.

Following boundary conditions are considered :

x = 0 $D\frac{\partial u}{\partial x} + K = 0$ (no moisture flow)
x = L $D\frac{\partial u}{\partial x} + K = -\Phi$ (with Φ = evaporation rate)

The modal technique is applied to this equation. As described by KOSHLYAKOV, SMIRNOV and GLINER, we have to look for a solution of the form

$$u(x,t) = \sum_{j=1}^{\infty} L_j(x) \, a_j(t)$$

The functions $L_j(x)$ are the eigenfunctions of the system. The differential equations in $a_j(t)$ are completely separated, and can be integrated independently.

DIRECT METHOD

If the diffusivity is time-independent, the first term of the right hand side of the equation

$$\frac{\partial u}{\partial t} = \frac{\partial}{\partial x}(D \frac{\partial u}{\partial x} + K)$$

is self-adjoint. This means that, if we have a solution of the form

$$u(x,t) = \sum_{j=1}^{\infty} a_j(t) \, L_j(x)$$

than $a_j(t)$ is integral-transformation of $u(x,t)$ with kernel $L_j(x)$. So

$$a_j(t) = \frac{1}{C_j} \int_0^L u(x,t) \, L_j(x) dx$$

The denomination C_j is the norm of $L_j(x)$ in the L_2 $(0,L)$-space. So

$$C_j = \int_0^L L_j^2(x) \, dx$$

In the case that the eigenfunctions are normalised, we get a set of beautiful results :

$$u(x,t) = \sum_{j=1}^{\infty} a_j(t) \, L_j(x)$$

$$a_j(t) = \int_0^L u(x,t) \, L_j(x) \, dx$$

The well-known Fourier-series expansion is a special case of this, by which the $L_j(x)$ are $\sin x$ and/or $\cos x$. In this case, the eigenfunctions satisfy the following differential equation :

$$\frac{d}{dx}(D \frac{dL_j}{dx}) + \lambda_j^2 \, L_j = 0$$

with boundary conditions

$x = 0$ $\quad\quad\quad \dfrac{dL_j}{dx} = 0$

$x = L$ $\quad\quad\quad \dfrac{dL_j}{dx} = 0$

The time-dependent parts of the series expansion of $u(x,t)$ satisfy

$$\dfrac{da_j(t)}{dt} + \lambda_j^2 a_j(t) + A_j = N_a - N_b$$

with $A_j = \int_0^L \dfrac{\partial K}{\partial x} L_j(x)\, dx$

$N_a = \dfrac{1}{C_j} \dfrac{1}{D_{x=0}} L_j(0) K_{x=0}$

$N_b = \dfrac{1}{C_j} \dfrac{1}{D_{x=L}} L_j(L)\, (K_{x=L} + \Phi)$

$C_j = \int_0^L L_j^2(x)\, dx$

In this case, the assumption is made that the conductivity is also time-independent.

The direct method has the following disadvantages :
1) The boundary conditions are not satisfied.
2) The computation of A_j requires a differentiation.

MODIFIED METHOD

Consider first the case of no evaporation. In this situation $u(x,t)$ has a steady-state curve. Then, u is only x-dependent, and so are the diffusivity D, and the conductivity K.

If K and D were constant (only x-dependent), it would be easy then to investigate the perturbation of $u(x,t)$ around the steady-state curve. One could take the steady-state curve as a reference curve :

$u' = u - u_{ref}$

satisfies

$$\dfrac{\partial u'}{\partial t} = \dfrac{\partial}{\partial x}\left(D\dfrac{\partial u'}{\partial x}\right)$$

with boundary conditions :

$$x = 0 \quad \frac{\partial u'}{\partial x} = 0$$

$$x = L \quad \frac{\partial u'}{\partial x} = 0$$

Therefore, consider K and D as being constants during the timestep Δt and compute u_{ref}

In the case with evaporation on the other hand, following reasoning can be made. Consider again K and D as being constants during the timestep Δt, and compute a u_{ref} which satisfies :

$$\frac{\partial u_{ref}}{\partial t} = \frac{\partial}{\partial x} (D \frac{\partial u_{ref}}{\partial x} + K)$$

Because of the evaporation, this u_{ref} will be also time-dependent. So :

$$u_{ref} = \alpha(x) + \beta(t)$$

and $\beta'(t) = \frac{\partial}{\partial x} (D\alpha'(x) + K) = C_1$

or $\beta(t) = C_1 t + C_2$

$D\alpha'(x) + K = C_1 x + C_3$

Furthermore, u_{ref} must satisfy the boundary conditions :

$$x = 0 \quad D \frac{\partial u_{ref}}{\partial x} + K = 0$$

$$x = L \quad D \frac{\partial u_{ref}}{\partial x} + K = -\Phi$$

The boundary condition at $x = 0$ makes $C_3 = 0$. At $x = L$ we find :

$$D \alpha'(x)_{x=L} + K = -\Phi$$

which gives :

$$C_1 = -\frac{\Phi}{L}$$

So

$$\beta(t) = -\frac{\Phi}{L} t + C_2$$

$$\alpha(x) = \int_0^x \frac{1}{D} (-\Phi \frac{\tau}{L} - K) \, d\tau + C_4$$

And
$$u_{ref} = -\frac{\Phi}{L}t + \int_0^x \frac{1}{D}(-\Phi\frac{\tau}{L} - K)d\tau + C$$

Since we don't have to specify the initial condition of u_{ref}, we can assign any value to C. So, let's take C equal to zero.

Now : $u' = u - u_{ref}$

satisfies

$$\frac{\partial u'}{\partial t} = \frac{\partial}{\partial x}(D\frac{\partial u'}{\partial x})$$

with boundary conditions :

$x = 0 \qquad \frac{\partial u'}{\partial x} = 0$

$x = L \qquad \frac{\partial u'}{\partial x} = 0$

The eigenfunctions $L_j(x)$ satisfy the following differential equation :

$$\frac{d}{dx}(D\frac{dL_j(x)}{dx}) + \lambda_j^2 L_j(x) = 0$$

and the boundary conditions are :

$x = 0 \qquad \frac{dL_j}{dx} = 0$

$x = L \qquad \frac{dL_j}{dx} = 0$

The time-dependent part of the series expansion of $u(x,t)$ satisfies :

$$\frac{da_j(t)}{dt} + \lambda_j^2 a_j(t) = 0$$

The initial condition of which is

$$a_j(0) = \frac{\int_0^L u'(x,0) L_j(x) dx}{\int_0^L L_j^2(x) dx}$$

The solution of which is :

$$a_j(t) = a_j(0) e^{-\lambda_j^2 t}$$

The solution of the original complete equation is :

$$u = u' + u_{ref}$$

and

$$u' = \sum_{j=1}^{\infty} a_j(t) L_j(x)$$

In this way, the two disadvantages of the direct method disappear. This is very clear, since instead of the preceeding quantity

$$A_j = \int_0^L \frac{\partial K}{\partial x} L_j(x) dx$$

we now have to compute u_{ref}, which is also an integration, but this time, there is no differentiation in the integrand. Furthermore, this integration is done on the analog computer, in the mean time that this computer is solving the differential equation for $L_j(x)$. So, the calculation of u_{ref} *consumes no time at all*, but only analog equipment.

The boundary conditions are now also exactly satisfied. Since at $x = 0$ and at $x = L$

$$\frac{dL_j}{dx} = 0$$

u' satisfies, at the same boundaries,

$$\frac{\partial u'}{\partial x} = 0$$

So : at $x = 0$ and at $x = L$

$$\frac{\partial u}{\partial x} = \frac{\partial u'}{\partial x} + \frac{\partial u_{ref}}{\partial x} = \frac{\partial u_{ref}}{\partial x},$$ and u_{ref} satisfies the boundary conditions

of the fundamental equation.

Following computerorganisation is used

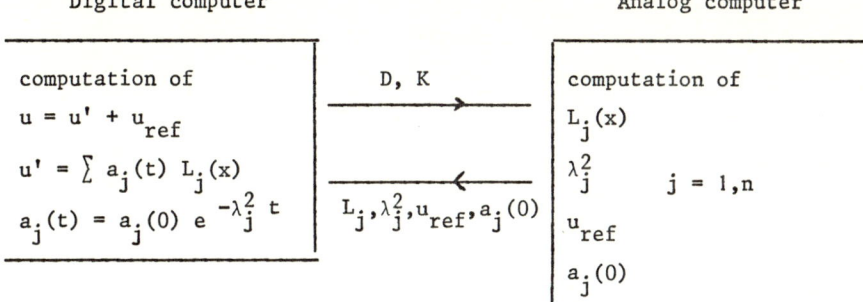

Starting from the initial condition for u(x,t), the digital partner computes the values of the diffusivity and of the conductivity.

It also computes the integrand in the formula for u_{ref}. Then, in *one single analog run,* the analog computer calculates the eigenfunction $L_j(x)$ and the eigenvalue λ_j^2, the reference-curve u_{ref}, $u - u_{ref}$ and also $a_j(0)$. Since all these calculations are simple algebraic ones and integrations, they can all be performed in the same analog run. Meanwhile, the digital computer supplies the analog with the values of D and takes samples of $L_j(x)$ and of u_{ref}.

All this is done for each eigenfunction. Then, the digital partner calculates $a_j(t)$, u' and finally $u = u' + u_{ref}$.

APPLICATION OF THE TECHNIQUE

Previous theory will now be applied to a practical example and it will be verified if the method leads to any valuable result or not.

Therefore experimental data were used from an infiltration of water in the soil. However, we had only data for horizontal infiltration at hand, thus the influence of gravity was left out of the fundamental equation. The water content has been measured and the diffusivity calculated.

These values of the diffusivity are used to simulate the experiment on the hybrid computer.

What is the merit of modal techniques, compared with all the other hybrid and digital methods ?

If the diffusivity were only x-dependent, the fundamental equation is perfectly self-adjoint, and the modal technique would need *no discretisation* at all, yielding the exact-solution, if we could make the summation

$$u(x,t) = \sum_{j=1}^{\infty} a_j(t) L_j(x)$$

until infinity.

Now, when D is also t-dependent (thus D = D(u)), we have to make discretisations, yet not in u, but in the diffusivity D, and in the conductivity K. So, we still have a solution which is continuous in time and continuous in space.

The main advantage, however, is the fact that the differential equation for $a_j(t)$ has an analytical solution, so this needs no integration. Furthermore, and even more important, these time-dependent functions are decaying exponentials. So, this method is unconditionally stable, since all errors are also decaying exponentially.

Let's now proceed to the pratical application of the modal technique.

EXPERIMENTAL MATERIALS AND METHODS

The soil used in these experiments is a silt-loam. The particle size distribution of the soil is given in Table 1.

Bulk density and water content of the soil columns were determined using the gamma-ray attenuation equipment described by VERPLANCKE H. (1973).

Table 1 : Particle size distribution of the soil : silt-loam

Fraction (µ)	%
0 - 2	24
2 - 10	10,2
10 - 20	17
20 - 50	39,6
50 - 100	8,8
100 - 200	0,2
200 - 500	0,1
> 500	0,1

Fraction	%
clay (0 - 2 µ)	24
silt (2 - 50 µ)	66,8
sand (> 50 µ)	9,2
% organic matter	0,17
% $CaCO_3$	0,25

Comments on Computer Modeling of a Moisted Soil

Briefly the equipment consists of a 100 millicurie Am^{241} source, a scintillation detector with a 2.54 cm diameter by 2.54 cm thick thallicum-activated NaI crystal with photomultiplier and an analyzer-scaler

$$\frac{\partial u}{\partial t} = \frac{\partial}{\partial x}(D(u)\frac{\partial u}{\partial x})$$

Boundary conditions

$u(x,t) = u_i \quad x > 0 \quad t = 0$

$u(x,t) = u_s \quad x = 0 \quad t > 0$

$u = f(\lambda) \quad \lambda = \frac{x}{\sqrt{t}}$

$u = f(\lambda(x,t))$

1. $u = u_i \quad \lambda = \infty$
2. $u = u_s \quad \lambda = 0$
3. $\frac{du}{d\lambda} = 0 \quad u = u_i$

$$\boxed{D(u_x) = \frac{1}{\left[\frac{du}{d\lambda}\right]_{u_x}} (-\frac{1}{2}) \int_{u_i}^{u_x} \lambda\, du}$$

BRUCE and KLUTE method for diffusivity measurement (ref. 2) spectrophotometer.

The gamma beam is collimated at the gamma ray source with a 5 cm diameter hole, 5 mm long in lead. The scaler is fully integrated to the operation of an automatic printing unit (teletype) that lists counts per unit time versus time.

Sieved air-dry soil (fraction < 0,3 mm) was carefully packed in a plexi cylinder. The columns have a length of 100 cm with a 5 cm inside diameter. A mechanical soil packer was used to pack the soil. The average bulk density of the soil is 1.1445 ± 0.0212 g.cm^{-3}.

The soil cylinders were used horizontally. Small holes were drilled 1 cm apart into the top of the cylinder to maintain atmospheric pressure in the entire sample at all times.

To determine packing uniformity, bulk density was measured along the entire sample at 1 cm intervals by using the gamma-ray equipment. Distilled water was then introduced at one end of the horizontal column, through a saturated, coarse (of negligible impedance) fritted glass plate 5 mm thick.

Water infiltrating into the soil columns through this glass plate was maintained at a slight negative head of 2 mbar (NIELSEN et all 1958).

The horizontal infiltration trials were run until the wetting front reached 30 cm from the origin of the soil column. The soil water content was measured along the entire sample at 1 cm intervals by using the gamma-ray equipment.

These measurements provided the water distribution along the soil column at a fixed time, which is needed for the Bruce and Klute method. Soil-water diffusivities were calculated from the data obtained during horizontal infiltration according to the method outlined by BRUCE and KLUTE(1956)

COMMENTS AND CONCLUSION

The described modeling technique applied to the silt-loam soil from table 1 results in the behavior of u(x) for several time increments as shown above. The simulation was, even when only five eigenfunctions (j = 5) were used, in agreement over the complete (t,x) range with experimental curves within measurement error bounds. The hybrid implemented model is used to study the influence on the moisture content through conditioning of the soil.

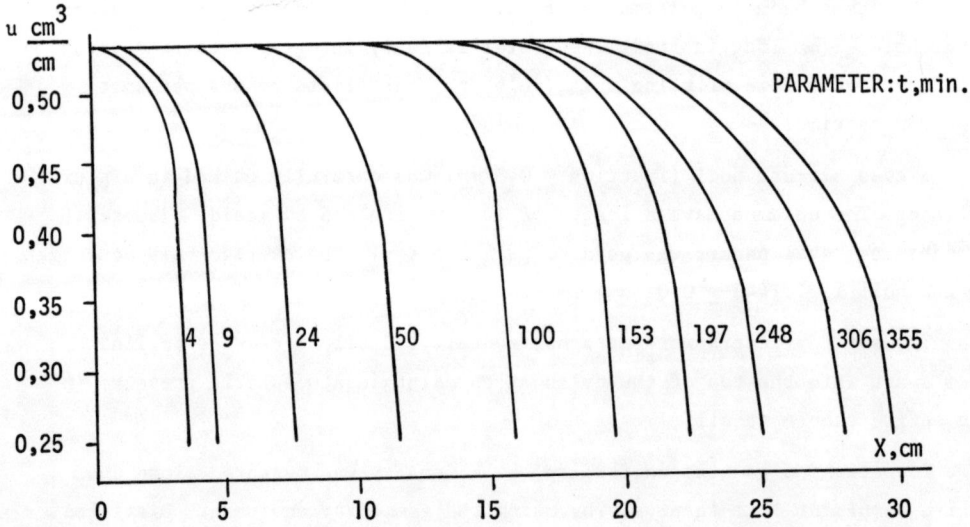

The effect of changing the diffusivity dependence on u can easily be studied a slight modification in the digitally stored D(u) table is readily made. Different soil types are in the same manner implemented on the computer. The necessity of mathematical models of higher complexity is currently under investigation.

REFERENCES

KOSHLYAKOV, SMIRNOV and GLINER, *Differential equations of mathematical physics*, Amsterdam, North-Holland Publishing Co. 1964.

BRUCE R.R., KLUTE A., *The measurement of soil moisture diffusivity*, Proc. of Soil Sci. Soc. 1956, 20, 458-462.

NIELSEN D.R., BIGGAR J.W., DAVIDSON J.M., *Exp. cons. of diff. analysis in unsatur. flow problems*, Proc. of Soil Sci. Soc. Amer. 1962, 26, 107-111.

VERPLANCKE H., *Doctoral thesis*, Rijksuniversiteit Gent.

VANSTEENKISTE G.C., *Hybrid Systems*, Von Karman Institute, Belgium, 1971.

SIMULATION OF THE THERMAL BEHAVIOUR OF BARE SOILS FOR REMOTE SENSING PURPOSES

ANDRIES ROSEMA

*Netherlands Interdepartmental Working Community for the
Application of Remote Sensing Techniques
(NIWARS), Delft, The Netherlands*

ABSTRACT

This paper reports the construction of a mathematical model, simulating the daily course of a.o. the surface temperature and the terms of the heat balance at bare soil surfaces. It is based on well known physical theories about moisture and heat transfer in the soil and the surface layer of the atmosphere. The behaviour of the solution in the top soil layer, and some simulation results are treated.

EXPLANATION OF SYMBOLS

C_H : soil heat capacity ($J/m^3.K$)
C_W : soil water capacity ($kg^2/J.m^3$)
D : vapour diffusivity (m^2/sec)
E : vapour flux in the atmosphere ($kg/m^2.sec$)
G : heat flux in the soil at the surface ($J/m^2.sec$)
H : heat flux in the atmosphere ($J/m^2.sec$)
I_n : net radiation ($J/m^2.sec$)
K_{VT} : soil vapour conductivity related to temperature gradients ($kg/m.sec.K$)
$K_{V\Psi}$: soil vapour conductivity related to potential gradients ($kg.sec/m^3$)
K_W : soil isothermal water conductivity ($kg.sec/m^3$)
K_{WT} : soil water conductivity related to temperature gradients ($kg/m.sec.K$)
$K_{W\Psi}$: soil water conductivity related to potential gradients ($kg.sec/m^3$)
L : vaporization heat of water (J/kg)
L' : Monin - Obukhov stability length
LE : latent heat flux in the atmosphere ($J/m^2.sec$)
M : molecular weight of water ($18\ kg/kmol$)
P : air pressure (N/m^2)
R : gas constant ($8.314 \times 10^3\ J/kmol.K$)
S : non dimensional wind shear parameter
T : temperature
W : moisture flux in the soil at the surface ($kg/m^2.sec$)
c : heat capacity of air ($J/kg.K$)
g : gravitation per unit mass ($9.8\ N/kg$)
h : height above the subsoil water level (m)
k : von Karman constant (0.4)

s : specific humidity ($=\rho_v/\rho$)
t : time (sec)
v : windspeed (m/sec)
v^* : friction velocity (m/sec)
x_m, x_o : volumetric content of minerals, resp. organic material
z : distance to the surface (m)
z_a : height of the boundary above the surface (m)
z_0 : aerodynamic roughness (m)

Ψ : matrix potential (J/kg)
α_E, α_H : ratio of eddy conductivities of vapour, resp. heat, to that of momentum
β : empirical turbulence parameter
γ : quantity denoting temperature dependence of the surface tension at the air-water interface
δ : quantity denoting the decrease of destillative vapour transport above a critical water content
ζ : dimensionless height
θ : volumetric water content (m^3 water/m^3 soil)
λ : soil heat conductivity (W/m.K)
λ_a, λ_a^* : conductivity of dry air and apparent conductivity of moist air (W/m.K)
υ : mass flow factor
ξ : ratio of temperature gradient in the air filled pore space to the average temperature gradient in the soil
ρ : air density (kg/m^3)
ρ_v : vapour density (kg/m^3)

Note on terminology: "vapour" is used for water in the vapour phase, "water" for water in the liquid phase, and "moisture" when referring to both, or when no special reference to the phase is made.

1. INTRODUCTION

NIWARS explores the possible applications of Infra Red Line Scanning. This is a method in wich radiation, emitted from the surface, of a suitably chosen wavelength band (e.g. 8-14 µm) is received by a device carried along in an aircraft. The data are reproduced either on photographic film or in the form of radiation temperature maps. The thermal resolution is about 0.1 K.

In the 8-14µm band the effect of emissivity on observed radiation temperatures is not neglegible. However, due to the temperature effect beïng usually considerably greater, study of the behaviour of the surface temperature and the factors that influence this will presumably supply relevant knowledge. The result of this study is a mathematical model.

A more detailed report is in preparation /1/.

2. TRANSPORT EQUATIONS AND BOUNDARY CONDITIONS

The model consists of a number of equations and a algorithm that numerically solves these equations. The equations that are used to describe moisture and heat transport as a function of depth and time are respectivily:

$$C_W \frac{\Delta \Psi}{\Delta t} = \frac{\Delta}{\Delta z} \left\{ (K_{W\Psi} + K_{V\Psi}) \frac{\Delta(\Psi + gh)}{\Delta z} \right\} + \frac{\Delta}{\Delta z} \left\{ (K_{WT} + K_{VT}) \frac{\Delta T}{\Delta z} \right\} \qquad (1)$$

$$C_H \frac{\Delta T}{\Delta t} = \frac{\Delta}{\Delta z}\left(\lambda \frac{\Delta T}{\Delta z}\right) \qquad (2)$$

The conductivities λ, $K_{W\Psi}$ etc., and the capacities C_W, C_H are functions of the principle variables: the matrix potential Ψ and the temperature T.

The boundary conditions are formed by assumed to be known values of Ψ and T at the subsoil water level, and by the heat and moisture balance at the surface. These can be written as:

$$HB = I_n + G + H + LE = 0 \qquad (3)$$
$$MB = \quad W + E = 0 \qquad (4)$$

The balance terms are assigned positive if they are directed towards the surface.

3. CONDUCTIVITIES AND CAPACITIES

Moisture conductivities are calculated using a theory developed by Philip and De Vries /2/ to describe moisture transport in porous media caused by potential and temperature gradients. Water transport due to temperature gradients is the consequence of temperature dependance of the surface tension at the air-water interface. Vapour transport can be imagined as a repeated process of evaporation, diffusion and condensation (distillation). It can be shown by means of thermodynamics that, at equilibrium between the liquid and vapour phase, vapourdensities can be expressed in Ψ and T. Based on such considerations the moisture conductivities can be expressed in terms of Ψ and T as follows:

$$K_{W\Psi} = K_W \qquad (5)$$
$$K_{V\Psi} = \delta \upsilon D \rho_V (M/RT) \qquad (6)$$
$$K_{WT} = K_W \gamma \Psi \qquad (7)$$
$$K_{VT} = \delta \upsilon \xi D \rho_V \frac{M\{L-(\Psi+gh)\}-RT}{RT^2} \qquad (8)$$

In which:

$$\upsilon = P/\left(P - \rho_V \frac{RT}{M}\right) \qquad (9)$$
$$\rho_V = \text{const } \frac{M}{RT} \exp \frac{M\{-L+(\Psi+gh)\}}{RT} \qquad (10)$$
$$D = 0.59 \times 10^{-5} \frac{T^{2.3}}{P} \qquad (11)$$

Each sort of soil exhibits a characteristic relationship between Ψ and the volumetric water content θ: the 'moisture characteristic'. In the same way there is a relationship between K_W and Ψ: the 'transport characteristic'. Rijtema /3/ gives these characteristics for twenty standard soils. Fig.3 shows for example the moisture characteristics of 'coarse sand', 'sandy loam' and 'basin clay'. In fig.1 and 2 the moisture conductivities for the same soils, and calculated according (5), (6), (7) and (8), are set out against Ψ, at 283 K.

The water capacity C_W can be derived from the moisture characteristic from its definition: $C_W = d\theta/d\Psi$. Moisture and transport characteristics are introduced in the model in the form of tables. The right values are found by interpolation.

De Vries /4/ developed a method to calculate the heat conductivity λ of a soil from the heat conductivities and volumetric contents of the constituent

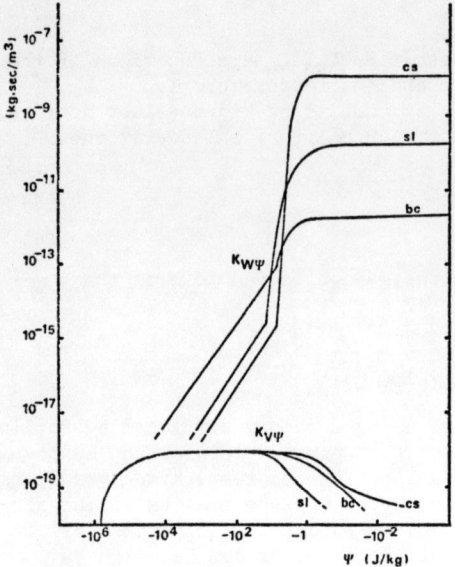

Fig. 1: Water and vapour conductivities related to potential gradients versus matrix potential, for 'sandy loam (sl),'coarse sand'(cs) and 'basin clay'(bc).

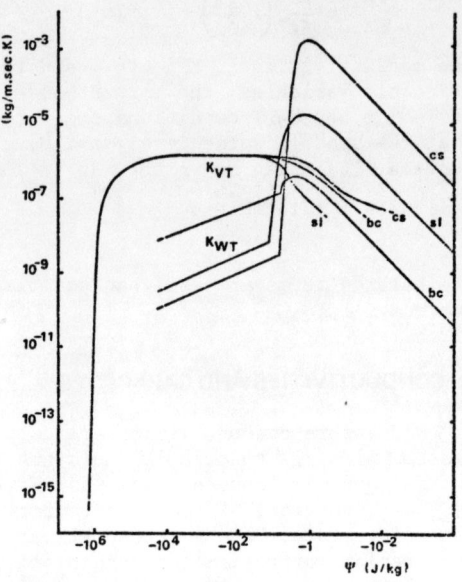

Fig. 2: Water and vapour conductivities related to temperature gradients versus matrix potential, for 'sandy loam' (sl), 'coarse sand' (cs) and 'basin clay' (bc).

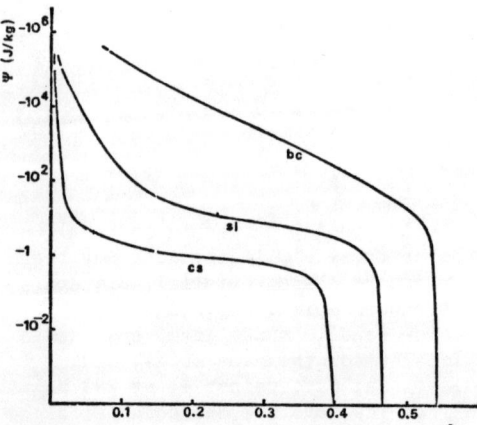

Fig. 3: Moisture characteristics of 'sandy loam'(sl), 'coarse sand' (cs) and 'basin clay'(bc).

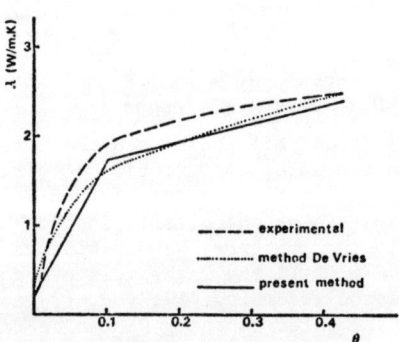

Fig. 4: Experimental and computed heat conductivities of quarz sand λ versus volumetric water content θ, at 293 K.

components. The heat conductivities of the solid components of the soil, and of water are weak functions of temperature. The previously mentioned distillation process causes transport of latent heat, which seems to increase the heat conductivity of the air in the pore space considerably. The apparent conductivity of the air can approximately be calculated from Ψ and T by an expression related to (8):

$$\lambda_a^* = \lambda_a + L \upsilon D \rho_v \frac{M\{L-(\Psi+gh)\}-RT}{RT^2} \qquad (12)$$

The procedure used by De Vries to calculate λ is applied here in a somewhat simplified manner. Fig.4 shows a result for comparison.
The heat capacity C_H of the soil is found from:

$$C_H = 4.19 \times 10^6 (0.46 \, x_m + 0.60 \, x_o + \theta) \qquad (13)$$

Although λ and C_H are computed as a function of θ, the latter can be reduced to Ψ via the moisture characteristic.

4. HEAT AND MOISTURE BALANCE TERMS

The fluxes of heat G and moisture W in the ground at the surface are calculated in terms of differences (the surface has index 0, the first grid point below has index 1), as follows:

$$G = \lambda_1 \frac{(T_1 - T_0)}{\Delta z_1} \qquad (14)$$

$$W = \left(K_{W\Psi 1} + K_{V\Psi 1}\right) \left(\frac{\Psi_1 - \Psi_0}{\Delta z_1} - g\right) + \left(K_{WT1} + K_{VT1}\right) \frac{T_1 - T_0}{\Delta z_1} \qquad (15)$$

The conductivities are assigned to the medium between the grid points 0 and 1, and are computed from the mean values of Ψ and T at these grid points.
The fluxes of heat H, vapour E and latent heat LE in the surface layer of the atmosphere are found on basis of the semi-emperical theory of turbulent transport as it was first satisfactorily described by Obukhov /5/ and subsequently developed further by Monin and Obukhov /6/, and others /7/. Due to turbulent transport in the surface layer being able to be treated as a semi-stationary process, the usual transport equations may be integrated with respect to height, and written as follows:

$$H = \alpha_H \rho c v_* k \frac{T(\zeta_a) - T(\zeta_0)}{F(\zeta_a) - F(\zeta_0)} \qquad (16)$$

$$E = \alpha_E \rho v_* k \frac{s(\zeta_a) - s(\zeta_0)}{F(\zeta_a) - F(\zeta_0)} \qquad (17)$$

$$LE = L \times E \qquad (18)$$

$$v_* = k \frac{v(\zeta_a)}{F(\zeta_a) - F(\zeta_0)} \qquad (19)$$

In which:

$$F(\zeta_a) - F(\zeta_0) = \int_{\zeta_0}^{\zeta_a} \frac{S(\zeta)}{\zeta} d\zeta \quad ; \quad \zeta_a = z_a/L' \, , \, \zeta_0 = z_0/L' \tag{20}$$

$$S - S^{-3} = \beta\zeta \tag{21}$$

The Monin-Obukhov stability length L' is defined by:

$$L' = \frac{v_*^3 \, \rho \, c \, \overline{T}}{Hgk} \tag{22}$$

$F(\zeta_a) - F(\zeta_0)$ is computed by numerical inversion of eq. (21) and subsequently numerical integration according to eq. (20). Fig.5 and 6 show the character of the functions $S(\zeta)$ and $F(\zeta_a) - F(\zeta_0)$ with values of 1 and 7 for β, the latter being the most realistic.

Air velocity, temperature and specific humidity at height z_a are taken to be given. α_H, α_E and β are also assumed to be known. H, E and LE are then functions of L', $T(z_0)$ and $s(z_0)$. It is subsequently supposed that the latter two may be equated T_0 and s_0 at the arbitrary soil surface. It is then further assumed that T_0 and s_0 can be identified as the temperature and specific humidity prevailing at the actual soil surface, and averaged over a given area. The specific humidity at the surface again is a function of Ψ_0 and T_0:

$$s_0 = \frac{const}{P} \exp \frac{M\{-L + (\Psi_0 + gh_0)\}}{RT_0} \tag{23}$$

For the net radiation I_n experimental values may be used. I_n may -for bright days and by approximation- also be computed. The net income of solar radiation is then computed from: time, latitude, albedo, atmospherical thickness and the back scattered fraction of sunlight /8/. The net income of thermal radiation may be calculated from the emissivity, the surface temperature, and the air temperature and humidity, using the well known formulae of Stefan-Bolzmann, and of Ångström or Brunt. This treatment is the one used in the examples of simulation to be cited later.

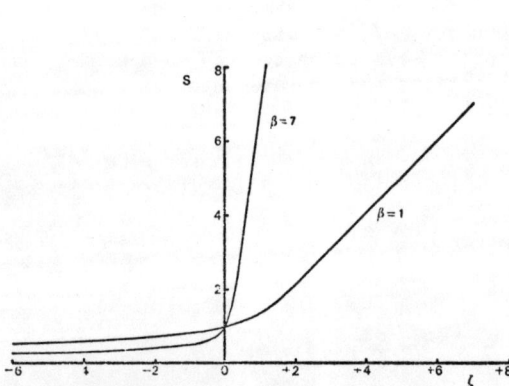

Fig. 5: The dimensionless wind shear parameter S as a function of the dimensionless height ζ, for two values of the empirical parameter β.

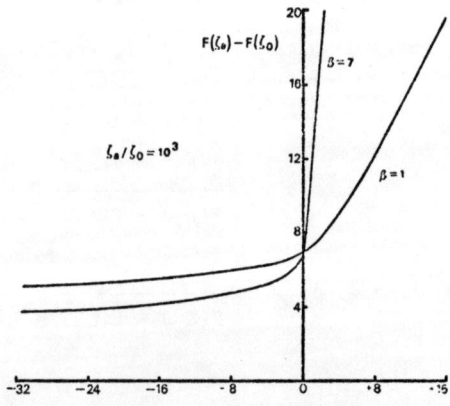

Fig. 6: $F(\zeta_a) - F(\zeta_0)$ as a function of ζ_a, for a ratio of $\zeta_a/\zeta_0 = z_a/z_0 = 10^3$ and two values of the empirical parameter β.

It is finally concluded that the moisture balance and heat balance terms together may be expressed in the variables Ψ_0, T_0 and L'.

5. SOLVING THE EQUATIONS

The transport equations are numerically solved using the method of Du Fort and Frankel /9/.
Therefore a grid is layd over the imaginary z,t plane.
According to this method Ψ and T at the grid point (n, t+Δt) are computed from Ψ(n-1, t), T(n-1, t), Ψ(n+1,t), T(n+1,t) and Ψ(n,t-Δt), T(n.t-Δt), as well as from the capacities at (n,t) and the conductivities at (n,t) and (n+1,t).(Fig7). The attractive feature of this method is that it has the advantages of an explicit method while being considerably more stable than the classical explicit method. High stability is necessary due to very small values of Δz having to be chosen in the region near the surface if the very steep gradients which occur there, in particular in Ψ, are to be simulated with sufficient accuracy. In order to limit the number of points, and consequently the computation time, an expanding grid is chosen, starting at the surface with small values of Δz (e.g. 0.5 mm) and doubling with each downwards step. Computation of Ψ and T at every grid point, and the corresponding heat and moisture balance terms is done stepwise with time as illustrated with the flow sheet in fig.8. After introducing the input data, which includes the initial values of Ψ and T at the grid coordinates of the soil profile, the first quantities to be calculated are the corresponding conductivities and capacities. Then, as previously stated, the values of Ψ and T at the grid coordinates of the profile beneath the surface are calculated for the next point in time. The sum of the heat balance components HB and the moisture balance components MB are considered as functions of Ψ_0 and T_0. The zero values of these functions are iteratively determined using the method of Newton-Ralphson. The values of Ψ_0 and T_0 at the previous point of time are taken as first approximations. L' and the conductivities are repeatedly revised during the iteration. If HB and MB are below given very small arbitrary values, the desired variables are produced in a certain way, and the entire procedure is repeated.Using an expanding grid, starting with 0.5 mm, with twelve grid points in the profile, and a time interval of 3 minutes, about 3 minutes computation time was required on the IBM - 360/65 computer for the simulation of one 24 hour cycle by a programme written in ALGOL.

6. SOME SIMULATION RESULTS

Figures (9), (10), (11) and (12) form a typical set of output data.
An investigation was carried out to determine the extent to which reduction of the distance between the grid points Δz and Δt leads to appreciable improvements in the resulting solution. In the course of these investigations, two striking phenomena were observed:
- oscillations having a period equal to the time interval,
- oscillations having a much longer period. (Fig. 10,11).

The first type of oscillation is a normal sign of instability and originates from choosing Δt too large with respect to Δz. The second variety of oscillation is more intriguing. Taking an equilibrium distribution of water in the gravitational field as an initial situation, a considerable drying out of the top layer occurs during the simulation, especially after sunrise. In this process the drying front gradually progresses downwards, causing a sequential

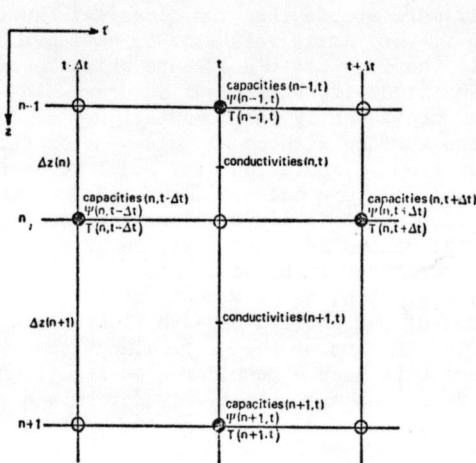

Fig. 7: Difference 'molecule' used in the method of Du Fort and Frankel, applying an expanding grid.

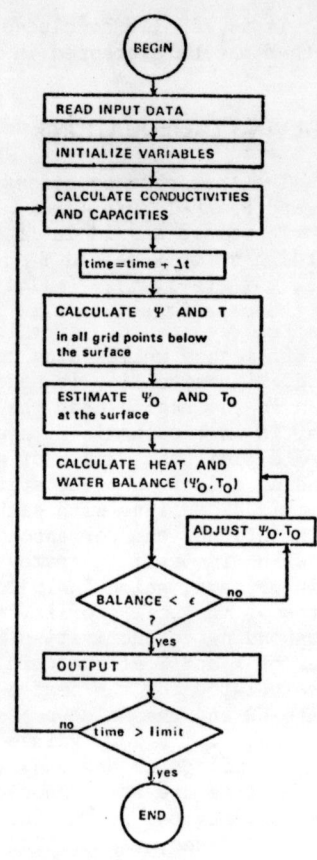

Fig. 8: Flow sheet of the simulation model.

rapid "drying" of grid points. It appears that these oscillations are caused by a periodic increase and decrease in evaporation in connection with the previously mentioned sequential and rapid "drying" of the grid points. This phenomenon can be seen as a failure of the solution to converge in the top soil layer. Even when reducing Δt and Δz (Δz to 0.1 mm), this phenomenon still remains. Due to the water phase usually being discontinuous, in the top layer of the soil, and with the various elements of the water phase having distances from each other of the same order of magnitude as the Δz chosen, there is no advantage to be gained in reducing Δz even further. It is concluded with some reservations, that the equation (1), used in describing unsaturated moisture transport, is in principle unsuitable for describing moisture transport near the soil surface, where very high gradients occur. One might however question whether in this way (admittedly wrong in principle) a phenomenon is being simulated which also occurs at real soil surfaces. The water phase near a natural surface is, after all, discontinuous, just as is the case with the grid. It is possible that the irregular evaporation data determined by Rose /10/ can be ascribed to a similar mechanism.

Fig. 9

Fig. 10

Fig. 11

Fig. 12

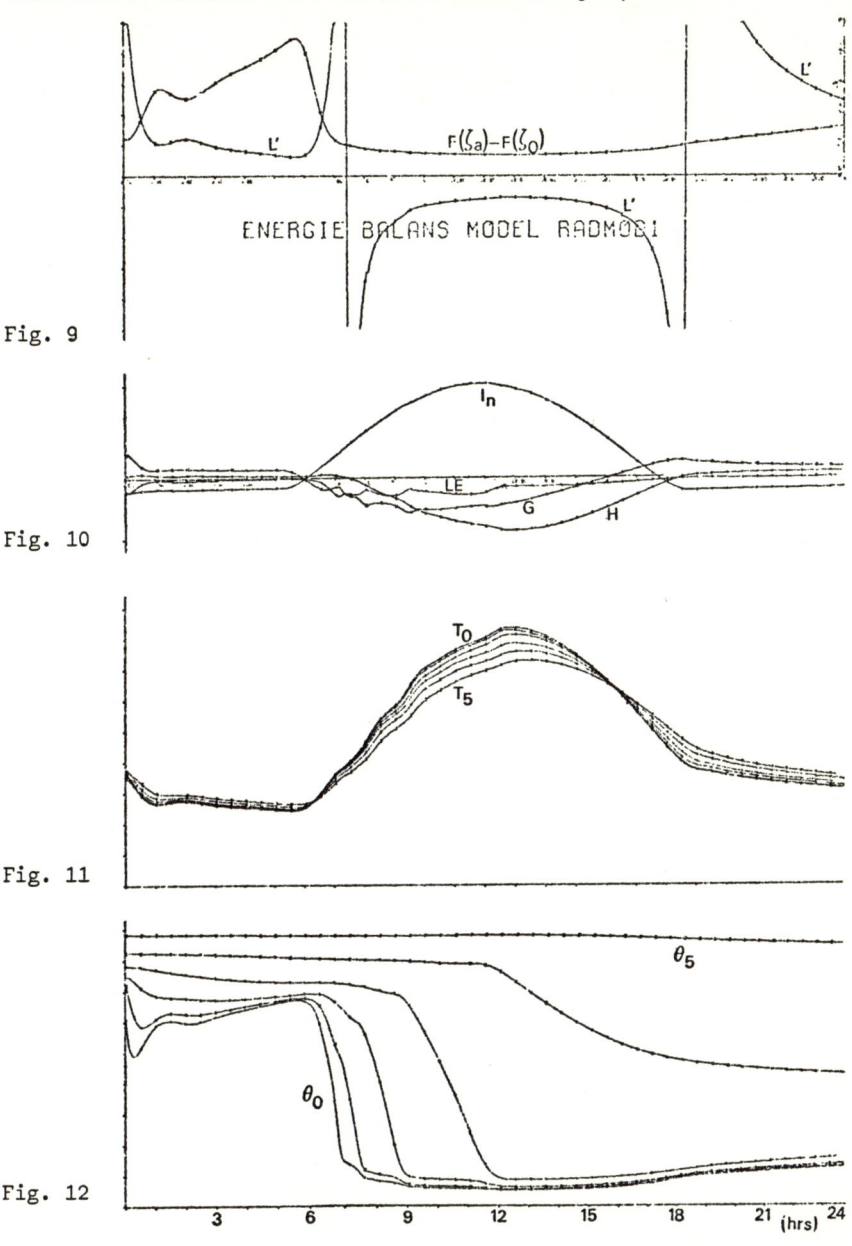

Fig. 9, 10, 11, 12 : Daily course of the Monin-Obukhov stability length L', the value of $F(\zeta_a)-F(\zeta_0)$, the terms of the heat balance I_n, G, H, LE, and the temperatures T and volumetric water content θ at six levels (0, 0.5, 1.5, 3.5, 7.5, and 15.5 mm beneath the surface), simulated and computer plotted for 'sandy loam'. Wind velocity 7.5 m/sec.

The sensitivity of the surface temperature to various variables has been investigated after choosing the most suitable grid point separation, this being a compromise between computing time and convergence of the solution. In general all variables appeared to have a relevant effect. Some examples will be given here, all based on the standard input mentioned in table 1.

A value 7 for β is in reasonable agreement with experimental results (see for example /11/). Fig. 13 shows the daily coarse of H and T_0 for $\beta=1,4$ and 7, simulated for 'sandy loam' and standard input. For the higher values turbulent transport is strongly suppressed at night, and increased during daytime, leading to faster cooling of the surface in the first, and slower warming of the surface in the latter case. This especially holds for clear days and low wind velocities. Then, since $H \approx LE \approx 0$ the nocturnal differential cooling of various soils is entirely due to differences in their thermal properties.

Fig. 14 provides an example of the daily course of the surface temperatures of 'coarse sand', 'sandy loam', and 'basin clay', for standard input. The low contrast by day between the first two is due to the high albedo of the sand. The effect of subsoil water level variations in 'sandy loam' is shown in fig. 15.

When interpreting IRLS-data, attention is paid rather to temperature contrasts then to the temperature itself. Fig. 16 and 17 show the simulated temperature contrast between 'sandy loam' and 'basin clay' for various subsoil water levels and wind velocities. Fig. 18 and 19 show these contrasts between 'sandy loam' and 'coarse sand'. They demonstrate that temperature contrasts are not necessarily maximal around noon, and that their character is quite dependent of variables not concerning soil type.

To analyse the behaviour of the moisture transport mechanism in the soil, fluxes caused by potential and temperature gradients, at the 5, 23 and 950 mm level beneath the surface of a 'sandy loam', are plotted separately in fig. 20, together with the evaporation. At night the net moisture flux at these levels is directed to the surface and exceeds the evaporation, causing an increase of moisture content near the surface. By day, during a certain period the net moisture flux is directed downwards at the lower two levels, which is caused by downward distillation of moisture to lower temperatures.

Table 1: Standard input values, used except if indicated otherwise.

	COARSE SAND	SANDY LOAM	BASIN CLAY
ALBEDO	0.30	0.20	0.15
EMISSIVITY	0.92	0.94	0.94
POROSITY	0.40	0.47	0.54
VOL. QUARTZ CONTENT	0.60	0.35	0.00
VOL. CLAY+FELSPAR CONTENT	0.00	0.18	0.46
VOL. ORGANIC MAT. CONTENT	0.00	0.00	0.00
MOISTURE AND TRANSPORT CHARACTERISTIC	(SEE FIG 1 AND 2)		

LATITUDE	+52 DEGR.
DAY	100 th. (APRIL 12 th.)
AERODYNAMIC ROUGHNESS z_0	0.01 m
HEIGHT OF BOUNDARY z_a	30 m
WINDSPEED	5 m/sec
AIR TEMPERATURE } AT z_a	280 K
AIR SPEC. HUMIDITY	0.006
DEPTH OF SUBSOIL WATER LEVEL	0.50 m
TEMPERATURE AT SUBSOIL WATER LEVEL	280 K
$\alpha_H = \alpha_E$	1
β	7

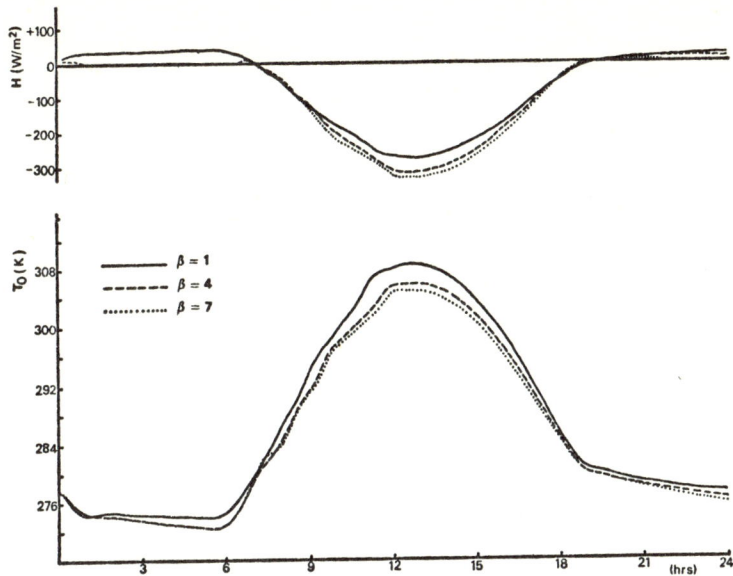

Fig. 13: Daily course of the surface temperature T_0 and the turbulent heat flux H for 'sandy loam' and three values of the empirical turbulence parameter β.

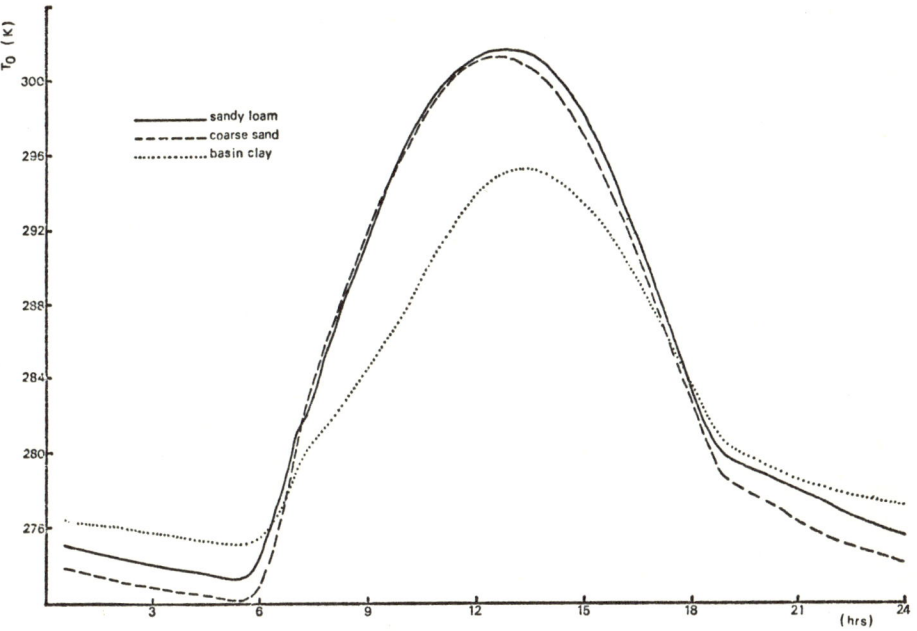

Fig. 14: Daily course of the surface temperature of 'sandy loam', 'coarse sand' and 'basin clay'.

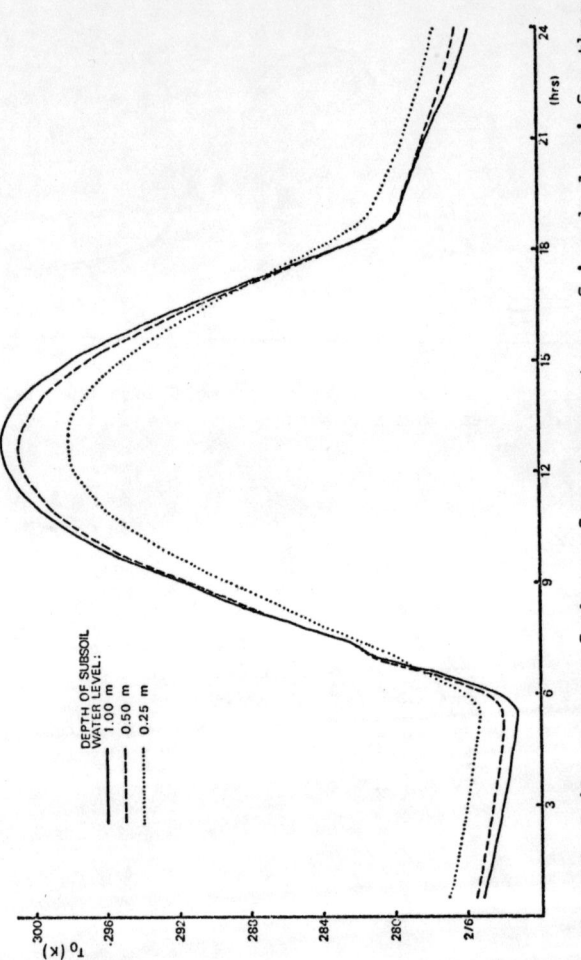

Fig. 15: Daily course of the surface temperature of 'sandy loam' for three different depths of the subsoil water level

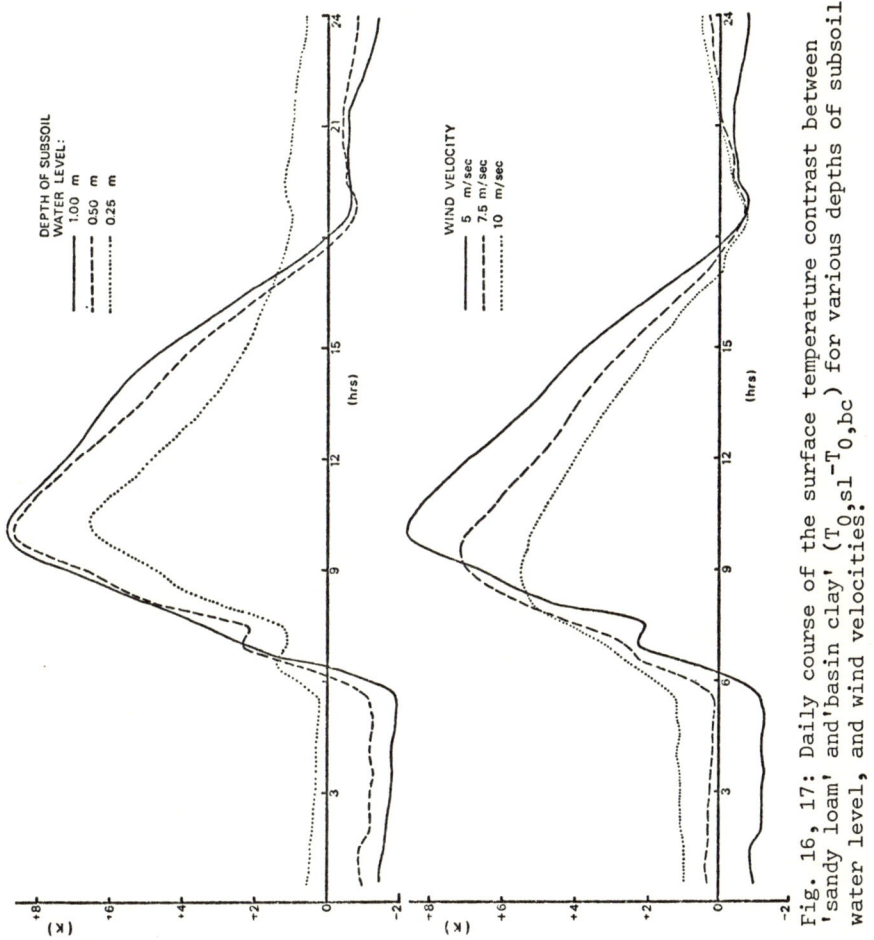

Fig. 16, 17: Daily course of the surface temperature contrast between 'sandy loam' and 'basin clay' ($T_{0,sl} - T_{0,bc}$) for various depths of subsoil water level, and wind velocities.

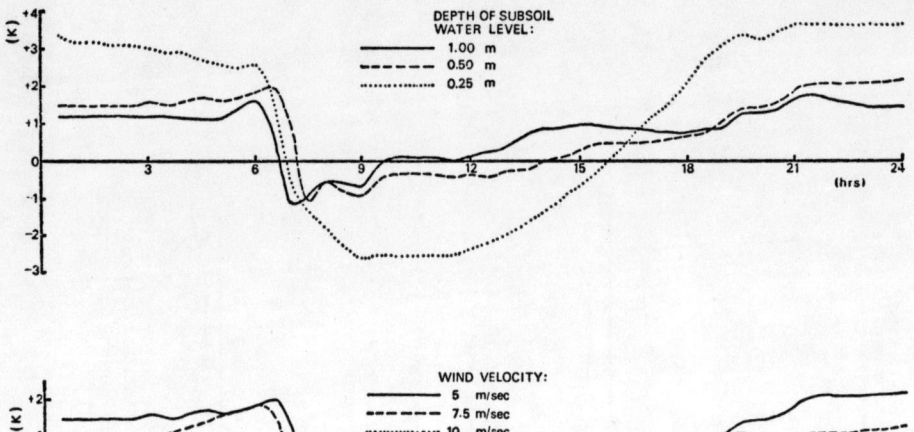

Fig. 18, 19: Daily course of the surface temperature contrast between 'sandy loam' and 'coarse sand' ($T_{0,sl}-T_{0,cs}$) for various depths of subsoil water level, and wind velocities.

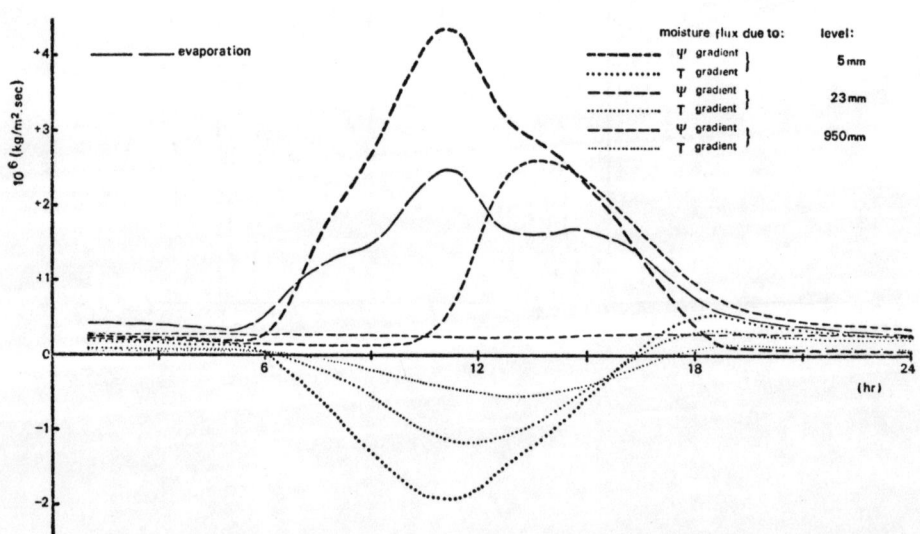

Fig. 20: Daily course of moisture fluxes due to potential and temperature gradients at three levels in 'sandy loam', together with evaporation.

CONCLUSIONS

Simulation results indicate that near the surface the usual transport equation does not describe moisture transport very well. Drying soils show oscillations in evaporation, that probably may have their counterpart in nature.

It has been shown that surface temperature contrasts are weather and subsoil water level dependent. Clear nights and low windspeeds may favour discrimination of soils from their thermal properties.

ACKNOWLEDGEMENT

Programming of the model was supported by the Information Processing Group of NIWARS. The author is especially indebted to S. Anthony. The manuscript was typed by Mady Hessels.

REFERENCES

/1/ Rosema, A., "A Mathematical Model for Simulation of the Thermal Behaviour of Bare Soils, Based on Heat and Moisture Transfer", NIWARS report no. 13, Delft, The Netherlands (in preparation).

/2/ Philip, J.R. and D.A. de Vries, "Moisture Movement in Porous Materials under Temperature Gradients", Trans.Am.Geoph.Union, Vol. 38, No. 2, 1957.

/3/ Rijtema, P.E., "Soil Moisture Forecasting", Institute for Land and Water Management Research, Wageningen, The Netherlands, 1970.

/4/ De Vries, D.A., "The Thermal Conductivity of Soil", Mededelingen van de Landbouwhogeschool te Wageningen, The Netherlands, 1952.

/5/ Obukhov, A.M., "Turbulence in an Atmosphere with a Non-uniform Temperature" 1946, Republished in: Boundary Layer Meteorology 2 (1971) 7-29.

/6/ Monin, A.S. and A.M. Obukhov, "Fundamental Regularities of Turbulent Agitation in the Ground Layer of the Atmosphere", Trans.Geoph.Inst.Ac.Sc. USSR 24, 163-187, 1954.

/7/ Rijkoort, P.J., "The Increase of Mean Wind Speed with Height in the Surface Friction Layer", Mededelingen en verhandelingen KNMI, no. 91, Staatsdrukkerij, 's-Gravenhage, The Netherlands, 1968.

/8/ Kondratyev, K.Ya., "Radiation in the Atmosphere", Academic Press, New York, London, 1969.

/9/ Du Fort, E.C. and S.P. Frankel, "Stability Conditions in the Numerical Treatment of Parabolic Differential Equations", Math.Tab.Wash. 7, 135-152, (237,239), 1953.

/10/ Rose, C.W., "Evaporation from Bare Soil under High Radiation Conditions", 9th. Int.Congr.Soil.Sc.Trans. Vol. I.

/11/ Businger, J.A. a.o., "Flux-Profile Relationships in the Atmospheric Surface Layer", Journ. of Atm. Sciences, Vol. 28, March 1971.

SURFACE PHENOMENA CONNECTED WITH EVAPORATING WATER AND CONDENSING WATER VAPOUR IN THIN CAPILLARIES

N.V. CHURAEV

*Institute of Physical Chemistry,
Moscow, USSR*

Evaporation of water from quartz capillaries less than 1 micron in radius has been experimentally tested. Studied are also the influence of the hydrophilization of the capillary surface on the evaporation rate and that of temperature on the thickness of the surface films of water. It has been shown that the film flow is able to render evaporation more intense only in fairly thin capillaries and at a high relative humidity of air. The effect of adsorption upon the diffusion of vapour into empty capillaries has been investigated.

NOMENCLATURE

- r : capillary radius
- v : evaporation rate
- ρ : liquid density
- x : distance from meniscus to the capillary end
- h : thickness of a liquid film
- p_s : pressure of saturated vapour
- p_m : pressure of vapour over the meniscus
- p_0 : pressure of vapour in the ambient medium
- p_x : critical pressure of vapour, corresponding to $\beta \to \alpha$ transition

Π : disjoining pressure of a liquid film
τ : time
v_m : molar volume of liquid
η : viscosity of liquid in a film
φ : relative humidity of air
R : gas constant
t : temperature, °C
T : absolute temperature
D : coefficient of diffusion of water vapour through air

1. INTRODUCTION

The transfer of moisture may occur, whether through diffusion of vapour or owing to the film flow in the liquid phase over the surface of particles. To carry into effect separate quantitative determinations of these flows, the evaporation of water from thin quartz capillaries was observed. Determined were the conductivity coefficients K included in the mass-transfer equation

$$v = \frac{K(p_m - p_o)}{\rho x} \qquad (1)$$

Experiments made with quartz capillaries $r > 1\mu m$ showed the following /1,2/. In the course of evaporation, a metastable β -film $(h > 200 Å)$ remains on the capillary surface after the retreating meniscus. At a certain distance from the meniscus, where the vapour pressure decreases to a critical one, this film is ruptured and passes over into a stabler state, i.e., into a thin α -film $(h < 80$ Å). Two types of water films coexisting on the capillary surface, this was made use of to explain a variation in the conductivity coefficients, which was experimentally observed during evaporation. The films of two different thicknesses correspond to two α - and β -branches of the disjoining pressure isotherm $\Pi(h)$ for water films on quartz surface /3/.

In fairly thin capillaries, condition $p_m \leq p_x$ may be fulfilled. In this case, only thin α -films remain on the quartz surface. The experiments made with capillaries of $r < 1\mu m$ /4/ showed that β -films did not form in the capillaries smaller than $0.25\,\mu m$ in radius. Hence, using the Kelvin equation as the basis, the value of critical pressure of vapour p_x/p_s was determined to be equal to 0.996. This value corresponds to the pressure of soil moisture of about 6.10^5 N/m^2. Consequently, within the region of the relative pressure of vapour $p/p_s < 0.996$, the mass transfer in the liquid phase may be carried into effect only owing to the flow of α -films.

The contribution of the film transfer increases as the capillary radius decreases. Thus, for the thinnest ones of the studied capillaries of $r = 0.1 \div 0.2 \mu m$, the evaporation rates measured at $p_o/p_s = 0.96$ were found to exceed by 5 to 6 times the vapour flow.

Comparing the experimental and theoretical values of evaporation rates enabled one to make the conclusion that the viscosity of α -films several times exceeds that of bulk water /4/. Under the same conditions, no appreciable variation in the viscosity of thin films was detected for nonpolar decane.

The stability of thin films of nonpolar liquids is determined by the effect of Van der Waals forces /5/. Thick β -films of water are found to be stable on silicate surface owing to the interaction of diffuse ionic layers /6/.

Analysis of various components of surface forces /3,6/ enabled one to conclude that stability of thin α -films on quartz surface is mainly conditioned by changes in the structure of water. If water would preserve its bulk structure, then, as calculations have demonstrated, the films should exhibit a considerably smaller thickness than that resulting from the experimental isotherms of disjoining pressure $\Pi(h)$.

Thin layers of water undergo structural changes probably under the effect of active centers on quartz surface, e.g., OH-groups. The increase in the viscosity of α-films detected may be regarded as one of the factors proving the structural changes of water.

2. EFFECT OF THE HYDROPHILIZATION OF THE CAPILLARY SURFACE AND TEMPERATURE UPON THE THICKNESS OF α-FILMS

If the assumptions were true to the effect of water structure varying in thin polymolecular α-films, their thicknes must be sensible both to the degree of hydrophilic property (hydrophility) of surface and to temperature.

The internal surface of thin capillaries was rendered hydrophilic by the following technique. On having filled up with water one-fifth of a capillary, its both ends were sealed. The sealed capillary was held for 24 hours in a quartz tube furnace at a temperature of $180^\circ C$. The interaction of hot vapour at a high pressure with quartz surface resulted in an increase in the surface density of OH-groups. After completing such a treatment, the capillary was extracted from the tube furnace, opened up, and dried at a temperature of $120^\circ C$. Another portion of the same capillary was not subjected to the treatment, and was used for carrying out comparative experiments.

Both the treated and untreated capillaries were filled up with water and placed side by side in a vacuum chamber to allow observing evaporation. Both the design of the chamber and the measurement technique were described in reference /4/. The shifting rates of menisci in the both capillaries were measured with the microscope through a plane-parallel window provided in the chamber.

Fig. I shows the experimantel results obtained with the capillaries of two radii of 0.25 and 0.2I μm at a relative pressure of vapour in the chamber $p_0/p_s = 0.96$. Since $v = dx/d\tau$, it will follow from equation (I) that at $p_m - p_0 =$ Const and K = Const:

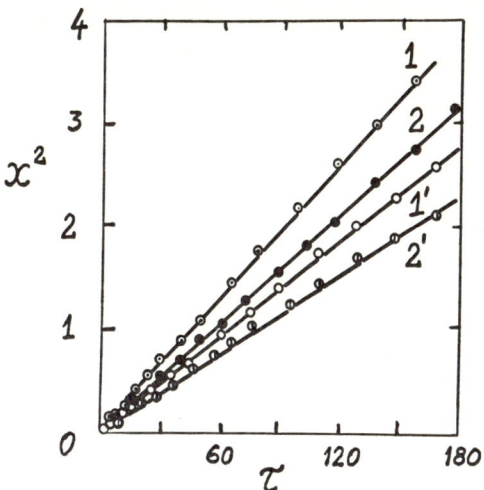

Fig. I. Results of observing rates at which water evaporates from the hydrophilic (I,2) and nontreated (I', 2') quartz capillaries. I and I' : r = 0.25 μm; 2 and 2' : r = 0.21 $\mu m (x, m; \tau, \text{min}; \ t = 27°C)$.

$$K = \frac{\rho}{2(p_m - p_o)} \cdot \frac{x^2}{\tau} \qquad (2)$$

Hence, the slope on the graph is proportional to the conductivity coefficients K.
As it appears from Fig. I, the values of K for the hydrophilic capillaries are approximately by 40% higher than those for the nontreated capillaries. Since the hydrophilic treatment could not have affected the flow of vapour $(r \gg h)$, an increase in the evaporation rate is rather connected with an increase in the thickness of α -films on the hydrophilic surface of quartz. Consequently, there is a correlation between the hydrophility of quartz surface and the thickness of α -films.

Investigated was also the effect of hydrophobization of the surface of capillaries with trimethylchlorosilane. After the treatment, the wetting angle was found to be equal to 98-I02°. Like in the earlier experiments /2/, the complete disappearance of β -films in the capillaries was found to have occurred. Unfortunately, one has not

succeeded in observing a variation in the thickness of
α -films after hydrophobization. Technical difficulties have arisen when trying to carry into effect the hydrophobization of very thin capillaries, their filling up with water and sealing.

Fig. 2 shows the relationship between the thickness of α -films of water (at $p/p_s \sim 1$) and temperature. An ellipsometric method was applied to measuring thickness by using polarized light /7/. The formation of films resulted from the adsorption of vapour on the plane polished surface of fused quartz. Special report /8/ sets forth in more detail the technique of these experiments. As it appears from the graph, an abrupt decrease in the thickness of α -films is observed within the temperature range from 10^0 to 20^0C. The practically complete ceasing of polymolecular adsorption is detected at t $\geqslant$ 65^0C.

As it may be supposed, the structural changes are transferred from the surface into the liquid owing to the effect of directed intermolecular H-bonds. An increase in

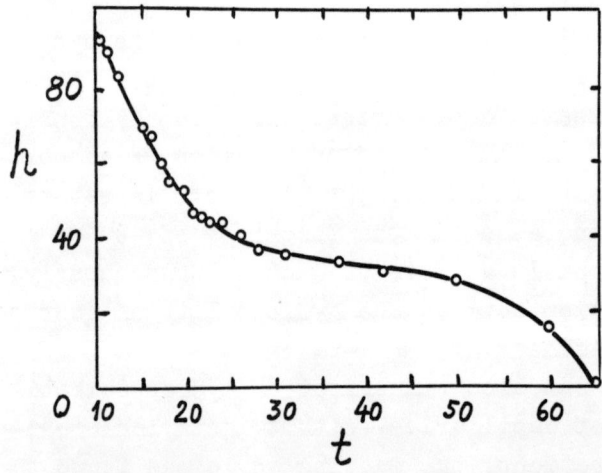

Fig. 2. Relationship between the thickness of α -films of water on quartz surface and temperature (h, Å; t °C).

the temperature of water results in the breaking of the network of H-bonds and in a decrease in the radius which the surface exerts its influence at. Consequently, the observations made corroborate the supposition that α - films of water exhibit a structure which is different from that of bulk water.

3. EVAPORATION OF WATER FROM THIN CAPILLARIES INTO HUMID AIR

The experiments, which have been described above, involved observing evaporation in evacuated chambers. It was of interest to find out whether the detected relationships hold when evaporating into humid air. To carry out these experiments required the development of a system for prolonged (e.g., during a month), continuous thermostating /9/. Different relative humidities of air were maintained by the use of saturated salt solutions.

In this case, the current of vapour may be calculated by using the Fick diffusion equation. Therefore, Eq. (1) may be converted as follows:

$$v = \frac{v_m D p_s (1 - \varphi_0)}{RTx} + \frac{2RT}{3\eta v_m rx} \int_{\varphi_0}^{\varphi_m} \frac{h^3}{\varphi} d\varphi = \frac{v_m D_x p_s (1 - \varphi_0)}{RTx} \quad (3)$$

Here D_x is the effective coefficient of diffusion, taking into account both the vapour current and the film flow. In order to find the film flow (the second term of the equation), adsorption isotherm $h(\varphi)$ will have to be preset.

Fig. 3 shows the experimental values of D_x/D as a function of the capillary radius. The calculation was done by using the slopes of the first portion of evaporation graphs as the basis, where both β-and α-films were assumed to participate in the surface transfer. This allows assessing the general contribution of the film flow.

The experiments have shown that at $\varphi_0 \leq 0.73$ there is practically no effect of the surface film flow identi-

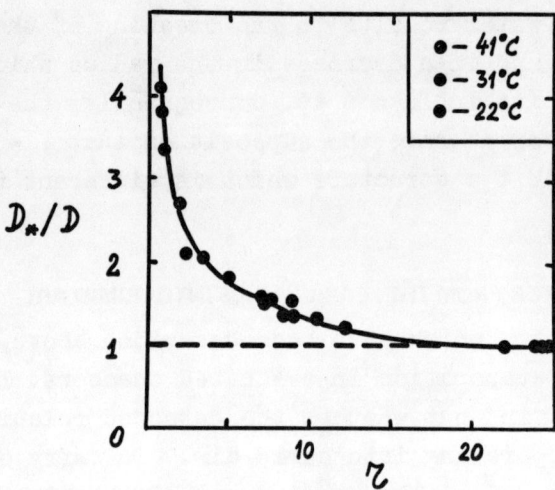

Fig. 3. Effect of the film flow upon the evaporation of water from quartz capillaries into the ambient air with relative humidity $\varphi_0 = 0.95 (r, \mu m)$.

fied. The evaporation rate is well described by Eq. (3) at $D_x = D$. The values of D obtained experimentally are close to the tabular ones; e.g., 0.25 cm^2/sec at 22°C; 0.26 cm^2/sec at 31°C, and 0.27 cm^2/sec at 41°C.

At $\varphi_0 = 0.86$, a contribution of the film transfer becomes noticeable only in the capillaries smaller than 5 μm in radius. At a still higher relative humidity of air $\varphi_0 = 0.95$, the effect of the film transfer is detected already in the capillaries up to 10 μm in radius. A variation in temperature within the range from 22°C. to 41°C. affected but slightly the contribution of film transfer.

The effect of film flow grows as the capillary radius decreases. So, for example, in the capillaries about 1 μm in radius, the surface film flow by 3 to 4 times exceeded the vapour flow, as appears from Fig. 3. The data thus obtained are in good agreement with the evaporation theory that was developed earlier /10/. Both the experiments and theory show the effect of the film flow is essential only in fairly thin capillaries and at a high (e.g., close to unity) relative humidity of ambient air.

In this case, also, comparing the experimental and theoretical data confirmed an increased value of viscosity of α-films of water /9/.

4. DIFFUSION AND CONDENSATION OF WATER VAPOUR IN THIN CAPILLARIES

Let us now discuss the reverse process – that is the diffusion of water vapour into an initially empty capillary. When a capillary is fairly thin, the velocity the vapour is entering therein at is determined not only by the diffusion, but also by the adsorption of vapour at the capillary surface. The mass flow increases owing to the condensation of vapour on the capillary walls and to the formation of α-films. The theory of diffusion of vapour in a capillary with the simultaneous polymolecular adsorption was developed earlier /11/. A solution was obtained for simplified isotherms of adsorption, consisting of several lengths of a broken line.

The experiments were made with quartz capillaries less than 1 μm in radius. The experimental technique consisted in observing the evaporation of a water column into the empty portion of a capillary /12/. Measuring the mass flow and using a decrease Δz in the length of the water column as the basis, the values of the effective diffusion coefficient De were calculated. The coefficient De takes into account the effect of adsorption upon the mass transfer:

$$\Delta z = \frac{2 v_m p_m}{RT} \left(\frac{D_e \cdot \tau}{\pi} \right)^{1/2} \tag{4}$$

The values of De were determined by the use of slope of experimental linear relationships $\Delta Z(\sqrt{\tau})$.

Fig. 4 shows how the relationship D_e/D depends upon the capillary radius. From the experiments made, it follows that the values of D_e in thin capillaries more than

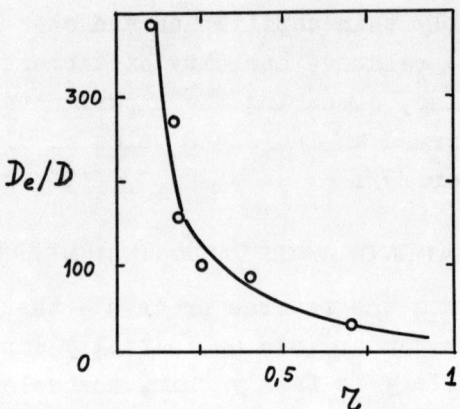

Fig. 4. Relationship between the effective coefficient of diffusion of water vapour into empty capillaries and their radius $r(\mu m)$.

by two orders of magnitude exceed the coefficients of diffusion of water vapour through air ($D = 0.26$ cm^2/sec). The difference of D_e from D decreases as the capillary radius increases. In the case of capillaries more than 50 μm in radius, the effect of polymolecular adsorption of water upon diffusion becomes insignificant. On the contrary, in thin capillaries $r \leqslant 1\mu m$, the polymolecular adsorption is the main mechanism determining the kinetics of the processes of mass exchange with the ambient medium.

In those cases where the effect of the film flow or film condensation is essential, the calculation of the mass-transfer processes should be based on the adsorption isotherms and those of the disjoining pressure of thin films. For instance, calculation of the mass-transfer coefficients, which has recently been carried out for a model porous medium composed of similar spherical particles /13/, may be referred to as an example of such an approach.

Consequently, the experiments were made with quartz cylindrical capillaries as the simplest model of pores, which enabled one to ascertain the region of the pore radii and that of the pressure of water vapour, where the assessing of surface phenomena is imperative.

REFERENCES

/1/ Derjaguin, B.V., Churaev, N.V., Ershova, I.G., Doklady AN SSSR, v. 182, 368(1968).

/2/ Churaev, N.V., Ershova, I.G., Kolloidn. zhur., v. 33, 782, 913(1970).

/3/ Derjaguin, B.V., Churaev, N.V., Doklady AN SSSR, v. 207, 572)1972).

/4/ Zorin, Z.M., Novikova, A.V., Petrov, A.K., Churaev, N.V., J. Coll. Interface Sci. (in Press).

/5/ Churaev, N.V., Kolloidn. zhur., v. 36, 318, 323(1974)

/6/ Derjaguin, B.V., Churaev, N.V., Structural Component of Disjoining Pressure. To be published in J. Coll. Interface Sci.

/7/ Derjaguin, B.V., Zorin, Z.M., Zhur. fisich. khimii, v. 29, 1755(1955).

/8/ Ershova, G.F., Zorin, Z.M., Churaev, N.V., To be published in Kolloidn. Zhur.

/9/ Churaev, N.V., Yashchenko, N.E., Soil science (in Russian), N9, 105(1973).

/10/ Derjaguin, B.V., Nerpin, S.V., Churaev, N.V., Kolloidn. zhur., v. 26, 301(1964).

/11/ Zolotarev, P.P., Churaev, N.V., Zhurn. fisich. khimii., v. 46, 1123(1972).

/12/ Zorin, Z.M., Sobolev, V.D., Churaev, N.V., Zhurn. fisich. khimii, v. 46, 1127(1972).

/13/ Churaev, N.V., Inzhenerno-fisich. zhurn., v. 23, 807(1972).

SECTION 2
LOWER ATMOSPHERE

AERODYNAMICS OF VEGETATED SURFACES

JOOST A. BUSINGER

Department of Atmospheric Sciences
University of Washington, Seattle, Washington, U.S.A.

ABSTRACT

The flux-profile relations over a uniform surface within the atmospheric surface layer are reviewed for the neutral and diabatic case. An attempt is made to describe roughness elements in terms of roughness length and displacement height by extending a simple formula proposed by Lettau (1969). A similar treatment is given for shelterbelts.

The flow regimes within a canopy are discussed briefly.

1. INTRODUCTION

An accurate description of the aerodynamics of vegetated surfaces is essential for the understanding of the turbulent transfer processes that take place near these surfaces. However, the complexities of the flows in and around plant canopies is such that we are forced to make some gross simplifications. The challenge therefore is to simplify the problem sufficient that it is tractable but that still valid and valuable conclusions can be made concerning the transfer processes. In this review I have made an attempt to do this.

Here and there it was tempting to digress and discuss an interesting phenomenon in detail such as, e.g., the fluttering of a poplar leaf. The interaction of wind and leaf results in an oscillation of the leaf and a vortex street in its wake.

This oscillation may have beneficial effects for the tree because it varies the exposure of the leaf to direct sunlight and allows other leaves to obtain also fluctuating amounts of direct sunlight. In direct sunlight the process of photosynthesis will be light saturated if the exposure is continuous, so by oscillating the amount of saturation will be reduced and leaves otherwise in the shade become more efficient. Another beneficial effect of the oscillation is that it may enhance the diffusion of CO_2 to the surface which is also necessary for the photosynthesis. The interaction of aerodynamics and plant physiology is such that one can stop and think about an almost infinite number of intriguing details, which carries us way beyond the scope of this review.

Therefore I have restricted myself to a discussion of the horizontally uniform case of flow above and within a plant canopy with special attention to development of roughness length and displacement height.

2. THE STRUCTURES OF THE FLOW OVER A UNIFORM SURFACE

The structure of the atmospheric surface layer (the lowest 10-20 m of the atmosphere) is relatively well known over uniform surfaces. Without going into detail concerning the various descriptions that have been proposed I will just summarize one of the simplest descriptions which gives sufficiently accurate results and which is being used for some numerical models of the atmospheric boundary layer.

a. The Neutral Case (No Heat Flux)

In the surface layer the fluxes may be considered independent of height, therefore the vertical momentum flux $\overline{\rho\, u'w'}$ measured in this layer is equal and opposite to the surface stress, τ ;

$$F_m = \overline{\rho\, u'w'} = -\tau \tag{1}$$

It is customary to introduce the friction velocity $u_* \equiv (\tau/\rho)^{1/2}$ as a convenient scaling velocity, and to introduce an eddy viscosity K_m to absorb the difficulties encountered in the correlation of turbulent fluctuations, i.e.,

$$u_*^2 = -\overline{u'w'} \equiv K_m \frac{\partial \overline{u}}{\partial z} \tag{2}$$

In order to obtain the wind profile from this equation it is necessary to make an assumption about K_m. Prandtl's (1932) mixing length model provided such an assumption. We can arrive at the same result by a simple dimensional argument. By observing that K_m has the dimensions of a velocity times a length, we have the natural choice of u_* as a scaling velocity and the height above the surface as the length. Therefore we assume

$$K_m \propto u_* z$$

or

$$K_m = k\, u_*\, z \tag{3}$$

where k is a constant of proportionality, the von Kármán constant. Combining (3) with (2) and integrating, yields the well known logarithmic profile

$$\frac{\overline{u}}{u_*} = \frac{1}{k} \ln\left(\frac{z+z_o}{z_o}\right) \tag{4}$$

where z_o is the roughness length. It may be interpreted as the size of the smallest turbulent eddy.

If we are dealing with a uniform vegetation, the height where the mean wind vanishes according to the logarithmic wind profile may not be equal to the surface of the earth but to some height above the surface because the vegetation acts as a new surface with respect to the wind. It is customary in this case to introduce a displacement height D, to indicate at which height the wind vanishes. Equation (4) is then written in the form

$$\frac{\bar{u}}{u_*} = \frac{1}{k} \ln \left(\frac{z-D+z_o}{z_o}\right) \tag{5}$$

This equation has three constants k, z_o and D which must be obtained from experiment because there is no theory which allows us to calculate them. This state of affairs is not quite satisfactory. Recently, Tennekes (1973) has provided an independent argument in support of the logarithmic profile. By matching the wind-profile near the surface to the velocity defect law near the top of the boundary layer the only solution for the matching is a logarithmic law. The logarithmic law is more basic than the similarity argument leading to (3) and (4) suggests and extends well beyond the surface layer up to heights of 100 m or so. If observations beyond the surface layer are available they may be helpful in determining the constants in (5).

The roughness length, z_o, and displacement height, D, have to be determined empirically from the wind-profile. Both quantities reflect characteristics of vegetated surfaces and will be discussed in some detail in Section 3.

"The von Kármán constant, pegged at 0.4 since the Depression, is now allowed to float any where between 0.33 and 0.40" (Tennekes, 1973). Indeed, since better surface layer measurements have become available, independent determinations of this constant have been made. Businger et al (1971) and Frenzen (1972) report 0.35. On the other hand Pruitt et al (1973) find values between 0.39 and 0.44. A theoretical analysis by Tennekes (1968) where he extrapolated wind tunnel data to very large Reynolds numbers yielded an asymptotic value of 0.33. In any case there is some evidence that the von Kármán constant is not quite constant but varies. Indications are that for rough surfaces the constant is smaller then for smooth surfaces contrary to Tennekes (1973) speculation on this point.

The fact that the logarithmic law is valid well above the surface layer under neutral conditions is helpful for describing the profile. It does not imply, however, as equations (2) and (3) suggest that the transfer of momentum is constant with height, in fact the momentum transfer decreases markedly over the transition layer. Consequently if u_* relates strictly to the surface stress the first equality of equation (2) and equation (3) are no longer valid above the surface layer.

In order to improve our description for K_m over the entire logarithmic layer we have to consider the entire steady state barotropic boundary layer. The equations of motion are

$$\frac{\partial}{\partial z} \overline{u'w'} = f(\overline{v} - v_g) \tag{6}$$

and

$$\frac{\partial}{\partial z} \overline{v'w'} = f(u_g - \overline{u}) \tag{7}$$

where $u_g = -\frac{1}{\rho f} \frac{\partial p}{\partial y}$, $v_g = \frac{1}{\rho f} \frac{\partial p}{\partial x}$, f is the coriolis parameter ($\simeq 10^{-4}$ sec^{-1} at mid latitudes). u_g and v_g are the components of the geostrophic wind in the x and y directions respectively. The x direction is determined by the mean surface wind.

Near the surface $\overline{v} \simeq 0$ and (6) may be integrated to

$$\overline{u'w'} = u_*^2 - f v_g z \tag{8}$$

and combined with the identity in (3)

$$K_m = u_* k z \left(1 - \frac{f v_g z}{u_*^2}\right) \tag{9}$$

This is only the asymptotic behavior of K_m near the surface, taking into account the first order term beyond the surface layer, and may be used as a reasonable approximation to a height of about 100 m.

A more general expression for K_m, which leads to wind profiles in the neutral boundary and is in excellent agreement with more sophisticated theoretical models, has been proposed by Businger and Arya (1974)

$$K_m = u_* k z \exp\left(\frac{-f v_g z}{u_*^2}\right) \tag{10}$$

Eq. (9) clearly is the first order approximation of (10). The advantage of assumption (10) over previous assumptions such as Blackadar's (1962) is that no arbitrary extra scaling length has been introduced.

b. The Diabatic Case

The neutral case is a special case because it requires that there is no heat flux from the surface to the atmosphere or vice versa. The more general situation with a heat flux present is called the diabatic case. The structure of the surface layer as well as the entire rest of the boundary layer is strongly modified by the fact that in this case buoyancy forces play a significant role. The turbulent energy equation indicates clearly the role of the heat flux in producing or consuming turbulent energy

$$- \overline{u'w'} \frac{\partial \overline{u}}{\partial z} + \frac{g}{\theta} \overline{w'\theta'} + D = \varepsilon \tag{11}$$

where g is the acceleration due to gravity, θ is the potential temperature, D is a transport term which includes transport of turbulent energy in the vertical and work done by pressure fluctuations, ε is the dissipation of turbulent energy due to a cascading process to smaller and smaller turbulent scales and ultimate viscous dissipation.

Richardson (1920) studied this equation and was especially interested in the situation where the turbulence would be suppressed. By taking the ratio of the first two terms of (11) and assuming that

$$\overline{w'\theta'} = - K_h \, \partial\overline{\theta}/\partial z , \qquad (12)$$

that
$$K_h = K_m , \qquad (13)$$

and using eq. (2) he introduced the Richardson number in the form

$$Ri = \frac{g}{\theta} \frac{\partial\overline{\theta}/\partial z}{(\partial\overline{u}/\partial z)^2} \qquad (14)$$

This number indicates the relative importance of the two production terms in eq. (11) and as such serves as a useful "stability" parameter.

The assumption (13) is not quite valid as indicated by experiment, therefore it is better to leave the two production terms as they appear in (11). The resulting ratio is called the flux Richardson number

$$R_f = \frac{g}{\theta} \frac{\overline{w'\theta'}}{\overline{u'w'} \, \partial\overline{u}/\partial z} \qquad (15)$$

This number has the drawback that it contains a mixture of covariances and mean profile information, which makes it more difficult to determine than Ri. Moreover both numbers have the drawback that they cannot easily be represented in simple scaling parameters such as a scaling velocity, temperature or height.

Using similarity arguments Obukhov (1946) introduced a scaling length

$$L = - \frac{\theta}{g} \frac{u_*^3}{\overline{w'\theta'}k} \qquad (16)$$

and a dimensionless height $\zeta = z/L$. (This dimensionless height is also obtained when in (15) $\partial\overline{u}/\partial z$ is replaced by u_*/kz). It is clear that L is approximately constant in the surface layer and therefore an important characteristic of this layer.

A great deal of effort has gone into developing relations between fluxes and profiles. We will not review these efforts in detail and only select the simplest formulation which does describe the data well. All efforts so far have been empirical and rigorous theory does not exist, so I felt free to present my personal bias.

In describing the diabatic profiles it is convenient to introduce the following quantities:

$$\phi_m \equiv \frac{kz}{u_*} \frac{\partial \bar{u}}{\partial z} \text{, a dimensionless wind shear} \tag{17}$$

$$\phi_h \equiv \frac{kz}{\theta_*} \frac{\partial \bar{\theta}}{\partial z} \text{, a dimensionless temperature gradient} \tag{18}$$

where $\quad \theta_* \equiv - \frac{\overline{w'\theta'}}{u_*} \quad$ is a scaling temperature $\tag{19}$

and $\quad \alpha = \frac{K_h}{K_m} \quad$ ratio of eddy transfer coefficients $\tag{20}$

With these definitions we recognize the following identities

$$Ri = \zeta \, \phi_h / \phi_m^2 \tag{21}$$

and $\quad \alpha = \phi_m / \phi_h \tag{22}$

Fleagle and Businger (1963) give a simple physical argument leading to a diabatic wind profile

$$\phi_m = (1 - \gamma \, Ri)^{-1/4} \tag{23}$$

for the unstable case, i.e. $Ri < 0$. γ is a constant to be determined by experiment ($\gamma \simeq 16$). Pandolfo (1966), Businger (1966) and Dyer (1966) from analyzing the Kerang

Aerodynamics of Vegetated Surfaces

observations concluded independently that for Ri < 0 a good approximation is given by

$$Ri \simeq \zeta \qquad (24)$$

a remarkably simple relation. This combined with (23) yields

$$\phi_m = (1 - \gamma \zeta)^{-1/4} \qquad (25)$$

and combined with (21) gives

$$\phi_h = (1 - \gamma \zeta)^{-1/2} \qquad (26)$$

consequently we find for (22)

$$\alpha = (1 - \gamma \zeta)^{1/4} \qquad (27)$$

The set of equations 24-27 describes the mean structure of the unstable boundary layer. Paulson (1970) integrated (25) and (26) in the form

$$\frac{\bar{u}}{u_*} = \frac{1}{k} \ln (\frac{z}{z_o} - \psi_1) \qquad (25a)$$

where $\psi_1 = 2 \ln [(1+x)/2] + \ln [(1+x^2)/2] - 2 \tan^{-1} (x+\pi/2)$
and $x = (1-\gamma\zeta)^{1/4} = \phi_m^{-1}$

and $\frac{\bar{\theta}-\theta_o}{\theta_*} = \frac{1}{k} (\ln \frac{z}{z_o} - \psi_2) \qquad (26a)$

where $\psi_2 = \ln [(1+y)/2]$ and $y = (1-\gamma\zeta)^{1/2} = \phi_h^{-1}$

The stably stratified surface layer is well represented by the log-linear law. Webb (1970) and Businger et al, (1971). In this case we have

$$\phi_m = 1 + \beta \zeta = \phi_h \qquad (28,29)$$

or integrated

$$\frac{\bar{u}}{u_*} = \frac{1}{k}(\ln \frac{z}{z_o} + \beta \zeta) = \frac{\bar{\theta}-\theta_o}{\theta_*} \qquad (28a, 29a)$$

where θ_o is the temperature at z_o.

For eqs. (25a, 26a, 28a, 29a) the same consideration of a zero plane displacement is valid as for eq. (5).

Also, similar to eq. (5), eqs. (25a, 26a, 28a, 29a) are valid well beyond the surface layer. This will enable us to obtain from the profiles the surface fluxes in a region where the fluxes are no longer constant with height. On the other hand if we try to obtain the fluxes directly with the eddy correlation technique at substantial heights which is sometimes needed in the case of tall vegetation, such as forests, a correction may be needed because the fluxes are already substantially less than the surface values.

Fig. 1 shows the agreement with observations of eqs. 25, 26, 28, and 29. The plots for α and Ri may be derived from this figure and are not given explicitly here.

The equations provide for an attractive scheme of similarity conditions. We only need to keep track of z/z_o and z/L. However, for practical purposes it makes a great deal of difference whether we consider a lawn or a forest. Above a lawn it is easy to require, say, that the measurements should be taken at $z \gg z_o$ and be still within the surface layer. A forest may have a $z_o \simeq 1m$, which makes the requirement $z \gg z_o$ difficult to meet and probably impossible to meet simultaneously with the requirement of a constant flux layer. Also, the requirement of horizontal uniformity is much sooner reached over a lawn than over a forest. Because of these logistical aspects there is an abundance of observations over relatively smooth surfaces (i.e. small z_o) and a serious scarcity of data over real rough surfaces such as forests.

Aerodynamics of Vegetated Surfaces

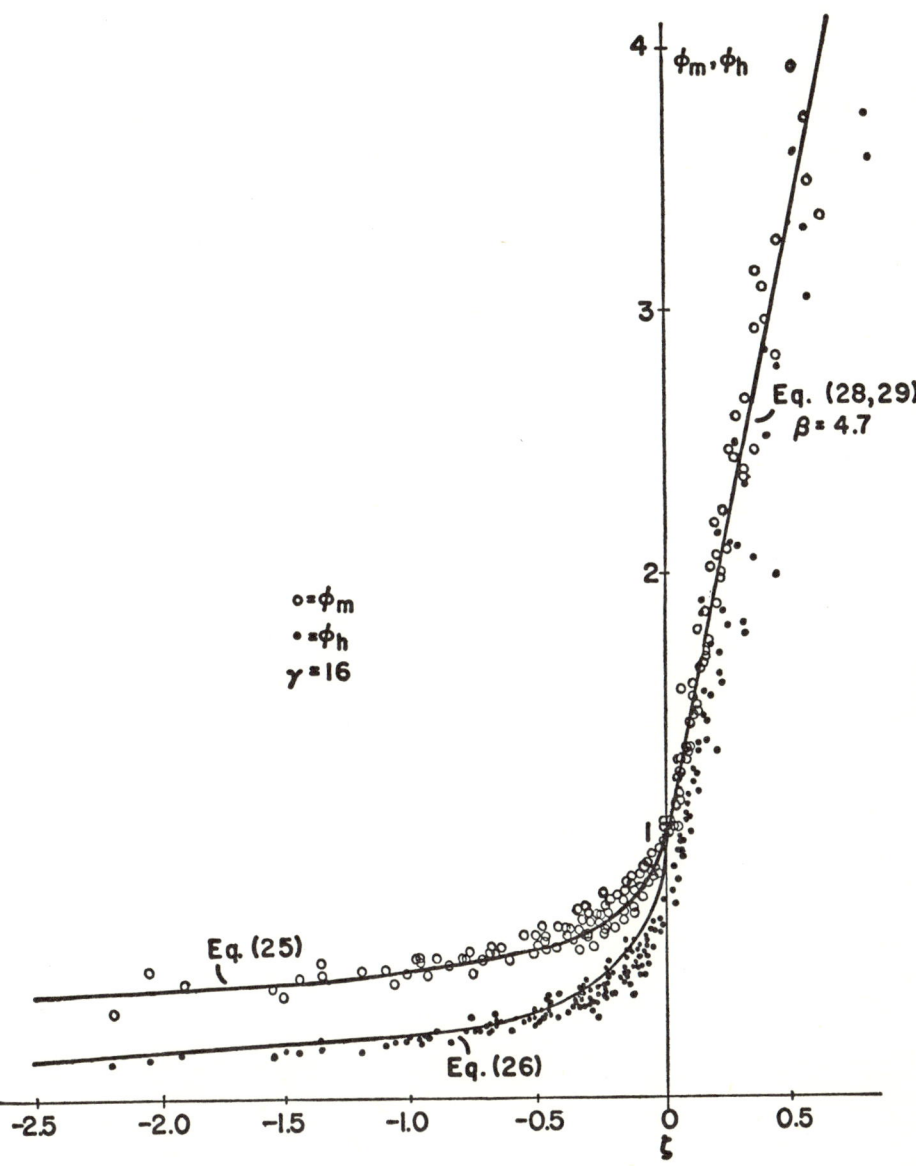

Fig. 1. Dimensionless wind gradient and temperature gradient.
Eqs. (25), (26), (28) and (29) are compared with the Kansas data presented in Businger et al (1971).
$\gamma = 16$ and $\beta = 4.7$.

3. ROUGHNESS, DISPLACEMENT HEIGHT AND SHELTERBELTS

a. In the preceeding section we have considered the average effect of a statistically large number of roughness elements. They may range from grains of sand and leaves of grass to tree tops. The requirement was uniformity. The roughness length and displacement height have been discussed for uniform conditions by Monin and Yaglom (1971, p. 284-295). What happens to the roughness if in the middle of our pasture we find a single tree standing? Does this increase the roughness of the pasture as a whole? Or is it better to consider the effect of the tree on the wind separately? Clearly it seems desirable to analyze the effects of single roughness elements. But it is not always clear what a roughness element is. Here we have to use a combination of common sense, intuition and some experience, not necessarily in that order. Whether or not a leaf (or other plant element) is a roughness element depends on whether or not its exposure to the mean wind is characteristic of a roughness element. Whatever a roughness element is, it can be characterized by a height, h, a cross section, S, to the wind and a porosity to the wind. If we are considering a population of more or less uniform roughness elements distributed over an area then we can also take into consideration the horizontal area , A, per roughness element. Lettau (1969), based on Kutzbach's (1961) bushel basket experiments over frozen lake Mendota formulated a remarkably simple empirical relation for the roughness length based on above mentioned characteristics assuming that porosity does not play an important role:

$$z_o = 0.5 \, h \, S/A \tag{30}$$

where 0.5 is an empirical coefficient.

This formula is particularly useful when $A \gg S$, i.e., when the roughness elements are fairly isolated. In fact Lettau used this relation to determine the increase in roughness due to micrometeorological equipment during a cooperative field experiment in Davis (April-May, 1967). The elements (masts, radiometers and various other instruments were scattered over the field and according to (30) contributed about 0.3 cm to z_o, which seemed quite reasonable if added to the roughness of the grass. It is, of course, not clear that roughnesses are additive. Eq. (30) tends to overestimate z_o when A becomes of the order of S. It is clear that when the elements are closely packed together the wind will respond to the tops of the elements as a new surface and z_o in eq. (30) corresponds probably more to the displacement height, and the actual roughness length has to be derived from the geometric configuration of the new surface. The individual roughness elements on which eq. (30) is based have lost their identity in this case.

Eq. (30) may be modified to take the above mentioned limitation into account. First we observe that usually S can be expressed as being proportional to h^2 ($S = \frac{\pi}{4} h^2$ for a circle; $S = \frac{\sqrt{3}}{3} h^2$ for an equilateral triangle etc.). Therefore (30) may be written

$$z_o = 0.5 \, C_1 \, h^3/A \qquad (31)$$

where C_1 is the geometrical constant between S and h^2. If now the displacement height, D, exists because there is too much overlap of the elements, the actual height of the roughness elements may be assumed to be $h' = h-D$. The cross section of the new elements will be proportional to h'^2 and eq. (31) may be written in the form

$$z_o = 0.5 \, C_1 \, \frac{(h-D)^3}{A} \qquad (32)$$

In this case C_1 may be a function of D/h because the geometry of the roughness elements changes with increasing overlap. On the other hand D/h itself will be a function of h^2/A again with a geometrical constant or function as a coefficient of proportionality. The simplest relation would be

$$\frac{D}{h} = C_2 \frac{h^2}{A} \qquad (33)$$

The right hand side must always remain less than one. Only in the case the roughness elements are cubes is it possible to pack them in such a way that a new surface is being generated so that $D=h$ and the z_o may become very small. Combining (33) with (32) yields

$$z_o = 0.5 \, C_1 \, h^3 \, (1-C_2 \, h^2/A)^3 \, A^{-1} \qquad (34)$$

If the geometry of the elements is specified eqs. (33) and (34) may be used to obtain a reasonable estimate of D and z_o. The coefficient C_1 relates to the geometry of the vertical cross section and C_2 to the horizontal cross section of the elements.

In the preceding discussion it has been assumed that the roughness elements were more or less uniformly distributed in the horizontal, this may to some extent include a random distribution provided the random variations are not too large.

Of special interest is the effect of shelterbelts. Eqs. (33) and (34) may be reduced to the one dimensional case when windbreaks are oriented perpendicular to the wind. The area may then be replaced by the distance, l, between the windbreaks and the cross section is then proportional to h. Eqs. (33) and (34) may be written as

$$\frac{D}{h} = C_2 \frac{h}{l} \qquad (35)$$

Aerodynamics of Vegetated Surfaces

and
$$z_o = 0.5 \, h^2 \, (1 - C_2 \tfrac{h}{l})^2 \, l^{-1} \tag{36}$$

because $C_1 = 1$ in this case.

The equation is only valid for $C_2 \tfrac{h}{l} \geq 1$. This leads to some difficulty in the unrealistic case the windbreak is very thin and one wants to consider large values of h/l, i.e., many windbreaks close together.

In Fig. 2 examples are given of eqs. (34) and (36) for given geometries of the roughness elements and shelterbelts. Eqs. 31-36 are purely empirical and only reflect the complicated aerodynamic effects roughness elements have on the airflow in the crudest senses. The justification for a purely geometrical treatment of these elements lies in the assumption that the mean flow is fully turbulent. Wind tunnel observations by Kawatani and Meroney (1970) suggest that eq. (34) underestimates z_0 and places the max effect at too high values of h^2/A. On the other hand, observations by Koloseus and Davidian (1966) as reviewed by Wooding et al (1973) suggest that with proper adjustment of the constants (34) may indeed be a useful interpolation formula.

The main purpose of a shelterbelt is to reduce the windspeed behind it over as large a distance as possible, so that a minimum of shelterbelts are needed to protect a given area. The maximum overall wind reduction will be obtained when z_o is a maximum, which according to (36) occurs for $l = 3 \, C_2 \, h$. This may be in many cases a closer distance than is economically feasible or desirable for other reasons.

An extensive study on the effect of shelterbelts with various degrees of porosity was carried out by Nägeli (1941) which is still frequently referred to. He found that very dense shelterbelts would reduce the wind strongly right behind the shelter but the wind would increase relatively rapidly farther away. A rather loose shelterbelt would reduce the wind less behind the shelter but its effect extends over a greater distance. Plate (1971) found an explanation of

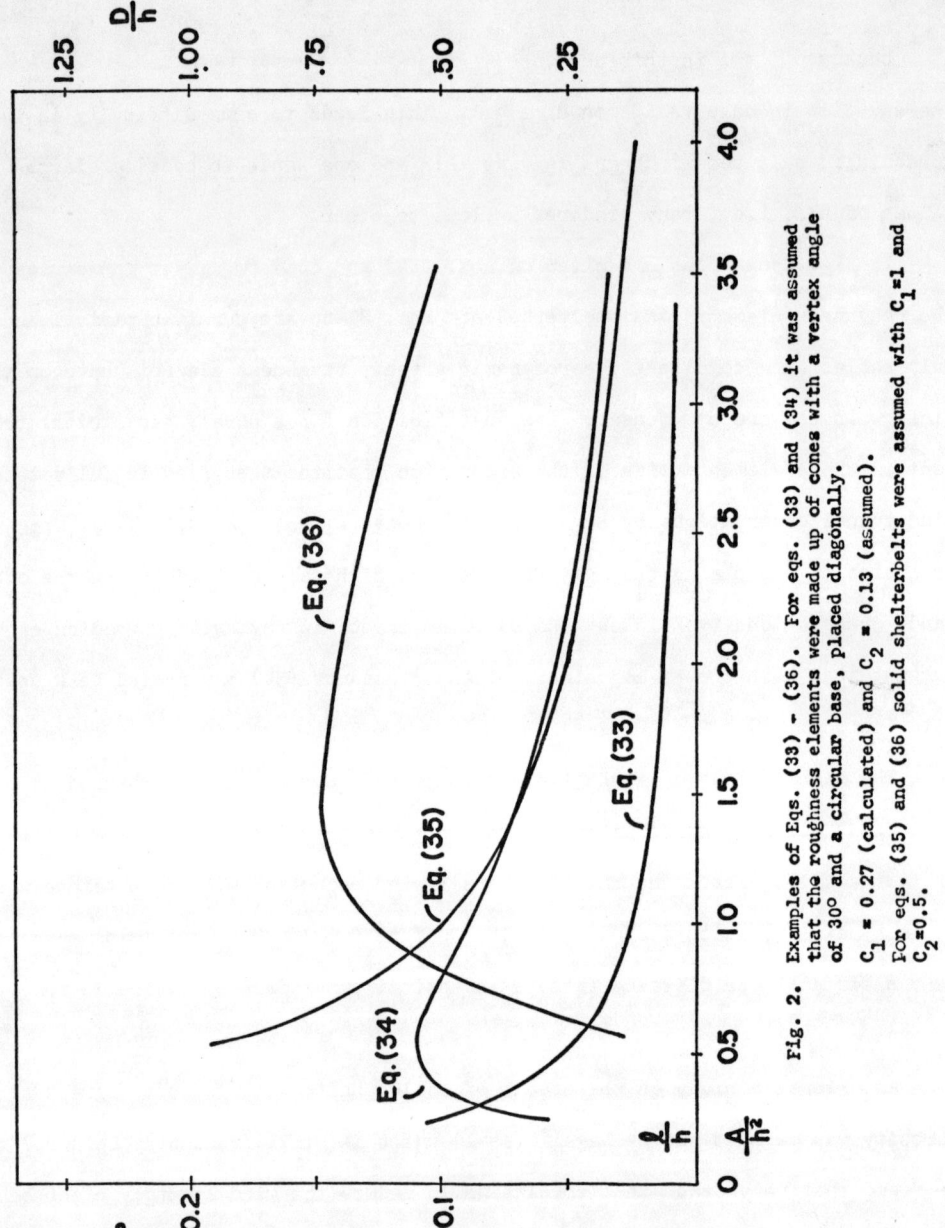

Fig. 2. Examples of Eqs. (33) - (36). For eqs. (33) and (34) it was assumed that the roughness elements were made up of cones with a vertex angle of 30° and a circular base, placed diagonally.

$c_1 = 0.27$ (calculated) and $c_2 = 0.13$ (assumed).

For eqs. (35) and (36) solid shelterbelts were assumed with $c_1 = 1$ and $c_2 = 0.5$.

this phenomenon. A dense shelter belt creates a relatively large pressure difference across it when exposed to the wind. At the height h the flow passing over the shelter will be separated from the air behind the shelter. However, the low pressure behind the shelter exerts a downward force on the separated flow and consequently curves the separation streamline downward, see Fig. 3. The curvature can be found approximately by equating the centrifugal force on an element following the streamline and the resultant pressure force. This yields

$$\frac{1}{R} = \frac{1}{\rho u^2} \frac{\partial p}{\partial z} \qquad (37)$$

where R is the radius of curvature, u is the windspeed at the streamline and ∂p/∂z is the pressure gradient across it. Eq. (37) expresses that the larger the pressure gradient the smaller the radius of curvature and consequently the stronger the curvature. Therefore the streamline bends down to the surface more rapidly behind a dense shelter than behind a more porous shelter and the re-attachment occurs sooner. This is called the "Coanda" effect.

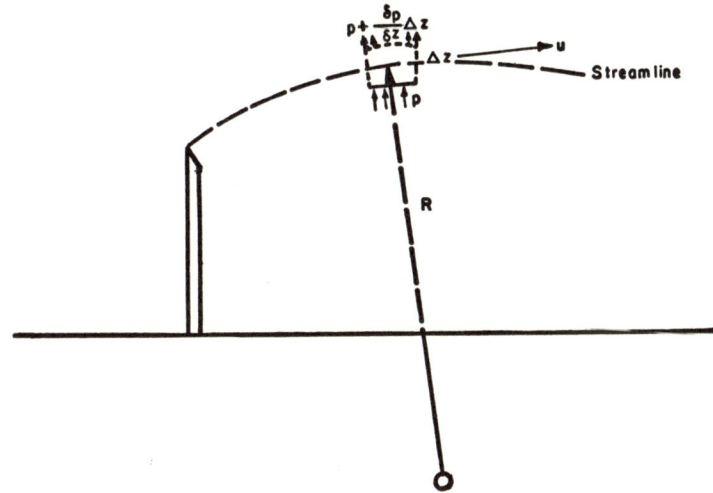

Fig. 3. Illustration of the Coanda effect (after Plate, 1971).

The complexities of the flow around a shelterbelt are illustrated schematically in Fig. 4. In zone 1 we have the undisturbed flow field of the surface layer. In zone 2 the flow field is displaced and distorted due to the presence of the shelterbelt. Beyond the shelter the lower boundary of 2 represents the shear zone related to the separation of the flow. When the shelter is solid back flow may occur (zone 3) with a reattachment point. Behind the reattachment point the boundary layer flow gradually develops again (zone 4).

The flow is even more complex when it encounters a series of shelterbelts. Much research can still be done on finding the optimum solutions as far as height and distance is concerned for various types of shelterbelts.

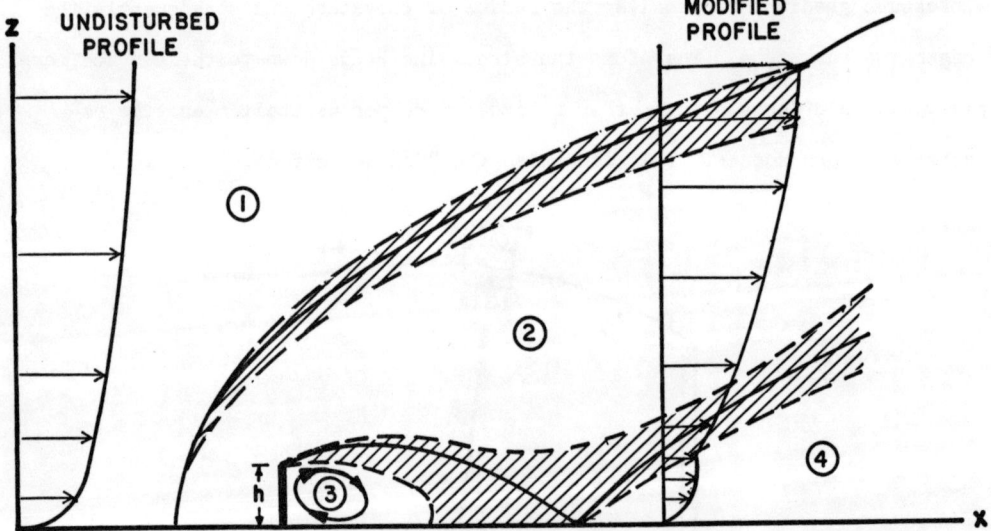

Fig. 4. The flow zones in a boundary layer disturbed by a shelterbelt.
    1. Undisturbed boundary layer flow.
    2. Region of influence of shelterbelt on pressure field.
    3. Region of flow separation (eddies).
    4. Re-established boundary layer.

The shaded areas are transition zones. (After Plate and Lin, 1965 with some modifications).

Aerodynamics of Vegetated Surfaces

4. THE FLOW WITHIN A CANOPY

When the roughness elements are packed so close that a new uniform surface (a canopy) is formed by their tops we again have a surface layer which is now displaced upward by the displacement height D. For the flow above this canopy at level D there might just as well have been a solid surface, it would have the same characteristics. However, this does not mean that below D there is no flow at all. A comparatively weak but complicated flow structure exists in this region which we call canopy flow.

The main difference between the flow inside the vegetation and immediately above it is that momentum is absorbed over a vertical cross section of the flow. Instead of eq. (6) we have now

$$\frac{\partial}{\partial z} \overline{u'w'} = -\frac{1}{\rho} \frac{\partial \overline{p}}{\partial x} - \frac{1}{2} C_D A \overline{u}^2 \tag{38}$$

where C_D is the drag coefficient of the plant elements and A is the effective aerodynamic surface area of the vegetation per unit volume. Both C_D and A are usually complicated functions of height, and C_D is also somewhat dependent on $\overline{u}$.

In well developed canopies we may distinguish 3 flow regimes;

I. The pressure term is negligible, which is true in the upper part of the canopy. Eq. (38) reduced to

$$\frac{\partial}{\partial z} \overline{u'w'} = -\frac{1}{2} C_D A \overline{u}^2 \tag{39}$$

II. The divergence of the momentum flux is negligible. This may be the case for the flow under a dense canopy, as in a well developed forest. In this case we have

$$\frac{1}{\rho} \frac{\partial \overline{p}}{\partial x} = -\frac{1}{2} C_D A \overline{u}^2 \tag{40}$$

with a significant wind shift from case I. The direction will be down the large scale pressure gradient.

    III. Very close to the surface we have again the same condition as above the canopy, i.e.

$$\frac{\partial}{\partial z} \overline{u'w'} = 0, \text{ with the}$$

corresponding logarithmic profile.

In the transition zones from I to II and from II to III all terms of (38) have to be considered.

Most studies of canopy flow deal with regime I. (Tan and Ling, 1961; Cionco 1962, 1965; Uchijima and Wright, 1964; Plate and Quraishi, 1965).

In the very simple case (Cionco, 1962) that $\frac{1}{\rho}\frac{\partial \overline{p}}{\partial x}$ is negligible in comparison to the last term in (30) and that $\frac{1}{2} C_D$ A is independent of height this equation may be integrated if we further assume that we have a constant mixing length ℓ such that

$$\frac{\partial}{\partial z} \overline{u'w'} = -\frac{\partial}{\partial z}(\ell \frac{\partial u}{\partial z})^2 = -\ell^2 \frac{\partial}{\partial z}(\frac{\partial u}{\partial z})^2 \tag{41}$$

The result is

$$u = u_h \exp\{a(z/h-1)\} \tag{42}$$

where h is the height of the canopy and $a = \{\frac{C_D A}{4\ell^2}\}^{1/3}$ h.
The simplifications leading to eq. (42) are such that one wonders whether this equation may be applied to real canopies. Surprisingly a number of observations do agree with (42) above expectations. The reason maybe that $\underline{a}$ changes relatively

slowly with C_D, A and ℓ. The agreement is best in the upper part of uniform canopies., such as wheat, corn and rice. (Inoue, 1963; Saito, 1964; Tan and Ling, 1961; Uchijima and Wright, 1964). Fig. 5 shows a comparison of (42) with some observations. It is, of course, possible to refine this model by introducing the variations of height of C_D and A as they occur in real canopies and conse-

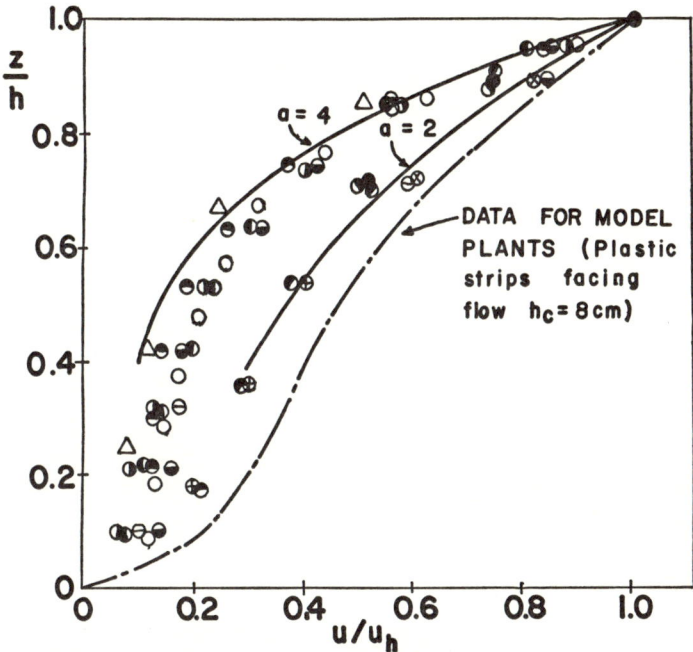

Fig. 5. Comparison of Eq. (42) with some observations. (Partly after Plate, 1971).

quently solve eq. (39) numerically. An effort in this direction was published by Cionco (1965).

If the vegetation is tall and the canopy is well developed the momentum from the mean wind above the canopy may be virtually absorbed in the upper part of the canopy. Farther down it is too shady for many leaves to develop and C_D A decreases. In this area it is possible that $\frac{1}{\rho} \frac{\partial p}{\partial x} \gg \frac{\partial}{\partial z} \overline{u'w'}$ and that eq. (40) is valid. It is clear that the velocity will increase with decreasing C_D A because the pressure gradient is independent of height. The wind direction tends to be perpendicular to the isobars whereas above the canopy the cross-isobaric angle is usually less than 40°. Consequently a significant windshift may be observed within the canopy. A collection of normalized wind profiles in various canopies is given in Fig. 6. Some of the characteristics discussed above may be recognized in these profiles but also some deviations from the simple conditions as discussed here are apparent.

CONCLUSION

Looking back at the previous discussion we must humbly conclude that we have barely scratched the surface. There are many problems that have been tackled which have not been discussed here and there are many more problems that have yet to be studied. A few comments should be made.

Many studies have been made on the effect of a sudden change of roughness on the boundary layer. A recent and very promising study by Rao et al,(1974) uses a second order closure model for the turbulence. Their results agree remarkably well with the observations by Bradley (1968). A good review of the literature is given by Plate (1971).

Windtunnel experiments with uniform roughness elements have been carried out by Marshall (1971), and also by Kawatani and Meroney,(1970) which might be used to test the simple geometric relations proposed in section 3.

Aerodynamics of Vegetated Surfaces

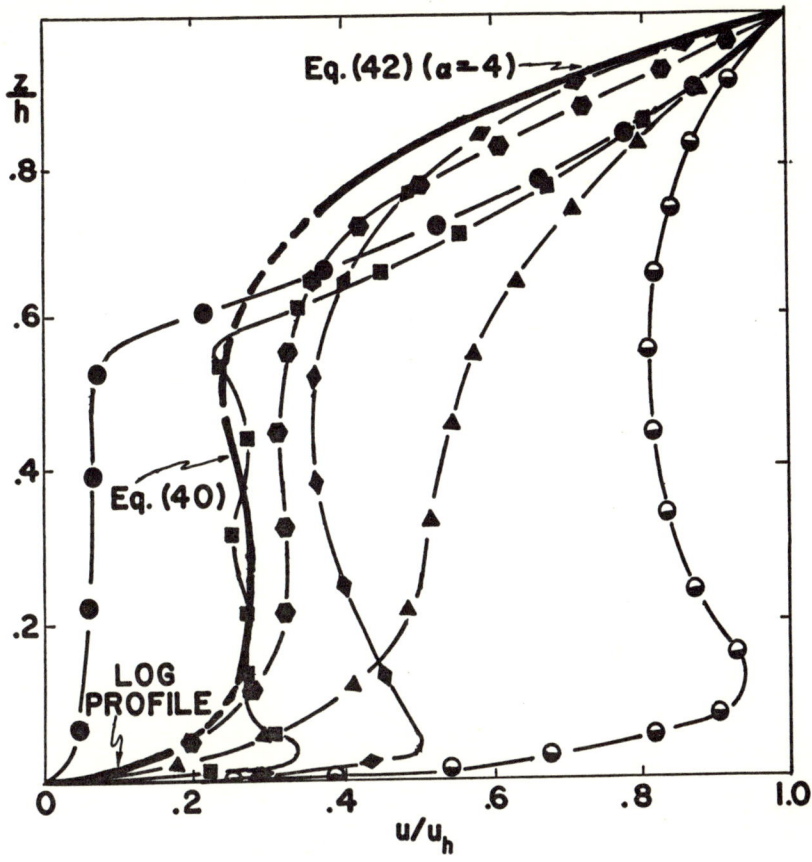

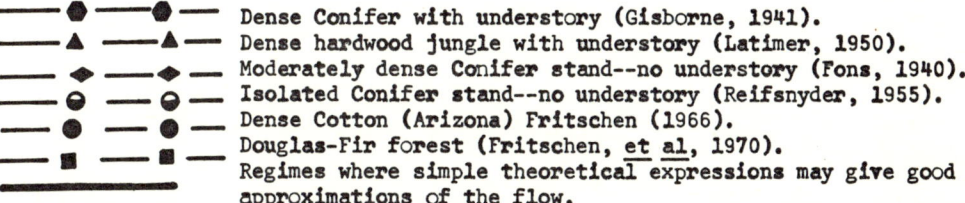
— ● — ● — Dense Conifer with understory (Gisborne, 1941).
— ▲ — ▲ — Dense hardwood jungle with understory (Latimer, 1950).
— ♦ — ♦ — Moderately dense Conifer stand--no understory (Fons, 1940).
— ◐ — ◐ — Isolated Conifer stand--no understory (Reifsnyder, 1955).
— ● — ● — Dense Cotton (Arizona) Fritschen (1966).
— ■ — ■ — Douglas-Fir forest (Fritschen, et al, 1970).
▬▬▬▬▬ Regimes where simple theoretical expressions may give good
approximations of the flow.

Fig. 6. Canopy flow in a variety of forests as well as in dense cotton. (After Fritschen et al, 1970).

The effects of advection have been studied extensively also. However, the aerodynamic problems associated with it are still formidable. The development of rational second order closure schemes for the turbulent transfer opens this area up for fruitful research.

The effects of static stability within the plant canopy are in many cases not negligible and require careful study especially of the turbulent intensity and the associated transfer of heat, water vapor, and CO_2.

The structure of turbulence within the canopy which is basic for problems of diffusion and turbulent transfer has been studied to some extent, e.g., (Baines, 1972; Cionco (1972) and Kawatani and Meroney (1970). It is still a wide open area with considerable potential for fruitful research.

Finally it is well known that the vegetation responds to the wind. Trees bend a wheat field waves and leaves flutter. The aerodynamic response of the vegetation to the flow interacts with the flow. For example: In a strong wind both the displacement height and the roughness length of a wheat field are less than in a light wind. The plants bend over and become more streamlined which reduces the stress.

REFERENCES

Baines, G.B.K., 1972: Turbulence in a wheat crop. Agr. Meteor., 10, 93-106.

Blackadar, A.K., 1962: The vertical distribution of wind and turbulent exchange in a neutral atmosphere. J. Geophys. Res., 67, 3095-3102.

Bradley, E.F., 1968: A micrometeorological study of velocity profiles and surface drag in the region modified by a change in surface roughness. Quart. J. Roy. Meteorol. Soc., 94, 361-379.

Businger, J.A., 1966: Transfer of momentum and heat in the planetary boundary layer. Proc. Symp. Arctic Heat Budget and Atmospheric Circulation, The RAND Corporation, Santa Monica, Calif. 305-322.

Businger, J.A. and S.P.S. Arya, 1974: Height of the mixed layer in the stably stratified planetary boundary layer. Advances of Geophysics. Academic Press, 73-92.

Businger, J.A., J.C. Wyngaard, Y. Izumi and E.F. Bradley, 1971: Flux profile relationships in the atmospheric surface layer. J. Atmos. Sci., 28, 181-189.

Cionco, R.M., 1962: A preliminary model for air flow in the vegetative canopy. Bull. Am. Meteor., 43, 319. (Abstract)

Cionco, R.M., 1965: A mathematical model for air flow in a vegetative canopy. J. Appl. Meteor., 4, 515-522.

Cionco, R.M., 1972: Intensity of turbulence within canopies with simple and complex roughness elements. Boundary Layer Met., 2, 453-465.

Fleagle, R.G. and J.A. Businger, 1963: An Introduction to Atmospheric Physics. Acad. Press, New York, 346 pp.

Fons, W.L., 1940: Influence of forest cover on wind velocity. J. For., 38: 481-486.

Frenzen, P., 1973: The observed relation between the Kolmogorov and v. Karman constants in the surface boundary layer. Boundary Layer Met., 3, 348-358.

Fritschen, L.J., 1966: Micrometeorological determination of water loss from various surfaces. In Atmospheric and Soil-Plant-Water Relationships. Technical Report ECON 2-66P-A. U.S. Army Electronic Command, Ft. Huachuca, Arizona.

Fritschen, L.J., C.H. Driver, C. Avery, J. Buffo, R. Edmonds, R. Kinerson, and Peter Schiess, 1970: Dispersion of air tracers into and within a forested area: 3. (Univ. of Wash., Seattle, Wash.) Technical Report ECOM-68-68-3. U.S. Army Electronic Comman, Ft. Huachuca, Arizona.

Gisborne, H.G., 1941: How the wind blows in the forest of northern Idaho. 11th Rocky Mt. For. Range Exp. Sta., 14pp.

Inoue, E., 1963: On the turbulent structure of airflow within crop canopies. J. Met. Soc. Japan, Tokyo, Japan, Ser. II 41, 317-326.

Kawatani, T. and R. N. Meroney, 1970: Turbulence and windspeed characteristics within a model canopy flow field. Agr. Meteor., 7, 143-158.

Koloseus, H.J. and Davidian, J., 1966: Free-surface instability correlations and roughness concentration effects on floro over hydrodynamically - rough surfaces, USGS Water-Supply Paper 1592-C,D.

Kutzbach, J.E., 1961: Investigations of the modification of windprofiles by artificially controlled surface roughness. Section 7 of studies of the three dimensional structure of the planetary boundary layer. Annual Report, Univ. of Wisconsin, 71-113. (DDC)

Latimer, W.M., 1950: General Meteorological Principles. In Handbook on aerosols, U.S. Atomic Energy Comm. (Chapt. 2). 147 pp.

Lettau, 1969: Note on aerodynamic roughness parameter estimation on the basis of roughness-element description.

Marshall, J.K., 1971: Drag measurements in roughness arrays of varying density and distribution. Agr. Meteor., 8, 269-292.

Monin, A.S., and A.M. Yaglom, 1971: Statistical Fluid Mechanics: Mechanics of Turbulence, Vol. 1 Cambridge, Mass., The MIT Press, 769 pp.

Nägeli, W., 1941: Untersuchungen über die Windverhältnisse im Bereich von Windschutzstreifen, Mitt. Schweis. Anst. Forstl. Versuchsw, 23: 221-276.

Obukhov, A.M., 1946: Turbulence in an atmosphere with a non-uniform temperature, Tr. Ahad. Nauk. SSSR Inst. Teoret. Geofis. No. 1. (Transl. in Boundary Layer Met., 2, 7-29).

Paeschke, W., 1937: Experimentelle Untersuchungen zum Rauhigkeits-und Stabilitaetsproblem in der freien Atmosphaere. Beitr. Phys. Atmos., 24, 163-189.

Pandolfo, J., 1966: Wind and temperature profiles for constant flux boundary layers in lapse conditions with a variable eddy conductivity to eddy viscosity ratio. J. Atmos. Sci., 23, 495-502.

Paulson, C.A., 1970: The mathematical representation of wind speed and temperature profiles in the unstable atmospheric surface layer. J. Appl. Meteor., 9, 857-861.

Plate, E.J., and A.A. Quraishi, 1965: Modeling of velocity distribution inside and above tall crops. J. Appl. Meteor., 4, 400-448.

Plate, E.J., 1971: Aerodynamic characteristics of atmospheric boundary layers. AEC Critical Review Series, 190 pp.

Plate, E.J., 1971: The aerodynamics of shelterbelts. Agr. Meteor., 8, 203-222.

Prandtl, L., 1932: Meteorologische Anwendungen der Strömumgs. lehre, Beitr. Phys. Atmos, 19, 188-202.

Pruitt, W.O., D.L. Morgan and F. J. Laurence, 1973: Momentum and mass transfers in the surface boundary layer. Quart. J. Roy. Met. Soc., 99, 370-386.

Rao, K.S., J.C. Wyngaard and O.R. Coté, 1974: The structure of a two-dimensional internal boundary layer over a sudden change of surface roughness. J. Atmos. Sci., 31, 738-746.

Reifsnyder, W.E., 1955: Wind Profiles in a Small Isolated Forest Stand. Forest Science 1: 289-297.

Richardson, L.F., 1920: The supply of energy from and to atmospheric eddies. Proc. Roy. Soc. London, A 97, 354-373.

Saito, T., 1964: On the wind profile within plant communities. Bull. Nat. Inst. Agric. Sci. Ser. A, No. 11, 67-73.

Seguin, B., 1973: Rugosité due paysage et evapotranspiration potentielle à l'échelle régionale (Land roughness and regional potential evapotranspiration). Agricultural Meteorology 11, 79-98.

Stoller, T. and E.R. Lemon, 1963: The energy budget of the earth's surface, Part II. Production Research Report No. 2, Agriculture Research Service, U.S. Dept. of Agriculture, 49 pp.

Tan, H.S. and S.C. Ling, 1961: A study of atmospheric turbulence and canopy flow. Therm Incorporated and Dept. of Agriculture, Ithaca, N.Y., TAR-TR611, Cooperative Research Program.

Tennekes, H., 1968: Outline of a second order theory of turbulent pipe flow. AIAA J., 6, 1735-1740.

Tennekes, H., 1973: The Logarithmic Windprofile, J. Atmos. Sci., 30, 234-238.

Uchijima, Z, and J.L. Wright, 1964: An experimental Study of airflow in a corn-plant layer, Bull. Nat. Inst. Agric. Sci., Japan, Ser. A, 11, 19-65.

Wooding, R.A., E.F. Bradley, and J.K. Marshall, 1973: Drag due to regular arrays of roughness elements of varing geometry. Boundary Layer Met., 5, 285-308.

HEAT AND MASS TRANSFER WITHIN PLANT CANOPIES

BRIAN LEGG* and JOHN MONTEITH†

*Rothamsted Experimental Station, Harpenden, Herts, U.K.,
†University of Nottingham School of Agriculture,
Sutton Boninton, Loughborough, Leics. U.K.

ABSTRACT

The foliage of vegetation is a complex source and sink of heat, water vapour, carbon dioxide, and other gases. The vertical transport of heat and mass in various crop stands has conventionally been estimated from the product of a one-dimensional transfer coefficient (K) and a vertical concentration gradient although there are serious theoretical and practical objections to this procedure. The properties of turbulence in canopies can be specified in terms of intensity, scale length, etc. More fundamental work is needed to relate specifications of turbulence to the diffusion of heat and mass in three dimensions.

OBJECTIVES

Investigations of heat and mass transfer in vegetation have two broad, complementary objectives, one physical and the other biological. The physical objective is to describe how parts of the earth's surface which are covered with grass, arable crops, swamps, or forests partition the radiant energy which they absorb, and in particular, to specify the input of heat and water vapour to the lower atmosphere. Realistic boundary conditions are an essential component of models of atmospheric behaviour and may eventually contribute to more accurate predictions of weather, both daily and seasonally. The biological objective is to understand how the growth, survival and reproduction of higher plants depends on the environment of foliage. Encouraged by a stimulus from ecology, micrometeorologists have attempted to measure, to analyse and to model processes of transfer which occur within plant communities and which are intimately related to physiological behaviour. This review is concerned with progress and with problems which inhibit progress. We shall try to determine whether micrometeorological methods of estimating heat and mass transfer in plant stands are worth pursuing, either in preference to other techniques or to provide complementary information.

SOURCES AND SINKS

Determinants

When sunlight strikes the surface of a leaf, part of the absorbed energy, usually a major part, is dissipated by the evaporation of water, and a much smaller part is stored in the products of photosynthesis. The leaf may therefore be regarded as a source of water vapour (or latent heat) and a sink of carbon dioxide. The mean temperature of the leaf adjusts to a value at which the net gain of heat is equal to the heat loss, i.e. the leaf may be a source or a sink of sensible heat depending on whether it is warmer or cooler than the surrounding air. At night, a leaf exposed to the sky loses heat by radiation and usually cools a few degrees (Celsius) below air temperature. Small amounts of heat are produced by respiration and by condensation when the temperature of the leaf is less than the dew-point of the ambient air

Proceeding from the simple heat balance of a single leaf to the much more complex system of a leaf canopy, it is necessary to take account of the following factors:

1) The spatial distribution and arrangement of foliage specified by a leaf area index L and by the elevation and azimuth angles of leaf surfaces (1). The leaf area index per unit depth of canopy $a(z)$, often referred to as the foliage area density, usually increases from zero at the top of the stand, height h, to a maximum value at approximately $2h/3$ and then decreases to a very small value near the ground (Fig. 1). Stephens (3) showed that the foliage in Pine forests follows a normal frequency distribution with respect to height and the foliage of many agricultural crops is arranged in a similar way. In consequence the layer of foliage between $2h/3$ and h is the 'active' layer for most types of mature vegetation, containing all the significant sources and sinks of heat, mass and momentum.
In some types of vegetation it is necessary to take account of the surface area of stems and other organs capable of transpiration and photosynthesis (2). The differentiation of green (active) and yellow (senescent) foliage is also relevant to the distribution of CO_2 sinks.

2) The spatial distribution of intercepted radiant energy which depends on the optical properties of the foliage and on its architecture (see contribution by Nilson).

3) The spatial distribution of potentials such as temperature, vapour pressure and CO_2 concentration. These potentials determine how the available radiant energy is partitioned between convection, transpiration and photosynthesis. There is a strong element of feedback in the system, however, so that the existence of sources and sinks of heat and mass, whose magnitude and distribution depends on a set of potentials, is responsible in turn for the evolution of potential gradients within the foliage and for diurnal variation of the potentials about mean values which are set by weather and soil conditions.

4) Resistances to the transfer of heat and mass, i.e. the potential difference between any two points in the foliage required to maintain unit flux density of heat or of mass between these points.
The resistances are of two main types:-
a) Physiological resistances to gaseous diffusion, in particular, the

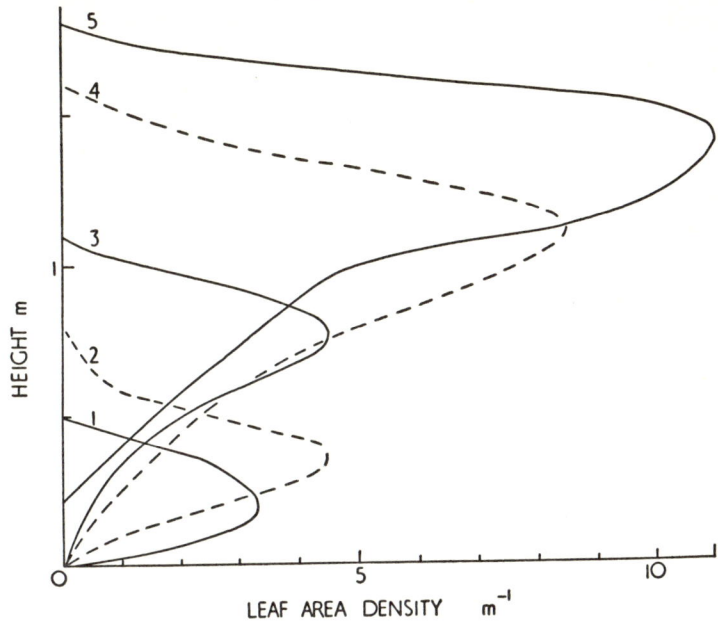

Figure 1. Vertical distribution of leaf area in a maize stand at Iygeva. 1963 (from Ross and Nilson (2)) 1, 12 July; 2, 25 July; 3, 3 August; 4, 13 August; 5, 20 August.

resistance of stomatal pores (discussed by Stigter). These resistances depend both on the physical environment of the foliage e.g. availability of light energy, temperature; and on biological factors such as age and genetic constitution.

b) aerodynamic resistances to heat and gaseous diffusion determined by the flow of air round individual foliage elements and through the canopy as a whole.

As the physiological resistances in a plant stand are often much larger than the corresponding aerodynamic resistances, they play a dominant role in determining the distribution of water vapour and carbon dioxide fluxes. However, as the main emphasis of this review is physical rather than biological, the following sections will be concerned mainly with the nature and specification of air flow and turbulence within canopies.

Aerodynamic regimes

A network of aerodynamic resistances extends from the surfaces of individual foliage element to the free atmosphere above the canopy, and transfer across this network may be considered a three-stage process. In the <u>first</u> stage, transfer occurs across the laminar or turbulent boundary layer of air in immediate contact with the surface. In principle, the rate of transfer can be expressed as a difference of potential between the surface and the air divided by the diffusion resistance of the boundary layer. Numerous complications arise in practice: the effect of hairiness and of other surface irregularities; and the effect of externally imposed turbulence on boundary layer transport, a topic discussed by Clark and Wigley. Pressure fluctuations are likely to influence the behaviour of boundary layers within canopies but experimental evidence is lacking.

The <u>third</u> stage is transfer in the atmosphere, essentially a three-dimensional process. Over extensive, uniform vegetation, however, horizontal transfer may be neglected and most experimental studies have concentrated on the relation between <u>vertical</u> fluxes and concentration gradients of heat and mass in a one-dimensional boundary layer. The coefficient of turbulent transfer $K(z)$, representing the flux per unit gradient at height z within this layer, can be estimated from a knowledge of windspeed and temperature profiles, and a corresponding resistance between two heights z_1 and z_2 can be found from $\int dz/K(z)$.

The first and third stages of transfer are therefore related to the behaviour of specific boundary layers - the boundary layer of individual foliage surfaces and the boundary layer of the whole canopy acting as an extended surface. The intermediate <u>second</u> stage is much harder to identify and to define and it has no accepted name. At a conference in Canberra some years ago it was appropriately referred to as the Sargasso Sea where marine vegetation proliferates in the clear swirling waters!

Homogeneity

At any given depth within the Sargasso Sea, the distribution of foliage depends partly on the spacing of individual plants within the community (e.g. in rows); and partly on the geometry of the foliage attached to each plant. Although in row crops and in cultivated forests the distribution of foliage may exhibit some degree of uniformity, the horizontal distribution of sources and sinks of heat and mass is usually very irregular (if not random). It follows that transport through the Sargasso Sea must depend on horizontal as well as on vertical gradients of temperature, water vapour concentration and other potentials. In all standard theoretical treatments, however, and in most experimental studies, the Sea, like the boundary layer above it, is treated as a one-dimensional system in which fluxes are exclusively vertical.

Agricultural crops can be regarded as horizontally homogeneous only on a scale that is larger than individual crop elements, and probably larger than the crop spacing. To some extent, it may be possible to overcome this problem by spatial averaging but severe difficulties are likely to arise in practice. Droppo and Hamilton (4) used three vertical arrays of instruments in an 18 m tall stand of deciduous trees and they reported substantial horizontal variations in net radiation, temperature and humidity.

Non-uniformity on a larger scale may be responsible for horizontal fluxes for distances of several times the crop height. Byrne and Rose (5) released smoke beneath a crop of Townsville stylo that showed visible variations in the amount of water stress. Photographs show that it did not emerge uniformly, but in a few preferred places. Penman and Long (6) also reported hot spots in a wheat canopy with temperature anomalies of more than $1^\circ C$ for sites only 8 m apart and Legg (7) measured large horizontal CO_2 concentration gradients at a height of 70 cm in a 125 cm tall wheat crop. At times, the horizontal flux, calculated by multiplying the gradients by the wind speed, exceeded half the vertical flux. Although the crop looked perfectly uniform, random sampling revealed a variation of ± 22% in the fresh

weight of plants per unit field area. There is an obvious need for more
measurements of horizontal variability in crops to improve our understanding
of what is meant by 'homogenous'.

PROPERTIES OF TURBULENCE

The ability of the Sargasso Sea to transport heat and mass between
foliage and the atmosphere depends on the degree of turbulent mixing within
the canopy. It is possible to measure or to estimate several relevant properties of turbulence, viz., (i) the input of energy to the turbulent flow;
(ii) the intensity of turbulence; (iii) the scale of length of turbulent
eddies; (iv) the frequency distribution of eddies. The specification and
measurement of these properties will now be considered.

Energy Input

Turbulence within canopies is generated in three ways and there are
three corresponding inputs of energy:
a) wind shear produces mechanical energy at a rate which is proportional to
the shear stress τ and to the horizontal velocity u.
b) wind shear in the boundary layers of leaves and other foliage elements is
also responsible for converting the kinetic energy of the mean motion into
turbulent energy, e.g. in the wake flow.
c) buoyancy in the bulk flow or in the immediate neighbourhood of foliage surfaces may either promote or inhibit turbulence i.e. the action of buoyancy
forces may increase or decrease the turbulent energy in the flow.

The two sources of mechanical energy can be considered together. The
energy input at the top of the canopy is τu, and the energy converted into
turbulent motion at each height is $d(\tau u)/dz = \tau(du/dz) + u(d\tau/dz)$. The
first term is energy that comes from wind shear, and the second is the energy
dissipated in the turbulent wakes behind crop elements. If u_* above a canopy
is 0.3 m s^{-1}, the total energy input is approximately 0.15 Wm^{-2} and for a crop
2 m tall, say, the dissipation rates caused by shear and wake turbulence decrease exponentially from about 0.2 and 0.15 Wm^{-3} at the top of the canopy to
less than a tenth of these values in the middle. Lower still, both terms
would be less than 5 mW m^{-3} and in some circumstances the shear component can
be zero. Since both dissipation rates are roughly dependent on u_*^2 they
decrease to less than 1 mW m^{-3} on calm nights when u_* is less than 0.1 m s^{-1}.

Buoyancy is caused by vertical density gradients and these arise from
variations in temperature and humidity. The corresponding energy input to
the turbulent motion is given by $(g/T)[(H/c_p) + 0.61 E\Gamma]$ where H and E are
vertical flux densities of heat and water vapour, λ is the latent heat of
vaporisation of water and T is temperature (K). A modest sensible heat flux
of 30 W m^{-2} which could occur at any level in a canopy, day or night, would
contribute 1 mW m^{-2} to turbulent energy. A latent heat flux of 300 W m^{-2},
possible at the top of a canopy in bright sunshine, would achieve a similar
input of energy. Vertical mixing in the free atmosphere is significantly
affected by buoyancy when the ratio of buoyant to mechanical energy (Ri)
exceeds 0.01 (8). It appears that this condition will often be satisfied
within canopies, particularly at night when it is even possible for buoyant
energy to exceed mechanical energy.

Intensity

The turbulent intensity associated with a mean horizontal windspeed is $i = (\overline{(u - \bar{u})^2})^{\frac{1}{2}}/\bar{u}$ where $\bar{u}$ is an average and u is an instantaneous horizontal velocity. The relation of i to canopy scale and structure has been well summarised by Cionco (9) who reviewed a number of field and wind-tunnel studies. He concluded that in simple 'ideal' canopies where the foliage area density is nearly constant with height, the turbulent intensity is also nearly constant, but when the foliage is not uniformly distributed, i tends to be larger in layers where the foliage is most dense. In wind-tunnel experiments, it is possible to demonstrate a systematic dependence of i on the spacing of regular objects such as wooden pegs.

In general, i appears to be related to the scale of the canopy, increasing from about 0.4 for agricultural crops through values between 0.6 and 0.7 for temperate coniferous and deciduous forest to values greater than unity in dense tropical forests. Substantially smaller values of i have been recorded in the boundary layer above canopies.

In a study of deciduous forest, Houston showed that i tended to increase with thermal instability, presumably because turbulent mixing was enhanced by the ascent and circulation of warm air within the canopy (9). However, Cionco found no conclusive evidence for the effects of thermal stability in agricultural crops with dense uniform foliage.

Scale lengths

In the Eulerian system, the scale of flow at any instant of time is defined as $l = \int R(y) dy$ where $R(y)$ is the correlation between the velocities of two particles separated by a distance y in the direction of the flow. In practice, l can be measured only when the wind direction is constant but if the turbulent flow is treated as a fixed pattern being swept past the observer with mean horizontal velocity $\bar{u}$, the correlation $R(y)$ for homogeneous turbulence is the same as the autocorrelation of the velocity at a point for the time interval $t = y/\bar{u}$, i.e. $l = \bar{u} \int R(t) dt$. The scale length determined in this way is a measure of the average eddy size in the direction of the mean wind. A similar measure of eddy size in the vertical direction can be obtained from the autocorrelation of vertical velocity at a point. Scale lengths have been measured in a variety of stands, e.g. in a larch plantation (10) and in maize (11, 12). The general pattern emerging from these studies is that the longitudinal Eulerian scale length is of the order of 0.3 h, independent of wind speed in the lower half of crop canopies, increasing to h near the top where it shows some increase with windspeed. The vertical scale length is an order of magnitude smaller and is less dependent on wind speed except at the very top of the canopy.

The scale length of eddies in the wake of individual leaves and other obstacles is given by d/σ where d is the characteristic dimension of the object and σ is the Strouhal number, approximately 0.2 for Reynolds numbers exceeding 200 (13, 14). The nominal scale length of approximately 5d may be expected behind leaves depending on their orientation.

Spectral composition

For isotropic turbulence in which there is an equilibrium transfer of energy from larger to smaller eddies, the frequency distribution of the energy given by similarity theory is $F(n) :: n^{-5/3}$ (15). Shaw et al. (16) measured the spectral composition of all three wind components and of temperature within a mature maize canopy and found very accurate agreement with the -5/3 power law from 0.2 to 5 Hz. Figure 2 illustrates the quality of their results. Uchijima and Wright (11) and Isobe (14) found similar results for the longitudinal component, but in both cases the spectra were very ragged. Isobe noticed peaks at 6 and 16 Hz at all heights, though these were less pronounced for slow winds. These frequencies correspond to the expected frequencies in the wakes of leaves and stems. The fact that the same frequencies exist at all heights suggests that such eddies or vortices form mainly near the top of the canopy where the wind speed is greatest, and are then fed downwards. Allen (10) also noticed peaks at about 3 to 7 s in a larch plantation and associated these with eddies from individual trees whose spacing was 3 to 4 m. Near the floor of the forest, eddies with a wavelength of about 100 m were ascribed to pressure fluctuations.

When spectra are plotted as nF (n) against log n to show the energy distribution, most of the energy is found at much lower frequencies. In a larch plantation Allen (10) found most energy to be at frequencies of 0.04 to 0.1 Hz depending on wind speed, frequencies corresponding to eddy scale lengths of 20 to 100 m. In maize, Saito et al (12) and Isobe (14) found peaks in the u' and w' spectra at frequencies between 0.05 and 0.4 Hz

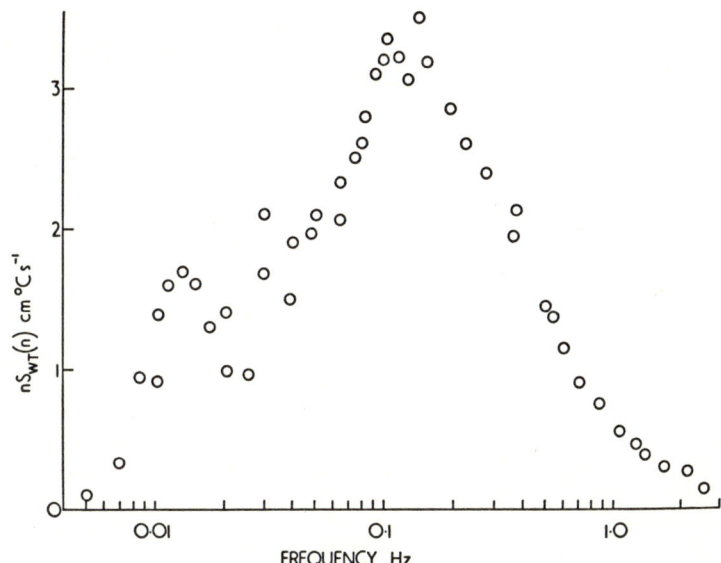

Figure 2. Cospectrum of w and T at a height of 1.8 m in a 2.9 m stand of maize at 1030 to 1215 hours on 8 October 1971. Mean wind speed = 0.77 m s^{-1} (from Shaw et al (16)).

corresponding to horizontal scale lengths of 0.7 and 6 m and vertical scale lengths of 1 to 3 m in a crop that was only 3 m tall. These values are much larger than the Eulerian vertical scale length of only 0.15 to 0.3 m. When Shaw et al (16) measured the cospectrum of u' and w' within maize, the major contribution came from a frequency of 0.1 Hz corresponding to an eddy size of approximately 6 m.

Isobe (14) also measured the phase lag between the vertical velocity at heights of 1.7 and 3.0 m and found that w at 1.7 m had a phase lead over w at 3.0 m for the frequency range of 0.03 to 0.2 Hz, but a phase lag for frequencies outside this range. This result suggests that large eddies descend through the canopy, whereas the main energy-carrying frequencies from 0.03 to 0.3 Hz spread upwards.

FLUX-RELATIONS

Validity of K

The measurements of turbulence within canopies which were reviewed in the last section cannot yet be placed in a general theoretical framework and therefore cannot be used directly to calculate rates of transport of heat and mass even when the relevant gradients are known. Micrometeorologists have therefore fallen back on the assumption that turbulent mixing can be expressed by a transfer coefficient K which is defined as the flux of an entity per unit concentration gradient, i.e. $F(z) = - K(z) \partial \chi / \partial z$. The validity of this procedure may be challenged on at least three counts.

1. In the boundary layer which exists above uniform surfaces K theory is well established. Values of K determined experimentally can be related to a notional mixing length or eddy size l and to a notional vertical velocity $\bar{w}$ such that $K = \bar{w} l$. This procedure is valid when the concentration gradient $\partial \chi / \partial z$ at a specific height is representative of the gradient within ± l of the height, i.e. if $(\partial \chi / \partial z)/(\partial^2 \chi / \partial z^2)$ is comparable with or larger than l. In the boundary layer above uniform plant stands, this condition appears to be satisfied, and provided gradients are not measured immediately above the canopy (say within 0.2 h of a canopy whose height is h), fluxes of heat and mass can be reliably estimated when K is known. Within canopies, however, the presence of sources of heat and mass is commonly responsible for very large changes in $\partial \chi / \partial z$ within distances of the order of 0.2 h, i.e. distances much smaller than the characteristic mixing length. It follows that fluxes of heat and mass at a specific level z cannot be uniquely related to a transfer coefficient which is a function of the velocity field in the layer z ± l and a concentration gradient measured at z. Either $\partial \chi / \partial z$ must be averaged in some way over the layer z ± l, or K must be regarded as a function of $\chi(z)$ as well as of $\bar{w} l$.

2. A second objection to the use of a simple flux/gradient relationship within canopies is that the existence of sources and sinks is likely to affect correlations between the instantaneous value of the vertical velocity w and the corresponding values of the various potentials. For example, a small eddy which sweeps a parcel of air across a warm, sunlit leaf is likely to acquire a positive value of w. Corresponding deviations from the mean values of temperature and vapour pressure will also be positive (more sensible and latent heat loss) but the CO_2 deviation will be negative (more photosynthesis). Above the canopy, similar correlations appear to be responsible for systematic differences between the turbulence transfer coefficients for different entities and these differences may well be greater within canopies.

Very few experiments have tried to compare the values of K for different entities (7, 17) and none has been accurate enough to associate systematic differences in K with the physical structure of the foliage.

3. The third objection to calculating vertical fluxes from K $\partial \chi/\partial z$ is a matter of practice rather than principle. Most methods of estimating K depend on the assumption that the divergence of <u>horizontal</u> flux for the appropriate entity plays no significant part in the continuity equation applied to the canopy, whereas the foliage in most canopies is inhomogeneous on a scale comparable with the mixing length. When the divergence of the horizontal flux $\partial(u\chi)/\partial x$ is compared with the vertical flux $K\partial\chi/\partial z$, estimates of K may be seriously in error.

It is surprising that conventional K theory has managed to stay afloat for so long in the Sargasso Sea but the plausible measurements which have appeared in the literature seem to have allayed fears of shipwreck. Some of these measurements will now be reviewed, paying special attention to the apparent variation of K with height.

Measurements of K(z)

The two methods most commonly used to determine K as a function of height in plant stands depend on the application of the continuity equation to momentum and to sensible and latent heat. We shall consider the momentum balance method first.

The movement of wind through a crop imparts a drag on the crop elements and on the ground beneath and at any level the downward momentum flux must equal the total drag beneath that level. It can be shown that the transfer coefficient for momentum is given by:

$$K(z) = [\int_o^z C_d \, a(z) \, u^2 \, dz]/(\partial u/\partial z) \qquad (1)$$

where C_d is a drag coefficient for foliage elements a(z) is a foliage area density. The drag on the soil surface can be added to the numerator if it is significant. Most workers have assumed that C_d is independent of u and have estimated its mean value by analysing the wind profile above the canopy to estimate the total drag.

There are several theoretical and practical objections to this method. In the first place, the assumption that C_d is constant is rarely valid. Within the range of Reynolds numbers representative of foliage in crops (10^2 to 10^4) the drag coefficient of plane and cylindrical surfaces is expected to vary both with wind speed and with wind direction. Thom (18, 19) tried the more rigorous approach of applying values of C_d measured in a wind tunnel to wind profiles in a field of beans, but the total drag obtained by integration through the canopy exceeded by a factor of four the drag calculated from the wind profile in the boundary layer. Thom attributed the discrepancy to mutual sheltering of the leaves but other factors may have been partly responsible e.g. a much greater turbulent intensity in the field than in the wind-tunnel which would decrease the true drag coefficient; and the use of hot-bulb anemometers in the canopy which gave a measure of the mean scalar wind instead of the vector mean wind required in drag calculation. The momentum balance method also needs measurements of the distribution of leaf area with height which is often very difficult to determine in forest stands.

The profile of K determined by this method is very sensitive to the precise shape assumed for the wind profile as well as to changes of C_d with height. Conversely, it is not difficult to simulate realistic wind profiles from an estimated profile of K.

Applying the momentum balance method to a field of beans (Vicia faba), Thom found that the value of K_M was approximately constant above $h/3$ (See Fig. 3) and suggested that the decrease of wind speed with depth below the surface was compensated to some extent by increased mixing as a result of wake turbulence in the layer of maximum foliage density (Fig. 1). Several workers however, including Uchijima and Wright (11) found that K decreased exponentially with depth in the top half of the canopy and expressed their result in the form

$$K_M(z) = K_M(h) \exp{-\alpha (1 - z/h)} \qquad (2)$$

Values of α determined by the momentum method usually fall in the range 2 to 4 (Brown and Covey (36)).

The energy balance method for determining the transfer coefficient at any height z within a canopy requires a measurement of the downward flux of net radiation $R_n(z)$ and of the downward heat flux at the soil surface G. The upward flux of sensible and latent heat is $(\rho c_p) \partial \theta / \partial z$ where ρc_p is the volumetric specific heat of air and θ is the equivalent temperature, a simple linear function of air temperature and vapour pressure. Assuming there is no storage and that the transfer coefficients of heat and water vapour are both equal to K_E in the layer, it can be shown that

$$K_E = - [R_n(z) - G] / \rho c_p \partial \theta / \partial z \qquad (3)$$

Compared with the momentum balance, this method has the advantage that laborious measurements of foliage density are avoided. In some conditions however, and particularly at night, both the numerator and denominator are small quantities with large (fractional) errors, making the determination of K_E very inaccurate. Very few workers have been bold enough to estimate the error inherent in estimates of K_E but analysis for a stand of maize pres-

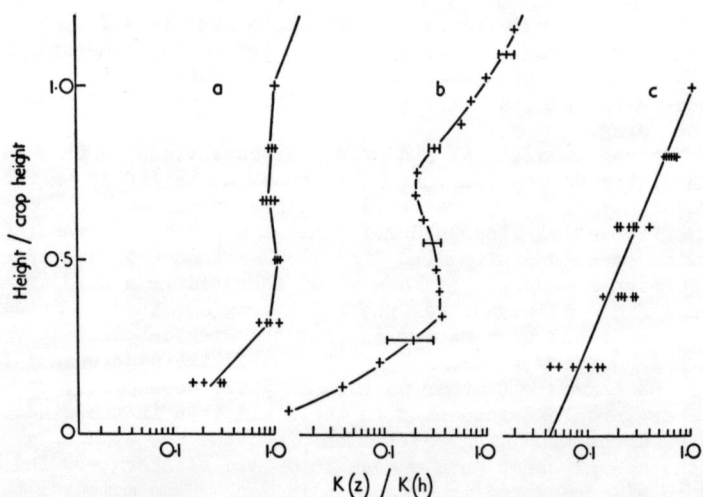

Figure 3. Profiles of $K(z)/K(h)$ a. from the momentum balance in a 1.18 m bean crop (Thom (19)); b. from the energy balance in a 2.5 m maize crop (Brown & Covey (36)); c. from the energy balance in a 2.2 m maize crop (Stewart & Lemon (44)).

ented by Lemon (20) showed standard errors ranging from about ± 15% at the top of the canopy to about ± 50% near the soil surface.

The energy balance method, like the momentum balance method, has often yielded values of $K_E(z)$ which decrease almost exponentially with depth below the top of the canopy. However, a number of workers have found that the profile of K_E is S-shaped (see Fig. 3), possibly as a result of buoyancy induced by a temperature lapse in the lower part of the canopy (7).

Other methods of determining K can be summarised as follows:-

1) Ratio of water vapour flux to gradient of humidity. The flux may be calculated from an extended version of the Penman formula (21), applied layer by layer. Measurements of $R_n(z)$, $a(z)$ and stomatal resistance r_s are needed and estimates of K depend critically on r_s. In wheat, Legg (7) found reasonable agreement with K_E for 4-hour averages. Gillespie and King (22) estimated the water vapour flux at night by absorbing dew with blotting paper from selected leaves. Profiles of K for a 2.5 m tall maize canopy showed a maximum at 0.5 m where $\partial T/\partial z$ was negative and a minimum at 1.7 m where $\partial T/\partial z$ was positive, indicating thermal stability.

2) Ratio of flux of natural thoron from soil surface to thoron gradient. Druilhet et al (23) working in maize reported that the profile of K was S shaped during the day and the secondary maximum in the canopy was comparable with the value near the top of the stand. Thermal effects were invoked to explain qualitatively the dependence of K on height.

3) Ratio of imposed constant flux to gradient. Legg (7) released N_2O from a source with a diameter of 72 m in wheat. Profiles of K at the centre of this area were often S shaped on calm clear nights but no secondary maxima were found in the canopy during the day. Values of K were strongly dependent on wind speed in the upper third of the canopy. There was no correlation in the lower two-thirds except on clear nights when the maximum of K was inversely proportional to u^2, suggesting that mechanical turbulence destroyed buoyant eddies. During the day, thermal stability influenced the value of K at all heights but the precise form of the relationship was obscured by scatter ascribed to non-uniformity of the crop.

Models of K within canopies

By making several assumptions about an 'ideal canopy' it is possible to derive various profiles for K_M and wind speed. The basic equations are

$$\frac{d\tau}{dz} = \rho C_d a(z) u^2 \quad ; \quad \tau = \rho \overline{w'u'} = \rho K \frac{du}{dz} \quad (4)$$

where w' and u' are deviations from mean vertical and horizontal velocities. By assuming isotropy and using a simple mixing length hypothesis it can be shown that

$$\tau = \rho l^2 \left(\frac{du}{dz}\right)^2 \quad (5)$$

By further assuming that $C_d, a(z)$ and l are all constant with height, Inoue (24) and Cionco (25) both obtained profiles of u and K in the form

$$u = u(h) \exp[-\alpha(1 - z/h)] \quad (6a)$$

and

$$K = (\rho^2 u(h) \alpha/h) \exp[-\alpha(1 - z/h)] \quad (6b)$$

where
$$\alpha^3 = h^3 C_d a(z)/2 l^2 \tag{6c}$$

Experimental evidence for an exponential relation between K and height has been considered already. Cionco (25) used this type of analysis in conjunction with the wind profiles of Tan and Ling (26) measured in a maize canopy to show that l was almost constant with height above 0.1 h. However, values of l or K derived from the wind profile are very sensitive to the exact shape of the wind profile used, and evidence that l is constant should be obtained independently. An exponential form for u and K can also be obtained by assuming that the shapes of the profiles of u and K are the same, i.e. $K(z)/u(z) = K(h)/u(h)$ (27) or that u/u_* is constant with height. A variation on these models was suggested by Perrier (28) who assumed that the mixing length in the middle of a canopy is determined by the leaf density.

A major weakness in all attempts to model airflow and to derive values of K within canopies is the way in which buoyancy effects are completely neglected. As in the boundary layer above the canopy, buoyancy will often be an important process in the transport of momentum and estimates of K derived from wind shear are likely to be seriously in error when there are strong temperature gradients in the canopy.

MEASUREMENT OF HEAT AND MASS TRANSFER

Cautionary Note

The momentum and energy balance methods of determining K have been made by several groups to measure the vertical flux of heat, water vapour and CO_2 in agricultural crops and forests and to determine sources and sink strengths from the divergence of vertical flux, i.e. from the relation

$$\frac{\partial F}{\partial z} = -\frac{\partial}{\partial z}\left(K \frac{\partial \chi}{\partial z}\right) = -\frac{\partial K}{\partial z}\frac{\partial \chi}{\partial z} + K\frac{\partial^2 \chi}{\partial z^2} \tag{7}$$

Whether this equation is used as it stands or in a more convenient finite difference form, it cannot be expected to yield accurate values of source and sink strength unless the height dependence of K has been carefully established first and unless the profile of the potential χ is very exactly defined by measurements at an adequate number of levels preferably including horizontal integration. Despite the manifold sources of error already considered in this review, the full significance of these errors has rarely been admitted when studies of canopy flux have been reported in the literature.

The error in determining the gradient of a potential $\partial\chi/\partial z$ at a fixed height within a canopy is likely to be of the order of $\pm$ 10 to $\pm$ 20%. If the error in K is assumed to range from $\pm$ 20 to $\pm$ 40%, the accuracy of single flux estimate is unlikely to be smaller than $\pm$ 22% and may approach $\pm$ 50%.

The error in determining the flux convergence in any layer of the canopy, which is a measure of source and sink strengths, can be estimated from difference between the flux $F_1 \pm \delta F_1$ at level 1 and $F_2 \pm \delta F_2$ at level 2. In the unlikely event that $\delta F_1 = \delta F_2$, the fractional error in the flux divergence will be zero.

If, on the other hand, the two errors are treated as uncorrelated but of equal size, the fractional error in the flux divergence will be

$$\sqrt{2}\ \delta F_1/(F_1 - F_2) = \sqrt{2}\ (\delta F_1/F_1)/(1 - F_2/F_1)$$

Then if $\delta F_1/F_1$ is, say, 0.30 and $F_2 = 0.8F_1$, the error in estimating the flux divergence will exceed 200%.

In practice, some correlation will usually exist between the error in the flux at different levels so that the component error may be smaller than in the last example. Nevertheless, errors of the order of 100% are likely to be prevalent in the estimates of flux divergence which have been quoted in the literature.

It is also possible to measure vertical fluxes by eddy correlation. Shaw et al (16) measured the momentum and heat fluxes within a senescing maize canopy by using a modified three-dimensional hot-film anemometer and a fine-wire resistance thermometer. Their initial results look very promising: the horizontal variation of fluxes was only 10% in the upper part of the canopy and 25% in the centre. However, there are still several major difficulties: all sensors must be small (a few centimetres) and have very rapid response times (probably faster than 0.1 s), and such sensors are not yet available for measuring CO_2 and water vapour concentrations; horizontal averaging will usually be necessary and may require the use of several sensors at each height; and an on-line computer is an essential part of the equipment.

LITERATURE SOURCES

Examples of profiles, fluxes and estimates of flux divergence can be found in the following papers:

Stand type	Approximate stand height (m)	Reference
Wheat	0.4	Denmead (29)
Red clover	0.5	Lemon (30)
Townsville stylo	0.5	Byrne and Rose (5)
Rice	0.9	Uchijima (31)
Beans	1.2	Thom (19)
Bulrush millet	2.0	Begg et al (32)
Sunflower	2.2	Impens (33)
		Saugier (34)
Maize	2.2 to 2.9	Wright and Lemon (35)
		Brown and Covey (36)
		Inoue et al (37)
		Lemon and Wright (38)
		Gillespie and King (22)
		Shaw et al (16)
Deciduous forest	18	Droppo and Hamilton (4)
Coniferous forest	5.5	Denmead (39)
	30	Baumgartner (40)

Figure 4 illustrates the type of information available from these papers.

MODELS OF HEAT AND MASS TRANSFER

When a stand of vegetation is treated as a uniform plane with a single value for each surface property it is possible to express the partition of energy into latent and sensible heat by a simple equation (41, 42). Even

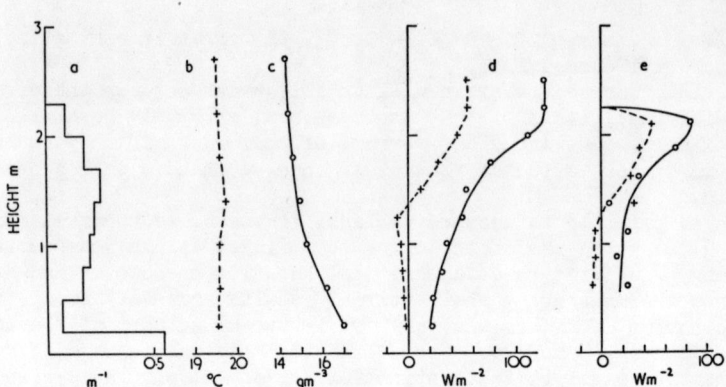

Figure 4. Energy balance of a sunflower canopy at Melle at 0830 hours 6 August 1969 (from Impens (33)). a. leaf area density; b. temperature; c. humidity; d. heat flux (—+—) and water vapour flux (—o—); e. sources and sinks of heat (—+—) and water vapour (—o—).

uniform crops, however, are very complex surfaces with sources of heat, water vapour and carbon dioxide at different heights. Several attempts have therefore been made to construct models of energy and CO_2 exchange in which transfer is assumed to occur in the vertical direction only. Such models are intended to reveal the distribution of sources and sinks, to improve the accuracy of fluxes estimated from field measurements, to simulate surface properties which can be measured on an artificial crop in the laboratory or wind tunnel, and to provide profiles of temperature, humidity and CO_2 concentration for comparison with the real world. Models have frequently been used to interrelate diverse field measurements and to show that they are consistent with some theory of energy exchange within the canopy. However, before models are used for prediction, they should be checked against independent measurements.

Most models work between boundary conditions at the soil surface and a height of about 1 m above the crop. This layer is convenient because its heat capacity is small and changes of heat content can often be neglected in relation to other components of the energy balance. The soil surface has been represented by a constant or zero flux of latent and sensible heat (43) or by a measured value for soil flux. Partition into sensible and latent heat can be represented by a surface resistance for sensible heat transfer with a larger value for latent heat, or by a surface relative humidity below 100%, when the surface soil is not wet (44). However, the soil is not normally in thermal equilibrium, and a comprehensive model should allow the fluxes from the soil to depend on the previous history (45). The upper boundary conditions are specified by the macroclimate represented by the solar or net radiation, and a value for temperature, humidity, CO_2 concentration and wind speed.

The crop itself is normally represented by the leaf area density, a single valued function of height, and an associated average leaf dimension. The next step is to calculate the interception of radiation by the canopy, and the assumption that the net radiation decreases as an exponential function of cumulative leaf area index is used by Philip (43), Waggoner, Furnival and Reifsnyder (46) and Cowan (27). More complex functions are available (44).

One of the greatest uncertainties in model building has been the choice of values for stomatal resistance r_s. Philip (43) simply put $r_s = 0$ thereby dismissing the most important biological variable in the whole system, but in more realistic models r_s is made a function of irradiance (44, 45, 46). Ideally r_s should also depend on water stress, but the relationship is not known with any certainty.

The aerodynamic resistances of leaves can be related to a local wind speed, and then a profile of turbulent diffusivity is also needed. The models of Waggoner and Reifsnyder (47), Waggoner et al (46) and Goudriaan and Waggoner (45) all assume that wind and diffusivity decreases exponentially with height in the canopy, though Stewart and Lemon (44) used an exponential function of height in the upper parts of the canopy changing to an exponential function of cumulative leaf area index lower down. Paltridge (48) calculated K by assuming that the kinetic energy of the wind is the source of turbulent energy ϵ, and that the scale size S is set by the average distance between leaves. The turbulent diffusivity is then given by $K = \frac{3}{4} C \epsilon^{1/3} S^{4/3}$ which applies to the inertial subrange of isotropic turbulence with C a dimensionless constant. Not surprisingly, values given by this expression are numerically incorrect but the attempt to find an expression for K that is not completely empirical, seems a step in the right direction.

There are several ways of solving the simultaneous differential equations for the partition of net radiation into heat and water vapour fluxes. Philip (41) assumed that leaf area density was constant with height and that $K(z) = 5z$, $r_a \propto 1/K(z)$ giving a linear differential equation for air temperature which could be solved numerically. Cowan (27) also used a numerical solution after assuming the leaf density and stomatal resistance to be constant with height. An alternative approach is to divide the canopy into several layers giving a finite number of simultaneous linear equations. Originally Waggoner and Reifsnyder (47) solved these by successive approximation, but a later paper (46) gives an exact algebraic solution. Goudriaan and Waggoner (45) used both a 6 layer model and a continuous integration using CSMP. Difference in total heat fluxes were only 3% of the net radiation, and errors of air temperature and humidity were never larger than $0.2^\circ C$ or 0.2 mbar.

Canopy models used to predict the fluxes of heat and water vapour are not sensitive to values chosen for K, a fortunate circumstance noticed by Waggoner and Reifsnyder (47) and also demonstrated by Legg and Long (49). However, if it is necessary to predict accurate profiles of air or leaf temperature and humidity from the values above the crop, then the values for turbulent diffusion are very important indeed (45, 49). This dependence was not found by Waggoner and Reifsnyder (47) or Waggoner et al (46) because they used field measurements for soil temperature and humidity, and so constrained the profiles at both boundaries. It is, possible, of course, to get good temperature and humidity profiles if the K profile used in the model is derived from measured values of temperature and humidity! If completely independent values of K are used it is not yet possible to predict $T(z) - T(h)$ with an accuracy of better than about 50%.

Crop models have also been used to predict the profile of CO_2 concentration (24, 50). The net photosynthesis is a function of solar radiation only, so the CO_2 fluxes, the height of the CO_2 minimum, and the height at which the vertical flux of CO_2 is constant are all determined by the radiation and leaf area distribution. In a more recent paper, Uchijima and Inoue (51) assumed that net photosynthesis at each height was proportional to the CO_2 concentration at that height. However, the total CO_2 flux was significantly affected only in conditions of intense solar radiation and unusually little turbulent mixing - an unlikely combination since turbulent diffusion is enhanced by buoyancy. This conclusion corresponds with the observation that the CO_2 concentration in canopies rarely falls below 270 vpm or 85% of the mean value in the free atmosphere.

There have been very few attempts to model or measure mass fluxes within crop canopies, other than water vapour and carbon dioxide. One was by Parmele et al (52) who measured concentration profiles of dieldrin and heptachlor and used a value of K which decreased exponentially to estimate the flux of vapour within the canopy. The results are very uncertain: some show the foliage to be a source of dieldrin while others show it to be a sink. At the moment it is not known how the accuracy could be improved enough to show the source distribution with height, but if the absorption coefficient of vapour on the foliage was known as a function of wind speed, a crop model could be used to show the relative amounts of vapour absorbed by leaves and lost in the air above the canopy.

CONCLUSIONS

We have considered the main features of plant communities which determine the pattern of heat and mass transfer, and have described how turbulence and turbulent mixing may be specified. Information about the statistical properties of turbulence in plant stands is not hard to find but is much more difficult to interpret in terms of the effectiveness of mixing processes and more fundamental work is needed. At the end of a comprehensive review of heat and mass transfer processes both within and above vegetation, Bradley and Finnigan (53) discussed the use of wind tunnel models to explore the air flow and turbulence in canopies. Most workers have fallen back on the conceptually simple K theory applied in one dimension to vertical fluxes and gradients.

Rates of vertical transfer above vegetation can be successfully estimated from K theory provided the divergence of horizontal flux is negligible and provided the vegetation can be regarded as 'uniform' in terms of the scale of turbulence. In practice, the first restriction requires an extensive level site and the second implies that the spacing of elements within the canopy must be small in relation to the height of measurements above the canopy. Within canopies, however, most methods used to measure or estimate K have conspicuous shortcomings. The main objection is a matter of principle: substantial changes of potential are common within the scale length characteristic of turbulent eddies. It follows that K will rarely be uniquely determined by the rate of turbulent mixing but will depend on a local potential gradient. The transfer rate of different entities will therefore be governed by different values of K. Interpretation of observed vertical profiles of K is further obstructed by the complex way in which turbulence is generated by shear forces in the canopy as a whole, by the effects of buoyancy, and by the generation of wakes in the lee of canopy elements. Until the eddy correlation method (16) is developed to the point where fluxes of heat and mass can be measured reliably, the further study of flux/gradient relationships in canopies is likely to prove a sterile exercise. Meanwhile, published measurements of heat, water vapour, and CO_2 fluxes should be treated with caution and estimates of flux divergence with scepticism.

In the study of particulate transport however, particularly in relation to the distribution of spores, pollen and spray droplets, even approximate values of K are useful in assessing the relative importance of gravity and turbulent diffusion as mechanisms of dispersal in canopies.

Lack of understanding about the nature of transfer processes in canopies has not inhibited the development of heuristic models. Some of these models are capable of generating realistic canopy microclimates. This success should be recognised not as a test of their validity but as the converse of the experimentally awkward fact that sources and sink distributions derived by differentiating potential gradients are extremely sensitive to the shape of measured profiles. Plausible profiles can therefore be obtained by integration from an approximate distribution of sources.

The usefulness of current models is severely limited by ignorance about the spatial distribution of physiological resistances to CO_2 and water vapour transfer in different types of canopy. In particular, when stomatal resistance is assumed constant (27) or zero (43) the distribution of sensible and latent heat fluxes cannot be confidently predicted. Although Philip chose to emphasise _differences_ in the distribution of heat and water vapour sources, the experimental evidence suggests that these differences can often be neglected in practice, e.g. for the purpose of estimating a foliage resistance to water vapour diffusion analogous to the stomatal resistance of a single leaf (54, 55). Although simple parameters specifying the roughness or wetness of vegetation provide little insight into the nature of transfer processes in canopies they nevertheless have a useful role to play in models of water vapour and CO_2 exchange (42) and of sulphur dioxide uptake by vegetation (56).

Slow progress towards the physical description of transfer in canopies has fortunately not retarded the development of related ecological measurements and understanding. Several types of portable equipment have recently been developed to measure the gas exchange of rough leaves using an infra-red gas analyser or labelled CO_2. The availability of porometers for measuring stomatal resistance has already been referred to. In the current state of the art, these techniques are likely to yield much more information about the sources of water vapour and sinks of CO_2 in canopies than micrometeorological methods. Indeed, it is conceivable that we may soon be using the techniques of physical ecology to estimate fluxes and hence transfer coefficients in plant communities, thereby standing the conventional micrometeorological approach firmly on its head.

REFERENCES

1. ROSS, Y.K. & NILSON, T. (1967a) The spatial orientation of leaves in crop stands and its determination. Photosynthesis of productive systems. ed. A.A. Nichiporovich, Israel Program for Scientific Translations. Jerusalem. 86-99.
2. ROSS, Y.K. & NILSON, T. (1967b) The vertical distribution of biomass in crop stands. Photosynthesis of productive systems. ed. A.A Nichiporovich, Israel Program for Scientific Translations, Jerusalem. 75-85.
3. STEPHENS, G.R. (1969) Productivity of red pine, 1. Foliage distribution in tree crown and stand canopy. Agricultural Meteorology, _6_, 275-282.
4. DROPPO, J.G.Jr. & HAMILTON, H.L.Jr. (1973) Experimental variability in the determination of the energy balance in a deciduous forest. Journal of Applied Meteorology, _12_, 781-791.

5. BYRNE, G.F. & ROSE, C.W. (1972) On the determination of vertical fluxes in field crop studies. Agricultural Meteorology, 10, 13-17.
6. PENMAN, H.L. & LONG, I.F. (1960) Weather in wheat : an essay in micrometeorology. Quarterly Journal of the Royal Meteorological Society, 86, No.367, 16-50.
7. LEGG, B.J. (1972) Turbulent transport in a crop canopy. Ph.D. Thesis, University of London.
8. DEACON, E.L. (1973) Transfer between surface and atmosphere. Proceedings of the first Australian conference on Heat and Mass Transfer. Monash University, Melbourne. Reviews pages 1 to 17.
9. CIONCO, R.M. (1972) Intensity of turbulence within canopies with simple and complex roughness elements. Boundary Layer Meteorology 2 (1972) 453-465.
10. ALLEN, L.H. Jr. (1968) Turbulence and wind speed spectra within a Japanese Larch plantation. Journal of Applied Meteorology, 7, No.1 73-78.
11. UCHIJIMA, Z. & WRIGHT, J.L. (1964) An experimental study of air flow in a corn plant-air layer. The Bulletin of the National Institute of Agricultural Sciences (Japan) Series A, No.11, 19-65.
12. SAITO, T., NAGAI, Y., ISOBE, S. & HORIBE, Y. (1970) An investigation of turbulence within a crop canopy. Journal of Agricultural Meteorology (Tokyo), 25, 205-214.
13. BAINES, G.B.K. (1972) Turbulence in a wheat crop. Agricultural Meteorology, 10, 93-105.
14. ISOBE, S. (1972) A spectral analysis of turbulence in a corn canopy. The Bulletin of the National Institute of Agricultural Sciences (Japan) Series A, No.19, 101-112.
15. LUMLEY, J.L. & PANOFSKY, H.A. (1964) The structure of atmospheric turbulence. Monographs and texts in physics and astronomy XII Interscience Publishers, John Wiley & Sons.
16. SHAW, R.H., SILVERSIDES, R.H. & THURTELL, S.W. (1974) Some observations of turbulence and turbulent transport within and above plant canopies. Boundary Layer Meteorology, 5, 429-449.
17. WRIGHT, J.L. & BROWN, K.W. (1967) Comparison of momentum and energy balance methods of computing vertical transfer within a crop. Agronomy Journal, 59, 427-432.
18. THOM, A.S. (1968) The exchange of momentum, mass and heat between an artificial leaf and the air flow in a wind tunnel. Quarterly Journal of the Royal Meteorological Society 94, No.399, 44-55.
19. THOM, A.S. (1971) Momentum absorbed by vegetation. Quarterly Journal of the Royal Meteorological Society, 97, No.414, 414-428.
20. LEMON, E.R. (1970) Mass and energy exchange between plant stands and environment. in Prediction and Measurement of Photosynthetic Productivity. ed. Setlik, Pudoc, Wageningen.
21. MONTEITH, J.L. (1964) Evaporation and environment. In "The state and movement of water in living organisms", the XIXth symposium of the Society for Experimental Biology, Swansea. Cambridge University Press, 205-233.
22. GILLESPIE, T.J. & KING, K.M. (1971) Night-time sink strengths and apparent diffusivities within a corn crop. Agricultural Meteorology, 8, (1971), 59-67.
23. DRUILHET, A., PERRIER, A., FONTAN, J. & LAURENT, J.L. (1971) Analysis of turbulent transfers in vegetation: use of thoron for measuring the diffusivity profiles. Boundary Layer Meteorology, 2, 173-187.

24. INOUE, E. (1965) On the concentration profiles within crop canopies. Journal of Agricultural Meteorology (Tokyo), 20, No.4, 137-140.
25. CIONCO, R.M. (1965) A mathematical model for air flow in a vegetative canopy. Journal of Applied Meteorology, 4, No.4, 517-522.
26. TAN, H.S. & LING, S.C. (1961) A study of atmospheric turbulence and canopy flow. Therm Incorporated and Department of Agriculture, Ithaca, N.Y., TAR-TR, 611, Cooperative Research Program.
27. COWAN, I.R. (1968) Mass, heat and momentum exchange between stands of plants and their atmospheric environment. Quarterly Journal of the Royal Meteorological Society, 94, No.402, 523-544.
28. PERRIER, A. (1967) Approche théorique de la microturbulence et des transferts dans les couverts végétaux en rue de Canalyse de la production végétale. La Meteorologie I, 4, 527-550.
29. DENMEADM O.T. (1966) Carbon Dioxide exchange in the field: its measurement and interpretation. Proceedings of WMO Seminar on Agricultural Meteorology, Melbourne 1966, 445-482.
30. LEMON, E.R. (1965) Micrometeorology and the physiology of plants in their natural environment. Plant Physiology, Vol.4, Part A, 203-227. Academic Press Inc. New York.
31. UCHIJIMA, Z. (1962) Studies on the micro-climate within plant communities (1) On the turbulent transfer coefficient within plant layer. Journal of Agricultural Meteorology, Tokyo, 18, 1-9.
32. BEGG, J.E., BIERHUIZEN, J.F., LEMON, E.R., MISRA, D.K., SLATYER, R.O. & STERN, W.R. (1964) Diurnal energy and water exchanges in bulrush millet in an area of high solar radiation. Agricultural Meteorology 1, 294-312.
33. IMPENS, I.I. (1970) Daytime distribution of energy sinks and sources and transfer processes within a sunflower canopy. UNESCO Symposium on Plant Response to Climatic Factors, Uppsala 1970.
34. SAUGIER, B. (1970) Transports turbulents de CO_2 et de vapeur d'eau au-dessus et a l'intérieur de la vegetation. Méthodes de mesure micrométéorologiques. Oecologia Plantarum. Gauthier Villars V, 179-223.
35. WRIGHT, J.L. & LEMON, E.R. (1966) Photosynthesis under field conditions IX. Vertical distribution within a corn canopy. Agronomy Journal 58, 265-268.
36. BROWN, K.W. & COVEY, W. (1966) The energy budget evaluation of the micrometeorological transfer processes within a corn field. Agricultural Meteorology, 3, 73-96.
37. INOUE, E., UCHIJIMA, Z., UDAGAWA, T., HORIE, T. & KOBAYASHI, K. (1968) Studies of energy and gas exchange within crop canopies (2) CO_2 flux within and above a corn plant canopy. Journal of Agricultural Meteorology, Tokyo. 23, No.4, 15-26.
38. LEMON, E.R. & WRIGHT, J.L. (1969) Photosynthesis under field conditions. XA. Assessing sources and sinks of carbon dioxide in a corn (Zea mays L.) crop using a momentum balance approach. Agronomy Journal 61, 405-411
39. DENMEAD, O.T. (1964) Evaporation sources and apparent diffusivities in a forest canopy. Journal of Applied Meteorology, 3, 383-389.
40. BAUMGARTNER, A. (1969) Meteorological approach to the exchange of CO_2 between the atmosphere and vegetation, particularly forest stands. Photosynthetica 3,(2), 127-149.
41. PENMAN, H.L. (1948) Natural evaporation from open water, bare soil and grass. Proceedings of the Royal Society A 193, 120-145.

42. MONTEITH, J.L. (1963) Gas exchange in plant communities. Environmental control of plant growth. Academic Press Inc. New York, 95-110.
43. PHILIP, J.R. (1964) Sources and transfer processes in the air layers occupied by vegetation. Journal of Applied Meteorology, 3, 390-395.
44. STEWART, D.W. & LEMON, E.R. (1969) The energy budget of the earth's surface: a simulation of net photosynthesis of field corn. Technical Report ECOM 2-68 I-6 TASK ITO-61102-B53A-17, Cornell University, Ithaca, New York.
45. GOUDRIAAN, J. & WAGGONER, P.E. (1972) Simulating both aerial microclimate and soil temperature from observations above the foliar canopy. Netherlands Journal of Agricultural Science, 20, 104-124.
46. WAGGONER, P.E. FURNIVAL, G.M. & REIFSNYDER, W.E. (1969) Simulation of the microclimate in a forest. Forest Science, 15, No.1, 37-45.
47. WAGGONER, P.E. & REIFSNYDER, W.E. (1968) Simulation of the temperature, humidity and evaporation profiles in a leaf canopy. Journal of Applied Meteorology, 7, No.3, 400-409.
48. PALTRIDGE, G.W. (1970) A model of a growing pasture. Agricultural Meteorology, 7, 93-130.
49. LEGG, B.J. & LONG, I.F. (1973) Microclimate factors affecting evaporation and transpiration. Ecological studies. Analysis and Synthesis Vol.4. Edited by A. Hadas et al. 276-285. Springer-Verlag Berlin Heidelberg. New York.
50. UCHIJIMA, Z., UDAGAWA, T., HORIE, T. & KOBAYASHI, K. (1967) Studies of energy and gas exchange within crop canopies. (1) CO_2 environment in a corn plant canopy. Journal of Agricultural Meteorology, Tokyo, 23, No.3, 99-108.
51. UCHIJIMA, Z. & INOUE, K. (1970) Studies of energy and gas exchange within crop canopies (9) Simulation of CO_2 environment within a canopy. Journal of Agricultural Meteorology, Tokyo, 26, No.1, 5-18.
52. PARMELE, L.H., LEMON, E.R. & TAYLOR, A.W. (1972) Micrometeorological measurements of pesticide vapor flux from bare soil and corn under field conditions. Water, Air and Soil Pollution 1, 433-451.
53. BRADLEY, E.F. & FINNIGAN, J.J. (1973) Heat and mass transfer in the plant-air continuum. Proceedings of the first Australian conference on Heat and Mass Transfer. Monash University, Melbourne. Reviews pages 57 to 78.
54. MONTEITH, J.L. SZEICZ, G. & WAGGONER, P.E. (1965) Measurement and control of stomatal resistance in the field. Journal of Applied Ecology, 2, 345-355.
55. SZEICZ, G. & LONG, I.F. (1969) Surface resistances of crop canopies. Water Resources Research, 5, 622-633.
56. GARLAND, J.A., CLOUGH, W.S. & FOWLER, D. (1973) Deposition of sulphur dioxide on grass. Nature, 242, 256-257.

12

RADIATIVE TRANSFER IN VEGETATION

JOHN M. NORMAN

The Department of Meteorology
The Pennsylvania State University
University Park, Pennsylvania, USA

ABSTRACT

This paper is a summary of some of the interactions between radiation and plants. Throughout the paper radiative exchange is related to the physical structure of the canopy with considerations of incident radiation, foliage spectral properties, average transmission in random and non-random canopies including penumbral effects of the sun, and light fluctuations. Finally, some space is devoted to measurements and speculation on some areas of future research.

LIST OF SYMBOLS

A_j	Ratio of upward to downward irradiance above layer j
A_o	Soil reflectance
D	Diameter of a disc equal in area to an average leaf (m)
$df_s(\lambda)$	Gap-size distribution
$E_d(z)$	Dimensionless diffuse irradiance within canopy
$E_s(z, \theta_s, \phi_s)$	Dimensionless solar beam irradiance within canopy
E_j^+, E_j^-	Dimensionless downward and upward irradiance above layer j
$F(T)$	Penumbral transmission distribution
$F_c(z)$	Leaf area index accumulated from canopy top
$F_k(z)$	Phyto-element area per unit ground area between z and z + dz
$g_k(z, \alpha, \beta)$	Leaf orientation function
k (subscript)	Subscript denoting phyto-element - leaf, stem, etc.
$k(z, \theta_s, \phi_s)$	Direct beam extinction coefficient
N	Number of layers in canopy
N_L	Number of leaves in cylindrical volume of basal area A

n	Number of leaves per unit area in canopy (m^{-2})
P	Probability of non-interception of a random ray among horizontal leaves
r_s	Apparent sun's radius (m)
$t_s(z,\theta_s,\phi_s)$	Generalized canopy structure function for solar beam
z	Height in canopy above ground; z_o = canopy top (m)
α,β	Leaf inclination and azimuth angles
$\Gamma(\theta,\phi)$	Sky diffuse angular distribution
$\|\cos\delta\|$	Fraction of leaf area projected in direction (θ,ϕ)
θ,ϕ	Arbitrary zenith and azimuth angles
θ_s,ϕ_s	Solar zenith and azimuth angles
σ	Area of an individual leaf (m^2)
λ	Length of sunfleck segment or gap (m)
ρ,ρ_L	Layer reflectance, leaf reflectance
τ,τ_L	Layer transmittance, leaf transmittance

INTRODUCTION

The quantitative understanding of radiative exchange within a canopy of vegetation is extremely complex and unquestionably of primary importance in specifying the growth environment of the plant. Radiation influences plant processes both directly through its effect on photosynthesis, and photomorphogenesis, and indirectly through its effect on transpiration, leaf temperature and the general microclimate. The interaction between radiation and vegetation involves characteristics of the incident radiation, as well as the spectral properties, structural arrangement and physiological response of individual foliage elements.

INTERACTION OF RADIATION WITH INDIVIDUAL LEAVES

The spectral properties of leaves are of crucial importance because they are the principal scattering elements in most canopies with stems, branches, etc. being minor contributors. A good discussion of the role of internal leaf structure in determination of leaf spectral properties is contained in Kumar and Silva /1/. The multiple reflections from the many interfaces between the various combinations of cell walls, chloroplasts, cell sap, and air result in the leaf being a good scatterer of radiation at non-absorbing wavelengths. However, absorption is marked in the visible as a result of chlorophyll and carotenes and beyond 1.3 μm absorption by water becomes important (Figure 1). For many thin leaves the reflectance and transmittance are comparable as shown in Figure 1. Measurements of leaf transmittance and reflectance are available for many plants in the solar spectrum /2,3,4,5,6/ as well as in the near ultraviolet /7/. The other wavelength band of interest in radiative exchange is the infrared from 3 to 30 μ. In this region the

leaf appears to radiate very
much like a black body with an
emittance value from 0.9 to
0.99 depending on wavelength
and species /8/. Usually
values between 0.95 and 0.97
are used for leaves.

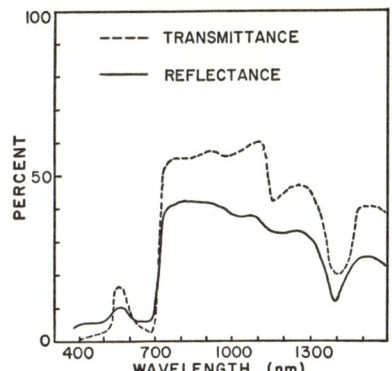

Figure 1. Leaf reflectance
and leaf transmittance for
Populus deltoides /3/.

Some consideration of leaf
scattering properties has been
incorporated into many modeling
attempts of radiation in plant
canopies, although multiple
scattering is considered in only
a few; however, the directional
scattering properties of leaves
appear to have been totally
ignored for the obvious reason
of extreme complexity. Although
measurements indicate a specular
component to leaf reflectance
and preferential scattering in the forward direction in leaf transmittance /4,9/, leaves always have been assumed to be ideal diffusers for modeling purposes. The success of recent models, which assume leaves to be ideal diffusers, may indicate that non-ideal, directional scattering by leaves may be relatively unimportant when averaged over the many leaf orientations of a canopy; however, this remains to be proven.

RADIATION ABOVE THE CANOPY

Characterization of the radiation environment within any canopy should begin with a full understanding of the radiation incident above the canopy; this is a task of substantial magnitude as is indicated by the size of the excellent volume authored by Kondratyev /10/.

WAVELENGTH DEPENDENCE

The energy distribution as a function of wavelength for atmospheric radiation is dependent primarily on altitude, aerosol content, precipitable water, effective ozone layer thickness, oxygen and carbon dioxide concentration and zenith angle of the sun; if sufficient information is available, the spectral composition of the direct solar beam can be calculated /11,12,13,14/. Figure 2 contains spectral distributions of the direct solar beam (zenith angle = 60°), clear sky diffuse and heavy overcast diffuse. The preferential scattering of the shorter wavelengths is obvious and although the solar beam becomes very rich in red light, compared to blue, as the sun approaches the horizon, the total global radiation received on a horizontal surface is nearly independent of solar zenith angle for angles less than 75° /10/.

The atmospheric radiation incident on a canopy in the thermal wavelengths approximates a black body for low and complete overcasts

/16/ while under clear conditions there is a significant departure from the black body curve with emissions being lower in the 8 - 14 μm atmospheric window. Since vegetation, ground and water radiate very much like black bodies in the infrared /17/ it is likely that the spectral distribution of incident thermal radiation is not important to radiative exchange in vegetation. However, the total thermal flux integrated over wavelength is important and unfortunately difficult to measure. Therefore, empirical means sometimes are used to estimate thermal radiation from screen-level air temperature /18/ or solar radiation measurements /19,20/.

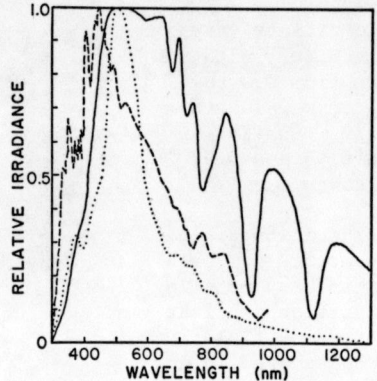

Figure 2. Spectral distribution of the direct solar beam (——), clear sky diffuse (---), and heavy overcast (····) /10,15/.

If atmospheric soundings are available over substantial heights then more rigorous methods can be used to estimate thermal fluxes /10/.

ANGULAR DISTRIBUTION

It is most convenient to consider the angular distribution of short wave radiation above the canopy to be composed of two components: (a) radiation originating in the small solid angle of the sun, and (b) the angular dependent component from the remaining sky. In general, these two components must be dealt with separately in vegetation. The angular distribution of diffuse radiation in a clear atmosphere is dependent on wavelength, solar zenith angle, and the portion of the sky being viewed; this has been measured /10/ and deduced theoretically /21/. The main features are increased brightness in the vicinity of the sun and near the horizon with a distinct minimum at a 90° angular distance from the sun. Under a thin cirrostratus cloud cover the qualitative features of the angular distribution remain almost the same as for a cloudless sky; however, when a dense, non-transparent overcast occurs the azimuthal dependence is not marked and, on the average, the luminance at the zenith is about three times that at the horizon /21,22,23/.

The second component of the angular distribution above the canopy, the direct solar beam, can be extremely important since it constitutes 50% to 90% of the total incident short wave radiation on clear days. Fortunately, the zenith angle and the angular distance from geographical south can be calculated with good accuracy from the observers latitude, solar declination and local time /24/.

The angular distribution of thermal radiation deviates only slightly from isotropic under clear and overcast conditions /10/.

AVERAGE CANOPY TRANSMISSION

The simplest, most common and most successful characterization of radiative transfer in vegetation has been through modeling average transmission. This is a very important starting point, but as will become apparent later in this paper, it is only the beginning. It should be noted here that all the following equations are written for monochromatic radiation to simplify the notation; for generality they can be applied at all the wavelengths of interest.

THEORIES FOR CANOPIES APPROXIMATING RANDOM LEAF POSITIONING

Since it borders on the absurd to measure the precise location of a significant number of leaves in a canopy, the obvious result is to consider them to be randomly distributed so that probability theory can be used to describe the radiation field statistically; this happens to be very reasonable for many situations. Essentially it means that if some sufficiently large, imaginary box is moved horizontally over any distance, the total leaf area, therein contained, remains unchanged.

The total irradiance at a level in the canopy has contributions from unintercepted direct beam and diffuse sky radiation, and reflected and transmitted portions of radiation intercepted by leaves (single and multiple scattering). Treatment of direct beam attenuation amounts to the geometrical problem of calculating the fraction of the foliage area projected in the direction of the sun; from this projected area the probability of a ray from a given angle intercepting an element can be defined. These results are then integrated over the hemisphere to provide estimates of diffuse radiation penetration.

One of the earliest light interception models, and still the one most often cited, was proposed by Monsi and Saeki /25/. They considered leaves to be the major radiation interceptors and determined the probability that a representative ray will miss N_L opaque, horizontal leaves each of silhouette area, σ, randomly located in a volume of basal area, A. The probability of non-interception is $P = (1 - \sigma/A)^{N_L}$ and in the limit as $A \gg \sigma$,

$$P = \lim_{A \to \infty} (1 - \sigma/A)^{N_L} = \exp(-N_L \sigma/A) = \exp(-F_c) \qquad (1)$$

where F_c is the silhouette leaf area per unit soil area in the layer above the level of interest (leaf area index accumulated from the top of the canopy downward). This is an expression for average direct beam transmission for randomly located, horizontal opaque leaves. It may also be interpreted as the fraction of horizontal area covered by gaps or sunflecks below a layer of leaves of thickness, F_c. To generalize this theory for inclined leaves it is necessary to specify the fraction of leaf area with its normal inclined at α to the horizontal and oriented at β [$g_k(z,\alpha,\beta)$]; in addition, the proportion of this leaf area, at

(α,β), projected in the direction of the sun, at (θ_s,ϕ_s), must be known $(|\cos \delta|)$. With this canopy structural information the unintercepted, dimensionless solar irradiance (normalized to unity above the canopy) can be written generally as

$$E_s(z,\theta_s,\phi_s) = \exp[-t(z,\theta_s,\phi_s)/\cos \theta_s] \tag{2}$$

$$t_s(z,\theta_s,\phi_s) = \sum_k \int_z^{z_o} F_k(z')\{\int_0^{2\pi}\int_{-\pi/2}^{\pi/2} g_k(z',\alpha,\beta)|\cos \delta|\sin \alpha \, d\alpha d\beta/4\pi\}dz' \tag{3}$$

where the subscript k refers to different plant elements (leaves, stems, etc.), $F(z')$ is the leaf area index between z' and $z' + dz'$, dz' is simply the dummy integration variable over height to distinguish it from the height of interest, z, and the canopy top, z_o, and $g_k(z,\alpha,\beta)$ is normalized so that the double integral in brackets in eq. (3) becomes unity if $|\cos \delta| = 1$ /26/. Since leaves intercept most of the radiation and the canopy can be divided into layers of equal leaf area index so that the integration over dz' can be replaced by the cumulative leaf area index down to height z, $F_c(z)$, the conventional extinction coefficient can be defined as

$$E_s(z,\theta_s,\phi_s) = \exp[-k(z,\theta_s,\phi_s) F_c(z)/\cos \theta_s] . \tag{4}$$

The extinction coefficient is the fraction of the total leaf area projected in the direction of the sun and the $\cos \theta_s$ factor accounts for the increased optical path length with increasing zenith angle. Some authors refer to $k(z,\theta_s,\phi_s) \sec \theta_s$, which is the fraction of leaf area projected onto the horizontal, as the extinction coefficient. The value of $k(z,\theta_s,\phi_s)$ can be calculated from the double integral in brackets in eq. (3); for flat leaves

$$|\cos \delta| = |\cos \theta_s \cos \alpha - \sin \alpha \sin \theta_s \cos (\beta - \phi_s)| \tag{5}$$

where θ_s is the angle of incidence (zenith angle) of the solar beam /25,27/. The canopy structure enters light extinction through $g_k(z,\alpha,\beta)$ and extinction coefficients have been calculated for horizontal leaves ($k(\theta_s) = \cos \theta_s$), azimuthally symmetric vertical leaves ($k(\theta_s) = 2 \sin \theta_s/\pi$), leaf area distributed as the surface area on a sphere ($k = 1/2$) and azimuthally symmetric leaves inclined at a single angle /25,27,28/. Canopies of azimuthally symmetric leaves with any leaf inclination distribution have been analyzed by deWit /29,30/. This excellent work of deWit has been applied to numerous canopies /31,32,33,34/ and extended to asymmetric azimuthal distributions /34,35/.

The detailed measurement of leaf inclination and azimuth angle distributions, $g(z,\alpha,\beta)$, is very difficult; several techniques have been developed /29,36,37,38,39/ and numerous measurements are available /34, 35,40/. Because of the great difficulties involved with these measurements, the simplified mathematical representations discussed above often are used; the most common being horizontal and spherical /28,29,36,41/. Somewhat less detailed measurements of mean leaf inclination angles have been obtained photographically /42/ and by inserting probes into canopies (point quadrats) /43,44,45,46/. The most important canopy structural measurement for radiation penetration is the total leaf area per unit soil area (leaf area index = $F(z)$) and it is determined by physically measuring the area of individual leaves /25,37/ or by subsampling area-to-weight ratios and measuring total leaf weight, or indirectly and non-distructively by point quadrats (with some assumptions).

The unintercepted penetration of diffuse sky radiation to a horizontal surface within a canopy can be derived by integrating the unintercepted beam penetration function, given by eq. (2) or (4), over the hemisphere.

$$E_d(z) = \pi^{-1} \int_0^{2\pi} \int_0^{\pi/2} \Gamma(\theta,\phi) \, E_s(z,\theta,\phi) \, \sin\theta \, \cos\theta \, d\theta d\phi \tag{6}$$

where $E_d(z)$ is the dimensionless diffuse flux (normalized to unity above the canopy) and $\Gamma(\theta,\phi)$ is the intensity distribution of the sky, which is normalized so that

$$\int_0^{2\pi} \int_0^{\pi/2} \Gamma(\theta,\phi) \, \sin\theta \, \cos\theta \, d\theta d\phi = \pi \, . \tag{7}$$

Solutions to eq. (6) have been considered by Cowan /28/ for horizontal, vertical and spherical leaf distribution canopies. Total canopy transmittance derived from eqs. (4) and (6), that is without scattering, agree well with measurements in visible wavelengths /47,48,49/.

The remaining contribution to canopy mean transmission, which is the most difficult to handle mathematically, arises from scattering of radiation that is intercepted by plant elements. The most rigorous treatment has been presented by Ross and Nilson /26/ and the numerous other models can be considered as special examples of their equations. Suits /50,51/ presents a very rigorous approach based on average horizontal and vertical leaf projections and arrives at canopy directional reflectance predictions useful in remote sensing. If the canopy is divided into layers, average leaf scattering properties can be used with simplified, single-scattering equations to arrive at once scattered contributions to irradiance /28,35/. Since all scattered radiation is diffuse, multiple scattering is estimated by applying diffuse penetration equations several times. This approach yields good agreement with measurements in the solar spectrum /35/. The same basic procedure is used by deWit /29/ except that computer iteration replaces the integrals of Cowan /28/, and although the diffuse angular distribution is assumed to be isotropic, this theory agrees well with measurements /31,32,33/. The scattering theories thus far discussed are based on canopy structural information. Several multiple scattering theories are empirically based because they require some radiation measurements within the canopy for their application /52,53,54/, or they are confined to horizontal leaves /55/. One model considers the multiple scattering between a single layered canopy and a turbid atmosphere above /56/.

If the direct beam and sky diffuse penetration is obtained from eqs. (4) and (6), approximate scattering corrections can be applied to these for horizontal, vertical and random leaf canopies of high leaf area index /37,57/. Some attempts have been made at using eq. (4) and replacing the actual extinction coefficient with one adjusted by the leaf transmissivity /25,58/. This may be reasonable for visible wavelengths but not in the near infrared.

The most simplified mathematical representation of average canopy transmittance is the empirical extinction coefficient derived from radiation measurements. These have been formulated in terms of the exponential, as Beer's Law, and the binomial /58,59,60/. Although they can be useful for some agronomic purposes, they are difficult to generalize for different kinds and densities of canopies, sky conditions and wavelength bands and they reveal little about interactions between canopy structure and radiation.

The scattering theories thus far discussed range from extremely complex and rigorous to very simple and limited in usefulness. Using the same basic approach as numerous authors cited above /28,29,52,54,55/, with a slightly different layer formulation, a very simple scattering model can be derived /61/ and is presented below. It was later presented in a two-layer form /62/ and then generalized to nearly any layer structure /63/. If the canopy is divided into N layers each with leaf density F/N, each with layer transmittance τ and layer reflectance ρ and all the radiation components have the same τ and ρ as with horizontal leaves, then

$$E^-_{j+1} = \rho E^+_{j+1} + \tau E^-_j$$
$$E^+_j = \rho E^-_j + \tau E^+_{j+1} \qquad (8)$$

where the fluxes above the jth layer are E^+_j for downward and E^-_j for upward. Simple manipulation of the above equations yields the ratio of upward to downward fluxes in a layer ($A_j = E^-_j/E^+_j = [(\tau^2 - \rho^2) A_{j-1} + \rho]/[1 - \rho A_{j-1}]$) and also the ratio of downward fluxes in successive layers ($E^+_j/E^+_{j+1} = \tau/(1 - \rho A_j)$). Starting at the soil surface with soil reflectance, A_o, the relation for A_j can be used to obtain the upward-to-downward flux ratios for all layers; with the A_j values and E^+_N, which is the downward flux from the sky, all the downward fluxes can be determined. For individual horizontal leaves of reflectance ρ_L and transmittance τ_L, the corresponding layer coefficients are $\tau =$

$\exp(-F/N) + [1 - \exp(-F/N)]\tau_L$ and $\rho = [1 - \exp(-F/N)]\rho_L$ for each of the N canopy layers. These equations consider all multiple scattering and have provided excellent agreement with spatially averaged measurements in visible and near-infrared wavelength bands /61/ while apparently being simpler than previous models of comparable rigor. In fact, if only a few layers are used, for example, N = 6 which often is sufficient, calculations can be done easily by hand with a slide rule. Beginning with these simple equations and the same basic procedure, highly sophisticated models can be constructed to incorporate leaves with very different transmittance and reflectance, any leaf angular distribution, any wavelength including thermal, some horizontal heterogeneities, inclusion of stems and branches, any sky light distribution and even reasonably approximate the angular distribution of diffuse light in the canopy /63/. Multiple scattering is treated with computer iteration in the layer model of deWit /29/ whereas the similar treatment of Norman et al. /61/ is analytical and thus somewhat easier to use.

The incorporation of plant parts other than leaves generally is not considered in radiation interception because of their minor role. Their treatment is analogous to that for leaves and some projected areas have been derived for randomly located vertical stems /64/ and for randomly oriented horizontal cylinders /63/.

THEORIES FOR CANOPIES APPROXIMATING NON–RANDOM LEAF POSITIONING

Any canopy that does not meet the statistical homogeneity requirements of randomness discussed at the beginning of the last section falls into this section. The most extreme example is widely separated individual plants where the radiation interception can be approximated by projected areas from individual opaque spheres, cylinders, cones /65,66/, or vertically displaced parallel discs /67/. The simplicity of these models makes it convenient to investigate climatic effects such as latitude, day length and cloudiness. Projected areas from arrangements of opaque shapes have been obtained for uniformly spaced vertical cylinders /68/, parallel horizontal cylinders (and equivalent slabs) /69/, and hedge rows of rectangular, triangular and trapezoidal cross sections /70,71/; the results indicate such climatic effects as the dependence of optimum row spacing on latitude.

A second means of characterizing radiation exchange in canopies with non-random foliage distributions is through semi-empirical relations derived from measurements of average direct-beam, visible transmission or from point quadrats /43/. These equations provide a direct link with random theory by multiplying the extinction coefficient, defined as in eq. (4), by a non-randomness factor that is greater than one for regular dispersion, equal to one for random and less than one for clumped foliage /41,72/. Exponential (eq. 4) and positive or negative binomial relations are used as well as a Markov model, which introduces structural dependency between layers /41/. Essentially this approach amounts to ascribing deviations between extinction coefficient calculations based on random theory and measurements to horizontal heterogenity; thus revealing little about the canopy structural elements and their arrangements.

A third method for considering the interaction of radiation with non-random foliage distributions derives entirely from the canopy structure by combining random theory with the regularily-spaced-opaque-obstacle approach; this might be termed a "stochastic-regular" or "weighted random" approach. Often it is possible to assume randomness at some level of the canopy structure and the essence of this approach is to choose those levels from knowledge of the plants organization. For example, needles are highly clumped on spruce trees in the form of shoots that have ratios of needle-surface-area to shoot-projected area between 2 and 10 depending on the angle from which the shoot is viewed /73/. From shoot photographs and measurements of needle area the shoot geometry can be defined; likewise whorl structure can be characterized by determining branch lengths and counting main stem nodes. The average canopy transmittance can be calculated with good accuracy by assuming shoots to be randomly distributed within the confines of whorls and whorls randomly distributed throughout the canopy. This grouping theory predicts significantly greater beam transmissions and agrees much better with measurements than random theory while preserving many of the advantages of random theories. Row crops have been treated in a similar stochastic-regular manner by allowing leaf area density to vary both in the vertical and in the cross-row horizontal within the confines of parallel hedgerows of rectangular cross section /74,75/. Although the results indicate the tremendous variability of light penetration across the row, the average transmission predictions do not agree particularily well with the measurements presented.

The consideration of scattering in canopies with non-random foliage distributions is rare. The analysis of Norman and Jarvis /63/ for sitka spruce is the only attempt that this author is aware of; their results clearly indicate that although less direct beam radiation is intercepted by clumped foliage /41,72,74/, increased multiple scattering within the clump results in less scattered radiation leaving the clump. The increased beam and decreased scattering effects are compensatory so that random theory (without stems and branches) provides reasonable estimates of mean transmission but incorrectly partitions it between beam and diffuse.

A very difficult task in dealing with non-random foliage distributions is deciding when they are necessary. With point quadrat measurements the theory of Acock et al. /72/ or Nilson /41/ may be used as a guide but often a close look at the true structure of the community is even more helpful. The sunfleck segment length theory presented in Miller and Norman /76/ provides a very sensitive and relatively simple method for detecting non-randomness in horizontal leaf canopies, such as sunflower or sumac, and assigning an approximate scale to it. Using a camera /76/ or metre stick /60/ this distribution can be measured and compared to the random theory.

WAVELENGTH CONSIDERATIONS WITHIN CANOPIES

The theories thus far discussed are rigorous only for monochromatic radiation with the exception of some empirical representations which are used in varying wavelength bands. The theoretical equations should be evaluated separately for each wavelength band where leaf properties differ and the total transmission obtained from the integral over wavelength of the narrow-band transmissions. For many purposes, such as photosynthesis, energy balance or transpiration estimates, it is sufficient to consider the solar radiation separately in visible and near-infrared wavelengths.

The presence of emission at thermal wavelengths requires somewhat different treatment than for solar wavelengths, but because it is only diffuse in nature and scattering is negligible, it is not difficult to handle /32,77,78/.

MEASUREMENTS OF RADIATION

Radiation penetration measurements in vegetation must involve an assessment of the wavelength and angular response of the sensors used, the requirements for adequate spatial sampling, and canopy structural description. Solar radiation sensors are basically of three types: solarimeters that are uniformly sensitive throughout the solar spectrum, narrow wavelength band sensors and limited wavelength band sensors with specific responses such as a quantum response for photosynthesis. Since the wavelength distribution of radiation in the canopy is similar to that transmitted through individual leaves /79,80/, and because photosynthesis and stomatal resistance are sensitive primarily to visible wavelengths, solarimeter measurements should be partitioned into visible and near-infrared wavelength bands with "sharp" cutoff filters provided by the manufacturer /37/. A suitable method for testing radiation models is to measure canopy radiation with narrow-response sensors sensitive in wavelength bands where leaf properties are known and differ greatly; if the model is successful then the canopy transmission can be calculated with considerable confidence at any wavelength where leaf properties are known /63/. Limited wavelength band sensors are useful mainly where photosynthesis is concerned; generally these sensors have a quantum response /81,82,83/. Light meters with the human-eye response and unfiltered photosensitive semiconductors should be avoided unless their response is suited to some special purpose. Thermal radiation measurements usually are made with net radiometers (downward minus upward irradiance) that are sensitive to solar and thermal wavelengths.

Any sensor intended for irradiance measurements in planes should be appropriately "cosine compensated" or at least its angular response should be known so that it can be corrected. In addition, radiation measurements above the canopy should be partitioned into their beam and diffuse components since they are intercepted differently by the canopy.

The appeal of narrow-band sensors becomes greatest when adequate spatial sampling is considered. Their small size, light weight, fast response and relatively large signal output results in their being much more suitable for scanning great distances than the larger and heavier broad-band solarimeters. Basically there are two methods for obtaining adequate sampling in a canopy: 1) spatial and temporal averaging of a number of sensors located at fixed points, and 2) spatial and temporal averaging of one or more sensors that are being scanned over substantial distances. The number of sensors required at fixed points for adequate sampling varies from as few as 5 to more than 200, in both forest /84,85,86/ and agricultural /87,88/ canopies, depending on canopy structure and averaging time. Scanning sensors in vegetative canopies appears to be more suitable /61,63,86/ and plots of mean transmittance error versus scanning distance can be most useful /63/. For example, in a 14-metre tall sitka spruce stand with trees spaced at 1.5 meters, a one-meter tubular solarimeter may indicate between one-half and two times the 14-metre mean at a leaf area index of 4. Norman and Jarvis /63/ suggest that the range may be more meaningful than the standard deviation since extreme values are more common than values near the mean. Sinclair et al. /89/ report that reliable mean canopy transmittance values can be obtained in corn with large time constant sensors if they are sampled at least as frequently as the sensor time constant; however, it should not be necessary to sample more often than once per 4 seconds for 12-to-23 metre scans.

The general usefulness of radiation measurements depend entirely on canopy structural information simultaneously presented. Leaf area index profiles and general canopy information such as plant spacing, height, etc. must be included; more complete information is desirable /29,32,63/.

Equidistant-projection hemispherical photographs have become quite common in recent years and this author believes their use should be encouraged. Not only do they allow a quick assessment of general canopy features but a considerable amount of quantative information can be extracted from them /37,47,84,90/.

TRANSMISSION AND GAP—SIZE DISTRIBUTIONS

A casual observation from any astute observer will reveal that the structure of a canopy involves not only the number density of leaves, their location and orientation, but also their size, shape and vertical distributions. Using leaf size, shape and number density Miller and Norman /76/ developed a theory that can predict the fraction of a very long horizontal transect that will be occupied by gaps longer than some length, λ, in canopies of randomly distributed horizontal leaves. This theory has been extended to randomly distributed compound leaf forms of variable size and non-random groupings of spruce shoots /63/. These theories are very complex in their general form; however, for many canopies an equivalent, horizontal disc-shaped leaf, equal in area to an average leaf, can be used as a reasonable approximation. If the equivalent-disc diameter is D, then the fraction of the sunlit length of a very long transect (which is composed of both sunlit and dark segments) that is occupied by sunfleck segments between λ and $\lambda + d\lambda$ in length is

$$df_s(\lambda) = (nD)^2 \lambda \exp(-nD\lambda)d\lambda \qquad (9)$$

where n is the number of equivalent discs per unit area and $f_s(\lambda) = E_s f(\lambda)$ where $f(\lambda)$ is the gap size distribution defined in Miller and Norman /76/. (These are geometric sunfleck segments that would arise from a point source sun.) This distribution has a maximum which is the "most probable gap" (λ_{max}) because it occupies the greatest fraction of the total sunlit segment length, and it can be conveniently used to characterize simple canopies ($\lambda_{max} = (nD)^{-1}$). An artificial canopy composed of 10-cm discs at a disc-area-index of 1.5 has $\lambda_{max} = 5$ cm, where as at a disc area index of 0.5, $\lambda_{max} = 16$ cm /61/. For a sunflower canopy of leaf area index 1.9, $\lambda_{max} = 4.5$ cm and in a spruce canopy for shoots at a leaf area index of 2, $\lambda_{max} = 3.6$ cm, at LAI = 4.2, $\lambda_{max} = 1.9$ cm, and at LAI = 7, $\lambda_{max} = 1.2$ cm. Crown diameters and the number of trees per unit area could also be used for specifying the spacing distribution of trees in a forest. For sitka spruce /63/ $\lambda_{max} \sim 1$ metre and the two canopies discussed by Reifsnyder et al. /84/ probably have $\lambda_{max} \sim 9$ metres for pine and $\lambda_{max} \sim 2$ metres for the hardwood; clearly this reflects the very different spatial sampling requirements for these canopies. Specification of this single parameter, λ_{max}, for leaves could allow reasonable penumbral calculations to be done for many canopies by simply counting leaves in a sample area and estimating their diameter. In fact, it could be specified for several scales, whorls and whole trees and with the theory reasonable gap-size distributions could be calculated.

The penumbral light transmission distribution, which arises from the finite solar disc /40,91/, is the fraction of the area exposed to beam light transmissions greater than E_s (neglecting diffuse for the moment); it is given by

$$F(E_s) = \int_0^\infty \lambda^{-1} r_s \overline{F}(\lambda, E_s)[df_s(\lambda)/d\lambda]d\lambda \qquad (10)$$

where $\overline{F}(\lambda, E_s)$ is from Table 2 of Miller and Norman /91/ and r_s is the apparent sun's radius which depends on the average height, h, of leaves above the level of interest ($r_s = 0.0047h$). The final transmission distribution obtains from multiplying the "E_s" of F in eq. (10) by the fraction of incident radiation above the canopy that is beam and adding on the background diffuse and scattered contributions at the level of interest. The transmission distribution from eq. (10) is cumulative; a plot of the slope of this distribution would yield the incremental distribution, which has been measured in several canopies for visible /61,92,93,95/ and near infrared wavelengths /61,95/ as well as net radiation /88,94/. At moderate leaf area indicies with the sun higher in the sky, the distributions are bimodal being composed of a low level background diffuse and bright sunflecks. The two modes tend to approach each other for higher leaf area indicies or sun angles near the horizon.

This distribution of intensities can have considerable importance for leaf processes which depend non-linearily on radiation such as photosynthesis and stomatal resistance; therefore, they are likely to be most important in the visible wavelength band.

LIGHT FLUCTUATIONS

Some of the interactions between radiation and plant structure are very subtle and fluctuating light is one of those. High frequency fluctuations (> 1Hz), which arise from the effects of wind, have been measured in several canopies /93,96,97/, and although mean transmittance apparently is not affected, sunflecks are redistributed over a greater number of leaves for shorter time periods. Depending on the intensity of background diffuse radiation, this could result in increased photosynthetic efficiency /97,98/. At somewhat lower frequencies near one minute stomatal resistance may be affected /99/, and longer period fluctuations, from minutes to hours, could result in reduced light efficiency because dark times may be long enough for induction phenomena to occur -- a "delayed start" of photosynthesis during the light portion of the period /100/. These periods might be associated with movement of the sun or clouds.

If fluctuating light is measured with a moving sensor than the sensor output will contain information about canopy structure, sensor speed and wind speed. Norman et al. /93/ have investigated canopy structure with such moving sensors and Desjardins et al. /96/ have separated the effects of canopy structure from wind.

RADIATIVE TRANSFER AND PLANT PHYSIOLOGY

The radiation incident on leaves from within or above the canopy can have profound effects on various physiological processes such as photosynthesis, photomorphogrnesis and transpiration. A detailed discussion of this topic is presented in another paper of this volume. At this point it is important to realize that the canopy radiation models thus far considered have been formulated to predict radiation conditions on a horizontal plane above or within the vegetation. Application of this information to estimates of photosynthesis requires a transformation to energy incident on (or absorbed by) the various leaves along with the relationship between incident (or absorbed) radiation and the physiological process of interest. The analytical form of this transformation from canopy-based to leaf-based coordinates is not trivial /101/. A relatively simple numerical procedure for obtaining the distribution of direct solar beam radiation over variously oriented leaves has been presented by deWit /29,30/. This procedure can be used to calculate the radiation incident on individual leaves for various physiological processes. The relation between radiation and plant physiology is discussed throughout the literature /25,29,59,95,102,103,104/.

REFLECTIONS ON FUTURE RESEARCH

One of the most arduous and self defeating tasks involved in developing, verifying, and understanding models of radiative transfer in vegetation is the measurement of canopy structure. Anyone who has attempted this knows that "there must be a better way" to do this than the head-on approach of physically sampling leaf by leaf and branch by branch. Even if this "brute-force" sampling approach works on small samples, the resulting sophisticated models will be useless on a large scale if their parameterization requires such data as inputs. However, in all fairness it must be recognized that many of the sophisticated models can be extremely useful in expanding our understanding of radiation interaction with vegetation while not being useful for the purpose of agronomic inventories or general practice; for example, they could be most useful in deciding what simplifications are suitable for realistic analysis of different plants in different climates /33/. This author feels that one of the high priorities of radiative transfer studies should be that of remotely sensing the canopy structure. Aside from being very valuable to those modeling radiative transfer, it also could be invaluable to every other branch of plant-environment studies from agronomic inventories to the most sophisticated holocoenotic modeler, because they all require some form of structural information. Some of the theory for this kind of "inversion" problem already has been developed in point quadrat theory; however, replacing the point quadrat probe with a sun beam is not trivial and much work remains to be done. Examples of such efforts are the determination of leaf area index from diffuse angular distribution measurements /63/ or measurements of the ratio of near-infrared to visible transmittance /105/. Work on this kind of "inversion" problem is likely to involve complex mathematics because that is the nature of it; however, some of the measurements may not be too difficult.

With the coming of aircraft and satellites as observing platforms, some attempts have been made at using radiation reflected from the canopy as an indicator of water stress or disease /50,51,106,107,108/ and Suits /51/ has suggested that the heiligenschein /109/ may be used to reveal some canopy structural characteristics from aircraft measurements. Study in this area has only begun and although the potential benefits of knowledge in this area could have tremendous impact on our societies, only one fact is certain -- that progress is going to be very difficult.

Another area deserving more research is the integration of radiative transfer knowledge with realistic treatments of heat and mass transfer so that the whole environment of variously exposed leaves can be used to assess physiological implications of that environment /32/. This involves both photosynthesis and photomorphogenesis. At the present time our knowledge of the radiation environment appears to greatly exceed our understanding of heat and mass transfer in the canopy or our understanding of the plants physiological response to the radiation. The great proliferation of "arm-chair" theoretical radiation models, without supporting measurements, definitely should not continue; from the viewpoint of plant productivity or manipulating crop yield, "arm-chair" radiation treatments are beyond their point of diminishing returns. Any radiation modeling in vegetation should be realistically evaluated with measurements before it is revealed to the scientific community. At the present time, relating light to the ongoing physiological processes appears to be much more important to plant productivity assessment than do more refined radiation models. However, models realistically simulating vegetation structure and incorporating radiation, heat and mass transfer with physiological processes are a very high priority. Environmental manipulation for higher yields is another closely related area deserving of more research /62,103/.

A third problem of interest, which has spurred debate for many years /110/ deals with the extrapolation of research from controlled environments to the field. Controlled environments include growth chambers (phytotrons) and glasshouses. The radiation environment arising from artificial sources in phytotrons has been characterized /81,102/ as have the other environmental parameters such as wind, temperature and humidity. Glasshouse environments also have been characterized /111, 112/. However, knowledge of these environments must be combined with physiological responses to appreciate fully the subtle differences that the plant is sensitive to. The use of comprehensive environmental models that consider radiation, heat and mass transfer in relation to physiological characteristics of plants will facilitate in understanding these subtle differences between controlled environments and the field.

REFERENCES

1. Kumar,R.;Silva,L.(1973) Appl. Optics. 12:2950-2954.
2. Birkebak,R.;Birkebak,R.(1964) Ecology. 45:646-649.
3. Gates,D.M.;Keegan,H.J.;Schleter,J.C.;Weidner,V.R.(1965) Appl. Optics. 4:11-20.
4. Wooley,J.T.(1971) Plant Physiol. 47:656-662.
5. Methy,M.(1972) Israel Jour. Agric. Res. 22:77-84.
6. Egan,W.G.(1970) Forest Science. 16:79-94.
7. McCree,K.J.;Keener,M.E.(1973) The effects of solar ultraviolet radiation on photosynthesis in agricultural crops. Final Report. U.S. Dept. Transport, NOAA Contract 1-35360, Biol. Dept. Texas A & M Univ., College Station, Texas.
8. Gates,D.M.;Tantraporn,W.(1952) Science. 115:613-616.
9. Breece,H.T.III;Holmes,R.A.(1971) Appl. Optics. 10:119-127.
10. Kondratyev,K.Ya.(1969) Radiation in the Atmosphere. Academic Press, N.Y.
11. Moon,P.(1940) Jour. Franklin Institute. 230:583-617.
12. Elterman,L.(1964) Atmospheric Attenuation Model, 1964, in the Ultraviolet Visible and Infrared Regions of Altitudes to 50 km. Rep. AFCRL-64-740 Air Force Cambridge Research Laboratory, Bedford, Mass.
13. Elterman,L.(1968) Visible and IR Attenuation for Altitudes to 50 km. Rep. AFCRL-68-0153, AFCRL. Bedford, Mass.
14. Guttman,A.(1968) Appl. Optics. 7:2377-81.
15. Henderson,S.T.;Hodgkiss,D.(1963) British Jour. Appl. Physics. 14:125-131.
16. Ashcheulov,S.V.;Kondratyev,K.Ya.;Styro,D.B.(1966) In Actinometry and Atmospheric Optics. (Ed. V. K. Pyldmaa). Israel Prog. Sci. Trans., 1971, Cat. No. 5835, pp. 43-53.
17. Ashcheulov,S.V.;Kondratyev,K.Ya.(1966) In Actinometry and Atmospheric Optics. (Ed. V. K. Pyldmaa). Isrl. Prog. Sci. Trans., 1971, pp. 24-27.
18. Idso,S.B.;Jackson,R.D.(1969) Jour. Geophys. Res. 74:5397-403.
19. Linacre,E.T.(1968) Agric. Meteorol. 5:49-63.

20. Nkemdirm,L.C.(1973) Agric. Meteorol. 11:229-42.
21. Grace,J.(1971) Jour. Appl. Ecology. 8:155-64.
22. Moon,P.;Spenser,D.E.(1942) Trans. Illum. Engng. Society, N.Y. 37:707-12.
23. Anderson,M.C.(1966) Jour. Appl. Ecology. 3:41-54.
24. List,R.J.(1966) Smithsonian Meteorological Tables. Smithsonian Miscellaneous Collections, Vol. 11. Pub. by the Smithsonian Institution.
25. Monsi,M.;Saeki,T.(1953) Jap. Jour. Bot. 14:22-52.
26. Ross,J.K.;Nilson,T.A.(1966) In Actinometry and Atmospheric Optics. (Ed. V. K. Pyldmaa). pp. 253-270. Translated from Russian by Israel Prog. Sci. Trans., Cat. No. 5835, 1971.
27. Warren Wilson,J.(1960) New Phytol. 59:1-8.
28. Cowan,I.R.(1968) Jour. Appl. Ecology. 5:367-79.
29. deWit,C.T.(1965) Photosynthesis of leaf canopies. Agric. Res. Reports No. 663, Inst. for Biol. and Chem. Res. on Field Crops and Herbage, Wageningen, The Netherlands, 57 p.
30. Idso,S.B.;deWit,C.T.(1970) Appl. Optics. 9:177-84.
31. Duncan,W.G.;Loomis,R.S.;Williams,W.A.;Hanau,R.(1967) Hilgardia. Vol. 38, No. 4.
32. Stewart,D.W.;Lemon,E.R.(1969) The energy budget at the earth's surface: Assimilation of net photosynthesis of field corn. Tech. Rept. ECOM2-68, I-6. (Avail: National Technical Information Center, Springfield, Va., USA.)
33. Gaudriaan,J.(1973) In Proceedings of the International Congress: "The Sun in the Service of Mankind." July 2-6, 1973, Paris.
34. Lemeur,R.(1974) Agric. Meteorol. 12:229-47.
35. Lemeur,R.(1970) Proc. Symposium on plant response to climatic factors, Uppsala. UNESCO, Paris.
36. Nichiporovich,A.A.(1961) Soviet Pl. Physiol. 8:428-35.
37. Anderson,M.C.(1971) In Plant Photosynthetic Production. Manual of Methods. (Eds. Z. Sestak, J. Catsky, P. G. Jarvis) pp. 412-66. Dr. W. Junk, N. V. The Hague.
38. Daynard,T.B.(1971) Agron. Jour. 63:133-5.
39. Lang,A.R.G.(1973) Agric. Meteorol. 11:37-51.
40. Loomis,R.S.;Williams,W.A.(1969) In Physiological Aspects of Crop Yield. (Eds. J. O. Eastin, et al.) Amer. Soc. Agron., Madison, Wisc. pp. 27-47.
41. Nilson,T.(1971) Agric. Meteorol. 8:25-38.
42. Smart,R.E.;Lang,A.R.G.(1973) Agric. Meteorol. 11:445-450.
43. Warren Wilson,J.(1963) Austral. Jour. Bot. 11:95-105.
44. Warren Wilson,J.(1965) Jour. Appl. Ecology. 2:383-90.
45. Philip,J.R.(1965) Austral. Jour. Bot. 13:357-66.
46. Philip,J.R.(1966) Austral. Jour. Bot. 14:105-25.
47. Anderson,M.C.(1966) In Light as an Ecological Factor. (Eds. R. Bainbridge, et al.) Blackwell Sci. Publ., Oxford. pp. 77-90.
48. Warren Wilson,J.(1967) Jour. Appl. Ecology. 4:159-65.
49. Anderson,M.C.;Denmend,O.(1969) Agron. Jour. 61:867-71.
50. Suits,G.H.(1972) Remote Sensing of Environ. 2:117-25.
51. Suits,G.H.(1972) Remote Sensing of Environ. 2:175-82.
52. Allen,J.H.;Brown,K.W.(1965) Agron. Jour. 57:575-80.
53. Allen,W.A.;Gayle,T.V.;Richardson,A.J.(1970) Jour. Opt. Soc. Amer. 60:372-76.

54. Allen,W.A.;Richardson,A.J.(1968) Jour. Opt. Soc. Amer. 58:1023-28.
55. Alderfer,R.G.;Gates,D.M.(1971) Ecology. 52:855-61.
56. Weinman,J.A.;Guetter,P.J.(1972) Jour. Appl. Meteorol. 11:136-40.
57. Anderson,M.C.(1969) Agric. Meteorol. 6:399-405.
58. Kasanaga,A.;Monsi,M.(1954) Jap. Jour. Bot. 14:304-24.
59. Monteith,J.L.(1965) Annals of Bot. N.S. 29:17-37.
60. Miller,P.C.(1969) Ecology. 50:878-85.
61. Norman,J.M.;Miller,E.E.;Tanner,C.B(1971) Agron. Jour. 63:743-748.
62. Fuchs,M.(1972) In Optimizing the Soil Physical Environment Toward Greater Crop Yields. Academic Press, New York and London. pp. 173-91.
63. Norman,J.M.;Jarvis,P.G.(1974) Photosynthesis in sitka spruce (Picea Sitchensis (Bong.) Carr.) IV. Radiation penetration theory and a test case. Jour. Appl. Ecology. Submitted.
64. Federer,C.A.(1971) Agric. Meteorol. 9:3-20.
65. Terjung,W.H.;Louie,S.S.F.(1972) Intl. Jour. Biometeorol. 16:25-43.
66. Jahnke,L.S.;Lawrence,D.B.(1965) Ecology. 46:319-26.
67. Pearman,G.(1968) Studies of Leaf Energetics. Ph.D. thesis. Faculty of Science, University of Western Australia.
68. Brown,P.S.;Pandolfo,J.P.(1969) Agric. Meteorol. 6:407-421.
69. Idso,S.B.;Baker,D.G.(1967) Agron. Jour. 59:13-21.
70. Jackson,J.E.;Palmer,J.W.(1972) Jour. Appl. Ecology. 9:341-57.
71. Smart,R.E.(1973) Sunlight interception by vineyards. Amer. Jour. Enol. Viticulture. 24: In press.
72. Acock,B.;Thornley,J.H.M.;Warren Wilson,J.(1969) In Prediction and Measurement of photosynthetic productivity. (Ed. I. Setlik) Proc. IBP/PP Tech. Meeting, Trebon, Czech. PUDOC. The Netherlands, pp. 91-102.
73. Norman,J.M.;Jarvis,P.G.(1974) Photosynthesis in sitka spruce (Picea Sitchensis (Bong.) Carr.) III. Measurements of canopy structure and interception of radiation. Jour. Appl. Ecology. In press.
74. Allen,L.H.(1974) Agron. Jour. 66:41-47.
75. Allen,L.H.;Sinclair,T.R.;Lemon,E.R.(1974) Direct-beam light penetration into erect-leaf and arch-leaf rows of corn. Agron. Jour. In press.
76. Miller,E.E.;Norman,J.M.(1971) Agron. Jour. 63:735-38.
77. Ross,J.K.;Tooming,H.G.(1966) In Actinometry and Atmospheric Optics. (Ed. V. K. Pyldmaa). Isrl. Prog. Sci. Trans., Cat. No. 5835, 1971. pp. 287-96.
78. Idso,S.B.(1968) A holocoenotic analysis of environment-plant relationships. Tech. Bull. 264, Agric. Exp. Station, Univ. of Minn., St. Paul. U.S.A.
79. Federer,C.A.;Tanner,C.B.(1966) Ecology. 47:555-60.
80. Scott,D.;Menalda,P.H.;Brougham,R.W.(1968) New Zeal. Jour. Bot. 6:427-49.
81. McCree,K.J.(1972) Agric. Meteorol. 10:443-53.
82. Norman,J.M.;Tanner,C.B.;Thurtell,G.W.(1969) Agron. Jour. 61:840-3.
83. Biggs,W.W.;Edison,A.R.;Brown,J.D.;Maranville,K.W.;Clegg,M.D.(1971) Ecology. 52:125-31.
84. Reifsnyder,W.E.;Furnival,G.M.;Horowitz,J.L.(1971) Agric. Meteorol. 9:21-27.
85. Gay,L.W.;Knoerr,K.R.;Braaten,M.O.(1971) Agric. Meteorol. 8:39-50.
86. Brown,G.W.(1973) Agric. Meteorol. 11:115-121.

87. Denmead,O.T.;Fritchen,L.J.;Shaw,R.H.(1962) Agron. Jour. 54:505-10.
88. Impens,I.;Lemeur,R.;Moermans,R.(1970) Agric. Meteorol. 7:335-7.
89. Sinclair,T.R.;Desjardins,R.L.;Lemon,E.R.(1974) Agron. Jour. 66:214-7.
90. Bonhomme,R.;Chartier,P.(1972) Isrl. Jour. Agric. Res. 22:53-61.
91. Miller,E.E.;Norman,J.M.(1971) Agron. Jour. 63:739-42.
92. Sinclair,T.R.;Lemon,E.R.(1974) Agron. Jour. 66:201-5.
93. Norman,J.M.;Tanner,C.B.(1969) Agron. Jour. 61:847-9.
94. Sinclair,T.R.;Lemon,E.R.(1972) Boundary Layer Meteorol. 3:246-54.
95. Sinclair,T.R.;Lemon,E.R.(1973) Solar Energy. 15:89-97.
96. Desjardins,R.L.;Sinclair,T.R.;Lemon,E.R.(1974) Light fluctuations in corn. Agron. Jour. In press.
97. Kriedemann,P.E.;Torokfalvy,E.;Smart,R.E.(1973) Photosynthetica 7:18-27.
98. McCree,K.J.;Loomis,R.S.(1969) Ecology. 50:422-8.
99. Gregory,F.G.;Pearse,H.L.(1937) Annals of Bot. 1:3-10.
100. Gardner,W.W.;Allard,H.A.(1931) Jour. Agric. Res. 42:629-51.
101. Paltridge,G.W.(1970) Agric. Meteorol. 7:93-130.
102. Prediction and Measurement of Photosynthetic Productivity (1970). Proc. IBP/PP Tech. Meeting, Trebon, Czech. PUDOC. The Netherlands.
103. Physiological Aspects of Crop Yield (1969) (Eds. J. D. Eastin, et al.) Amer. Soc. Agron., Madison, Wisc. U.S.A.
104. Duncan,W.G.(1971) Crop Science. 11:482-5.
105. Jordan,C.F.(1969) Ecology. 50:663-666.
106. Knipling,E.B.(1970) Remote Sensing of Environ. 1:155-59.
107. Ausmus,B.S.;Hilty,J.W.(1972) Remote Sensing of Environ. 2:77-81.
108. Guasman,H.W.;Allen,W.A.;Myers,V.I.;Cardonas,R.;Leamer,R.W.(1970) Remote Sensing of Environ. 1:103-7.
109. Tricker,R.A.R.(1970) Introduction to Meteorological Optics. Amer. Elsevier Pub. Co. Inc., New York and Mills and Boon, London.
110. Evans,L.T.(1963) In Environmental Control of Plant Growth. (Ed. L. T. Evans). Academic Press, N.Y., London. pp. 421-37.
111. Morris,L.G.(1972) Isrl. Jour. Agric. Res. 22:85-97.
112. Kimball,B.A.(1973) Agric. Meteorol. 11:243-60.

GENERAL PRINCIPLES OF NATURAL EVAPORATION

FRANK KREITH and W.D. SELLERS

University of Colorado, Boulder, Colorado, USA,
University of Arizona, Tucson, Arizona, USA

ABSTRACT

This article presents an overview of the mechanisms governing evaporation in nature from a macroscopic and a microscopic point of view. The microscopic viewpoint presents a steady-state model of the water loss through the stomata in the leaves and the surrounding boundary layer whereas, the macroscopic approach considers a time dependent model to calculate evaporation from an entire ecosystem, such as a field, into the atmosphere.

1. INTRODUCTION

Evaporation from natural surfaces, especially vegetation is a very complex process which almost defies mathematical description. There are numerous variables and interactions involved which are not only difficult to quantify, but sometimes even difficult to define. Nevertheless, there are certain basic physical processes involved which must be considered in any realistic approach to the problem and which, when considered together, lead to reasonably consistent and reliable predictions of experimental results.

Evaporation in nature has been studied by micro-meteorologists, plant physiologists, agricultural engineers, and atmospheric scientists for over a century. The literature on the subject is extensive and some excellent review papers have been prepared by Monteith /1/, Webb /2/, Zelitch /3/, Lemon /4,5/, and Denmead /6/, while books and monographs on the subject have been written by Slatyer /7/, Kozlowski /8/, Konstantinov /9/, and Sellers /10/. We will give here only an overview of natural evaporation from a microscopic and a macroscopic viewpoint, respectively. The presentation is on a general level and for detailed derivation of equations, discussion of analytical methodologies, experimental measurements, and other viewpoints the reader is referred to the bibliography at the end of the article. The macroscopic viewpoint is presented by W. D. Sellers and the microscopic view by F. Kreith.

2. MACROSCOPIC ANALYSIS

No attempt will be made here to present a comprehensive review of methods used to estimate evapotranspiration from vegetation, crops, and other natural surfaces. This has been done several times, most impressively by A. R. Konstantinov in his well-known monograph, "Evaporation in Nature" /9/. Rather, in order to illustrate those factors which are of primary importance to evaporation on a scale larger than a single leaf or a plant, a simple time-dependent model of the evapotranspiration process will be described and applied to the special case of a small field of short grass with variable rooting, depths in an arid environment.

This model is based on the energy and water budgets of the particular site. It is assumed that a number of basic meteorological variables are known for each hour of the day. Included are the incoming solar radiation (Q), the cloud cover, the surface albedo (α), the air temperature (T), the relative humidity (r), and the wind speed (u) at a height of 1 or 2 m above the surface. Given this information, and the composition of the surface cover and soil, the model yields hourly values of the net radiation (R); the energy used for evaporation (LE), for warming the air (H), and for warming the soil (G); and the surface temperature (T_s).

A more general, but also more complicated model has been described by Sasamori /11/ who obtained time-height distributions of temperature, wind velocity, the eddy diffusion coefficient, and humidity up to a height of 1 km in the atmosphere and of temperature and humidity to a depth of 1 m in the soil.

2A THEORY

a. The energy balance equation

The basic relationship used for this analysis is the energy balance equation for a soil column extending from the ground surface or top of the vegetation to that depth in the soil at which the vertical transfer of heat is negligible. For the daily cycle this depth averages 50 to 100 cm /10, p.136/. The energy balance equation may be written,

$$R = H + LE + G, \qquad (1)$$

where R is the net radiation or radiation balance; H is the sensible heat flux between the surface and air (positive when directed upward); LE, the product of the latent heat of evaporation L and the evaporation rate E, is the latent heat flux between the surface and air (positive when the surface is losing water by evaporation); and G is the sensible heat flux into the soil column. G is directly proportional to the rate at which the column is warming (G positive) or cooling (G negative).

The purpose of the following paragraphs will be to show how each term in eq (1) may be written in terms of a single unknown, either the surface temperature T_s or the temperature difference ΔT between the surface and the air. Having done this, the terms will be recombined to give the basic prediction equation used in the next section.

b. The net radiation, R

The net radiation is defined as the difference between the incoming and outgoing streams of solar and longwave sky or terrestrial radiation. Thus,

$$R = Q(1-\alpha) + I_\downarrow - \varepsilon\sigma T_s^4, \tag{2}$$

where Q is the incoming solar radiation; α is the surface reflectivity or albedo; $I_\downarrow$ is the long-wave sky (counter) radiation emitted downward by the overlying atmospheric constituents (primarily water vapor, carbon dioxide, ozone, and clouds) and absorbed by the surface; ε is the surface infrared emissivity, and σ is the Stefan-Boltzmann constant (8.1×10^{-11} cal cm^{-2} min^{-1} K^{-4}). The albedo varies greatly, depending on the surface and ranging from as high as 0.95 for fresh snow to 0.05 for a black top road or a dense coniferous forest. Most natural surfaces have an infrared emissivity ε lying between 0.90 and 0.96.

The infrared counter radiation $I_\downarrow$ in eq (1) can be estimated fairly accurately by any one of a number of empirical equations relating it, usually, to the air temperature, moisture content, and cloud cover. Some of these expressions are reviewed by Morgan, Pruitt, and Lourence /12/. In the absence of clouds, they obtained good results using

$$I_\downarrow = \varepsilon\sigma T^4 (0.605 + 0.048\sqrt{e}), \tag{3}$$

which was first suggested by Brunt /13/. In this equation T and e are, respectively, the air temperature (°K) and vapor pressure (millibars) at a height of about 2 m.

It will be appropriate later to rewrite eq (2) in an approximate form in order to eliminate the fourth power of T_s. Thus, adding and substracting $\varepsilon\sigma T^4$, we get approximately

$$R = Q(1 - \alpha) + I_\downarrow - \varepsilon\sigma T^4 - 4\varepsilon\sigma T^3(T_s - T). \tag{4}$$

c. The sensible heat flux, H

The sensible heat flux is directly proportional to the temperature difference between the surface and air. Thus

$$H = \rho c_p D_h (T_s - T) = \rho c_p D_h \Delta T \tag{5}$$

where ρ is the air density, c_p is the specific heat of air at constant pressure (0.24 cal gm^{-1} K^{-1}), and D_h is nebulous transfer coefficient. When the air temperature does not vary greatly with height, as is often the case near sunrise and sunset or under cloudy skies, the transfer coefficient may be estimated from

$$D_h = D_m = 0.16u(\ln\frac{z}{z_o})^{-2}, \tag{6}$$

where u is the horizontal wind speed at the height z and z_o is the roughness length, a parameter reflecting the physical characteristics of the surface. z_0 increases with the average height of the vegetation of surface obstacles. It averages about 0.003 cm for water, 0.03 cm for bare soil, and 0.75 cm for 6-cm grass /10, p.150/.

When the temperature varies with height, D_h does not equal D_m and eq (6) is no longer valid. The proper relationship to use in these situations is not known. It is known, however, that D_h may be several times larger than D_m under very unstable conditions (rapid temperature decrease with height and light winds) and only a fraction of D_m under stable conditions. On the basis of observations made by the author /14/ over dry bare soil in the Tucson area, a relationship of the form

$$D_h = D_m(1 + 14\,\frac{\Delta T}{u^2 z})^{1/3}, \tag{7}$$

where u is the wind speed (m sec^{-1}) at a height of about 2 m, is not inappropriate, at least under unstable conditions when ΔT is positive. Under stable conditions, the expression

$$D_h = D_m(1 - 14\,\frac{\Delta T}{u^2 z})^{-1/3} \tag{8}$$

seems to fit the data better. Equations (6), (7), and (8) will be used in the following analysis, even though the reliability of the latter two is uncertain.

d. The latent heat flux, LE

The latent heat flux depends on the vapor pressure difference between the surface and the air and is given by

$$LE = \rho L D_w(q_s - q) = 0.622\rho L D_w(e_s - e)/p, \tag{9}$$

where e_s and e are the surface and air vapor pressures, respectively, q_s and $\bar{q}$ are the surface and air specific humidities, respectively, p is the air pressure, and D_w is a transfer coefficient for water vapor. If the surface is wet enough so that its vapor pressure is equal to the saturation value at the temperature T_s, then the vapor pressure difference may be expressed as a function of ΔT:

$$e_s - e = a\Delta T + e_{sa} - e \text{ and}$$

$$LE = 0.622\rho L D_w a\Delta T/p + 0.622\rho L D_w(e_{sa} - e)/p, \tag{10}$$

where e_{sa} = the saturation vapor pressure at the air temperature,

$$a = \frac{0.622 L e_{sa}}{R_d T^2}, \text{ and}$$

General Principles of Natural Evaporation

R_d = gas constant for dry air (0.0686 cal gm^{-1} K^{-1}).

Equation (10) defines what is called potential evapotranspiration E_o; that is, the evapotranspiration that occurs in the presence of a non-limiting water supply. If the soil column is partially dry so that moisture is limiting, it is usually assumed, with considerable accuracy, that the actual evapotranspiration rate is proportional to the potential rate. The factor of proportionality may be defined as the relative soil-moisture content m. It depends on the amount w of moisture in the active soil layer of depth d and on the moisture-holding capacity w_m of the soil. Following Sellers /10 p. 176/, we assume that

$m = 1.0$ when $w \geq 0.75w_m$ and

$m = w/0.75w_m$ when $w < 0.75w_m$.

The depth of the active soil layer coincides roughly with the root zone for vegetation-covered surfaces. For bare soil d depends on the soil type and ranges from 1 to 5 cm.

Forms similar to eqs (7) and (8) can be used for D_W in eq (10) if the value of the constant 14 in these equations is reduced, on the basis of theoretical considerations, to 10.5.

e. The sensible heat flux into the soil column, G

There are a number of ways of estimating the sensible heat flux into a soil column. The method used here is derived from the standard theory of heat transfer in a homogeneous medium. Assuming, as a first approximation, a sinusoidal variation of the surface temperature, the result obtained is

$$G = \frac{\sqrt{2}}{2}(\omega C \lambda)^{1/2}(1 + \frac{2}{\omega \Delta t})\Delta T - \frac{\sqrt{2}}{2}(\omega C \lambda)^{1/2}[(\overline{T}_s - T) + \frac{2}{\omega \Delta t}(T_{s_1} - T)], \quad (11)$$

where ω is the frequency of the oscillation ($\pi/12$ hr^{-1}), C and λ are the heat capacity and thermal conductivity respectively, of the medium, t is the time interval involved (1 hr), $\overline{T}_s$ is the mean 24-hr surface temperature, and T_s is the surface temperature at the start of a given hour. The quantity $(C\lambda)^{1/2}$ is sometimes referred to as the thermal property of the medium. It ranges from about 0.01 cal cm^{-2} K^{-1} sec$^{-1/2}$ for dry peat moss to 0.06 cal cm^{-2} K^{-1} sec$^{-1/2}$ for wet clay.

f. The working equations

If for simplicity, the transfer coefficients are determined from values of ΔT and u existing at the start of the hour, eqs (1), (4), (5), (10), and (11) can be combined to give

For moist soil with an unlimited water supply: $\Delta T = B_1/B_2$. (12)

For partially dry soil: $\Delta T = B_3/B_4$. (13)

where $B_1 = Q(1 - \alpha) + I_\downarrow - \varepsilon\sigma T^4 - 0.622\frac{\rho L}{p}D_w(e_{sa} - e)$

$$+ \frac{\sqrt{2}}{2}(\omega C\lambda)^{1/2}[(\overline{T}_s - T) + \frac{2}{\omega\Delta t}(T_{s_1} - T)] \qquad (14)$$

$$B_2 = 4\varepsilon\sigma T + \rho c_p D_h + 0.622\frac{\rho L}{p}D_w a + \frac{\sqrt{2}}{2}(\omega C\lambda)^{1/2}(1 + \frac{2}{\omega\Delta t}) \qquad (15)$$

$B_3 = Q(1 - \alpha) + I_\downarrow - \varepsilon\sigma T^4 - mLE_o$

$$+ \frac{\sqrt{2}}{2}(\omega C\lambda)^{1/2}[(\overline{T}_s - T) + \frac{2}{\omega\Delta t}(T_{s_1} - T)] \qquad (16)$$

$$B_4 = 4\varepsilon\sigma T^3 + \rho c_p D_h + \frac{\sqrt{2}}{2}(\omega C\lambda)^{1/2}(1 + \frac{2}{\omega\Delta t}) \qquad (17)$$

The surface temperature and the moisture content of the soil at some initial time, say, midnight, and the potential evapotranspiration rate E_0 (obtained from eqs 12 and 10) must be specified or predetermined. T_s for 0030 can then be determined and used to obtain T_{s_1} at 0100 from

$$T_{s_1}(0100) = 2T_s(0030) - T_{s_1}(0000). \qquad (18)$$

The process is repeated, hour-by-hour, adjusting the relative moisture content m downward as evaporation occurs, for as many hours or days as desired. In eqs (14) and (16), $\overline{T}_s$ should be constantly changed as new values for T_s are obtained.

2B APPLICATIONS

The model has been tested against observations obtained by Vehrencamp /15/ in a dry lake bed near El Mirage, California, and by Van Bavel, Fritschen, and Reginato /16/ and Van Bavel and Fritschen /17/ at Phoenix, Arizona. In both cases, however, the soil-moisture content was held constant (at either zero or field capacity) for the short period for which data are available. Although good results were obtained, the model should be tested further under more variable conditions.

To illustrate one possible application of the model it has been used to estimate evapotranspiration from a hypothetical small plot of grass with variable rooting depth (and active soil layer) at Tucson, Arizona, during 10 identical clear days in the month of June. The input data are given in Table 1. The air vapor pressures used are compatible with relative humidities ranging from 9 percent in the late afternoon to almost 30 percent near sunrise. With such dry air, large evaporation rates can be expected. The potential evapotranspiration values given in the last column of the table indicate this.

They were obtained from the model assuming a wet surface, using an albedo of 0.15, a roughness length of 0.75 cm, and a thermal property of 0.05 cal cm^{-2} K^{-1} sec$^{-1/2}$.

The daily total potential evapotranspiration is just over 12 mm, quite typical for small wet surfaces in an arid environment in summer. The energy involved is about 50 percent greater than that supplied by radiation, implying strong advection of dry air from the arid environment and a considerable heat transfer to the cool surface from the overlying warm air. The 24-hour average surface temperature for this case is 20.0°C or 8.5°C cooler than the average air temperature. The maximum temperature difference between grass and air, 12.1°C, occurs in the late afternoon when the air is still warm and wind speeds are strongest. Highest evapotranspiration rates, however, occur earlier in the afternoon when the total energy supply is greater.

With the input data of Table I and appropriate initial conditions obtained from the potential evapotranspiration analysis, hour-by-hour estimates of evapotranspiration were made for short grass for a period of 10 days, starting with a soil column containing the maximum amount of water, assumed to equal 250 mm of water per meter depth of soil. The calculations were carried out for active soil layer depths of 5, 10, 20, 40, and 80 cm (equivalent to initial available moisture contents of 12.5, 25, 50, 100, and 200 mm, respectively) and are summarized in Table 2.

The more shallow soils, with the root zone down to only 5 or 10 cm, dry out quickly. After the fifth day evapotranspiration is negligible, most of the available water being used up. (In practice there may be an upward diffusion of water, liquid and vapor, from below the active layer. This process could increase the moisture available for evapotranspiration, but, in most cases, apparently not by a substantial amount.) As the soil layer dries out and warms up (positive G) the net radiation decreases and the sensible heat flux changes from large negative (air warmer than the grass) to large positive (grass warmer than the air). In the 5 cm case, after the third day the air was warmer than the grass only from 8:30 to 11:30 P.M.

For the deeper soils evapotranspiration proceeds at close to the potential rate (12.05 mm day^{-1}) for several days. After 5 days only a relatively small amount of the total water available is lost (30 percent for the soil with a rooting depth of 80 cm). It is possible in some favorable cases to maintain evapotranspiration at close to the potential rate for nearly a month, even in arid regions. See, for example, Van Bavel /18/ who attributes the eventual drop of the evapotranspiration rate below the potential rate after 17 days to an increase in the resistance of the canopy, in his case alfalfa, to water loss.

CONCLUSIONS

Modeling the macroscopic evapotranspiration process is a relatively complex problem, difficult to solve in all of its ramifications. Yet it is possible to achieve reasonably good results using a fairly simple, but consistent, model which takes into account most of the major variables involves. The basic model, described in this paper, may be modified in order to apply it to almost any specific problem. For example, the investigation of the microclimate of a desert plain partially covered with tall grass or brush would require incorporating into the problem the energy and water budgets of the plant-occupied layer. Two equations and two unknowns, the soil surface temperature, and the radiating temperature of the vegetation, would result. A solution to this problem is discussed by Foster /19/. Even more complex models may be developed to handle, for example, the lateral diffusion and transport of heat and moisture, precipitation, and the CO_2 budget. Instead of specifying the incoming solar radiation, the surface wind speed, and the air temperature and humidity, these could be derived from numerical models of the large-scale atmospheric circulation. In any case, however, the general procedure should not differ greatly from that outlined here.

3. MICROSCOPIC ANALYSIS

The microscopic view of evapotranspiration from vegetation presented in the second part of this lecture gives a comparison between analytic predictions of water consumption from various models and experimental data. Emphasis is placed on modeling evapotranspiration from the leaves of plants and trees with the aid of measurements of physiological and geometrical parameters. For a general and exhaustive discussion of plant water consumption the reader is referred to /7 and 8/.

3A WATER TRANSPORT AND STOMATAL PHYSIOLOGY

Water is an essential constituent of all vegetation and is involved in all physiological processes and metabolic activities of living plants. The paths and the corresponding resistances for the transport of water from the soil through vegetation into the atmosphere are shown schematically in Fig. 1 by a simplified electrical analog /20/. Water is first absorbed from the soil by the root system of the vegetation. In the roots, water transport occurs either from cell to cell or through the cell walls along gradients of decreasing water potential until it reaches the xylem vessels of the roots. From the roots water is transported to the leaves of a plant primarily in the xylem. Once it reaches the leaves, liquid water moves through the mesophyll cells to evaporating sites located within the cell walls in the leaves' stomatal cavities. From the stomatal cavities water is transported in the vapor phase through stagnant air to and through the stoma and finally through the external boundary layer into the atmosphere. According to the cohesion theory, transpiration reduces the leaf water potential and develops a pressure gradient which is transmitted by cohesion between water molecules through a continuous liquid system from the leaf's xylem into the roots. However, Van den Honert /21/, Kozlowski /8/, and Slatyer /7/ have shown that the vapor phase transfer of water, called transpiration, from the interior of the leaf to the outside air is the rate limiting step which determines the rate of evaporation as long as the soil moisture is sufficient to satisfy the transpirational demand.

General Principles of Natural Evaporation

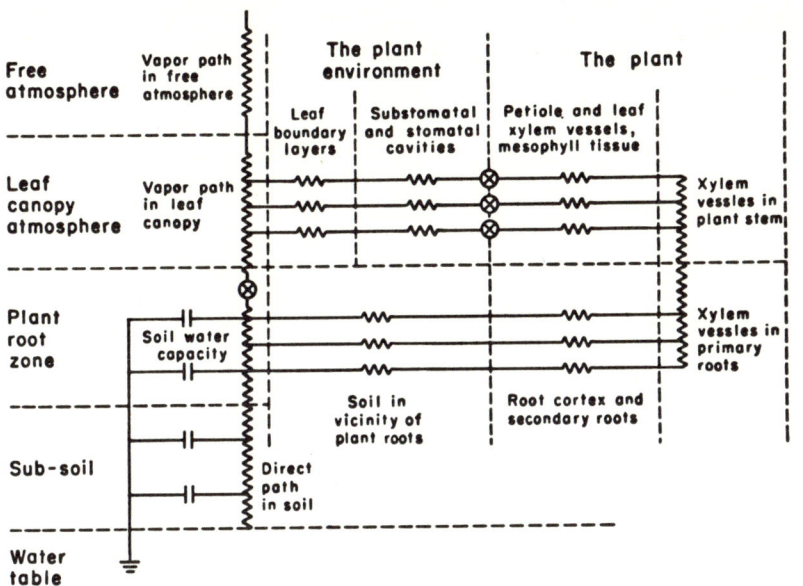

Figure 1

1. Electric Analogue of soil - plant - atmosphere system for water transport.

Water enters the leaf via the petiole and midrib and flows through the network of subsidiary veins and individual vessels until it finally moves through the cell walls and/or protoplasts to the evaporating surfaces located within cell walls bordering on the substomatal cavities. There is much uncertainty and few qualitative data concerning the relative magnitude of the resistances in the several parts of this pathway. However, the fact that severance of the main veins has little effect on transpiration over appreciable periods and that a leaf exposed to high insolation will wilt while a shaded adjacent leaf remains turgid, suggests that the resistances of the veins are relatively small compared to the resistance in the wall to protoplasts in the pathway.

Figure 2 shows the cross-sectional view of a typical dicot leaf. Since the outer wall of the epidermal cells are covered with a waxy substance, called, cutin, which is almost impervious to the transport of water, the epidermis and cuticular layer constitute a highly effective barrier to evaporation. Consequently, most evaporation occurs through the stomata, tiny openings in the leaf /22/. The size of a stoma is determined by the shape and relative position of the two kidney shaped guard cells surrounding it /23/. A typical stoma is about 15 micron long, with width up to 5 micron. Foliar stomates occupy between 1 and 3% of the total leaf area in most green leaf plants.

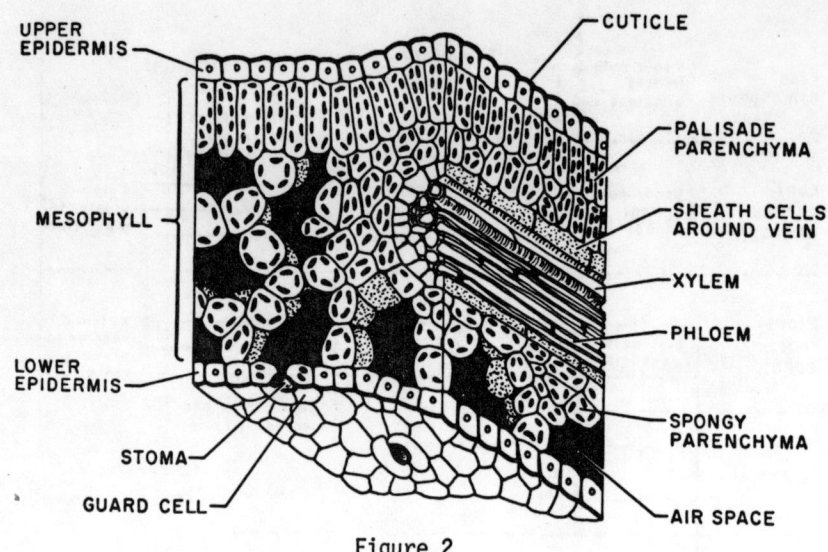

Figure 2
2. Cross-section of Dicot leaf.

Stomatal openings increase when the guard cells become more turgid, swell and bend outward. The degree of opening of the pore is a function of the turgor pressure difference between guard cells and the surrounding epidermal cells. Numerous factors, both external and internal, are known to cause stomatal responses /23/: light, carbon dioxide concentration, water stress, evaporative demand, and temperature. For example, an increase in light intensity causes an opening of the stoma whereas a decrease in the available water supply closes the stomata. Generally the balance between photoactive opening and hydroactive closing determines the size of the stoma opening. Recent evidence indicates that the mechanism underlying stomatal opening in light is active uptake of K and Cl ions in osmotic amounts by guard cells whereas the efflux of K and Cl ions from guard cells causes closing /24, 25/. The free energy for the active transport of these ions is provided by the hydrolysis of Andenocine 5' - Triphosphate [ATP] which is the direct product of photophosphorylation /26, 27, 28/.

Evidence about the effect of temperature on stomatal movement is conflicting /23/. It appears that within the range of 10 - 25°C, temperature has little effect on stomatal apperture. Hôwever, temperatures above 30°C appear to cause closure of the apperture.

3B DIFFUSION RESISTANCE

There are two general methods of approach for studying the diffusion resistance of leaves. One approach measures the rate of evaporation from a small area of the leaf by placing a vapor cup over it and observing the increase in relative humidity as a function of time. From this information one can then deduce the rate of evaporation from the leaf surface into the vapor cup. Excellent reviews of the equipment for this methodology are given by Stigter et al. /29/, Slatyer and Jarvis /30/, and Jarvis and Slatyer /31/. The latter developed also an ingenious extension of the vapor-cup method which permits the evaluation of the vapor diffusion resistance of the stoma as well as the cell wall resistance at the water-air interface which, as will be shown, can be an important parameter/ 32/.

The other approach is to measure the number and sizes of the stomata in the leaf and calculate by means of the diffusion equation and an appropriate analytical model the diffusion resistance of the leaf or the leaves on a plant. If the plant is weighed periodically, the rate of transpiration can be calculated and if the size and the number of leaves are known, the measured evaporation rate can be compared with the rate predicted analytically.

If, as the evidence indicates, the transpiration process from the leaves occurs mainly through the stomatal opening and is the rate limiting process which determines the rate of evaporation, it should in principle be possible to calculate rates of evaporation from a knowledge of the environmental conditions and stomatal sizes, distribution, and shapes. This approach towards an analysis of evapotranspiration on a microscopic level was initiated by the pioneer study of Brown and Escombe /29/ at the turn of the century and is still continuing. To illustrate the method of approach followed in these analyses, Fig. 3 shows schematically the resistances of a leaf in the evapotranspiration process. There are two parallel diffusion paths in the leaf: the cuticular and the stomatal. But since the resistance in the cuticular path is 10 to 20 times larger than in the stomatal path, its effect on the internal resistance is negligible. Thus, in constructing an analytical model one needs to consider only the stomatal path.

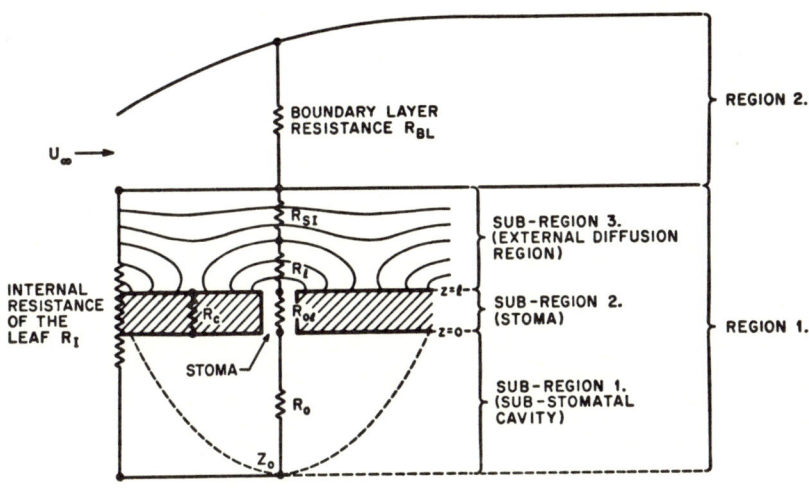

Figure 3

3. Schematic sketch showing resistances in a leaf.

3C AN ANALYTICAL MODEL FOR PREDICTING TRANSPIRATION FROM A LEAF

The rate of evapotranspiration from a leaf depends primarily on the number, the sizes, and shapes of the stoma in the leaf and the difference between the partial pressure of the water vapor in the sub-stomatal cavities and in the atmosphere. If one idealizes the process for a single stoma by dividing the water vapor diffusion path into three distinct regimes, the inlet to the stoma, the stomatal passage (or stoma), and the exit from the stoma, with diffusion resistances R_o, R_ℓ, and R_ℓ, respectively, the total resistance of a single stoma to diffusion is (see Fig. 3)

$$R_t = R_o + R_\ell + R_\ell \tag{19}$$

Photographs indicate that the flow cross-section of a typical stoma is approximately elliptical. Assuming that any cross-section of a stoma is elliptical, but a function of distance from the entrance, the total resistance of a single stoma can be expressed in the form

$$R_t = \frac{K(\varepsilon_o)}{\mathcal{D}\, 2\pi a_o} + \int_0^\ell \frac{dz}{\mathcal{D}_N \pi a(z) b(z)} + \frac{K(\varepsilon_\ell)}{\mathcal{D}\, 2\pi a_\ell} \tag{20}$$

where a_o is the semi-major axis of the stoma inlet opening,

$\mathcal{D}$ is the diffusivity of the water vapor in air,
ℓ is the length of the stoma from inlet to outlet,
a_e is the semi-major axis of the pore exit
b_e is the semi-minor axis of the pore exit, and
$K(\varepsilon)$ is the elliptic integral of the first kind whose argument is the eccentricity of the ellipse $\varepsilon = (a^2 - b^2)/a^2$.

Simplified assumptions for the shape of the stomatal cross-section can be used to evaluate R_s as shown below:

1. Uniform elliptic cross-section (if circular: $a=b$, $\varepsilon=0$, $K(\varepsilon)=1$)

$$R_t = \frac{\ell}{\mathcal{D}\,\pi a b} + \frac{K(\varepsilon)}{\mathcal{D}\,\pi a}$$

2. An elongated slit of length a and width b where $\varepsilon \to \infty$ and $K(\varepsilon) = \ln(4a/b)$

$$R_t = \frac{\ell}{\mathcal{D}\, ab} + \frac{\ln(4a/b)}{\mathcal{D}\,\pi a}$$

For a leaf having a stomatal density of N stomates per square centimeter, the average internal diffusion resistance $\bar{R}_I$, is obtained by summing the internal resistance of the individual stoma in parallel. Assuming that all stomata are of the same size and shape

$$\overline{R}_I = \frac{1}{\mathfrak{D}N} \left[\frac{K(\varepsilon_0)}{2\pi a_0} + \int_0^\ell \frac{\ell \, dz}{A(z)} + \frac{K(\varepsilon_\ell)}{2\pi a_\ell} \right] \quad (21)$$

If the stomata do not have the same sizes and shapes and a sample of n stomates is available, the statistically averaged resistance is the sum of the inverse of the average of the n individual stomatal conductances, or

$$\overline{R}_I = \frac{1}{N} \overline{\sum_{i=z}^{n} \frac{1}{(R_t)_i}} \quad (22)$$

In addition to these three geometrical stomatal resistances, there can be an additional resistance due to the interaction of adjacent stomata, the interference resistance R_{si}, and the cell wall resistance, R_W, which can reduce the vapor pressure in the sub-stomatal cavity.

Table 3 summarizes the most significant models used to predict the internal stomatal resistance from stomatal geometry and density. Ever since the pioneer study of Brown and Escombe in 1900 /33/ investigators have been trying to improve the basic model in an effort to explain why the predicted stomatal resistance was smaller than many measured values. Among the most important contributions are those of Penman and Schofield /34/ and Bange /35/ who recognized the importance of the shape of the pore passage and introduced realistic shapes into the model, and that of Milthorpe and Penman /36/ who found that the diffusion process of water vapor through wheat stomata was not pure molecular diffusion because the dimensions of a stoma are of the same order of magnitude as the mean fill path of the water molecules. To account for this phenomenon they introduced an intermediate zone diffusivity, $\mathfrak{D}_N$, in the model, whose value can be calculated from relations developed by Carman in 1956 /37/ from work with porous catalysts. Figure 4 gives the ratio of the intermediate zone diffusivity to the bulk diffusivity as a function of the mean hydraulic radius of the passage for various ratios of hydraulic radius to pore length as predicted by Carman. In 1970 Parlange and Waggoner /38/ extended the model to realistic elliptical shapes and included also the interference effect between adjacent stomata which had originally been mentioned qualitatively by Brown and Escombe, and been placed on an analytical basis by Keller and Stein /39/ for equally spaced cylindrical pores diffusing into a stagnant environment. A similar analysis of the interference effect for diffusion into a moving boundary layer has recently been completed by Weinberg and Kreith /40/. It showed that interference effects become appreciable when the ratio of inter-stomatal distance to hydraulic radius of the flow cross-section is of the order of ten or less. Figure 5 shows the ratio of the Sherwood Number for a plate with finite sources corresponding to the stomatal opening with interference divided by the Sherwood Number for a plate at uniform potential times the percent porosity (the ratio of total stomatal outlet

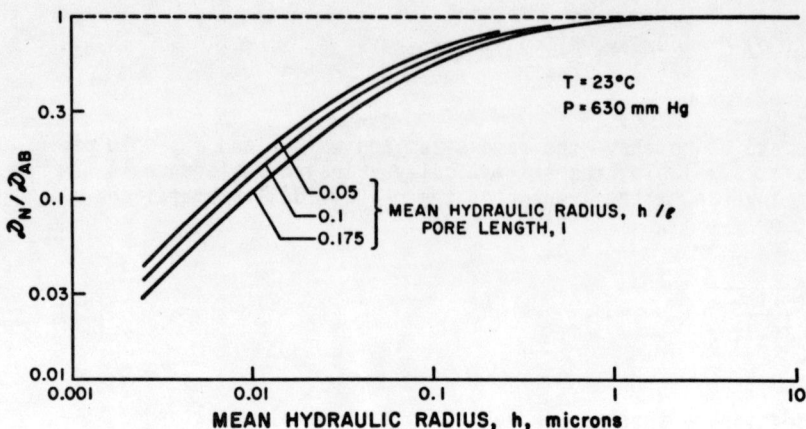

Figure 4

4. Ratio of the intermediate zone diffusivity to bulk diffusivity as a function of the mean hydraulic radius.

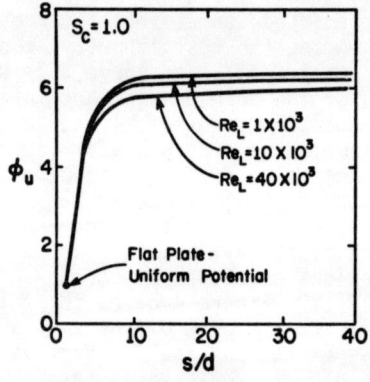

Figure 5

5. Effect of stomatal interaction on the Sherwood Number in laminar flow over a leaf.

area to the plate surface area) as a function of the ratio of stomatal hydraulic radius to inter-stomatal distance. It is apparent that interference effects can increase the boundary layer resistance appreciably at small inter-stomatal distance.

After passing through the stoma the water vapor must diffuse through the boundary layer outside the leaf before reaching the atmosphere. The resistance offered by the boundary layer can easily be measured experimentally by replacing the leaf with a thoroughly wetted piece of blotting paper cut to exactly the same shape as the leaf and measuring the change in weight during a specified time. The resistance of the boundary layer will depend on the flow conditions, i.e., whether the flow is laminar or turbulent and free or forced convection. Its value can be calculated approximately by a method proposed by Kreith et al. /41/. The results are summarized in Table 4. The effect of flapping has been studied by Parlange and Waggoner /42/.

The microscopic evaporation model has been tested under carefully controlled conditions in an ecological wind tunnel by Cannon /43/ and the correspondence between experimental results, including interference, from a simulated leaf and experimental data are shown in Figure 6. Since in the perforated plates used in Cannon's work to simulate a leaf no cell wall resistance was present, his work proved the proposed model is valid under these conditions.

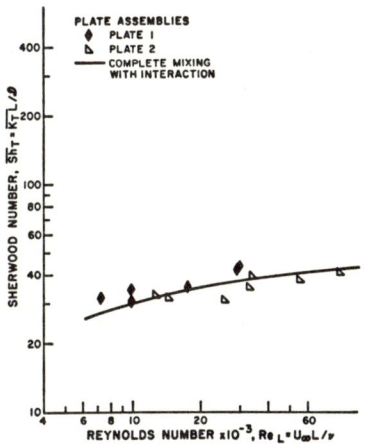

Figure 6

6. Experimental results for evapotranspiration from a perforated plate resembling stomatal geometry.

Subsequent work by Taori /44/ with living plants in the same ecological wind tunnel used by Cannon and in vivo showed good agreement between theory and experiments under low irradiation and free convection, two environmental variables conducive to eliminate or reduce cell wall resistance to a negligibly small value. In order to achieve correspondence between theory and experiments in living plants Taori determined the actual shapes of stoma from microtone sections and the sizes and density of the stomata from leaf impressions. He then used statistical methods to predict the diffusion resistance by means by eq 22. His results are compared with those of previous investigators in Table 3. The analytical predictions of Taori's statistical model are compared with experimental results in Fig. 7 where the ratio of the predicted to the measured diffusion resistance is plotted against the experimentally measured resistance.

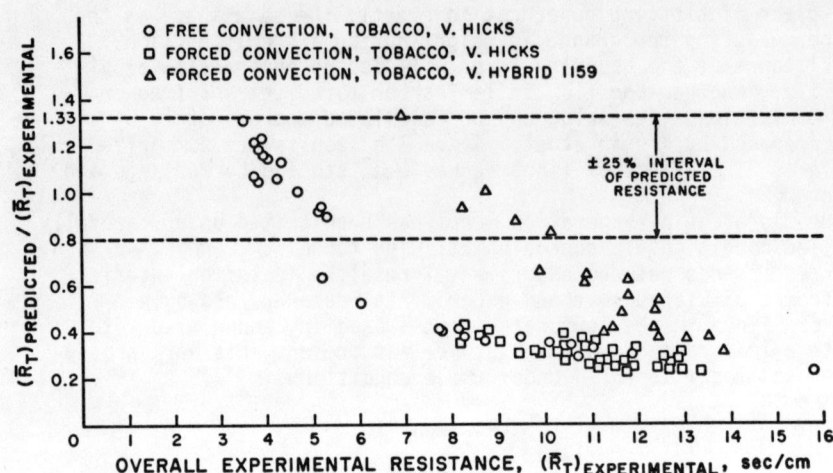

Figure 7

7. Comparison of experimental diffusion resistance with analytical prediction for evaporation from tobacco leaves.

The results show that the analysis is satisfactory for free as well as forced convection conditions as long as the evaporation rate and the cell wall resistance are not too large. A more complete discussion of these conditions, as well as of the analysis, is forthcoming /45/.

3D CONCLUSION

In setting up an evaporation model, the choice of the location for the liquid-vapor interface is important. Water is transported in the liquid phase to the evaporation sites located within cell walls bordering on sub-stomatal cavities, but there are few quantitative data on the diffusion resistance in that region. This causes no problems, provided the sub-stomatal cavities are filled with water vapor saturated at the leaf temperature. Then, the internal resistance resides entirely in the stomata and a knowledge of their sizes and density distribution suffices to calculate the evaporation rate. There is, however, considerable evidence that the liquid-vapor interface is located within cell walls which are covered by a thin waxy layer that resembles a permeable membrane. Electron photomicrographs show this layer extending around the guard cells into the sub-stomatal cavities and on walls bordering intercellular spaces. In a wheat leaf this layer is about 0.06μ and in some dicotyledons it may be as thick as one micron. The thickness of the layer over the cell walls of the sub-stomatal cavity is generally much less than on the guard cells. One can only speculate how the diffusion resistance of cell walls varies, though indirect evidence /31/ suggests that it increases with transpiration rate. Values between 0.22 sec/cm and 1.66 sec/cm have been reported. It is therefore evident that correspondence between the water loss in nature and the evaporation predicted from a diffusion model based on physical parameters of the stomata, the leaf, and the external evironment can only be achieved under conditions when cell wall resistance is negligible, unless an analytical method can be found to predict and incorporate this resistance into the model.

Table 1   Input data used with the evapotranspiration model

Hour (Apparent Solar Time)	$Q(\text{cal cm}^{-2} \text{hr}^{-1})$	$u(\text{cm sec}^{-1})$	$T(°C)$	$e(\text{mb})$	$E_0 (\text{mm hr}^{-1})$
2330-0030	0	313	24.4	6.88	0.15
0030-0130	0	291	23.4	6.76	0.13
0130-0230	0	291	22.5	6.81	0.13
0230-0330	0	291	21.6	6.84	0.13
0330-0430	0.1	291	20.8	6.88	0.13
0430-0530	2.6	313	20.6	7.16	0.14
0530-0630	13.2	313	21.7	7.40	0.15
0630-0730	30.8	335	24.2	7.55	0.24
0730-0830	48.6	358	26.7	7.53	0.42
0830-0930	64.0	358	28.9	7.37	0.61
0930-1030	75.4	380	30.8	6.88	0.84
1030-1130	82.5	425	32.2	6.25	1.05
1130-1230	85.2	469	33.4	5.66	1.16
1230-1330	82.5	514	34.4	5.44	1.20
1330-1430	75.4	536	35.0	5.62	1.13
1430-1530	64.0	559	35.0	5.34	1.09
1530-1630	48.6	559	35.0	5.06	0.90
1630-1730	30.8	604	34.7	5.53	0.78
1730-1830	13.2	626	33.6	5.98	0.61
1830-1930	2.6	514	32.0	6.42	0.35
1930-2030	0.1	425	30.2	6.87	0.23
2030-2130	0	380	28.6	6.85	0.19
2130-2230	0	335	27.2	6.85	0.16
2230-2330	0	313	25.8	6.98	0.14

Table 2   Summary of results

$d(\text{cm})$	5		10		20		40		80	
$w_m(\text{mm})$	12.5		25		50		100		200	
day no.	E	$\bar{T}_s$	E	$\bar{T}_s$	E	$\bar{T}_s$	E	$\bar{T}_s$	E	$\bar{T}_s$
1	9.06	22.3	11.33	20.5	12.05	20.0	12.05	20.0	12.05	20.0
2	2.55	28.3	6.58	25.1	10.52	21.3	12.05	20.0	12.05	20.0
3	0.66	30.9	3.41	28.7	7.61	24.7	11.32	20.6	12.05	20.0
4	0.17	31.5	1.77	30.4	5.50	27.3	9.63	22.6	12.05	20.0
5	0.05	31.6	0.92	31.1	3.97	28.9	8.20	24.7	11.73	20.2
6	0.01	31.6	0.48	31.4	2.87	29.8	6.97	26.3	10.84	21.3
7	0	31.6	0.25	31.5	2.08	30.4	5.93	27.4	10.00	22.6
8	0	31.7	0.13	31.6	1.50	30.8	5.05	28.2	9.23	23.8
9	0	31.7	0.07	31.6	1.08	31.0	4.30	28.9	8.51	24.8
10	0	31.7	0.03	31.6	0.78	31.2	3.66	29.4	7.85	25.6
1-10	12.50		24.97		47.96		79.16		106.36	

E in mm day^{-1}; $\bar{T}_s$ in °C

Table 3

COMPARISON OF SUBSTOMATAL CAVITY PORE AND
EXIT RESISTANCE CALCULATED BY VARIOUS MODELS
(Courtesy of Dr. A. Taori, private communication)

Description of Model	Inlet Resistance R_i, sec/cm	Stoma Resistance R_s, sec/cm	Exit Resistance R_e, sec/cm
1. Elliptic pore cross-section with microtome measured, statistically correlated pore shape (Taori Andersen, & Krieth, 1974) /45/	0.53	4.42	0.31
2. Elongated slit-like pore cross-section with microtome measured, statistically correlated pore shape (Taori, 1973) /44/	0.52	4.42	0.31
3. Elliptic pore cross-section with exponential pore shape (Parlange & Waggoner, 1970) /38/	0.31	3.82	0.31
4. Rectangular slit pore cross-section with exponential pore shape (Milthorpe and Penman, 1967) /36/	0.38	1.84	0.25
5. Equivalent circle pore cross-section with unsymmetric hourglass pore shape (Bange, 1953) /35/	0.34	2.47	0.25
6. Elliptic pore cross-section with hyperbolic pore shape (Penman & Schofield, 1951) /34/	0.19	2.70	0.19
7. Circular pore cross-section with cylindrical pore shape (Brown & Excombe, 1900) /33/	0.34	1.73	0.34

Table 4 Summary of Analytical Relations for Calculating Diffusion Resistances of Leaves /41,43/44/

I. Forced Convection Laminar Flow
($Re_L < 1.5 \times 10^4$)

$$\overline{R}_T = \overline{R}_I + 1.7(\frac{L}{}) Re_L^{-1/2}$$

$$L = \frac{\int_0^D w(y)\, dy}{\int_0^D w(y)^{1/2}\, dy}$$

II. Force Convection Turbulent Flow
($Re_L > 1.5 \times 10^4$)

$$\overline{R}_T = \overline{R}_I + 31.(\frac{L}{}) Re_L^{-4/5}$$

$$L = \int_0^D w(y)\,dy / \int_0^D (wy)^{0.8}\,dy$$

III. Free Convection Laminar Flow
($Gr_L/Re_L^{2.5}) > 1.0$
$Gr_L < 1.3 \times 10^7$

$$\overline{R}_T = \overline{R}_I + K(L/) Gr_L^{1/4}$$

$$\{K \begin{array}{l} = 2.0 \text{ for upper surface} \\ = 4.0 \text{ for lower surface} \end{array}$$

$$L = \int_0^D w(y)\,dy / \int_0^D w(y)^{0.75}\,dy$$

IV. Free Convection Turbulent Flow
$Gr_L > 1.3 \times 10^7$
$(Gr_L/Re_L^{2.5}) > 1.0$

$$\overline{R}_T = \overline{R}_I + 8.2(L/) Gr_L^{-1/3}$$

(for upper surface, no date for lower surface)

$$L = \int_0^D w(y)\,dy / \int_0^D w(y)^{2/3}\,dy^3$$

$Gr_L = \rho_\infty L^3 g(\rho_0 - \rho_\infty)/\mu_\infty^2$ (dimensionless)

$Re_L = V\rho L/\mu$ (dimensionless)

$\overline{R}_I$ see Eq. 21 in text

BIBLIOGRAPHY

1. Monteith, J. L., 1965 "Evaporation and Environment", Symp. Soc. Exptl. Biol., 29: 205-234.
2. Webb, E. K., 1965. Aerial microclimate. Met. Monogr. 6: 27-58.
3. Zelitch, I., 1967. Control of leaf stomata - their role in transpiration and photosynthesis. Am. Sci. 55: 472-486.
4. Lemon, E. R., 1963. Energy and water balance of plant communities. In L. T. Evans (ed.): Environmental control of plant growth. Academic Press, New York: 55-78.
5. Lemon, E., 1969. Gaseous exchange in crop stands. In J.D. Eastin, F.A. Haskin, C.Y. Sullivan, and C.H.M. van Bavel (eds.): Physiological aspects of crop yield. Am. Soc. Agron., Madison: 117-137.

6. Denmead, O.T., 1971. "Relative Significance of Soil and Plant Evaporation in Estimating Evapotranspiration," Proc. UNESCO Symp. Plant Response to Climatic Factors.
7. Slatyer, R.O., 1967. *Plant-Water Relationships*, Academic Press, New York.
8. Kozlowski, T.T., 1968. *Water deficit and Plant Growth Development, Control and Measurement.* Academic Press, New York.
9. Konstantinov, A.R., 1966. *Evaporation in Nature.* Israel Program for Scientific Translations, Jerusalem.
10. Sellers, W.D., 1965. *Physical climatology.* Univ. of Chicago Press, Chicago.
11. Sasamori, T., 1970. "A numerical study of atmospheric and soil boundary layers," J. Atmos. Sci. 27(8): 1122-1137.
12. Morgan, D.L., W.O. Pruitt, and F.J. Lourence, 1971. "Estimation of atmospheric radiation," J. Appl. Meteor. 10(3): 463-468.
13. Brunt, D., 1932. "Notes on radiation in the atmosphere," Quart. J. Roy. Meteor. Soc. 58(247): 389-420.
14. Sellers, W.D., 1967. "An investigation of heat transfer from bare soil," Final Report, Grant No. DA-AMC-28-043-66-G27, Institute of Atmospheric Physics, Univ. of Arizona, Tucson, Arizona, April, 1967.
15. Vehrencamp, J.E., 1953. "Experimental investigation of heat transfer at an air-earth interface," Trans. Amer. Geophys. Union. 34(1): 22-30.
16. van Bavel, C.H.M., L.J. Fritschen, and R.J. Reginato, 1963. "Surface energy balance in arid lands agriculture 1960-61," Production Research Report No. 76, U.S. Water Conservation Laboratory, Phoenix, Arizona.
17. van Bavel, C.H.M. and L.J. Fritschen, 1964. "Energy balance studies over sudangrass, 1962," Interim Report, DA Task 1-A-0-11001-B-021-08, U.S. Water Conservation Laboratory, Phoenix, Arizona.
18. van Bavel, C.H.M., 1967. Changes in canopy resistance to water loss from alfalfa induced by soil water depletion. Agricult. Meteor. 4(2): 165-176.
19. Foster, K.E., 1972. Mathematical analysis of soil temperatures in an arid region. Ph.D. Dissertation, Dept. of Watershed Management, Univ. of Arizona, Tucson, Ariz.
20. Cowan, I.R., 1968. Mass, heat, and momentum exchange between stands of plants and their atmospheric environment. Quart. J. Roy. Meteor. Soc. 94: 523-544.
21. van den Honert, T.H., 1948. "Water Transport in Plants as a Catenary Process," *Dist. Far. Soc.*, 3, 146-153.
22. Meidner, H., 1965. "Stomatal Control of Transpirational Water Loss," *Symp. Soc. Exptl. Biol.*, 19, 185-188.
23. Meidner, H., and T.A. Mansfield, 1968. *Physiology of Stomata*, McGraw-Hill Book Co., England.
24. Raschke, K., and M.P. Fellows, 1971. "Stomatal Movement in *Zea Mays*: Shuttle of Potassium and Chloride Ions Between Guard Cells and Subsidiary Cells," *Planta* (Berl.) 101, 296-316.
25. Pallaghy, C.K., 1971. "Stomatal Movement and Potassium Transport in Epidermal Strips of *Zea Mays*: The Effect of Co_2," *Planta* (Berl.), 101, 287-295.
26. Pallas, J.E.,Jr., and R.A. Dilley, 1972. "Photophosphorylation

can Provide Sufficient Adenosine 5'-Triphosphate to Drive K+ Movements During Stomatal Opening," <u>Plant Physiol.</u>, 49, 649-650.
27. Thomas, D.A., 1970. "The Regulation of Stomatal Aperture in Tobacco Leaf Epidermal Strips, I. The Effect of Ions," <u>Aust. J. Biol. Sci.</u>, 23: 961-79.
28. Thomas, D.A., 1971. "III. The Effect of AIP," <u>Aust. J. Biol. Sci.</u>, 24: 698-707.
29. Stigter, C.J., J. Birnil, and B. Lammers, 1973. Leaf diffusion resistance to water vapour and its direct measurement, II. Design, calibration, and pertinent theory needed. Landbouwhogeschool Wageningen 73-15.
30. Slatyer, R.O. and P.G. Jarvis, 1966. A gaseous diffusion parameter for continuous measurement of diffusive resistance of leaves. Science. 151: 574-576.
31. Jarvis, P.G. and R.O. Slatyer, 1970. The role of mesophyll cell wall in leaf transpiration. <u>Planta</u> Berl. 90: 303-322.
32. Scott, F.M., 1966. Cell wall surface of higher plants. Nature 210: 1015-1017.
33. Brown, H.T., and F. Escombe, 1900. Static diffusion of gases and liquids in relation to the assimilation of carbon and translocation in plants. Phil. Trans. Roy. Soc., London, Ser. B. Biol. Sci. 193: 223-291.
34. Penman, H.L. and R.K. Schofield, 1951. Some physical aspects of assimilation and transpiration. Symp. Soc. Exp. Biol. 5: 115-129.
35. Bange, G.G.J., 1953. "On the Quantitative Explanation of Stomatal transpiration," Acta. Bot. Neer. 2(3): 255-297.
36. Milthorpe, F.L. and H.L. Penman, 1967. "The diffusive conductivity of stomata of wheat leaves," J. of Exp. Bot. 18(56): 422-457.
37. Carman, P. C., 1956. Flow of gases through porous media. Butterworth Scientific Publication, London.
38. Parlange, Jean-Yves and Paul E. Waggoner, 1970 "Stomatal Dimensions and Resistance to Diffusion", Plant Physiol., 46, 337-342.
39. Keller, K. H. and T. R. Stein, 1967. A two dimensional analysis of porous membrane transport. Math. Bio. Sci. 1:421-437.
40. Weinberg, B., and F. Kreith, 1974. Evaporation by convection from a plate with finite sources. To be presented at the 5th International Heat Transfer Conference, Tokyo, Japan, Sept. 1974.
41. Parkhurst, D. F., P. R. Duncan, D. M. Gates, and F. Kreith, 1968 "Wind-Tunnel Modelling of Convection of Heat Between Air and Broad Leaves of Plants", Agr. Meteorol., 5, 33-47.
42. Parlange, Jean-Yves, P. E. Waggoner, Gary H. Heichel, 1971 "Boundary Layer Resistance and Temperature Distribution on Still and Flapping Leaves", Plant Physiol., 48, 437-442.
43. Kreith, F., J. Cannon, and D. Naot, 1974. A model study of transpiration from plant leaves. Paper presented at the AIAA-ASME Thermophysic and Heat Transfer Conference, Boston.
44. Taori, A., 1973. Mathematical modelling of transpiration from leaves. Ph.D. Thesis, Univ. of Colorado, Boulder, Colorado.
45. Taori, A., J. Anderson, and F. Kreith, 1974 "Evaporation from Tobacco Leaves - Comparison between a Statistical Model and Experiments", to be published.

METHODS OF OBSERVATION OF HEAT AND MASS TRANSFER IN THE LOWER ATMOSPHERE AND IN PLANT CANOPIES

A. PERRIER

INRA Station Centrale de Bioclimatologie Agricole, Versailles

In order to develop an understanding of transfer processes in canopies, general descriptions of vertical heat and mass transfer relations are formulated. These relations draw attention to appropriate and measurable properties of the canopy, the atmosphere and the soil surface. That could be used predictively in canopy modelling. Conventional methods for obtaining data on these parameters are discussed in terms of their validity and range of application. A number of problems that requires more investigation are mentioned.

GENERAL SYMBOLS:

ΔX, finite variation of a value X for an interval Δz
$\bar{x}$, x', Mean value of x, and its fluctuation, respectively
$X'(z)$, $X''(z)$, first and second derivative functions of $X(z)$ according to z
dx, Differential operator on quantity x

Subscripts

R, Refers to radiation transfer
S, Refers to sensible heat
L,   "   " Latent heat
M,   "   " Momentum exchange
H, Refers to Enthalpy flux
C,   "   " CO_2 exchange
SC, "   " conduclive heat flux in the soil
a,   "   " air characteristics
o,   "   " surface level
v, V "   " vegetation characteristics

LIST OF SYMBOLS

t, time (s)
z, height (m)
z_r, h, reference level and height of the canopy (m)
T, T_d, T_w, temperature, dew point, and wet bulb temperature of the air (K)
q, Concentration of water vapour in atmosphere (kg m^{-3})
C, Concentration of CO_2 in atmosphere (kg m^{-3})
$\phi_x(z)$, Flux density for the entity x (x m^{-2} s^{-1})
$\tau(z)$, Momentum flux density (kg m^{-1} s^{-2})
u, w, v, three velocity components (horizontal, vertical, horizontal) (m s^{-1})
u_*, Friction velocity for the canopy (m s^{-1})
τ_o, shearing stress for the canopy (kg m^{-1} s^{-2})
d_o, displacement height of the zero plane (m)
ZO, ZO_o, roughness parameter of the canopy and soil surface (m)
l, mixing length (m)
$L(z)$, Length scale function (m)
$Z(z)$, particular function of wind speed in a canopy
k, Kármán's constant (~0.4)
i, Turbulent intensity
R, R_u, R_w, Auto-correlation coefficients
ζ, Dimensionless height (z/L) where L is the Obukhov length /1/
$K_x(z)$, Turbulent diffusivity for entity x (m^2 s^{-1})
h_x, Exchange coefficient for entity x (ms^{-1})
D_x, Drag coefficient for entity x
$f(z)$, Surface density of the vegetation (m^{2}m^{-3}) (one face only)
$v(z)$, Spatial density of the vegetation (m^{3}m^{-3})
l_o, Characteristic length of a single leaf (m)
$\alpha(z)$, function of exchange parameters of a single leaf
$E(z)$, Mean irradiance at a level z for active photosynthetic wave bands (wm^{-2})
$Y(z)$, Air deficit of saturation function (°C)
Σr, sum of resistances (sm^{-1})
r, r_s, r_m, r_c, resistance and resistances of stomates, mesophyll carboxylation (sm^{-1})
b, Maximum efficiency of light energy conversion (mol CO_2 fixed/Einstein abs.)
$\Phi_p(z)$, Photosynthetic function for $C(z_r)$
ρ_x, Density of phase x (kg m^{-3})
cp_x, Specific heat at constant pressure of phase x (J kg^{-1} °C^{-1})
ν, Kinematic viscosity of air (m^2 s^{-1})
γ, Psychrometric constant (P°C^{-1})
S, Slope of the saturation curve of water vapour in air (P °C^{-1})
$P(T)$, Water vapour saturation function (P)
L, Latent heat of vaporisation of water (J kg^{-1})
h_1, Energy equivalent of CO_2 fixation (J kg^{-1})
R, universal gas constant (J K^{-1} mol^{-1})
M, Molar mass of H_2O (kg mol^{-1})
Ψ_o, Water potential at soil surface (J kg^{-1})
P_r, Prandtl number (ratio between kinematic viscosity and heat diffusivity in air)
Sc_x, Schmidt number for entity x (as Pr but with Mass diffusivity)

INTRODUCTION

Measurement of aspects of heat and mass transfer in a well-defined system should be performed not only to obtain illustrative data and to understand the system, but also to predict the effects on it of variations in its principal parameters.

Most biological systems involve complex mixtures of gas and liquid (water) within a complicated solid framework. These are very difficult situations to describe, but it may be possible to simplify them by careful selection of a restricted number of important, measurable, physical characteristics: physical properties of each phase, exchange properties of the interfaces, their spatial distribution, and appropriate boundary conditions. The choice of the boundaries depends on the particular application, but generally it is desirable to set them at such a level that they are not influenced by the system but depend essentially on external conditions.

In order to relate the exchanges of heat or mass to biological processes, it is necessary to choose, if possible, those parameters which are influenced only by the structure of the system and not by interactions between the system and the external environment. This choice is, however, always difficult not only because parameters may provide a valid description over a limited range of variation only, but also because it may not be strictly independent of the external environment. For example, the definition of resistance used to describe gas flow through the surface of a leaf becomes unsatisfactory when it is applied to both leaf surfaces because their porosities and roughnesses differ /2/. The notion of surface resistance is also quite invalid when considering evaporation of a drying soil (in this case r is linked not only to the water potential at the soil surface, but also to the flux itself /3/. Finally, the concept of crop resistance is also erroneous when it is applied to a whole canopy to predict evapotranspiration because it depends on wind speed, net radiation, canopy structure etc. /4/.

Each parameter chosen must thus be clearly defined as to its range of validity and its independence of other parameters: the latter must be checked continiously by experimental measurements.

The particular case to be considered is that of a plant community which extracts solutes and water from the soil and exchanges heat and mass with air. The problem analysed here concerns only exchange between the surfaces (soil and plant) and the air. With well-established conditions, and for short times, the boundaries of the system will be:
- The soil surface, at which the fluxes are small and relatively constant for a short time;
- A reference level in the air (z_r) above which the variation of air properties can be expected to be determined not by the system itself, but by general surrounding conditions.

The boundary conditions are then practically independent of the system and are nearly constant if we choose a suitable time interval (e.g. 10 min.). This interval is defined with reference to air turbulence, the most variable parameter of the system, and allows us to work with mean values obtained by integration over periods for which the mean value can be considered as a constant.

These particular conditions generally permit simple analyses of basic processes of transfer and the definition of the most important parameters.

The problem in this form is, nevertheless, over-simplified. The continuous diurnal and seasonal changes in physical parameters, coupled with biological changes, lead in reality to a permanently unsteady state which might approach a steady state only during very stable conditions. Furthermore, the time lag in many biological processes prolongs this unsteady state, even though the atmospheric perturbations such as cloudiness or gustiness, may pass quickly.

All measurements and methods must be adapted to cope with these complex conditions.

GENERAL RELATIONS CONCERNING HEAT AND MASS TRANSFER IN THE LOWER ATMOSPHERE AND IN PLANT COMMUNITIES

In the system considered here, involving exchanges with air only, the relations are examined generally at two levels:
- first, at any level, z, in the air between the soil surface and the reference level z_r.
- second, at soil and vegetation surfaces. Vegetation surfaces are generally restricted to leaves but must include stems, inflorescences and so on, in some cases.

1. BASIC RELATIONS

1.1 Energy balance

The main relation relies on the conservation of energy. The balance of any layer $(o - z)$ of the system is given by the algebraic sum of all energy fluxes, incoming or outgoing:

$$\phi_R(z) + \phi_{SC}(0) + \phi_S(z) + \phi_L(z) + \lambda [\phi_C(z) - \phi_C(0)] = \int_o^z \rho_v(z)\, cp_v(z)\, v(z)\, [dT_v(z)/dt]\, dz \quad (1)$$

The flux densities are taken to be positive, if they represent a gain to the system (layer $o - z$) and vice versa. This expression may be used inside the canopy for a thin layer $(z, z + \Delta z)$, viz:

$$\Delta\phi_R(z) + \Delta\phi_S(z) + \Delta\phi_L(z) + \lambda\Delta\phi_C(z) = \rho_v(z)\, cp_v(z)\, v(z)\, [dT_v(z)/dt]\, \Delta z \quad (2)$$

The differentiation with respect to height of equation (1) or the limit of equation (2) $(\Delta z \to o)$ describes the source and sink strengths which may be related to leaf surfaces by the relation:

$$-\phi_R'(z) = \phi_S'(z) + \phi_L'(z) + \lambda\phi_C'(z) = 2f(z)\, [\phi_{SV}(z) + \phi_{LV}(z) + \lambda\phi_{CV}(z)] \quad (3)$$

Generally, the storage term is negligible in steady conditions, e.g. less than 1% of the net radiation in a corn canopy; in dense vegetation, however, such as forest, and in transient conditions, this term may become significant, e.g. a variation of 100% in net radiation may result in a variation of 20% in the storage term during the first minutes of exchange. CO_2 fixation is also included as an energy term in the general equation and is often neglected. It may, however, be significant (0.5 to 2% for wheat at Versailles /5/ and as great as 8% for corn /6/ while values greater than 10% are given by Lemon /7/).

If we neglect these two terms, we should either demonstrate experimentally that they are unimportant, or include an appropriate correction.

1.2. Convection expressions

All convective transfer processes are strongly dependent on air movement and, in particular, on the turbulent characteristics of the air.

1.2.1. *Expression of fluxes in the air*

Given that the general movement of the air is essentially horizontal above all surfaces, including the canopy, changes in the slope of the ground and strong gustiness in the wind give a general three-dimensional structure to the problem. If, however, the effect of horizontal variation in wind direction is neglected, the general problem becomes two-dimensional.

Many authors have discussed the appropriate upwind fetch which leads to a one-dimensional problem in which the horizontal fluxes are negligible (Philip /8/ and Dyer /9, 10/ for heat and mass fluxes; Bradley /11/ for momentum fluxes). Generally, this fetch is of the order of some hundreds of metres, and, with a plant community of classical pattern, i.e. not strongly differentiated, one-dimensional analysis approaches reality.

In steady conditions the turbulence of the air may be characterised by mean values ($\bar{u}$, horizontal velocity: $\bar{w}$ vertical and $\bar{v}$ the third component) and by their fluctuations u', w', v'. The most common index used for turbulence measurements is the intensity of the turbulence /12/:

$$i = [\overline{(u')^2}]^{1/2} / \bar{u} \quad \text{or} \quad [\overline{(w')^2}]^{1/2} / \bar{u} \tag{4}$$

but a more complete description is given by the auto-correlation coefficient R (R_u or R_w) /12/:

$$R_u = \overline{u'(t)\, u'(t+\Delta t)} / [\overline{u'(t)^2}\; \overline{u'(t+\Delta t)^2}]^{1/2} \tag{5}$$

In a one-dimensional problem, where only vertical turbulent fluxes are considered, and with our sign convention for flux densities, we can write /13/:

$$\tau(z) = \rho_a \overline{u'w'} \tag{6}$$

$$\phi_S(z) = -\rho_a\, cp_a\, \overline{T'w'} \tag{7}$$

$$\phi_L(z) = -L\, \overline{q'w'} \tag{8}$$

$$\phi_C(z) = -\overline{C'w'} \tag{9}$$

and by analogy with the classical flux equation $\Phi_x = K(z)\,(dX/dz)$, where $K(z)$ is the turbulent diffusivity of the entity X /13/:

$$\tau(z) = -\rho_a\, K_M(z)\, [du(z)/dz] \tag{6a}$$

$$\phi_S(z) = \rho_a\, cp_a\, K_S(z)\, [dT(z)/dz] \tag{7a}$$

$$\phi_L(z) = L\, K_L(z)\, [dq(z)/dz] \tag{8a}$$

$$\phi_C(z) = K_C(z)\, [dC(z)/dz] \tag{9a}$$

It should be noted that equations 6 to 9 can also be written in vector form for a two-dimensional problem and two-dimensional case analogous to equations 6a to 9a can also be developed.

The most important fact is the similarity argument, which assumes that the transfer coefficients K_x are essentially identical. It is generally admitted /14,15/ that K_S is close to K_M in near neutral and moderately stable conditions /16,17/ but can differ markedly in strong instability where K_S/K_M may be as much as 2 or 2.5 /18/. Moreover, Businger et al /16/ have shown that, on average, the ratio K_S/K_M is closer to 1.3 than to unity. But generally the ratio K_C or K_L/K_S does not seem to differ very much from unity /19/. Similarity considerations can thus be used in a general relationship with the best approximation provided by the introduction of a function $g(\zeta)$ ($K_S = g(\zeta) K_M$) which depends on the stability parameters ζ /20/.

1.2.2. *Expression of fluxes at a surface level*

Generally, in the vicinity of small isolated obstacles, such as leaves or other small parts of plants, significant gradients are confined to a very thin boundary layer (0.1 to a few mm) /21/. The flux density of a particular entity from such surfaces is proportional both to the total gradient of the entity through the boundary layer and to an exchange coefficient; viz

$$\tau_{pv} = \rho_a h_M (u - u_o) \quad \text{with } u_o = 0 \tag{10}$$

$$\phi_{SV} = \rho_a c p_a h_S (T - T_o) \tag{11}$$

$$\phi_{LV} = L h_L (q - q_o) \tag{12}$$

$$\phi_{CV} = h_C (C - C_o) \tag{13}$$

All these exchange coefficients h_x (ms^{-1}) can be expressed as the product of a dimensionless coefficient and the wind speed,

$$h_x = D_x u \tag{14}$$

For h_M, the coefficient D_M is the classical drag coefficient for one face of a leaf /22/. Blasius /23/ and Pohlhausen /24/ showed that for a thin plate at zero incidence in laminar flow:

$$D_M = D_o (u l_o/\nu)^{0.5} \; ; \; D_S = D_M Pr^{-2/3} \; ; \; D_L = D_M Sc_L^{-2/3} \; ; \; D_C = D_M Sc_C^{-2/3} \tag{15}$$

The coefficient D_x differ only slightly between themselves at zero incidence, but when leaf incidence increases, the value of the total drag coefficient D_M may change /25,26/, although little effect may be observed in the mean values of heat and mass transfer coefficients (for the two leaf surfaces).

T_o, q_o and C_o are the values at the leaf surface, but q_o and C_o are unknown. The value q_i inside the stomatal cavity is thus chosen as a parameter, because q_i is the value at saturation for the temperature T_v (practically equal to T_o /26/) and C_o is inferred from carboxylation reactions and mesephyll resistances /24/. The relations 12 and 13 may then be rewritten:

$$\phi_{LV}(z) = L [h_L/(1 + h_L r_s)] [q(z) - q_i(z)] \tag{12a}$$

$$\phi_{CV} = [b E(z) C(z)] / [bE(z)\Sigma r + C(z)] \tag{13a}$$

in which b is the maximum efficiency of light energy conversion, $E(z)$ the irradiance for the active wave bands, and Σr the total resistance through chloroplast, mesophyll, epidermis, and air ($\Sigma r = r_c + r_m + r_s + 1/h_c$).

2. GLOBAL EQUATION FOR A CANOPY

To solve the problem of heat and mass transfer above and inside the canopy, it is necessary to combine the previous relations and to integrate them as functions of height.

2.1. Wind speed and diffusivity profiles

If, in a general way, a length scale which depends only on the geometry of the system, is introduced /28/ ($l = k\, L(z)$), the conventional expressions for the momentum flux (6a) and momentum balance inside the canopy (from 10), may then be rewritten:

$$\tau(z) = \rho_a k^2 L(z)^2 (du/dz)^2 \qquad (16)$$

$$d\tau(z)/dz = \rho_a D_M(z) u^2(z) f(z) \qquad (17)$$

The combination of 14 and 15 gives the general differential equation:

$$u'' u' + [L'(z)/L(z)](u')^2 - [D_M(z) f(z)/2 k^2 L^2(z)] u^2 = 0 \qquad (18)$$

or
$$Z' + [2/L(z)] Z^{3/2} - [D_M(z) f(z)/k^2] = 0 \qquad (18a)$$

with
$$Z(z) = [L(z)\, u'(z)/u(z)]^2 \qquad (19)$$

Introducing the boundary conditions $u(o) = 0$ and $u(z_r)$, the wind and diffusivity profiles are given by:

$$u(z) = u(z_r)\, \exp\left[-\int_z^{z_r} Z(z)^{\frac{1}{2}} L(z)^{-1} dz\right] \qquad (20)$$

$$K(z) = k^2 L^2(z) (du/dz)_z = k^2 L(z)\, u(z)\, Z(z)^{\frac{1}{2}} \qquad (21)$$

The classical parameters of the whole canopy are the following:

$$u_* = [k\, u(z_r)\, Z(h)^{\frac{1}{2}}] / [1 + Z(h)^{\frac{1}{2}} \int_h^{z_r} L(z)^{-1} dz] \qquad (22)$$

$$d = [z_r - h\, \exp(k\,[\Delta u]_h^{z_r}/u_*)] / [1 - \exp(k\,[\Delta u]_h^{z_r}/u_*)] \qquad (23)$$

$$z_0 = (h - d)\, \exp(-k u(h)/u_*) \qquad (24)$$

This type of solution is more general than previous solutions /29,30, 31,26,4/; all profiles are defined in terms of canopy structure and boundary conditions only. The sole assumption made in the first instance is that $L(z)$ is independent of wind speed.

2.2. Temperature, mass and flux profiles

If $K(z)$ and $u(z)$ are known, or measurable, general equations describing the microclimate inside and above the canopy, can be derived from relations 1, 2, 7a, 8a, 11, 12a, 15 for heat and H_2O, and 9a, 13a for CO_2.

2.2.1. Heat and water vapour

To combine the equations for heat and H_2O in a simple way, we first transform gradients of H_2O concentration to gradients of dew point temperature ($\Delta q(z) = (M/RT) S \Delta T_d(z)$). Then first, from 1, 7a, 8a we write the enthalpy flux equation $\phi_H(z)$ /32/, and second, from 2, 7a, 8a, 11, 12a, 15 the deficit of saturation ($Y(z) = T(z) - T_d(z)$):

$$\rho_a c p_a K(z) \phi_H' + \phi_R(z) + \phi_{SC}(0) = 0 \tag{25}$$

$$K(z)Y'' + K'(z)Y' - h_L(z) f(z) \alpha(z) Y + (1 - \alpha(z))\phi_R'(z)/\rho_a c p_a = 0 \tag{26}$$

with $[\phi_H(z)]_{z_1}^{z_2} = [(S + \gamma)/\gamma][T_w(z)]_{z_1}^{z_2} = [T(z) + T_d(z) S/\gamma]_{z_1}^{z_2}$, $\alpha(z) = [1 + (\gamma/(S + \gamma))h_L r_s]^{-1}$ (26a)

The boundary conditions for these two equations are given by data at the reference level ($T(z_r)$ and $T_d(z_r)$) and also for the second equation by the water potential Ψ_0 at the soil surface:

$$Y(0) = [1 - \exp(-M\psi_0/RT)] [d \log P(T)/dT]_{T \sim T(0)} \tag{27}$$

The second differential equation is easily solved throughout the canopy for the two boundary conditions because it is linear. In this way, a complete understanding of the canopy is obtained, and solutions for $T_d(z)$, $T(z)$ $\phi_{LV}(z)$, $\phi_{SV}(z)$, $\phi_L(z)$, $\phi_S(z)$, and $T_0(z)$ are easily attained.

2.2.2. Photosynthesis and CO_2 profile

Since the variation of $C(z)$ is small relative to the value of $C(z_r)$ at the reference level, a very good linear approximation of Eq (13a) can be used:

$$\phi_{CV}(z) \sim \phi_p(z)[1 - \phi_p(z) \Sigma r/C(z_r)] + \Sigma r [\phi_p(z)/C(z_r)]^2 C(z) \tag{28}$$

in which $\phi_p(z) = [b E(z) C(z_r)]/[b E'(z) \Sigma r + C(z_r)]$ (28a)

Equation (28) together with Eq. (9a) then provides a more general solution than those given by Inoue /33/ and Uchijima et al /34/; this solution depends only on the elementary photosynthetic function of leaves $\phi_p(z)$, on canopy structure $f(z)$, and on $D(z)$ and $L(z)$ through the term $K_M(z)$ (21):

$$K(z)C'' + K'(z)C' - f(z)\Sigma r [\phi_p(z)/C(z_r)]^2 C(z) - \phi_p(z) [1 - \phi_p(z) \Sigma r/C(z_r)] f(z) = 0 \tag{29}$$

This equation, which is similar to relation (26), is solved in the same way, with the two boundary conditions $C(z_r)$ and $\phi_c(0)$.

DESCRIPTION OF USEFUL ENTITIES

Today, sophisticated techniques provide increasingly accurate measurements of parameters, but to understand canopy changes a general model describing all links between climatic conditions and the vegetation is needed.

The earlier parts of this paper gave some relations derived in an attempt to create this kind of model. The flow chart summarizes the principal parameters described later and the links which determine the profiles and fluxes above and within the canopy. Some simplifying approximations which accord quite well with reality have been introduced, but a number of secondary effects, such as those of temperatures, have been ignored.

1. LEAF AND CANOPY PARAMETERS

Generally, leaves represent the bulk of the canopy; however, exchanges also occur at stems, and inflorescences, such as the ears of wheat and corn.

- Knowledge of spatial distribution of vegetation elements in a plant community /35/ is fundamental in calculating reliable values of net radiation penetration $\phi_R(z)$ /36/ and also of light penetration /37,38/, particularly photosynthetically-active wave-lengths. The different techniques used to obtain these data, e.g. direct measurement, photographic techniques, or point quadrat analysis, have been described /39/; but in order to define E(z) more precisely, additional details are required, e.g. radiative properties of leaves as functions of wave length, incidence angle and leaf morphology.

This description of canopy structure is also directly useful in describing momentum transfer above and inside the canopy. Cionco /40/ and Inoue /30/ give approximate diffusivity profiles as a function of leaf area index, and Perrier /28/ used a calculation of mixing length based on canopy structure. Several authors have used various of these parameters to obtain general flow properties; in particular, the roughness parameter and the displacement height /41/.

- The shape, the dimension and orientation of a single leaf have very important effects on the drag coefficient. The latter must be established, as a function of wind speed, leaf orientation and dimension and age, to permit the most accurate prediction of the momentum exchange /42/. The flutter of leaves may also change the drag coefficient /43/, and valid wind tunnel estimates of drag coefficient can be obtained only if it is possible to reproduce natural turbulence and to take into account bluff body effects.

Usually sensible and latent heat coefficients are very similar, differing markedly from momentum exchange which is more affected by the angle of incidence. Some measurements of the exchange coefficients as function of wind speed in controlled conditions /26,44/, or directly in canopies /2/ give more satisfactory solutions to the general equation than does the use of the drag coefficient. Generally, they are considered in terms of $u^{0.5}$ (Laminar flow) /45/, but for fully-turbulent flow, experiments have shown that better predictions over a large range of wind speeds are obtained using $u^{0.75}$ /26/. Denmead and Bradley /46/ found a wind speed dependence intermediate between these two relations at wind speeds of $0.4 - 0.6$ ms^{-1}.

- The following parameters are more closely linked with the biology of the plant and are more variable. However, some of them relating to maximum efficiency, or to internal resistances, such as carboxylation and mesophyll resistances, are fairly constant, changing only minimally with the physiological age of the leaves. Chartier /47/ gives a method of investigation and some values for the changes in these parameters along a profile in a canopy and assesses the importance of these variations to total photosynthesis /6/.

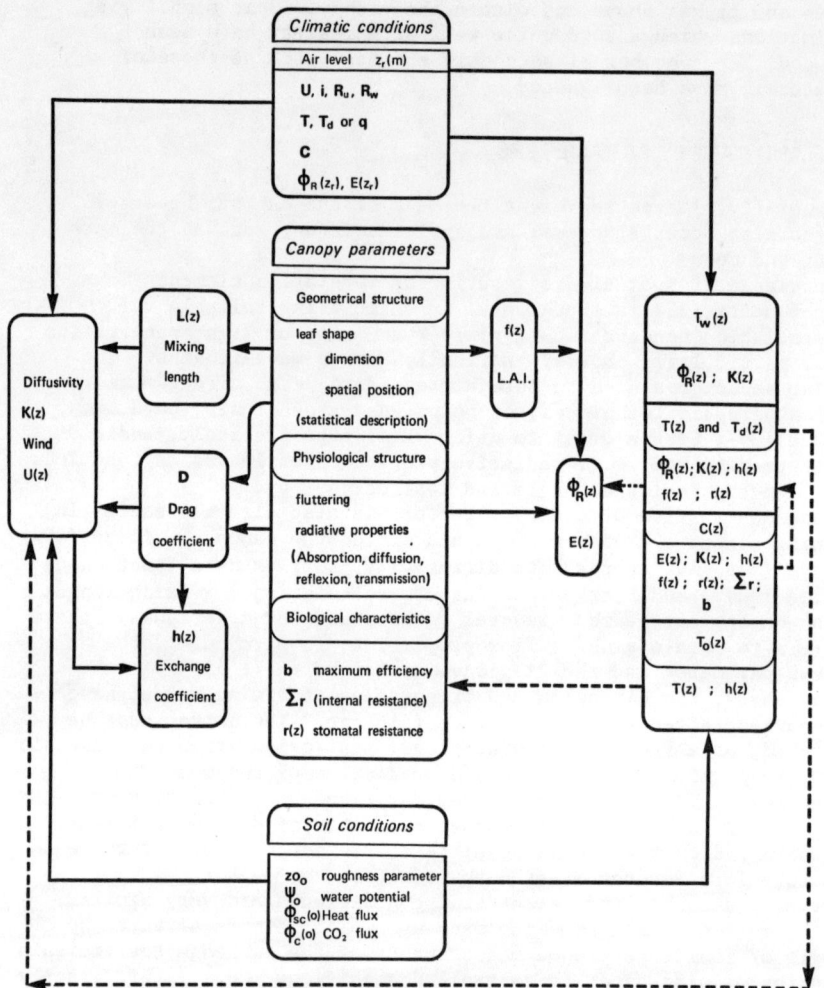

Flow chart: Solid arrows indicate the direct action, on resulting profiles of essential entities, of boundary conditions and canopy parameters. Broken arrows draw attention to some of the main feedback reactions. The parameters which determine the profiles $T_w(z)$, $T(z)$ or $T_d(z)$, $C(z)$, and $T_o(z)$ are listed below each of them.

The physiological parameter with the strongest influence on the partitioning of sensible and latent heat and also on photosynthetic rate, is the stomatal resistance. It is quite difficult to predict accurately this complex parameter, although some classical forms of variation, particularly in response to irradiance and age of leaves /48/ and sometimes less defined relationships with the saturation deficit of the air and the water potential have been obtained.

De Parcevaux and Perrier /2/ also point to the difficulty of using the resistance concept for aerodynamically or physiologically asymmetric leaves; the canopy must then be characterized by a double profile of upper- and lower- leaf stomatal resistances ($\alpha(z) = \alpha_u(z) + \alpha_1(z)$, Eq. 26). So, measurements for two faces at several heights and at various times throughout the day are required basic data for a model.

The last parameter which can be associated with the canopy is the leaf surface temperature. The determination of its variation with height in the canopy gives the final term of the energy balance at each level and provides the most useful test for adapting the model to reality. This parameter is also difficult to obtain and its measurement is quite critical /49/, but a continuous record at several levels inside the canopy is a good method for understanding energy changes of each level. Furthermore, in this way, it is possible to determine the value of the exchange coefficient and of the stomatal resistance in situ /2/.

2. RADIATION BALANCE PARAMETERS

Many techniques and methods of calculation or measurement are presented by Anderson /39/ and some well-adapted sensors for photosynthetic light can be used /50/. However, net radiation is always more difficult to determine, and even with special tube solarimeters misleading representations of mean average values for a crop are often obtained. The best method is to follow point values of net radiation with a sufficiently fast small sensor moving along a transect inside the canopy /51/. In this way, the length may be adapted to the horizontal heterogeneity and problems concerning the relation of the linear net radiometer output to the numbers and sizes of sunflecks are avoided. Hence more suitable mean values and a complete description of the spatial heterogenity are obtained.

3. ATMOSPHERIC PARAMETERS

Generally, all entities must be determined at some reference level; usually that is not difficult to do although a suitable reference level may be not very well defined. Theoretically the ideal level is the top of the boundary layer where the relevant parameter have values imposed by the surrounding environment. Comparison between profiles upwind of the canopy and above the canopy, given a reasonable fetch /10/ permits determination of this height: it is the level where the two profiles become similar. This level is not the same for all entities, but a mean height may be used as a convenience.

In order to check a general model, or to apply a particular method, it is generally necessary to determine mean gradients on profiles; wind speed /30,50,51,52/, temperature and humidity /53/, and CO_2 concentration /54,55,56,57/ are the most commonly measured profiles, but tracers have been used /58,59/.

Because of links between turbulence and diffusivity, particular attention must be paid to the definition of turbulence characteristics and stability parameters above and within the canopy. Hot-wire and sonic anemometers, which provide essentially three-dimensional measurements, give the only satisfactory data for investigations of turbulence structure. So, relationships between profiles and fluxes as functions of stability parameters (Richardson number or Obukhov length) are quite well documented /12,60,18/. Also, Cionco /61/ presents many measurements of turbulent intensity; this is the best parameter for determining the significance of mean velocity. In addition, many auto-correlation coefficients for the vertical and horizontal components of wind fluctuation and temperature fluctuation have been measured /61,62/ in an attempt to determine a mixing length scale of the form $[\overline{(w')^2}]^{\frac{1}{2}} \int_o^t R_w dt$ and to correlate this length scale with the diffusivity.

4. SOIL SURFACE PARAMETERS

- The heat flux at the soil surface, can be measured repetitively without difficulty, or can be approximated by some sinusoidal function of time based on physical properties of the soil surface /63/ and the net radiation receipt $\phi_R(0)$. For short periods of time, this term becomes less important as the ratio $\phi_R(0)/\phi_R(h)$ decreases, or as the leaf area index increases.
- The water potential at the soil surface. This varies considerably after rain and also during the day as a consequence of redistribution following periods of low evaporation and condensation at night. Generally, however, it does not vary very much beneath a canopy, but it may be estimated by a general theory of water movement in soil /64/.
- The CO_2 flux at the soil surface. This is certainly the most difficult term to obtain. It may be estimated from concentration measurements near the soil, but the low value of diffusivity is accompanied by great variation in concentration and as a result in the calculated flux. Monteith et al /65/ describe a variant of the Lundegärdh method, but it provides only mean values over several hours. Perrier (unpublished) used the principle of the assimilation chamber, covering the soil for periods of 2 or 3 minutes with a ventilated box held at the exact ambient pressure. In this way, changes in CO_2 concentration of only 10 or 20 ppm occurred, so that the concentration gradient between soil and air was not greatly affected, and an instantaneous estimation of CO_2 soil flux was obtained. Its value may change by 20% or more during the day, and by as much as 100-200% during and after periods of rain.
- The roughness parameter of the soil surface ZO_o. This does not greatly affect the momentum balance inside the canopy. At the very low wind speeds that occur near the soil surface this parameter is practically indistinguishable from the mean height of the roughness elements.

GENERAL METHOD OF DETERMINATION

1. DIRECT MEASUREMENT

With a plant community, some variables can be determined directly but only in terms of an integral over the whole canopy.

1.1. Evapotrauspiration

The direct measurement of evapotrauspiration is the basic measurement of heat and mass transfer of a canopy. Lysimeters /66/ have been improved in order to obtain greater accuracy /67/ and continuous recording, these permitting measurements over small time intervals /68,69/. Some weighting lysimeter installations have been made with sufficient accuracy to permit measurements over periods of half an hour with surfaces of a few m^2, thus providing good checks on indirect measurements /70/.

1.2. Shearing stress

The direct measurement at the surface is a technique originated by Sheppard. Since then other workers have used this technique to investigate shearing stress /71,72/ τ_o, the drag coefficient of the surface D_M, and the friction velocity U_*:

$$D_M = (u_*/u)^2 = \tau_o/\rho_a u^2 \qquad (30)$$

This technique, however, used also by Bradley /11/ to analyse the variation of surface drag with change in surface roughness, is difficult to use in all canopies, particularly tall ones.

1.3. CO_2 and water vapour fluxes

These measurements are based on the general principle of gas metering. They may be applied to the whole canopy, or only to a part, such as a single plant or leaf /43,74/. The principal problem in this type of measurement, is the disturbance of the natural environment (wind speed, turbulence, surface temperature, radiation) created by the inevitable presence of the chamber walls. However, even through many restrictions apply, much useful information can be obtained quite easily if precautions are taken to match climatic factors inside and outside the chamber, and care is exercised in interpretation of results /75/.

2. FLUCTUATION METHOD

The measurement of eddy fluxes from measurement of fluctuation both of the entity and of the vertical wind speed over a surface layer is becoming quite widely applied. This calculation /66/ requires compact and fast-response sensors with sophisticated recording equipment. Generally, the lack of suitable sensors for use inside a canopy (particularly for $CO2$) restricts the method. Extensive application of this method has been carried out by Dyer et al /76/ using the fluxatron and evapotron. Inoue et al /77/ also used the method to make a complete analysis of a canopy (sensible and latent heat fluxes and CO_2 flux). When the surface is not horizontal and when heterogeneities occur, it is necessary to introduce the mean vertical component and to replicate sufficiently over the surface.

3. SIMILARITY APPROACH: TURBULENT DIFFUSIVITY CALCULATION

Most information concerning fluxes above and inside canopies is obtained using this approach. From the known flux ϕ_x and gradient C_x of an entity x, its diffusivity K_x is calculated (the flux is taken as positive if it is a gain for the surface, i.e. a downward flux. Eq. 6a to 9a). The application of this diffusivity to the transport of any entity, assuming that the turbulent diffusivities are identical, is only valid for the mean level considered z. The second and most important hypothesis is the conservation of flux between the level where it is measured (generally soil or canopy surface) and the level z where the gradient is measured.

Three kinds of entity are commonly examined in this way:
- Momentum flux → Aerodynamic method
- Enthalpy flux → Energy balance method
- Tracer flux → Tracer method

3.1. Momentum flux

This is the most commonly used approach /78,79,7,65,60/ because the classical theory of the logarithmic profile in neutral conditions gives a relation between momentum flux and wind speed gradient or wind speed. As a result, only one measurement (wind speed as a function of height) is necessary to determine the diffusivity. The two classical relations for above and inside the canopy are:

$$\tau_o = \rho_a u_*^2 = \rho_a k(z-d) u_* [du(z)/dz] \quad (31)$$

$$\tau(z) = \tau(0) + \rho_a \int_0^z D_M(z) f(z) u^2(z) dz \quad (32)$$

For the air layer above the canopy, if we make the assumptions that the turbulent diffusivity for momentum is identical with the entity x, that the momentum flux is constant with height above the surface, and that neutral conditions exist (section 1.2.1), we have for the flux of x:

$$\phi_x = k^2 [\Delta u]_{z_1}^{z_2} [\Delta C_x]_{z_3}^{z_4} / [\text{Log}(z_2/z_1) \text{Log}(z_3/z_4)] \quad (33)$$

Alternatively, if we consider the drag expression for the surface we have

$$\phi_x = D_M(z_1) u^2(z_1) [\Delta C_x]_{z_1}^{z_2} / [\Delta u]_{z_1}^{z_2} \quad (34)$$

where $D_M(z_1)$ is a function of z_1 (near the surface), ZO the roughness parameter, and d the height of the zero plane displacement ($D_M = k/\text{Log}(z_1-d)/\text{ZO})^2$); D_M may be determined directly from a drag-plate, or by calculation of ZO and d from the wind profile /7/.

Of course, the most important variation is caused by non-neutral conditions. Corrections have been proposed based on stability parameters /12,60,18/. The second correction comes from the difference between the momentum diffusivity and the mass or heat diffusivity (section 1.2.1).

Inside the canopy the same difficulties exist and they are exaggerated by low wind speeds which increase the relative error according to a square law (Eq. 17 or 32) /80/. Also, the common assumption that the shearing stress at the soil surface is zero; introduces another significant error, particularly if the foliage density is very small near the soil.

3.2. Enthalpy flux

The energy balance and the enthalpy flux equations (Eq. 1, 15, 26a) show that the latter is directly proportional to the wet bulb temperature and linked to the net radiation by the diffusivity. Thus, only measurements of the enthalpy gradient (very close to the wet bulb gradient) and net radiation profile are needed for calculation of the turbulent diffusivity for enthalpy ($K_L \sim K_S$). For each level, the following exact equation can then be written

$$K(z) = [\phi_R(z) + \phi_{SC}(0) - \epsilon] / [\rho_a c_{p_a}(dT/dz) + L(dq/dz) + \lambda(dC/dz)] \quad (35)$$

This relationship, when applied inside the canopy, can be imprecise because of the difficulties in obtaining accurate gradient measurements, but it has been widely used /80,34,56/. More details concerning error sources and methods of measurement are given by some authors /80,81,82/. One of the principal problems is to obtain an accurate mean value in time and horizontal space at a given level.

Above the canopy, the usefulness of this method is well established. McIlroy /83/ used a direct evaporation recorder (E.P.E.R.) with an analogue system. Perrier et al /84/ use a numerical system based on the calculation of $K(z)$ to obtain hour-by-hour measurements of the fluxes (ϕ_R, ϕ_L, ϕ_S, ϕ_{SC}); these are calculated from thirty measurements each hour and continuous inversion of the sensors avoids some errors in gradient measurement.

3.3. Tracer method

In theory, this is a very simple method, because the flux density of the tracer is generally known and constant with height and the diffusivity is simply calculated from the gradient measurements. However, to obtain homogeneity between the soil and the reference level in the presence of advection, the flux density of the tracer must be large and uniform over a considerable surface area /85,58/.

Fontan et al /86/ used as a tracer radioactive Thoron emitted naturally from the soil. In this case, by using the equation for the decay of the gas, an estimation of the flux is calculated directly from the profile of Thoron concentration as a function of height. This kind of measurement has been applied successfully above and inside the canopy /59/.

3.4. Combination methods

Different combinations of the previous methods can be used, depending on the data available. One method commonly used is to combine direct measurement (e.g. weighting lysimetry) of evapotranspiration with profile measurements to obtain the turbulent diffusivity of water vapour, which is then applied to CO_2 or sensible heat flux:

$$\phi_C = (\phi_L/L)([\Delta c]_{z_1}^{z_2} / [\Delta q]_{z_1}^{z_2}) \quad (36)$$

Another classical combination is based on the Penman calculation of evapotrauspiration /79/, in which the exchange coefficient of the saturation vapour deficit term is determined from drag plate measurements (Eq. 34), or from an aerodynamic analysis (Eq. 33) /60,66,81,87/. Thus:

$$\phi_L = [s/(s + \gamma)][\phi_R(h) + \phi_{SC}(0)] + \rho_a \, cp_a \, \bar{h}_s (T_a - T_d)$$

($\bar{h}_s$, exchange coefficient for a whole canopy $\sim D_M \, U(z_r)$ or $k^2 \, u(z_r)^2 / \text{Log}^2 [(z_1 - d)/Z0]$).

4. SIMULATION IN WIND TUNNEL

It is obviously fascinating to compare controlled conditions with unsteady natural conditions and the complications due to heterogeneities in space and time that occur in the field. But two important difficulties arise:

Firstly, similitude between the various length scales cannot be achieved in all directions, and some reasonable range of variation about the exact solution must be tolerated.

Secondly, the simulation of all atmospheric conditions, particularly atmospheric turbulence /88/ and a negative long wave balance, induce many difficulties, even before the difficulty of conducting a model experiment in which three kinds of energy exchange (ϕ_K, ϕ_L, ϕ_S) occur. Of course such experimentation is easier to perform on a simple model, such as a leaf, than on a whole canopy.

An exhaustive analysis of mean leaf properties as functions of height in the canopy appears to be one of the most useful studies. It may thus be possible to obtain useful information for a general model. This kind of study has been undertaken for a number of years /43,44,89, 90/ but rarely in a way sufficiently general to determine, as functions of incidence angle, turbulence, and wind speed /25/, drag coefficients and exchange coefficient profiles for a canopy.

It is more ambitious to attempt to simulate exchange processes in the whole canopy, although for aerodynamic studies /91,42/, or for systems in which only heat or mass exchange is occuring /92/, information approximating reality has been obtained. Wind tunnel studies are particularly suitable for examination of drag coefficients, roughness parameters, and general exchange characteristics of the whole canopy. But in fact because of the two important restrictions presented above, the greatest advances will occur when there is constant feedback between wind tunnel data and experimental field data.

5. MATHEMATICAL MODELS AND SIMULATION

It is important to dissociate purely statistical and numerical models from analytic ones. In the first kind, the model may describe accurately but give little clue to the mechanisms involved and to the links between factors and parameters. In the second kind, the analysis seeks to discover from elementary physical theory, the links and interrelations between factors, and, in simpler cases, to give analytical solutions. This kind of model has been proposed by many authors /32, 31,40,28,4,93/. Some numerical models, based on experimental profiles of wind speed and diffusivity, have been elaborated also /94,95/.

In general, an analytical model provides the simplest way of understanding the operation of canopy processes, of simulating the effects of a single parameter on profiles and fluxes, and of interpreting the evolution of some general entity such as evapotranspiration, CO_2 flux, or leaf surface temperature.

CONCLUSIONS

This analysis of heat and mass transfer within and above canopies has attempted to show that methods of observation must be based on an understanding of canopy processes. Our understanding of these is now sufficiently complete to permit some economies in experimental measurement.

However, these reflexions are only as valid as the hypotheses on which the analysis has been based, and new efforts must certainly be directed more and more towards testing the validity of the hypotheses.

The first problem is one of variation in space. A one-dimensional analysis, or sometimes a two-dimensional one for advection problems, is generally accepted. The heterogeneity of most canopies in three dimensions, even superficially regular ones, and the overall three dimensionality of turbulence may in certain circumstances lead to gross errors of interpretation.

The second, and perhaps most important problem is variation in time, a reality which must be considered. This continuous change with time is often approximated by successive steps, and the important thing is to know what errors are introduced by the assumption of successive steady states. For this reason, studies concerning the time constants of processes in the whole canopy and in its different parts will be very important for improving our present knowledge of canopy functioning. Certainly, the most interesting changes to be studied are those induced by changes in wind speed and particularly in radiation. For example, even without dramatic change resulting from the passage of clouds, the regular passage of the sun induces movement of sunflecks along a leaf, resulting in a transient state. On average, this state would not induce changes in heat and mass transfer if the biology of the system remained constant, but biological time responses, with effects on stomatal regulation and photosynthetic reaction, generate variation. In such ways, canopy heat and mass transfer may vary considerably.

This leads us to devote more attention to the biological parameters and to try to determine both their instantaneous time responses and their variation with age of the plant.

ACKNOWLEDGMENT

The author thanks his colleagues, Drs. O.T. Denmead and D. Smiles of the Division of Environmental Mechanics, CSIRO, Canberra, Australia, for the benefit of discussion and for making the text more comprehensible.

REFERENCES

/1/ Monin, A.S. and Obukhov, A.M. 1954. Trudy Geofis. Inst. Akad. Nauk SSSR 24, 163-87.
/2/ de Parcevaux, S. and Perrier, A. 1970. In 'Plant Response to Climatic Factors'. Proc. Uppsala Symp., 1970. pp.117-35 (Unesco: Paris)
/3/ Perrier, A. 1973. Etude physique de l'evaporation dans les conditions naturelles: part II - Expressions et parametres donnent l'evapotranspiration reelle d'une surface 'mince'. Ann. Agron. (in press)
/4/ Cowan, I.R. 1968. Q.J.R. Metcorol. Soc. 94, 523-44.
/5/ Baldy, C. 1973. Ann. Agron. 24, 1-31.
/6/ Chartier, P. 1972. In 'Crop Processes in Controlled Environments'. (Eds. A.R. Rees et al). Proc. Symp. Littlehampton, 1971. pp. 203-16. (Academic Press: London).
/7/ Lemon, E.R. 1960. Agron. J. 52, 697-703.
/8/ Philip, J.R. 1959. J. Meteorol. 16, 535-47.
/9/ Dyer, A.J. 1963. Q.J.R. Meteorol. Soc. 89, 276-80.
/10/ Dyer, A.J. 1968. Unesco Nat. Resourc. Res. 5, 493-8.
/11/ Bradley, E.F. 1968. Q.J.R. Meteorol. Soc. 94, 361-79.
/12/ Lumley, J.L. and Panofsky, H.A. 1964. 'The Structure of Atmospheric Turbulence'. 239 pp. (Wiley Interscience: New York).
/13/ Priestley, C.H.B. 1959. 'Turbulent Transfer in the Lower Atmosphere'. 130 pp. (University of Chicago Press: Chicago).
/14/ Panofsky, H.A. 1963. Q.J.R. Meteorol. Soc. 89, 85-94.
/15/ Swinbank, W.C. 1968. Q.J.R. Meteorol. Soc. 94, 460-7.
/16/ Webb, E.K. 1969. Q.J.R. Meteorol. Soc. 96, 67-90.
/17/ Taylor, P.A. 1971. Q.J.R. Meteorol. Soc. 97, 326-9.
/18/ Businger, J.A., Wyngaard, J.C., Izumi, Y. and Bradley, E.F. 1970. J. Atmos. Sci. 28, 181-9.
/19/ Swinbank, W.C. and Dyer, A.J. 1967. Q.J.R. Meteorol. Soc. 93, 494-500.
/20/ Obukhov, A.M. 1946. Trudy Inst. Teor. Geofiz. Akad. Nauk SSSR 1, 95-115.
/21/ Ramdas, L.A. and Paranjpe, M.K. 1936. Current Sci. 4, 642-6.
/22/ Schlichting, H. 1955. 'Boundary Layer Theory'. (Pergamon Press: New York and London).
/23/ Blasius, H. 1908. Z. Angew. Math. Phys. 56, 1.
/24/ Pohlhausen, E. 1921. Z. Angew. Math. Mech. 1, 115.
/25/ Thom, A.S. 1968. Q.J.R. Meteorol. Soc. 94, 44-55.
/26/ Perrier, A. 1968. Rev. Gen. Therm. 79/80, 721-40.
/27/ Chartier, P. 1966. C.R. Hebd. Seances Acad. Sci. 263, 43-6
/28/ Perrier, A. 1967. La Meteorologie 1, 527-50.
/29/ Uchijima, Z. 1962. J. Agric. Meteorol. 18, 1-10.
/30/ Inoue, E. 1963. J. Meteorol. Soc. of Jap. II, 317-21.
/31/ Denmead, O.T. 1964. J. Appl. Meteorol. 3, 383-9.
/32/ Philip, J.R. 1964. J. Appl. Meteorol. 3, 390-5.
/33/ Inoue, E. 1965. J. Agric. Meteorol. 20, 137-40.
/34/ Uchijima, Z., Udagawa, T., Horie, T., and Kobayashi, K. 1967. J. Agric. Meteorol. 23, 99-108.
/35/ Warren Wilson, J. 1959. New Phytol. 58, 92-101

/36/ Impens, I. and Lemeur, R. 1969. Arch. Meteorol. Geophys. Bioklimatol. Ser. B 17, 403-12.
/37/ Ross, Y. and Nil'son, T. 1968. Issled. Fiz. Atmos. 11, 5-54.
/38/ Chartier, P. 1966. Ann. Agron. 17, 571-602.
/39/ Anderson, M.C. 1971. In 'Plant Photosynthetic Production: Manual of Methods'. (Eds. Z Sesták, J. Catský, and P.G. Jarvis) pp. 412-66. (W. Junk: The Hague).
/40/ Cionco, R.M. 1965. J. Appl. Meteorol. 4, 517-22.
/41/ Lettau, H. 1969. J. Appl. Meteorol. 8, 828.
/42/ Thom, A.S. 1971. Q.J.R. Meteorol. Soc. 97, 414-28.
/43/ Shuepp, P.H. 1971. Boundary-Layer Meteorol. 2, 263-74.
/44/ Linacre, E.T. 1967. Plant Physiol. 42, 651-8.
/45/ Monteith, J.L. 1965. In 'The State and Movement of Water in Living Organisms'. pp. 705-34. (Cambridge University Press).
/46/ Denmead, O.T. and Bradley, E.F. 1973. Paper presented to 1st Australasian Conf. on Heat and Mass Transfer, Monash University, Melbourne, 1973. 6 pp.
/47/ Chartier, P. 1970. In 'Prediction and Measurement of Photosynthetic Productivity'. Proc. IBP/PP Tech. Meet., Trebon, Czechoslovakia 1969. pp. 307-15 (Centre for Agricultural Publishing and Documentation: Wageningen).
/48/ Djavanchir, A. 1971. Mise au point d'une chambre à transpiration et son application à l'etude de la régulation stomatique. Thesis, Faculty of Orsay, June 1971.
/49/ Perrier, A. 1971. In 'Plant Photosynthetic Production: Manual of Methods'. (Eds. Z Sesták, J. Catský, and P.C. Jarvis). pp. 632-69. (W. Junk: The Hague)
/50/ Takeda, K. 1964. J. Agric. Meteorol. 20, 1-6.
/51/ Stoller, J., and Lemon, E.R. 1964. USDA Prod. Res. Rep. 72. (The energy budget at the earth's surface, Part II).
/52/ Bradley, E.F. 1969. Agric. Meteorol. 6, 185-93.
/53/ Begg, J.R., Bierhuizen, J.F., Lemon, E.R., Misra, D.K., Slatyer, R.O., and Stern, W.R. 1964. J. Agric. Meteorol. 1, 294-312.
/54/ Monteith, J.L. and Szeicz, G. 1960. Q.J.R. Meteorol. Soc. 86, 205-14.
/55/ Inoue, E., Uchijima, Z., Udagawa, T., Horie, T., and Kobayashi, K. 1968. J. Agric. Meteorol. 13, 165-7.
/56/ Denmead, O.T. 1970. In 'Prediction and Measurement of Photosynthetic Productivity'. Proc. IBP/PP Tech. Meet., Trebon, Czechoslovakia, 1969. pp. 149-64. (Centre for Agricultural Publishing and Documentation: Wageningen).
/57/ Chartier, P., Perrier, A., and Verbrugghe, M. 1971. Ann. Agron. 22, 367-81.
/58/ Legg, B.J. 1970. Annual Report Rothamsted Exptal. Sta., Harpenden, England. p.58.
/59/ Druilhet, A., Perrier, A., Fontan, J., and Laurent, J.L. 1971. Boundary-Layer Meteorol. 2, 173-87.
/60/ Webb, E.K. 1965. Agric. Meteorol. Meteorol. Monogr. 6(28), 27-58.
/61/ Wright, J.L. and Lemon, E.R. 1965. Agron. J. 58, 255-61.
/62/ Saito, T., Nagai, Y., Isobe, S. and Horibe, Y. 1970. J. Agric. Meteorol. 25, 205-14.
/63/ vanWijk, W.R., Ed. 1963. 'Physics of Plant Environment. 382 pp. (North-Holland Publishing Co.: Amsterdam).

/64/ Philip, J.R. 1957. J. Meteorol. 14, 354-66.
/65/ Monteith, J.L., Szeicz, G., and Yabuki, J. 1964. J.Appl.Ecol. 1, 321-37.
/66/ Slatyer, R.O. and McIlroy, I.C. 1961 'Practical Microclimatology'. 310 pp. (Unesco and CSIRO: Melbourne).
/67/ Popov, O.V. 1959. Publication No. 49 Ass.Int. Hydrol. Sci. 26, 37.
/68/ Hallaire, M. and Bouchet, R.J. 1965. C.R. Acad. Agric. 51, 1024-5.
/69/ vanBavel, C.H.M. and Reginato, R.J. 1965. In 'Methodologie de l' ecophysiologie vegetale'. Actes du Colloque de Montpellier. 1962. Arid Zone Res. XXX, 129-35.
/70/ Perrier, A., Archer, P., and Blanco de Pablos, A. 1974. Etude de l'evapotranspiration reelle et maximale de diverses cultures: dispositif et mesures. Ann. Agron. (in press).
/71/ Goddard, W.B. 1963. Introductory measurements of shear-stress across rye grass sod. Univ. Calif. Dept. Agric. Eng. and Irrig. Final rep. 149 pp.
/72/ Bradley, E.F. 1968. Q.J.R. Meteorol. Soc. 94, 380-7.
/73/ Lange, O.L., Koch, W. and Schulze, E.D. 1969. Ber. Deut. Bot. Ges. 82, 39-61.
/74/ Eckardt, F.E. 1967. Oecol. Plant. 2, 367-93.
/75/ Jarvis, P.G. and Catsky, J. 1971. In 'Plant Photosynthetic Production: Manual of Methods'. (Eds. Z. Sesták, J. Catský, and P.G. Jarvis) pp. 49-104. (W. Junk: The Hague)
/76/ Dyer, A.J., Hicks, B.B. and King, K.M. 1967. J. Appl. Meteorol. 6, 408-13.
/77/ Inoue, E., Uchijima, Z., Saito, T., Isobe, S. and Uemura, K. 1969. J. Agric. Meteorol. 25, 165-71.
/78/ Thornthwaite, C.W. and Holzman, B. 1942. USDA Tech. Bull. No. 817.
/79/ Penman, H.L. 1948. Evaporation in nature. Report on Progress in Physics 11, 366-88.
/80/ Lemon, E.R. 1968. In "Plant Environment and Efficient Water Use'. Proc. Symp. Copenhagen, 1966. pp. 381-9. (Unesco: Paris).
/81/ Denmead, O.T. and McIlroy, I.C. 1971. In 'Plant Photosynthetic Production: Manual of Methods'. (Eds. Z. Sesták, J. Catský, and P.G. Jarvis). pp. 467-514. (W. Junk: The Hague).
/82/ Perrier, A. and Seguin, B. 1970. In 'Techniques d'Etude des Facteurs Physiques de la Biosphere'. pp.425-45 (INRA: Paris).
/83/ McIlroy, I.C. 1971. Agric. Meteorol. 9. 93-100.
/84/ Perrier, A., Blanco de Pablos, A., and Itier, B. 1973. Methode du bilan d'energie et son application: utilisation d'un bilan d'energie automatique. Ann. Agron. (in press).
/85/ Millington, R.J. and Peters, D.B. 1969. Agron. J. 61, 815-9.
/86/ Fontan, J., Birot, A., Blanc, D., Bouville, A. and Druilhet, A. 1966. Tellus 18, 623-32.
/87/ Bradley, E.F. and Finnigan, J.J. 1973. Proc. 1st Australasian Conf. on Heat and Mass Transfer, Melbourne, 1973. Reviews, pp. 57-78 (Monash University, Melbourne).
/88/ Arya, S.P.S., and Plate, E.J. 1969. J. Atmos. Sci. 26, 656-65.
/89/ Takechi, O. and Haseba, T. 1962-63. J. Agric. Meteorol. 18, 92-3 and 19, 7-10.
/90/ Parkhurst, D.F., Duncan, P.K., Gates, D.M. and Kreith, F. 1968. Agric. Meteorol. 5, 33-47.

/91/ Armitt, J. and Counihan, J. 1968. Atmos. Environment 2, 49-71.
/92/ Moreney, R.N. 1968. J. Appl. Meteorol. 7, 780-8.
/93/ Thom, A.S. 1972. Q.J.R. Meteorol. Soc. 98, 124-34.
/94/ STewart, D.W. and Lemon, E.R. 1969. The energy budget at the earth's surface: a simulation of net photosynthesis of field corn. Interim report on microclimate investigations, USDA Cornell 69-3.
/95/ Waggoner, P.E. and Reifsnyder, W.E. 1968. J. Appl. Meteorol. 7, 400-9.

SIMULATION OF FLOW ABOVE FOREST CANOPIES

WILLY Z. SADEH

Department of Civil Engineering, Colorado State University, Fort Collins, Colorado, USA

The similarity criteria for achievement of meaningful wind-tunnel simulation of neutral flow above forest canopies are discussed. A model forest canopy consisting of miniature plastic trees in an adequate meteorological wind tunnel was utilized to investigate the upper-canopy flow. Mean velocity distributions are presented. Comparison with field data indicates that satisfactory wind-tunnel simulation can be accomplished.

INTRODUCTION

Wind movement above forest canopies is of paramount importance in analyzing the various transport processes which occur within this zone. Forest canopies, regardless of their specific features, affect considerably the surface-layer wind and its accompanying turbulence. Consequently, the flow field and the turbulent exchange processes are significantly different from those which would prevail were the canopy absent. To start with, the properties of the flow and momentum transport are determined by the interaction between the oncoming turbulent wind and the canopy. Then, the mass and energy transports are considerably influenced. For instance, diffusion and dispersion of pollutants are basically established by the resulting turbulent wind. Estimations of carbon dioxide and water vapor exchange rates depend upon knowledge of flow properties and, particularly, of turbulent exchange coefficients within the layer where they are being transported. Snow accumulation, seed dispersion, fire spread rates, soil erosion, stocking norms of trees are affected by the prevailing turbulent wind. Knowledge of local flow conditions is of prime significance in regard to efficient planning of cutting patterns, clearings and forest roads. It is, further, worthwhile to note that the canopy induced effects on the flow vary considerably with changing oncoming wind direction and properties for the very same forest canopy.

Theoretical analyses of the turbulent boundary layer above a canopy and accompanying turbulent transport processes are exceedingly difficult due to the complexity of the canopy-wind interaction. To attain an understanding of this intricate flow, detailed experimental investigations are imperative. Field measurements, which are highly important, are inherently limited since realizations of adequate controlled flow situations are constrained by the weather vicissitudes. Additionally, they are exceedingly expensive. On the other hand, wind-tunnel simulation, which has proved its effective usefulness in innumerable flow problems, can be effectively utilized to study canopy flows. Physical simulation is of utmost significance not only in the light of its wide applications but also for achievement of a fundamental understanding of these compound flows. Whenever a particular canopy flow is simulated, the flow conditions in a wind tunnel can be maintained unchanged over sufficiently long time periods to warrant adequate investigations of the mean wind distribution, turbulence characteristics and transport processes above the canopy. The flow condition can be systematically controlled and varied for checking available theoretical models and developing new ones. Furthermore, the relative significance of the various factors affecting the flow can be methodically ascertained. Moreover, considerable savings are achievable through wind-tunnel simulation of canopy flows.

The forest canopy flow can be divided into three distinct but interrelated zones: (1) the flow within the trunk spacing or the sub-canopy flow; (2) the flow within the tree crown area usually called canopy flow; (3) the flow above the canopy or upper-canopy flow. Since these three flow zones are strongly interdependent their separate analyses and subsequent matching are most difficult. Insofar as the turbulent transport processes, knowledge of the properties of the turbulent boundary layer above the canopy is of prime concern. A brief account of efforts to simulate the upper-canopy flow in a meteorological wind tunnel with emphasis on the mean velocity variation is presented in this paper.

SIMILARITY CRITERIA

Simulation of flow above forest canopies by means of meteorological wind-tunnel flow depends upon fulfillment of geometric, kinematic, dynamic and thermal similarity conditions. In addition, similitude of the surface boundary conditions including the approaching wind are to be met. The similarity criteria can be readily deduced from the fundamental conservation equations of mass, momentum and energy [1]. Discussion of the various problems related to wind-tunnel modeling of atmospheric flows can be found in References 1, 2 and 3. Since this background literature is readily accessible only the specific aspects concerning simulation of canopy flow are briefly examined hereafter.

To start with, the results presented herein are confined to thermally neutral flow and, therefore, the thermal similitude criteria are disregarded. Achievement of satisfactory similitude of neutral flow is contingent on the capabilities of the meteorological wind tunnel being used. By and large, the physical size and flow characteristics of existing meteorological wind tunnels are such that the similarity criteria for modeling of the turbulent wind within the small-scale atmospheric motion can be reasonably approximated. The small-scale region encompasses flow over forest canopies and agricultural fields, over relatively small topographic features and over cities. This important region lies within the high frequency domain of the micrometeorological range of the horizontal turbulent wind energy spectrum, i.e., within frequencies larger than about 0.03 Hz [3]. Essentially, the small-scale atmospheric motion is associated with fluctuations generated primarily by the surface roughness, e.g., forest canopy, urban roughness. The properties of the turbulent wind are determined by the surface friction and upward transport of turbulent shear stresses through the turbulent momentum exchange. The greater the surface roughness, the higher are the turbulent shear stresses and the upward turbulent momentum exchange.

Geometric similarity can be accomplished using a scaled down model of the tree elements, i.e., the roughness elements. In selecting the geometric scale it is necessary to ensure the immersion of the model canopy into a fully developed turbulent boundary layer. This boundary layer simulates the atmospheric surface layer above a forest canopy. Its depth varies from about 100 to 300 m depending on the specific surface roughness. Additionally, the blockage effect induced by the model canopy must not exceed 5% to obliterate any undesirable flow pattern distortions. To obtain a fully developed canopy flow it is further essential to model the canopy over a sufficiently long horizontal fetch, i.e., to model the surface roughness distribution (surface boundary conditions). Thus, the geometric scale depends upon the size of the wind-tunnel cross section and the length of its test section. Due to the large variation of tree height this scale must be determined for each particular forest canopy being modeled. Based on preliminary investigations it appears that a geometric scale of 1:50 to 1:150, depending upon the field tree height, yields satisfactory modeling [4].

Kinematic similarity can be achieved by simulating the mean velocity and turbulence intensity distributions. The rotation of the wind with height, due to the Coriolis effect, is negligible within the atmospheric surface layer [5]. To obtain similar mean velocity and turbulence intensity variations it is indispensable to adequately simulate the wind upstream of the model canopy. This approaching velocity should vary accordingly to the logarithmic law characteristic to the lower atmosphere above a smooth surface. Consequently, the terrain upstream of the canopy, i.e., surface boundary conditions, must also be modeled. Since the model canopy should be fully immerged into a turbulent boundary layer, artificial thickening of the wind-tunnel boundary-layer thickness is necessary. Generally, a boundary layer thickness at least 4 times greater than the canopy height is desired. The thickening is accomplished by roughness installed at the wind-tunnel entrance section. This roughness also acts as a turbulence generator yielding the needed approaching turbulent wind. It is, further, important to remark that the growth of a sufficiently thick turbulent boundary layer above the model canopy is practically effected by the available length of the wind-tunnel working section.

Dynamic similarity for neutral flow requires equal Rossby and Reynolds numbers for the simulated and field flows. The Rossby number $Ro = U/L\Omega$ (U is the velocity relative to the rotating earth, L is a measure of the distance over which the velocity varies and Ω designates the earth angular velocity) expresses the effect of Coriolis force. Unless the wind tunnel were to be rotated the Rossby-number equality condition cannot be satisfied. Fortunately, the Rossby-number criterion can be discarded whenever the horizontal extent of the forest canopy is less than about 150 km [6]. Then, the Coriolis acceleration is negligible compared with the convective and/or local accelerations. This is true when the Rossby number is greater than 10 and, hence, relaxation of the Rossby-number condition does not affect significantly the flow.

The Reynolds number $Re = UL/\nu$ (U and L are characteristic velocity and length, respectively, and ν is the air kinematic viscosity), which represents the effect of viscous forces, is of prime significance in the case of laminar flow. By and large, the Reynolds-number equality condition cannot be met for canopy flows. The Reynolds number of the wind-tunnel simulated flow is smaller by a few order of magnitude than its field counterpart. On the other hand, flow above a rough surface, such as a forest canopy, is highly turbulent and, hence, predominantly governed by the inertial forces. This situation occurs whenever the Reynolds number is of order of 10^5 or larger. The viscous damping action is constantly counteracted by the turbulence produced by the tree elements which act as vortex generators [7]. Recent experimental results, which are referred to in Ref. 4, supply acceptable evidence that the drag coefficient and wake features of a tree are essentially independent of the Reynolds number within the above range. Consequently, the Reynolds-number equality criterion can be set aside provided that the simulated flow Reynolds number is sufficiently high to ensure the invariance of the flow pattern. The condition of equal Reynolds number is then replaced by a minimum Reynolds number for the wind-tunnel modeled flow. Its value should be at least of order of 10^5. However, where the turbulence characteristics are compared extreme caution must be exercised insofar as the modeling of turbulence scales and simulation of its energy spectra.

An additional similarity requirement is vanishing streamwise pressure gradient. In the atmosphere the pressure over longitudinal fetches of interest is practically constant. This condition can be easily attained by sectional adjustment of the wind-tunnel ceiling.

EXPERIMENTAL PROCEDURE

Model Forest Canopy

A model forest canopy 1100 cm long and 183 cm wide was employed. Miniature plastic evergreen boughs of a height of 18 cm with a stem 5 cm high modeled the trees. The plastic trees were inserted into 5 mm diameter holes drilled at intervals of 1.27 cm on the face of the canopy base. A schematic diagram of the model canopy and plastic tree including all important dimensions is displayed in Fig. 1. These plastic trees were selected based on preliminary measurement of the drag coefficient and wake growth for a single tree [8]. By and large, both exhibited a reasonable agreement with available live tree data [4,8,9]. Based on a real tree of 20 m height, say, the scale was about 1:110. Hence, the model canopy corresponded to a 1200x200 m forest.

The model trees were randomly distributed such that no distinct rows were discerned. Since about 4360 plastic boughs were utilized, the canopy areal density was approximately 2 trees/100 cm^2, i.e., a closed canopy. The volumetric density number, i.e., the ratio of the volume occupied by the crowns to the total canopy volume, was about 0.26. A photograph of the model forest canopy installed in the wind tunnel is provided by Fig. 2.

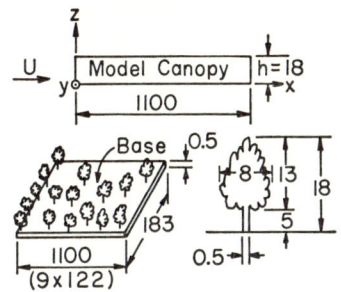 

Fig. 1 Sketch of the model canopy and tree element.

Fig. 2 Overall view of the model forest canopy installed in the meteorological wind tunnel.

METEOROLOGICAL WIND TUNNEL

In the light of the similarity conditions, the modeling of the flow above a forest canopy and of atmospheric flow, in general, depends upon the capabilities of the wind tunnel being utilized. Essentially, a wind tunnel with a large cross section, a sufficiently long working section and with a sectionally adjustable ceiling is necessary. A facility possessing these characteristics was completed at the Fluid Dynamics and Diffusion Laboratory, Colorado State University, in 1963 [10]. The experimental investigation of the flow in the upper-canopy was conducted in this meteorological wind tunnel. This wind tunnel is of closed circuit type with a 29 m long test section and a 1.83x1.83 m cross section. Its axial fan is driven by a 25 hp DC motor. The air velocity can be changed continuously up to about 20 m/s by varying the pitch of the fan blades and/or the motor speed. Boundary layers 90 to 100 cm thick can be obtained at the downstream position of the working section. The wind-tunnel contraction ratio is 9:1, and the background turbulence intensity is about 0.05%. Consequently it is possible to generate controlled turbulence by means of adequate devices. The wind-tunnel ceiling can be sectionally adjusted to produce a longitudinal vanishing pressure gradient. Its side panels are made of glass to allow visualization of the flows. In addition, thermally stratified flow can be produced since the wind-tunnel flow can be heated and/or cooled over a stretch of 13 m.

A sketch of this meteorological wind tunnel, which also shows the model canopy and the system of coordinates used, is depicted in Fig. 3. The leading edge of the model canopy, which stretched over a distance of 11 m, was located 15 m downstream of the test section entrance. This upwind fetch simulated a smooth approaching terrain and allowed the growth of a thick enough turbulent boundary layer. Initial thickening of the boundary layer and production of an adequate level of turbulence were effected by gravel installed in the contraction section and by flexible plastic strips regularly spaced along the first 3 m of the wind-tunnel working section.

EXPERIMENTAL TECHNIQUE

The experiment was conducted under approximately zero streamwise pressure gradient at a constant free-stream velocity (oncoming wind) of 6 m/s. Along the wind-tunnel ceiling the static pressure was monitored at eight pressure taps located 2.4 m apart as shown in Fig. 3. The ceiling slope was adjusted for a maximum pressure difference of 0.002 mm Hg with respect to the pressure at about 5 m upstream of the canopy. The oncoming wind was constantly monitored by means of a Pitot-static tube located 1 m upstream of the canopy outside of the boundary layer. All pressure measurements were performed using an Pitot-static tube located 1 m upstream of the canopy outside of the boundary layer. All pressure measurements were performed using an electronic pressure meter (Trans-Sonics Inc., Equibar Type 120A). This is a differential micromanometer with a range of 20 mm Hg and a resolution of 0.0001 mm Hg.

The mean velocity and turbulence characteristics were measured with a single hot-wire anemometer. In this experiment the velocities ranged up to about 6 m/s while the turbulence intensities varied from low levels above the canopy to extremely high levels near the canopy top and inside it. A novel hot-wire anemometer system conceived, designed and built at Colorado State University particularly appropriate for measurement of low velocity with large fluctuations was utilized [11]. Calibration of the hot wire was performed before and after each run using an accurate calibrator (Thermo-Systems Inc., Calibrator Model 1125). In all cases the hot-wire 1/2-power law, i.e., $E^2 \sim U^{1/2}$ where E is the hot-wire anemometer output voltage and U is the total velocity normal to the sensor, was reasonably satisfied. The reproducibility of the calibration curves was within 1 to 3%.

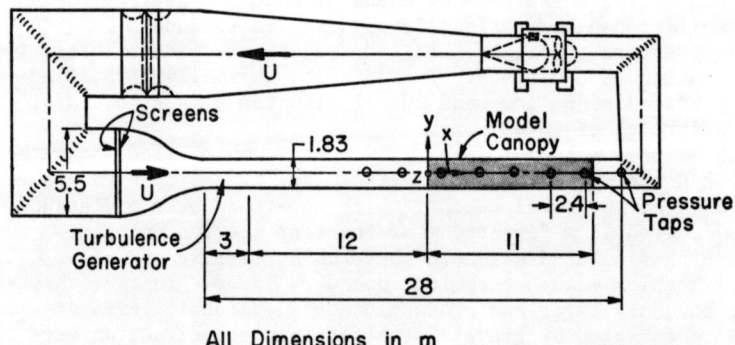

All Dimensions in m

Fig. 3 Overall view of the meteorological wind tunnel.

RESULTS

Velocity Survey

To study the feasibility of wind-tunnel simulation of upper-canopy flow, i.e., the flow above a forest canopy, detailed measurements of mean velocity and turbulence characteristics were performed. However, only the main results concerning the mean velocity are presented hereafter because of space limitations. The results of the turbulence survey will be reported in a forthcoming paper.

The system of coordinates employed in the presentation of the results is shown in Figs. 1 and 3. Its origin is at the geometrical center of the canopy leading edge, i.e., on the wind-tunnel centerline. All the results, unless mentioned otherwise, are presented in dimensionless form. Dimensionless variables are denoted by a tilde placed over the symbols. The height of the canopy is used as the characteristic length of the flow. Consequently, the dimensionless coordinates are defined by

$$\tilde{x}, \tilde{y}, \tilde{z} = z/h, y/h, z/h, \qquad (1)$$

where h = 18 cm. The experiment was conducted at a constant free-stream velocity, i.e., the velocity at the outer edge of the boundary layer, U_∞ = 6 m/s. This velocity was monitored continuously at x,z = -1,1 m. At this station the local thickness of the boundary layer was 72 cm. It is worthwhile to remark that this velocity is within the nonaeroelastic range which extends up to about 13 to 15 m/s. Hence, the dimensionless velocity was defined by

$$\tilde{U} = U/U_\infty . \qquad (2)$$

All the velocity measurements were performed along the centerline of the canopy, i.e., along the $\tilde{x}$-axis in the plane $\tilde{y} = 0$. Essentially, the flow in the neighborhood of the canopy centerline can be assumed as being two dimensional. A check of the velocity and turbulence intensity variations within a distance of ±30 cm off the centerline, ($\tilde{y}$ = ±1.67) revealed a transversal change smaller than 5%.

The mean velocity along the approaching terrain varied accordingly to the logarithmic law characteristic to atmospheric flow on smooth surfaces [12]. At $\tilde{x}$ = -5.55 the estimated values of the friction velocity and roughness length were 36 cm/s and 0.093 cm, respectively. The upstream terrain, according to suggested values of roughness length for natural surfaces [12], corresponds to a smooth area. Thus, the approaching velocity was adequately simulated.

Measurements of the mean velocity within the upper-canopy were carried out at 13 locations along the canopy centerline from 1 m upstream to 1 m downwind of it, i.e., over the distance $\tilde{x}$ = -5.55 to 66.66. At each position the measurements in the vertical direction, i.e., along the z-axis, were performed at 17 stations over a height of 90 cm. Thus, the mean velocity was monitored over a depth of about 4 canopy heights above the canopy.

The measured mean velocity variation along 6 selected isoheights is depicted in Fig. 4. This representation permits examination of the

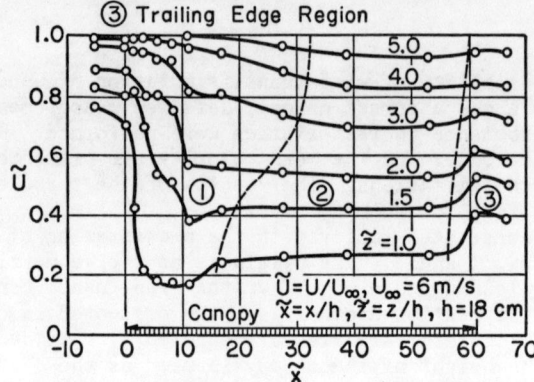

Fig. 4 Velocity variation along isoheights above the canopy, i.e., in the upper-canopy zone.

streamwise velocity evolution ensuing from its interaction with the canopy. To begin with, the canopy frontal area causes an upward deflection of the flow. As a result, a drastic flow retardation arises close to the canopy top which is clearly discerned in Fig. 4. Most of the velocity deceleration occurred over a longitudinal distance of about 15 to 20 h ($\tilde{x}$ = 15 to 20) within a depth extending approximately up to one canopy height ($\tilde{z}$ = 2). With increasing depth a larger adjustment range to the new surface roughness conditions is needed. The region throughout which most of the deceleration occurs can be defined as a transition region. On the other hand, the fully developed flow domain is defined, at any depth, as the region where the mean velocity deviates by less than 5% from its value at $\tilde{x}$ = 45. Within this domain the flow reaches a state of relative equilibrium. According to the foregoing definitions the boundary between the transition and fully developed flow regions is delineated by the broken line in Fig. 4. The transition region stretched roughly up to $\tilde{x}$ = 15 to 32 over a height range $\tilde{z}$ = 1 to 5. Basically, the extent of the transition region is determined by the momentum loss induced by the canopy-wind interaction. This loss depends, in turn, upon the canopy areal density. The higher the density, the greater is the momentum loss and the shorter is the transition region. This transition region, furthermore, affects the development of the internal boundary layer produced by the canopy. Hence, the flow properties in the upper-canopy are basically established by the characteristics of this domain. Toward the trailing edge of the canopy a slight velocity acceleration is observed over a fetch of about 5h. Within this trailing edge region, which is also outlined by a broken line in Fig. 4, the acceleration is due to the flow readjustment to the smooth surface leewind of the canopy.

Based on the vertical velocity variation the canopy effect on the growth of the boundary-layer thickness was estimated. Its thickness was defined as commonly done, as the distance from the wall (canopy ground) where $U/U_\infty = 0.99$. The boundary-layer thickness revealed a smooth increase from about 4h at the canopy leading edge to beyond 5.5h within the fully developed flow region, i.e., an increase of approximately 40%. This growth of the boundary layer is induced by the upward flow displacement and the development of an internal boundary layer. It is further, important to define a characteristic Reynolds number of the upper-canopy flow based on the local mean velocity and boundary-layer thickness. It was found that the average Reynolds number (streamwise averaged) was approximately 190,000. Since the characteristic Reynolds number was of order of 10^5, the simulated flow met the Reynolds number relaxation condition.

Comparison With Field Data

To ascertain the validity of simulating the upper-canopy flow it is imperative to compare wind-tunnel and real forest canopy velocity distributions. This comparison was carried out utilizing existing data for a spruce forest about 20 m high [14]. The field measurements were performed under approximately thermally neutral conditions within the transition region and up to a height of 1.5h. Since for the spruce forest no free-stream velocity was monitored, the velocity defect

$$\tilde{U}_o - \tilde{U} = (U_o - U)/U_{o1} \qquad (3)$$

was used in examining the velocity similarity. In this equation U_o is the velocity upwind of the canopy at $\tilde{x} = -5.55$ and U_{o1} designates the velocity at canopy top, viz., at $\tilde{z} = 1$, at very same upstream station. It is interesting to remark that U_{o1} was roughly the same in both cases, viz., about 5 m/s. The velocity defect variations for both flows at $\tilde{x} = 10$, i.e., in the transition region, are shown in Fig. 5. *A striking similar change is clearly discerned.* At other stations within the transition domain and upwind of the canopy a similar congruent behavior was found. Even inside the canopy the velocity defect variation exhibits an acceptable agreement as observed in Fig. 5. The results concerning the flow inside the canopy will be presented in a future paper.

It is, further, important to examine the streamwise changes of the stream function and, hence, the streamlines pattern for both cases. The dimensionless stream function, whose value is constant along a streamline, is defined by

$$\tilde{\psi} = \psi/U_{o1}h, \qquad (4)$$

where the stream function is

$$\psi = \int_0^z U \, dz. \qquad (5)$$

The streamlines for the wind-tunnel and spruce forest flows are portrayed in Fig. 6. *A similar longitudinal variation is distinguished* even for the limited available field data. Generally, the streamlines

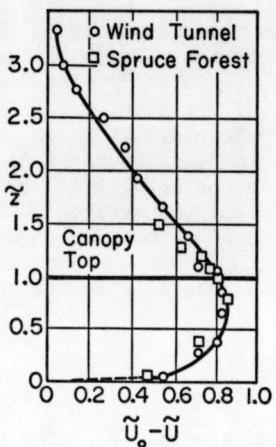

Fig. 5 Wind-tunnel and field forest velocity defect variation with height at $\tilde{x} = 10$ in the transition region.

configuration shown in Fig. 6 clearly reveals the canopy effects on the wind. The outward flow deflection and local drastic velocity diminution within the canopy frontal area and its transition region are evidenced by the significant upward displacement of the streamlines. In the fully developed flow region the almost parallel streamlines substantiate the attainment of a streamwise state of equilibrium. The velocity increase within the trailing edge region is evinced by the downward shift of the streamlines. Thus the longitudinal velocity variation and the division of the upper-canopy flow into three regions are corroborated by the streamlines pattern. The properties and extents of these three regions depend upon the canopy density and oncoming wind.

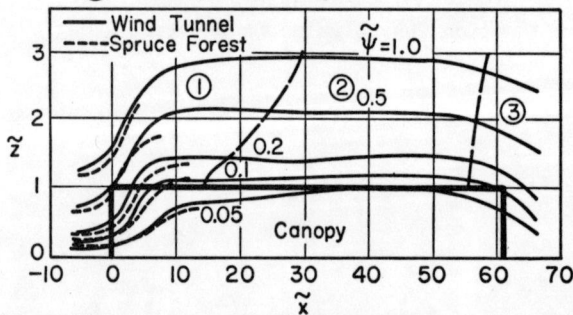

Fig. 6 Streamlines pattern for the wind-tunnel simulated flow and field forest wind.

The momentum loss brought about by the canopy is represented by the boundary-layer momentum thickness change. Inasmuch as most of the momentum loss occurs within the transition domain it is of considerable significance to observe the longitudinal variation of the momentum thickness relative to its upwind value. The dimensionless momentum thickness is defined by

$$\tilde{\theta} = \theta/\theta_o , \qquad (6)$$

where the momentum thickness is

$$\theta = \int_0^\infty \frac{U}{U_\infty}(1 - \frac{U}{U_\infty})dz, \qquad (7)$$

and θ_0 denotes its value upstream of the canopy at $\tilde{x} = -5.55$. In computing the momentum thickness for the spruce forest flow, the free-stream velocity was approximated by extrapolation of the measured velocity at $\tilde{x} = -5.55$. It was found to be roughly $1.22U_{01}$, i.e., about 6.1 m/s. Recall that the free-stream velocity for the wind-tunnel flow was 6 m/s. The longitudinal variation of the momentum thickness $\tilde{\theta}$ over a stretch extending up to 20h in the transition region for both cases is displayed in Fig. 7. A *satisfactory similar change is discerned.* They differ by ±20% at the most. Thus, the momentum loss is reasonably simulated by the wind-tunnel flow. The important aspects of the foregoing results is *the achievement of an adequate wind-tunnel modeling of upper-canopy flow characteristics.*

CONCLUDING REMARKS

The results presented in this work indicate clearly that the neutral mean flow above forest canopies can be satisfactorily simulated in an adequate wind tunnel. Its simulation is strongly substantiated by the remarkable similar variation of mean velocity, streamlines pattern and momentum thickness for the wind-tunnel modeled flow and real forest wind.

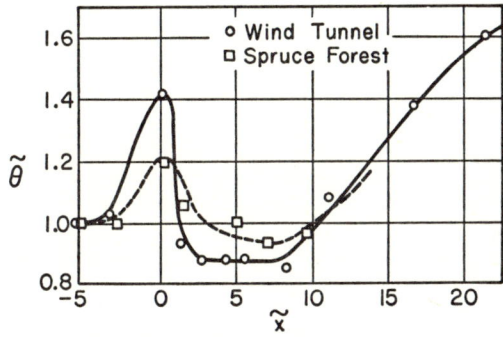

Fig. 7 Momentum thickness streamwise change within the transition region.

To simulate upper-canopy flow under neutral conditions, which pertains to small-scale atmospheric motion, it suffices to meet geometric and kinematic similitude criteria. Since this flow is inertia dominated the equal Reynolds-number condition may be generally relaxed. However, the Reynolds number of the wind-tunnel simulated flow must be of order of 10^5 or larger to warrant the flow invariance. In addition, the Rossby-number criterion can be disregarded whenever the forest canopy horizontal extent is smaller than about 150 km. Satisfactory wind-tunnel modeling is achieved when a 1:50 to 1:150 scaled down model of a forest canopy completely immersed into a fully developed turbulent boundary layer, which simulates the atmospheric surface layer, is utilized. Additionally, it is necessary to carry out the simulation under vanishing streamwise pressure gradient and to model the approaching upwind terrain.

Wind-tunnel simulation is currently the best practical, reliable and, undoubtedly, economical method of determining the forest canopy effects on the flow characteristics. Properly conceived simulation studies can provide the turbulent wind data needed to establish the properties of the various exchange processes. Additional immediate benefits concerning pollution and forest planning can be readily derived from such wind-tunnel studies. Moreover, due to the complexity of upper-canopy flow, the development of realistic theoretical models depends strongly on the experimental data being amassed.

A wind-tunnel model study of upper-canopy flow should furthermore be accompanied by a general survey of the local meteorological conditions. By and large, information of natural winds above and within forest canopies is in need of augmentation to permit adequate verification of results obtained through simulation investigations.

ACKNOWLEDGMENT

The support in preparing this paper by National Science Foundation Wind Engineering Grant (GK39266K) is gratefully acknowledged.

REFERENCES

[1] Cermak, J. E., "Laboratory Simulation of the Atmospheric Boundary Layer," AIAA J., 9, 9, 1746-1754 (1971).
[2] Cermak, J. E. and Arya, S. P. S., "Problems of Atmospheric Shear Flows and Their Laboratory Simulation," Boundary-Layer Meteor., 1, 40-60 (1970).
[3] Sadeh, W. Z. and Cermak, J. E., "Simulation of Roughness Effects on Atmospheric Flow," paper presented at the 10th Anniversary Meeting of Society of Engineering Science, North Carolina State University, Raleigh, N.C., 5-7 Nov. 1973.
[4] Sadeh, W. Z., Cermak, J. E. and Kawatani, T., "Flow Over High Roughness Elements," Boundary-Layer Meteor., 1, 321-344 (1971).
[5] Pasquill, F., "Wind Structure in the Atmospheric Boundary Layer," Phil. Trans. Roy. Soc. London, A.269, 1199, 439-456 (1971).
[6] Cermak, J. E., et al., "Simulation of Atmospheric Motion by Wind-Tunnel Flows," Dept. of Civil Eng., Fluid Dynamics and Diffusion Lab., Colorado State University, Fort Collins, Colorado TR CER66JEC-VAS-EJP-GHB-RNM-SI17 (1966).

[7] Rotta, J. C., "Turbulent Boundary Layers in Incompressible Flow," Progress in Aeronautical Sciences,(Ferri, A., et al., eds.) Pergamon Press, New York, N. Y., Vol. 2, 1-219, 1962.
[8] Hsi, G. and Nath, J. H., "A Laboratory Study of the Drag Force Distribution Within Model Forest Canopies in Turbulent Shear Flow," Dept. of Civil Eng., Fluid Dynamics and Diffusion Lab., Colorado State University, Fort Collins, Colorado, TR CER67-68GH-JHN50 (1968).
[9] Raymer, W. G., "Wind Resistance of Conifers," NPL Aero Report/1008 (1962).
[10] Plate, E. J. and Cermak, J. E., "Micro-Meteorological Wind Tunnel Facility: Description and Characteristics," Dept. of Civil Eng., Fluid Dynamics and Diffusion Lab., Colorado State University, Fort Collins, Colorado, TR CER63EJP-JEC9 (1963).
[11] Sadeh, W. Z. and Finn, C. L., "A Dual-Amplifier Hot-Wire Anemometer System," to be published.
[12] Sutton, O. G. Micrometeorology, McGraw-Hill Book Co., New York, N.Y., 1953.
[13] Deacon, E. L., "Vertical Profiles of Mean Wind in the Surface Layers of the Atmosphere," Geophys. Mem., 11, 91 (1953).
[14] Shinn, J. H., "Steady-State Two-Dimensional Air Flow in Forests and the Disturbance of Surface Layer Flow by a Forest Wall," U.S. Army Electronics Command, Fort Monmouth, N.J., TR ECOM-5383 (1971).
[15] Schlichting, H., Boundary-Layer Theory, McGraw-Hill Book Co., New York, N.Y. 6th ed., 1968.

ENERGY AND MASS TRANSFER IN VEGETATION BY ELECTROCHEMICAL ANALOG

PETER H. SCHUEPP and KEN D. WHITE

Macdonald Campus of McGill University, Montreal, Canada

ABSTRACT

Questions relating to the transfer of energy and mass in crops, such as boundary layer resistances near the ground, entrance effects of the flow into the crop, effects of plant spacing, row spacing and row orientation on the transfer, and limitations of the one-dimensional model for the flow field in the crop, are being studied in electrochemical model experiments. Application of the model to air is discussed.

LIST OF SYMBOLS

x	downstream displacement in the crop, [cm]	j	mass flux density [g.cm^{-2}.s^{-1}]
y	horizontal displacement perpendicular to x, [cm]	d	probe diameter [cm]
z	vertical displacement perpendicular to x, [cm]	ℓ	characteristic eddy size [cm]
u	mean velocity along x [cm.s^{-1}]	Re	Reynolds number $(\frac{u \cdot d}{\nu})$ [—]
w	mean velocity along z [cm.s^{-1}]	Sc	Schmidt number $(\frac{\nu}{D})$ [—]
D	molecular diffusivity of mass [cm^2.s^{-1}]	δ_0	thickness of viscous sub-layer [cm]
ν	kinematic viscosity [cm^2.s^{-1}]	δ	thickness of diffusion sub-layer [cm]
K_e	eddy diffusivity [cm^2.s^{-1}]	a, β	damping constants (defined by equation (1) in the text)
h	mass transfer coefficient [cm.s^{-1}]		
u^*	friction velocity [cm.s^{-1}]	z_ℓ	height where crop starts to limit eddy size [cm]
c	concentration [g.cm^{-3}]		
α	$h \cdot K_e^{-1}$ [cm^{-1}]	χ	von Karman's constant [—]

1. INTRODUCTION

Meaningful experimental studies of heat and mass transfer in vegetation require the simultaneous monitoring of so many parameters as to make such field programs too expensive and too complex for their wide application. Mathematical solutions on the other hand, are impossible to obtain for all but the most unrealistically simplified conditions. It is the aim of this paper to suggest that under these circumstances electrochemical model experiments may provide a worthwhile new approach, offering rapid and controlled variation of such parameters as flow velocity, crop structure and turbulence. The technique is not new; its applications to micrometeorology are. It is not supposed to replace field programs, but to offer a desirable versatile extension of such experiments, as well as an analytical tool for the interpretation of their results.

The studies reported here should be regarded as a first step toward establishing validity and limits of the technique. The emphasis is on controlled variation of parameters rather than on a 'true' simulation of the atmospheric flow field.

2. THE ELECTROCHEMICAL TECHNIQUE

The transfer of ions in an electrolytic solution is governed by diffusion, convection and migration of ions in the electric field. By adding to the reacting electrolyte a non-reacting (neutral) electrolyte in much larger concentration, it is possible to eliminate the electric field gradient in the solution, except for the immediate vicinity of the electrode surfaces, so that the transfer of ions is essentially determined by diffusion and convection. Measurements of mass transfer coefficients are thus reduced to measurements of electric current densities on the model (test electrode), for variations in model geometry and electrolyte flow.

The technique was pioneered in the early 1950's [1,2]. Applications during the last 20 years to the simulation of engineering heat and mass transfer problems have been too numerous for detailed reference. Lately it has been applied to the study of free and forced convection transfer of leaves [3,4].

The system used in the present experiments is the cathodic reduction of 0.025 N ferri-cyanide in the presence of 2 N sodium-hydroxide. Exact values for the molecular diffusivities are known for this system [5]. The ions are discharged, but not deposited, at the electrode surfaces, so that surface geometry remains unchanged by the transfer. The measurements were carried out in a closed-circuit electrolytic tunnel with a volume of 200 litres and a measuring cross-section of 20x20 cm^2. Test electrodes (models) were made of nickel, the anode plates along the tunnel wall of stainless steel. Free stream turbulence is of the order of 2%. Flow velocities were varied between 30 and 60 $cm.s^{-1}$. Velocity and turbulence measurements were made with DISA hot-film anemometer and Bruehl-Kjaer frequency analyzer. Systematic errors in velocity calibration, current measurement and molecular diffusivities are estimated at < 10%.

The molecular diffusivity of ferri-cyanide ions in the given system is very low, resulting in a (very high) Schmidt number of

3. SIMILARITY CRITERIA

Sc ~ 2300, compared with a value of ~ 0.7 for the diffusion of water-vapor and heat in air. Extrapolation of electrochemical data to air is therefore very critical in regions where molecular mechanisms govern the transfer.

In the present model it is assumed that
a) the Reynolds number Re in the model is high enough so that molecular diffusivities are negligible compared to eddy diffusivities except at solid surfaces, i.e. the non-matching of the Schmidt (or Prandtl) number is not serious as far as the transfer in the crop is concerned;
b) molecular viscosity determines the thickness δ_0 of the momentum sublayer (viscous sublayer) and molecular diffusivity D the thickness δ of the diffusion sublayer at solid surfaces. The relation between δ and δ_0 is given by the mode of damping of turbulent motion within the viscous sublayer. If the latter can be determined, extrapolation of measured transfer rates to air should be possible;
c) the width of the turbulent energy spectrum in the model is sufficient, so that effects of viscosity on the viscous dissipation of turbulent kinetic energy does not affect those eddies mainly responsible for the transfer. Fortunately the transfer is mainly determined by eddies with space scales comparable to the geometry of the model, i.e. by the low frequency end of the spectrum.
d) mass transfer results can be applied to heat transfer due to the basic similarity of the two transfer mechanisms.

Under these conditions liquid models are quite suitable for the simulation of atmospheric flow and transfer problems [6].

4. THE MODELS

The models were designed for adaptability to both fluid dynamics and mass transfer studies. Two were built: A rectangular one (22 cm x 16 cm) allows changes in plant and row spacing, with 6 rows and a maximum of 15 plants per row. The flow is down the rows. A square one (11 cm x 11 cm) with 6 rows and 7 plants per row can be rotated to study orientation effects. The plants were made from 0.1 cm nickel wire, wound around a 0.2 cm stem; they are 5 cm high and schematically shown in Figure 1.

The given model represents plants with approximately constant leaf-area index with height. Few plant elements are protruding into the free space between rows, possibly a serious departure from reality.

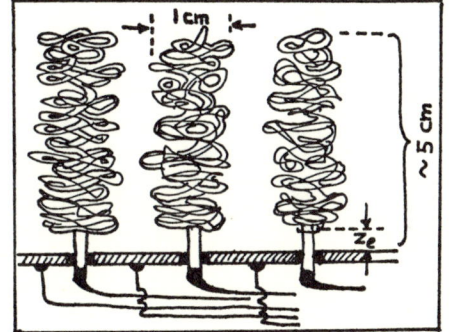

Fig. 1. Schematic view of 'plants' used in the electrochemical model.

The plants have a diameter of about 1 cm, so that a row spacing of 2 and 3 cm leaves 1 and 2 cm respectively of free space along the rows.

All measurements were taken along the centerline of the model. The transfer to the ground was recorded on 10 adjacent plates, 3.5 cm wide and 2.2 cm deep, also arranged along the centerline. A vertical sectionalized cylindrical probe with a diameter of 0.4 cm was inserted into the space between the rows to measure local transfer coefficients.

5. RESULTS AND ANALYSIS

5.1. Transfer to the ground:

Mean transfer coefficients h to the ground were measured for 4 combinations of plant and row spacing at free-flow velocities of 36, 43, 52 and 58 $cm.s^{-1}$. Figure 2 shows typical results. For comparison the friction velocities $u*$ at the ground, as determined from the measured velocity profiles, are also shown in Figure 2.

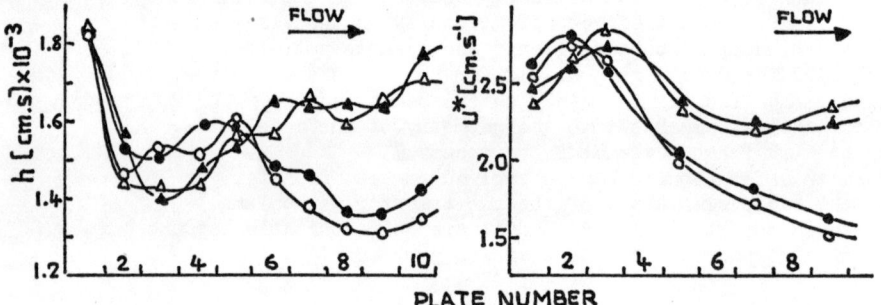

Fig. 2. Mass transfer coefficient h and friction velocity $u*$ at the ground at free-flow velocity of 43 $cm.s^{-1}$, for 3 cm row spacing with 2.25 cm (▲) and 1.5 cm (△) plant spacing and 2 cm row spacing with 2.25 cm (●) and 1.5 cm (○) plant spacing respectively.

Plant spacing has a very small effect on both parameters. A positive correlation between h and $u*$ appears to exist in the downstream half of the model. The strong two-dimensionality of the flow, particularly in the entrance region (see section 5.4) masks any such correlation in the upstream half.

Although the use of mixing length theory may be questionable outside one-dimensional shear flow, an attempt was made to interpret the data from the 4th plate down in terms of a model similar to that used by Levich [7] to describe the transfer across highly turbulent boundary layers. Three zones are distinguished:
- I: the thoroughly mixed crop zone, where the concentration c is approximately constant ($= c_o$) as far as the transfer to the ground is concerned, and where eddy size is limited by crop geometry;
- II: a zone where eddy size is determined by the distance from the wall and where momentum loss due to foliage becomes negligible. It is bounded on the lower side by the viscous boundary layer, and on the upper one by the height $z = z_\varrho$, where

the crop starts affecting momentum flux and eddy size. z_ℓ is about 0.5 cm in our model (see Figure 1). The velocity profile is approximately logarithmic in this region;
- III: the viscous sublayer, its thickness defined as the height $z = \delta_0$ where eddy diffusivity equals molecular momentum diffusivity ν. Within this sublayer the turbulent motion is damped until at $z = \delta$ the eddy diffusivity equals the molecular diffusivity of mass D.

The crucial factor in this model is the <u>assumption made about the damping of turbulence in the viscous sublayer</u>. The characteristic eddy size ℓ within the viscous sublayer is assumed to be

$$\ell = a.z^\beta \quad \dots\dots\dots\dots\dots\dots(1)$$

where 'a' and damping coefficient β are to be determined. They must satisfy the boundary conditions at the top of the viscous and diffusion sublayers. Considering that the downward flux density j must be constant with height, the concentration profiles could then be calculated for the additional boundary-conditions $c(z_\ell) = c_0$ and $c = 0$ at the wall:

in zone II : $\quad c_{II}(z) = \dfrac{j}{\chi u^*} \ln\left(\dfrac{z}{z_\ell}\right) + c_0 \quad \dots\dots\dots\dots(2)$

where χ is the von Kármán's constant (~ 0.4) and in zone III for $\delta < z < \delta_0$:

$$c_{III}(z) = c_0 + \dfrac{j}{u^*}\left[\dfrac{1}{\chi}\ln\left(\dfrac{\delta_0}{z_\ell}\right) + \dfrac{\delta_0^{2-2\beta}}{a^2(2\beta-1)} - \dfrac{\delta_0}{a^2(2\beta-1)}z^{1-2\beta}\right] \dots(3)$$

Replacing j by the diffusional flux D.dc/dz within the diffusion sublayer we get:

$$c(\delta) = c_0 \bigg/ \left\{1 + \dfrac{D\delta^{-2\beta}\delta_0}{a^2(2\beta-1)u^*} - \dfrac{D}{\delta u^*}\left[\dfrac{1}{\chi}\ln\left(\dfrac{\delta_0}{z_\ell}\right) + \dfrac{\delta_0^{2-2\beta}}{a^2(2\beta-1)}\right]\right\} \dots(4)$$

from which the concentration at the top of the diffusion sublayer can be calculated. For example, taking quadratic damping ($\beta = 2$), as recommended by Levich [8] for turbulent boundary layers, we get

$$a^2 = \chi \cdot \delta_0^{-2} \quad \text{and} \quad \delta = \delta_0 \, Sc^{-\frac{1}{4}} \quad \dots\dots\dots\dots(5)$$

which gives

$$c(\delta) = c_0 \bigg/ \left\{1 + \dfrac{1}{3} + Sc^{-\frac{3}{4}}\left[\ln\left(\dfrac{\delta_0}{z_\ell}\right) + \dfrac{1}{3}\right]\right\} = A_2 c_0 \sim 0.77 c_0 \quad \dots(6)$$

Similarly concentrations $c(\delta)$ for different damping terms were calculated. The results can be used to <u>deduce the relationship between mass transfer coefficient h and friction velocity u^*</u> in the model:

e.g. if $\beta = 2 \quad h = A_2 . \chi . Sc^{-3/4} . u^* \quad \dots\dots\dots(7)$

if $\beta = 2.5 \quad h = A_{2.5} . \chi . Sc^{-4/5} . u^* \quad \dots\dots\dots(8)$

etc.

The linear relationship between h and u^* predicted from equations (7) and (8) are plotted in Figure 3, together with the measured data. It is clear that the <u>direct proportionality between h and u^* could only be considered a first approximation</u>, that the damping is somewhat stronger than quadratic, particularly for the wide row spacing, where the damping appears to be velocity-dependent. On the whole, the correlation between h and u* is remarkably independent of velocity and position in the crop for any given plant and row spacing.

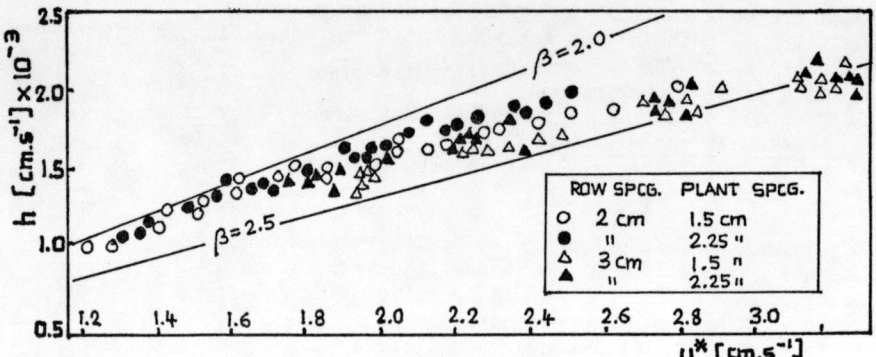

Fig. 3. Calculated and measured relationship between mass transfer h and friction velocity u* at the ground.

The observed approximate proportionality between h and u* leads to speculation about the proposed <u>linear relationship between the transfer coefficient at a solid wall and the eddy diffusivity K_e</u> 'in the vicinity of the wall' [8, 9, 10]:

$$h = \alpha \, K_e \quad \ldots\ldots\ldots\ldots\ldots\ldots(9)$$

The value of α near the ground can be calculated from the model described above. Defining (arbitrarily) the eddy diffusivity at $z = z_\ell$ as the eddy diffusivity 'near the wall', working with damping coefficients between 2.0 and 2.5 (see Figure 3), and estimating K_e as $K_e \sim \chi u^* . z_\ell$, the ratio of h/K_e becomes:

$$\alpha_{(model)} \sim (3.5 \pm 0.5) \times 10^{-3} \text{ cm}^{-1} \quad \ldots\ldots(10)$$

Applying the same analytical model to the transfer in air, with $Sc \sim 0.7$, so that $\delta \sim \delta_0$, the concentration at the top of the viscous sublayer becomes:

$$c(\delta_0) \sim c(\delta) = c_0 \Big/ \left\{1 - Sc^{-1} \ln\left(\frac{\delta_0}{z_\ell}\right)\right\} = A_{(air)} c_0 \quad \ldots(11)$$

where z_ℓ again is the height at which the crop begins to extract momentum and to limit eddy size.

Estimates of transfer in air according to this model require realistic values for u* and z_ℓ. With $u^* \sim 10$ cm.s^{-1} and z_ℓ assumed between 5 and 20 cm in a mature corn crop, A_{air} varies between 0.1 (z=20 cm) and 0.13 (z=5cm), and therefore

$$\alpha_{(air)} \text{ between } 7 \times 10^{-3} \text{ and } 4 \times 10^{-2} \text{ cm}^{-1} \quad \ldots\ldots\ldots(12)$$

Eddy diffusivities in crops near the ground are typically of the order of 10 to 100 cm^2.s^{-1} [11]. With the given value for α this would correspond to transfer coefficients of between 0.1 and 1 cm.s^{-1} or boundary layer resistances of between 10 and 1 s.cm^{-1}. These values appear reasonable for the flow down the rows, when compared with measured resistances of about 1.2 s.cm^{-1} in evaporation from drying (bare) soil [12].

It would be useful if this kind of relationship could allow successful scaling of transfer coefficients from model to air: h should scale as the product of α and K_e. The former seems to be 5 ± 3 times larger in air than in the model. K_e could be expected to scale with u^* and z_ℓ near the ground, i.e. with about $5 \times 20 = 100$, so that measured u^* values of 2 cm.s^{-1} correspond to in-air diffusivities of about 40 cm^2.s^{-1}. As a first approximation it is therefore proposed that measured transfer coefficients should be multiplied by 500 ± 300 to get appropriate coefficients in air.

It should be noted that the boundary layer concept is used in this model not so much as the 'true' picture of a boundary layer expected in real crops, but rather as an equivalent boundary layer, offering on the average a resistance to the transfer comparable to that of the very fluctuating actual one.

5.2. Eddy diffusivities: Several methods were tried to compute eddy diffusivity profiles in the model, all with little success: Estimating the shear stress at any given height by working down from the shear layer at the canopy top, taking into account the foliage as momentum sinks [13], did not work satisfactorily because it requires a one-dimensional shear layer where the gradients du/dz must be determined only by the distribution of the sinks of horizontal momentum. In our model vertical velocity components distorted the velocity profiles, causing zero velocity gradients at several positions, which leads to infinite diffusivities in the one-dimensional model.

Similarly, simultaneous solution of the continuity and momentum equations to obtain K_e requires knowledge of the horizontal pressure gradients dp/dz which we did not possess.

Neglecting effects of vertical motion, friction velocities u_T^* at the canopy top were determined from the measured velocity profiles. Zero displacements z_1 were generally between 3.0 and 3.5 cm, with roughness heights z_0 between 0.5 and 1 cm. For flow velocities between 30 and 60 cm.s^{-1} the u_T^* values ranged between 8 and 12 cm.s^{-1} for $z_0 = 0.5$ cm, and between 11 and 18 cm.s^{-1} for $z_0 = 1$ cm respectively. The top of the measuring section (10 cm above the canopy top) constrained the velocity profiles sufficiently to make it impossible to distinguish clearly between z_0 of 0.5 and 1 cm on the basis of the best-fit regression of measured velocities. Eddy diffusivities $K_e \sim \chi\, u_T^* .(z-z_1)$ could therefore be anywhere between 6 and 9 cm^2.s^{-1} at the 4.5 cm level. Scaling K_e with friction velocity and z, i.e. with a factor of $\sim 5 \times 40 = 200$, this would correspond to eddy diffusivities in air of between 1200 and 1800 cm^2.s^{-1} slightly below the canopy top, which is not unreasonable [10, 11].

Longitudinal turbulence intensities and frequency spectra were also measured, but the difficulty in correlating time and space scales and the lack of information about anisotropy of the turbulence made it impossible to use these data to obtain quantitative information on eddy diffusivities.

5.3. Transfer coefficients in the crop Transfer coefficients from the cylindrical probe were measured between rows at heights of 1.5, 3.0 and 4.5 cm. The results for 1.5 and 4.5 cm are shown in Figure 4 as a function of local Reynolds number, together with the

relationship
$$\frac{h \cdot d}{D} = 0.68 \, Re^{0.466} \, Sc^{0.33} \quad \ldots\ldots\ldots(13)$$

derived from the heat transfer of cylinders for $40 < Re < 4000$ and negligible turbulence [14], assuming analogy between heat and mass transfer.

It shows, perhaps not surprisingly, a close correlation between h and mean local flow velocity, particularly for the wide row spacing. The high turbulence intensity ($\sim 20\%$ between rows) only seemed to affect the transfer in narrow rows at the 4.5 cm level.

Knowing at least approximately the eddy diffusivities near the canopy top, it is possible to calculate the value $\alpha = h/K_e$ for the cylindrical probe, as was previously done for the ground plates: Mean transfer coefficients at the 4.5 cm level for all flow velocities are 3.9 and 3.5 x 10^{-3} cm.s^{-1} for

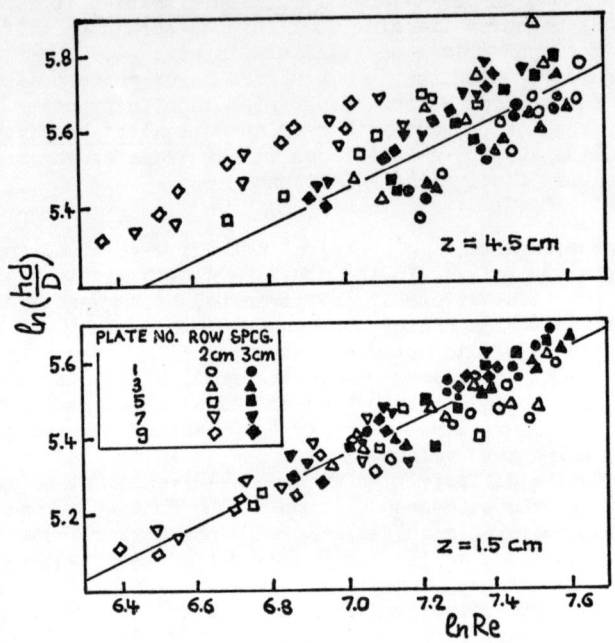

Fig. 4. Measured non-dimensionalized transfer coefficients to a cylindrical probe at various positions in the crop, together with values calculated from the heat transfer of cylinders.

close and wide row spacing respectively. Mean eddy diffusivity is approximately 7.5 cm^2.s^{-1} (see section 5.2). The value of α therefore becomes

$$\alpha = h \cdot K_e^{-1} \sim (5 \pm 2) \times 10^{-4} \quad \ldots\ldots\ldots\ldots(14)$$

It is interesting to compare this value with α values to foliage in crops, originally suggested to be $\sim 1 \times 10^{-3}$ [8], later improved to 0.7×10^{-3} 'with large scatter' [9], based on the data by Brown and Covey [10].

It would be premature to draw far-reaching conclusions from the relative closeness of these values, obtained with great inaccuracies in completely different systems. But if more evidence can be gathered about the correlation between h and K_e, it might well lead to an extremely useful parameter in the study of canopy heat and mass exchange.

5.4. The two-dimensionality of the model: As outlined in section 5.2, the quantitative computations of fluxes in the model was severely hampered by the fact that dp/dx, du/dx and consequently dw/dz were not zero. Along the centerline of the model the derivatives d/dy may be assumed to be negligible, so that the model is two-dimensional. By finite differencing between 450 gridpoints, and subsequent numerical integration, the continuity equation was solved for w(z) and u and w values were mapped out for different combinations of flow velocity and row spacing, as shown in Figure 5, for a flow velocity of 36 cm.s^{-1}.

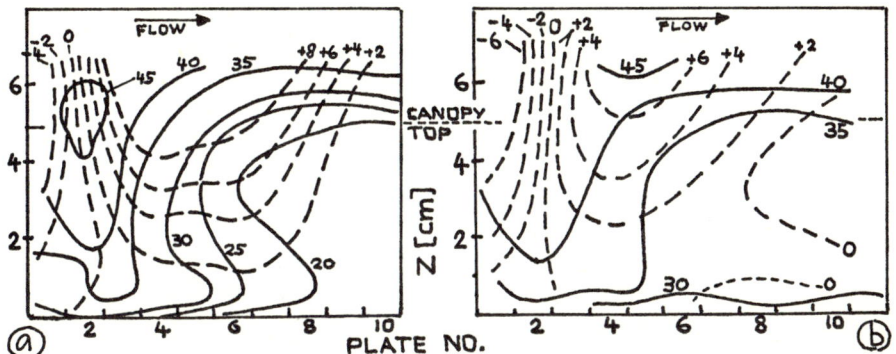

Fig. 5. Horizontal (solid lines) and vertical (dashed lines) flow velocities in the model for close (a) and wide (b) row spacing.

It shows the considerable gradients du/dx, regions of updraft (+w) due to the pressure gradients dp/dx, as well as some downdraft (-w) in the entrance region.

Perhaps this two-dimensionality should be viewed less as a flaw than as an opportunity to intensify studies relating to field situations with strongly pronounced two-dimensionality, whether it be induced in a 'normal crop' by gusty winds - or in the modelling of wind-breaks with the associated strong pressure gradients.

5.5. Orientation effects: Figure 6 shows the effect of orientation on velocity profiles and transfer coefficients of the cylindrical probe for a free-flow velocity of 36 cm.s^{-1}.

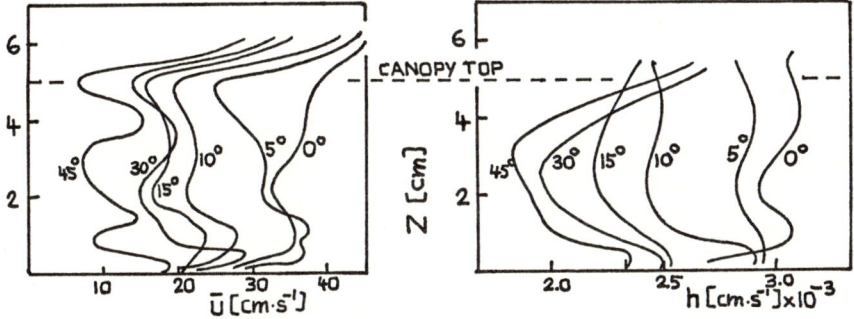

Fig. 6. Effect of orientation on flow velocity and transfer.

Similar graphs were obtained for higher flow velocities. The effect on both parameters is very strong, particularly in the center of the canopy, where the transfer coefficient may be diminished by 30% or more in a 45 degree rotation. At the ground the effect is relatively weak ($\sim 15\%$).

It might be noteworthy that in the region of maximum turbulence intensity the correlation between flow velocity and transfer coefficient is rather weak.

CONCLUSIONS

Within the present considerable uncertainities, extrapolation of electrochemical results to air appears possible, but it is unlikely that eddy diffusivities in the model will ever be determined with accuracy whereas transfer coefficients can easily be measured. For this reason the connection between eddy diffusivities and transfer coefficient should be further explored.

REFERENCES

(1) Lin C.S.,Denton E.R.,Gaskill H.S. and G.L.Putnam,1951:"diffusion controlled electrode reactions", Ind.Eng.Chem.,$\underline{43}$,2136-2143.
(2) Tobias C.W.,Eisenberg M. and C.R.Wilke,1952:"diffusion and convection in electrolysis - a theoretical review", J.Electrochem. Soc. $\underline{99}$, 359C-365C.
(3) Schuepp P.H.,1972:"studies of forced convection heat and mass transfer of realistic fluttering leaf models",Boundary-Layer Meteorol., $\underline{2}$,263-274.
(4) Schuepp P.H.,1973:"Model experiments of free-convection heat and mass transfer of leaves and plant elements", Boundary-Layer Meteorol., $\underline{3}$,454-467.
(5) Bazan J.C. and A.J.Arvia,1965:"the diffusion of ferro- and ferricyanide ions in aqueous solutions of sodium hydroxide" Electrochim. Acta,$\underline{10}$,1025-1032.
(6) Snyder W.H.,1972:"Similarity criteria for the application of fluid models to the study of air pollution meteorology", Boundary-Layer Meteorol.,$\underline{3}$,113-134.
(7) Levich V.G.,1962:"Physiochemical Hydrodynamics", Prentice-Hall Inc. 29 pp. and 144 pp.
(8) Philip J.R.,1964:"Sources and transfer processes in the air occupied by vegetation", J.Appl. Meteorol., $\underline{3}$,390-395.
(9) Uchijima Z.,1966:"Micrometeorological Evaluation of integral exchange coefficients at foliage surfaces and source strength within a corn canopy", Bull.Natl.Inst.Agr.Sci.(Japan). Serial A.,$\underline{13}$,81-92.
(10) Brown K.W. and W.Covey,1966:"An energy budget evaluation of the micrometeorological transfer processes within a corn field", J.Agric. Meterol.,$\underline{3}$,73-96.
(11) E.R.Lemon,1969:"Mass and energy exchange between plant stands and environment",Proceedings of the IBP/PP Technical Meeting, Třeboň Sept.14-21. Summary Section 2,199-205.
(12) Fuchs M.,Tanner C.B.,Thrutell G.W. and T.A.Black,1069:"Evaporation from drying soils by the combination method",Agron.J.,$\underline{61}$,22-26.
(13) Druilhet A.,Perrier A.,Fontan J. and J.L.Laurent,1971:"Analysis of turbulent transfers in vegetation:Use of thoron for measuring the diffusivity profiles". Boundary-Layer Meteorol., $\underline{2}$,173-187.
(14) Parker J.D.,Boggs J.H. and E.F.Blick,1969:"Introduction to fluid dynamics and heat transfer", Addison-Wesley Inc. 259 pp.

17

MICROCLIMATIC MODELING OF THE DESERT

JOHN MITCHELL, WILLIAM BECKMAN, RALPH BAILEY and WARREN PORTER

University of Wisconsin, Madison, Wisconsin, USA

ABSTRACT

A model for predicting the diurnal variations in the microclimate for dry, sparsely vegetated areas is developed. Predictions are made of the temperature distributions in the air and the soil and the velocity profile. The predictions are verified through comparison with data obtained at three desert sites. The predicted soil and air temperatures agree with the measurements within about 2 C over the course of the day.

NOMENCLATURE

A_f - frontal area of single vegetation element (m^2)
c_p - specific heat (kJ/kg-C)
h - heat transfer coefficient ($w/m^2 C$)
$\bar{H}$ - average height of vegetation (m)
k - thermal conductivity (w/m-C)
K - Karman constant (0.4)
L - Monin-Obukov length (m)
q_{cond} - conduction heat flux in ground (w/m^2)
q_{conv} - convection heat flux between soil surface and air (w/m^2)
q_{evap} - evaporation heat flux between soil surface and air (w/m^2)
q_{rad} - radiation heat flux between soil surface and sky (w/m^2)
q_{sol} - solar radiation flux absorbed by soil surface (w/m^2)
q_{sun} - solar flux incident on soil surface (w/m^2)
S - vegetation lot size (m^2/element)
St - Stanton number
T_{air} - air temperature at the reference height (C)
T_s - soil surface temperature (C)
T_{sky} - sky temperature (C)
u - air velocity (m/min)

u_* - shear velocity, $\sqrt{\tau_0/\rho}$ (N/m^2)

z - height above soil surface (m)

z_a - reference height (m)

z_o - surface roughness height (cm)

α_s - soil surface solar absorptivity

ε - soil long wavelength emissivity

ε_h - eddy diffusivity for heat (m^2/min)

ε_m - eddy diffusivity for momentum (m^2/min)

ρ - density (kg/m^3)

σ - Stefan-Boltzmann constant

τ - shear stress in the air (N/m^2)

τ_0 - shear stress at the soil surface (N/m^2)

INTRODUCTION

Recently, engineering techniques have been applied to the simulation of the thermal response of animals to their natural environments [1,2]. These simulations for the microclimate depend on accurate relations for the solar radiation, velocity profiles, temperature profiles, and heat flows in the vicinity of the ground surface. In order to readily simulate different areas, a minimum number of physical parameters and data should be required. Ideally, existing Weather Bureau data should provide the basis for simulations of a desired environment.

In this paper, the microclimate model developed in References 1 and 2 for dry desert-like areas with sparse vegetation will be reviewed. Experimental measurements made at three desert sites with different amounts of vegetation will be compared to model predictions for temperature profiles in the air and soil. Velocity measurements made at heights below that of the surrounding vegetation will be analyzed, and new relationships for the velocity profile near the surface will be discussed.

The basic relation for the microclimate model is the thermal energy balance for the soil surface. This equates the incoming solar energy to that transferred from the surface by radiation, convection, conduction, and evaporation. In equation form

$$q_{sol} = q_{rad} + q_{conv} + q_{cond} + q_{evap} \tag{1}$$

Meteorologists have extensively studied, both experimentally and analytically, the various components of the energy balance, e.g. [3,4,5]. However, though the various components have been studied in detail, few simulations of diurnal changes have been made. In part, this may stem from the lack of accurate relations for the velocity and temperature profiles near the soil surface. Consequently, few of the existing relations contain surface temperature explicitly.

The development and verification of relations containing soil surface temperature depends on accurate experimental measurements of surface temperature.

The ground is, in general, rough due to rocks, sand grains, grass, etc., and the position of the surface is not at all well defined. Nevertheless, the interface does exist, and it should be possible to develop predictive relations utilizing surface temperature.

The velocity profile near the surface and within the vegetation layers is also not well defined. In many experimental studies on sparsely vegetated areas, care is taken to insure that the measurements are made above the level of the surrounding vegetation or roughness elements and most of the expressions for velocity profiles are valid only for this regime, e.g. [3,4,5]. The analytical relations resulting from these experiments are not applicable below the height of the roughness elements since the basic assumption of constant shear stress in the air is not satisfied, and thus the relations cannot be extrapolated to the ground surface. Velocity profiles within relatively uniform and dense crop canopies have been determined both experimentally and analytically (eq. 6). However, little information exists for profiles in moderately vegetated areas. Simulations of the microclimate in crop canopies have been carried out by Lemon and Stewart [6], and Murphy and Knoerr [7], but there appears to be no models developed for bare or sparsely vegetated areas. The simulation of heat flow in the soil has been carried out in a classical fashion by de Vries [8], and more recently, by numerical techniques [9,10]. In all of these studies, the air and soil temperatures near, but not at, the soil surface are modeled and/or measured.

MICROMETEOROLOGICAL MODEL

Model Development

In this section, the micrometeorological model employed in [1,2 and 11] will be briefly outlined, and the parameters and input variables necessary for performing simulations will also be discussed. The model combines mechanism equations for the different modes of heat transfer with the energy balance expression, eq. (1). The mechanism equation for the solar energy flux absorbed by the soil surface is

$$q_{sol} = \alpha_s q_{sun} \qquad (2)$$

The incident solar energy flux is a measured input variable.

The mechanism equation for long wavelength radiant exchange between the surface and the sky is given by

$$q_{rad} = \varepsilon\sigma(T_s^4 - T_{sky}^4) \qquad (3)$$

where T_{sky} is the equivalent sky temperature for radiant energy exchange. For clear skies several relations have been proposed to relate T_{sky} to the air temperature, with those developed by Swinbank and Brunt [12] being most commonly used. The correlation proposed by Swinbank is of a simple form and is given by

$$T_{sky} = 0.0552 \, T_{air}^{1.5} \qquad (4)$$

where T_{air} is in degrees Kelvin. Equations (3) and (4) were combined to compute the radiant energy exchange between the ground surface and the sky.

The convective heat flow between the ground surface and the air is given by

$$q_{conv} = h(T_s - T_{air}) \tag{5}$$

where h is the convection heat transfer coefficient. Heat transfer relations have been established for air flow over bare ground in the absence of vegetation. It is assumed that the heat flux and shear stress are constant in the surface layer and that the velocity profile is that for adiabatic flow [4]

$$(u/u_*) = (1/K)\ln(z/z_0 + 1) \tag{6}$$

The heat transfer coefficient can then be written as

$$h = K^2 \rho c_p u / \ln(z_a/z_0 + 1) \tag{7}$$

where u is the air velocity at the reference height z_a. Modifications to eqs. (6) and (7) have been proposed to account for the distortion of the velocity profile due to the heat flow from the surface, and the consequent effect on the heat transfer coefficient, but will not be included here due to lack of detailed knowledge on the flow pattern inside the vegetation layer.

In contrast to the situation for bare soils, relations for sparsely vegetated surfaces, such as the desert sites under study, have not been formulated. In order to provide modeling relations for the current study, it was assumed that the heat transfer coefficient was given by an equation of the form of eq. (7)

$$h = K^2 \rho c_p u / [\ln(z_a/z_{0_{effective}} + 1)] \tag{8}$$

where $z_{0_{effective}}$ is an effective roughness height that accounts for the distortion of the profile due to the vegetation elements. The value of $z_{0_{effective}}$ in eq. (8) must be determined empirically for the site under consideration. Separate experiments were undertaken to determine the value of the effective z_0 [11], and it was found the effective value of z_0 is very close to that for the sand surface.

The heat conducted into the ground is modeled using a finite difference representation for transient heat conduction through the soil. In the present work, it was found most accurate and efficient of computer time to have the node size increase with depth. The first node, which contains the soil surface, is 1.25 cm thick, with a thickness of 10 cm for the last node at a depth of 0.6 m. Thermal conductivity and the density-specific heat product were assumed constant. It was assumed that soil water content was constant with time and evaporation had a negligible effect on the energy flows. Thus q_{evap} is zero in eq. (1).

REQUIRED PARAMETERS AND DATA

The required physical properties are the soil solar absorptivity, long wavelength emissivity, surface roughness height, soil thermal conductivity, and soil density-specific heat product. For a given site these values may be determined by direct measurement or taken from the literature. In the comparisons between predictions and data presented later, soil thermal property values were determined from the literature [8]. The long wavelength emissivity of the soil surface was taken to be unity. The soil surface roughness height, z_0, was determined through measurements using convection plates [11]. As discussed in [11], vegetation causes an increase in shear stress with height and thus the roughness length obtained from velocity profile data does not reflect the substrate roughness. It is important to use the soil roughness height to accurately model the surface heat transfer coefficient.

The data that need to be supplied are measured values of the incident solar radiation, air velocity and temperature at a reference height, and soil temperature at a reference depth. These may be obtained as functions of time from the Weather Bureau. The soil temperature at 0.6 m does not change diurnally and the average air temperature for the month may be used to satisfactorily approximate this value. [8,p.109,131]. In the present study, the simulation is performed using data taken at each site.

EXPERIMENTAL PROCEDURE

During July, 1972, experiments to verify the model were conducted at three desert sites in the Southwestern United States. Las Cruces, New Mexico, and Rock Valley, Nevada, were sparsely vegetated with small bushes more-or-less uniformly distributed over the desert surface. Tucson, Arizona was heavily vegetated with a variety of different species ranging from small annuals to Palo Verde, ironwood trees and suarro cactus. The average height, diameter, and lot size (desert surface area per plant) for these sites are given in Table 1.

At each site, measurements were made of incident solar radiation, soil and air temperatures, and wind speed. The instrument locations are indicated on the figures. An Eppley pyroheliometer was used for direct solar measurements with the output continuously recorded. The unit was inverted periodically to obtain

TABLE 1 Vegetation Characteristics and Predicted and Experimental Roughness Heights.

	LAS CRUCES	TUCSON	ROCK VALLEY
$\bar{H}$ (m)	0.4	1.90	0.44
$\bar{D}$ (m)	0.49	0.60	0.63
A_f (m²/bush)	0.20	1.62	0.28
S (m²/bush)	2.6	1.6	1.4
z_0 used in the simulation (cm)	0.02	0.02	0.02
z_0 for 0.2-0.4-0.6 m (cm)	0.2	0.034	0.9
z_0 for 0.8-1.2-1.6-2.0 m (cm)	4.76	2.62	3.67
z_0 for 2.0-4.15-8.3 m (cm)	-	13.95	-
$\bar{z}_0$ for all heights (cm)	1.95	10	3.29
z_0 Eq. (10) (cm)	1.5	6.5	4.3

the reflected energy and to allow estimation of the surface solar reflectivity and absorptivity. Shielded thermocouples were used for air temperature measurement. Soil temperatures were determined with bare thermocouples buried at various depths to 0.6 m. The soil surface temperature was determined by placing a thermocouple between two sheets of aluminum foil (approximately 3 cm x 3 cm), laying the assembly on the soil surface, and gently covering the aluminum foil with a fine layer of sand. Temperatures obtained in this manner have been compared with those obtained using a radiometer and agree within about 1 C. All temperatures were measured with a portable hand balancing potentiometer accurate to ±0.5 C. Wet and dry bulb temperatures were taken periodically with a sling psychrometer. Wind speeds were measured at heights ranging from 0.2 to 8.3 m using cup anemometers with the outputs (counter readings) recorded by hand.

Measurements were made one to two days after the instrumentation was placed to allow induced thermal disturbances to decay. Temperature measurements were made hourly over the daylight period of the tests and periodically at night. Wind speeds were measured at half-hour intervals during the day and also periodically at night.

RESULTS AND DISCUSSION

Velocity Profiles

The velocity profile data were analyzed to determine the applicability of the velocity profile given by eq. (6) and to provide an estimate of the roughness length z_0. From the shape of the measured profile, it is apparent that a single relation of the form of eq. (6) does not adequately represent the data for all heights. Vegetation creates additional drag so that the shear stress, and consequently u_*, is not constant but increases with height. This yields an increase in the computed value of z_0 with height.

In order to estimate the variation in shear stress with height, the velocity profile data were divided into groups of three or four neighboring heights. The data were analyzed assuming that the shear stress is constant over these smaller ranges. While this is not really valid, the stress does vary less than over the larger ranges, and thus some estimate of the variations of shear stress and z_0 with height may be obtained.

For each subset of velocity measurements, the unknown values of u_* and z_0 must be determined. The adiabatic velocity profile, eq. (6), is not strictly valid for all heights under conditions of strong surface heating. Below 1 m, the adiabatic form is a good approximation [3]. Above this height, equation (6) is increasingly in error due to the distortion of profile by increased thermal mixing. However, since the form of the profile inside the vegetation layer is not known, it was felt that sufficient insight into the variation of shear stress and z_0 with height could be obtained using the adiabtic form, eq. (6).

Equation (6) was re-written as

$$u = (u_*/k)\ln(z/z_0+1) \qquad (9)$$

Microclimatic Modeling of the Desert

The computational procedure was to fit the data from neighboring heights to eq. (9) using least squares techniques. This allowed determination of both z_0 and u_* for the height interval for each set of velocity measurements. For each site and each height interval, an average z_0 for all velocities was then obtained. A single profile for all heights based on a constant z_0 equal to the average z_0 over all heights was also determined for comparison.

The nondimensional velocity profile, u/u_*, is presented for each site as a function of height in Figs. 1a,b and c. The velocity profile for the different height intervals is discontinuous since the value of u_* for each height interval is different. As height increases, u_* increases and the value of u/u_* con-

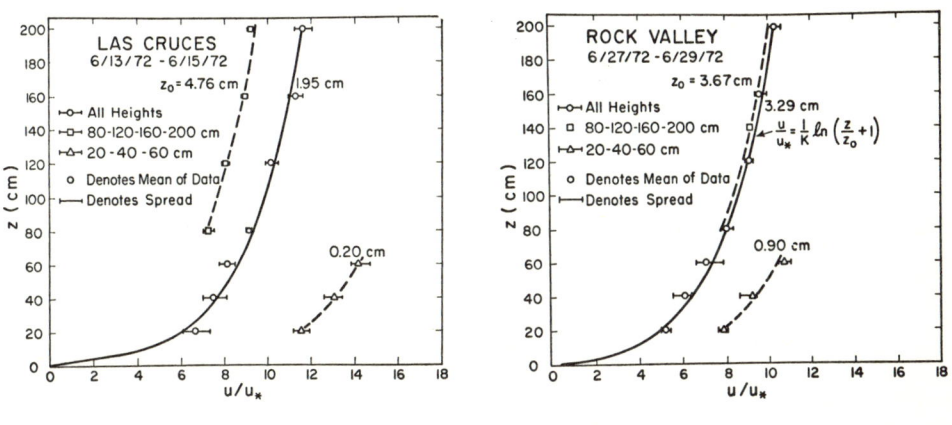

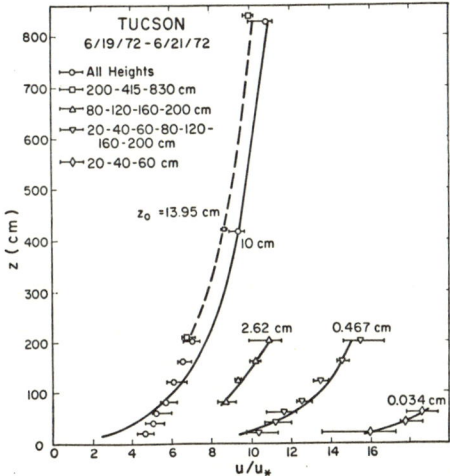

FIGS. 1a,b,c Nondimensional Velocity Profile.

sequently decreases. It is seen that the profiles based on the overall average z_0 do not fit the form of the data as well as do the sequential profiles for each height interval. The values of z_0 for each interval are given in Table 1.

A value of z_0 for a site may also be estimated from the vegetation geometry [14] by

$$z_0 = 0.5 \bar{H} A_f/S \tag{10}$$

The values of z_0 computed from this relation are also given in Table 1 in comparison to the values of the apparent z_0 for each height interval. It is seen that the average z_0 for all heights is close to that estimated from eq. (10). However, the value of z_0 close to the ground, which is the important parameter for the evaluation of the convection heat flow, differs considerably from both the results given by eq. (10) and by the overall velocity profile. This ground value is up to two orders of magnitude smaller than the other values.

The measured velocity profiles were also used to estimate the height variation of shear stress by assuming that the Prandtl mixing length theory is valid in the vegetation layer. Thus the shear stress is given by [4]

$$\tau = \rho K^2 z^2 (\partial u/\partial z)^2 \tag{11}$$

The values of the velocity gradient were determined graphically from a smoothed curve drawn through the average velocity profile for each site. The values of the shear stress were then computed from eq. (11), and are shown as a function of height in Fig. 2 for the three sites. Within the vegetation, the increase of shear stress with height can be approximated by

$$\ln \tau = \ln \tau_0 + az \tag{12}$$

or

$$\tau = \tau_0 \, e^{az}$$

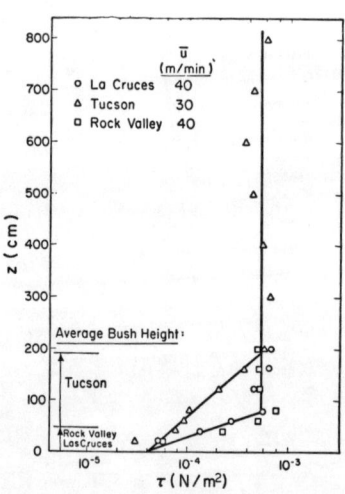

FIG. 2 Shear stress profile

The substitution of eq. (12) into eq. (11) and consequent integration yields an essentially linear velocity profile that fits the data for all heights considerably more accurately than does eq. (6). The shear and velocity relations also differ from those developed for dense crops which are based on a constant drag coefficient in the crop. These considerations indicate that further study is needed into the basic mechanisms governing the velocity profile in lightly vegetated areas.

THERMAL RESULTS

The model described earlier was used to simulate the data taken at each site. The measurement locations have been described earlier, and the parameter values are given in Table 2.

The simulation results are compared to the data in Fig. 3a, 3b and 3c. The simulations are not restarted with new initial conditions each day, but use the last predicted values from the previous day. The predictions for soil surface temperature, subsurface temperatures, and the 2.5 cm air temperature generally agree within about 2 C of the observed values. The temperature gradients in the soil near the surface are on the order of 2-5 C/cm during the day, and thus small errors in thermocouple placement result in large apparent differences between the predicted and observed soil temperatures. The close agreement between the measured and simulated air temperatures at 2.5 cm provides further verification of the convection heat transfer relations.

One of the major sources of inaccuracy in the modeling results from inhomogemeties in soil properties due to the presence of water. In developing the model, the goal was to achieve reasonable accuracy with minimum complexity, and thus the soil was modeled as a uniform medium with constant properties. This approximation was tested at Las Cruces where heavy, unseasonable rains fell the night before the test period commenced. However, the use of a different conductivity each day to represent the effects of soil drying allowed a reasonable prediction of soil temperatures.

The sensitivity of the predicted surface temperature to a 10 percent increase in the value of each parameter is given in Table 2. Since it is the logarithm of z_0 that is important in determining the convective heat transfer (Eq. 7), the sensitivity for z_0 is given in terms of a 100 percent increase (doubl-

TABLE 2 Soil Parameter Values Used in the Micrometeorological Simulation

PARAMETER	LAS CRUCES	TUCSON	ROCK VALLEY	SENSITIVITY
α_s	0.61	0.72	0.70	4 C/10% increase
z_0 (cm)	0.02	0.02	0.02	-1 C/100% increase
k (w/m-C)	0.56	0.63	0.35	-7 C/10% increase
ρc_p (kJ/kg-C)	2.1	2.1	2.1	-1 C/10% increase
ε	1.0	1.0	1.0	-

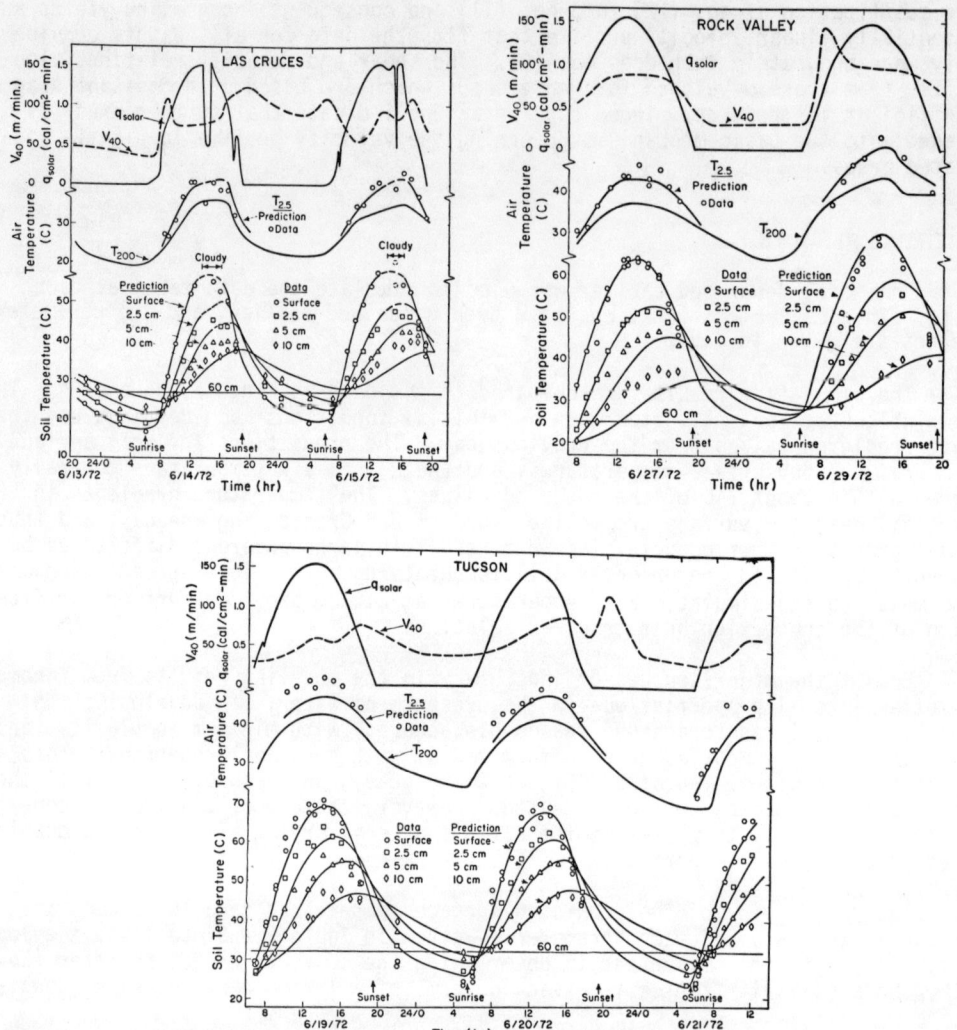

FIGS. 3a,b,c Comparison of Model Simulation and Data.

ing) of z_0. These values are obtained from simulations and are average effects over midday. It is seen that rough estimates for z_0 and ρc_p are acceptable, while α_s and k must be determined more accurately.

SUMMARY AND CONCLUSIONS

A microclimate model has been developed for predicting the temperatures in dry, desert-like environments. The data required are the incident solar radiation, an air temperature, a wind speed, and a deep soil temperature as functions of time. The required physical parameters are the soil solar absorptivity, long wavelength emissivity, thermal conductivity, density-specific heat product, and surface roughness. The model predictions have been verified through comparison with data from three different desert sites, and are seen to predict temperatures near the surface to within about 2 C.

Velocity profile measurements have shown that different roughness regimes exist with height in lightly vegetated areas. Near the surface, the flow is characterized by the roughness of the surface. As height increases. the profile becomes influenced by the presence of the vegetation. The roughness height determined from profile measurements is not useful in modeling the heat transfer from the soil surface.

Weather Bureau data may be used as an input to simulate a microenvironment. Hourly values of air temperature at 2 m, solar radiation, and wind speed are available. Wind speed is commonly measured high enough above the ground surface to avoid the influence of surrounding obstacles, and thus supplementary information is necessary to provide an adequate expression for the velocity profile. The deep soil temperature may be inferred from the mean air temperature for the month. Parameter values may be obtained from the literature. With this information, an estimate of the diurnal variations in temperature and wind speed near a desert surface may be obtained.

ACKNOWLEDGEMENT

The authors would like to acknowledge the assistance of the Desert Biome Group of the International Biological Program (IBP), and in particular Dr. Whitford, Las Cruces; Dr. Thames, Tucson; and Dr. Bamberg, Rock Valley. The efforts of Pat Wathen, Jim Tuschon, Wynn Wacker, Susan Beck, Carol Mitchell, John C. Mitchell, Chris Mitchell, and Laura Mitchell in support of the experimental study are appreciated. Portions of this research were supported by NSF Grant GB-31043.

REFERENCES

1. Beckman, W.A., J.W. Mitchell and W.P. Porter, Thermal Model for Prediction of a Desert Iguana's Daily and Seasonal Behavior, *Trans. ASME*, Series C, 95: 257-262, 1973.
2. Porter, W.P., J.W. Mitchell, W.A. Beckman and C.B. DeWitt, Behavioral Implications of Mechanistic Ecology, *Oecologia 13*: 1-54, 1973.
3. Sellers, W.D., *Physical Climatology*, The University of Chicago Press, 1965, p. 152-154.
4. Sutton, O.G., *Micrometeorology*, McGraw Hill Book Company, 1953, pp. 77-83.
5. Lettau, H.H. and B. Davidson, ed., *Exploring the Atmosphere's First Mile*, Pergamon Press, New York, Vol. 1, 1957, p. 307.
6. Stewart, D.W. and E.R. Lemon, The Energy Budget at the Earth's Surface, Interior Report. 69-3, U.S. Department of Agric. and Cornell Univ., 1964, pp.83,84.
7. Murphy, C.E. and K.R. Knoerr, A General Model for the Energy Exchange and Microclimate of Plant Communities, 1970 Summer Computer Simulation Conference, Denver, Colo., 1970.
8. DeVries, D.A., Thermal Properties of Soils, in *Physics of Plant Environment*, W.R. Van Vijk, ed., North Holland Publishing Co., 1966.

9. Wierenga, P.J. and C.T. deWit, Simulation of Heat Transfer in Soils, *Soil Science Soc. of Am.*, *34*:845-848, 1970.
10. Hanks, R.J., D.D. Austin and W.T. Ondrechen, Soil Model-Heat, Water and Salt Flow. (Submitted to Soil Science of Am.)
11. Bailey, R.T., J.W. Mitchell and W.A. Beckman, An Experimental Method for Determining Convective Heat Transfer from a Desert Surface, ASME Paper 73-WA/HT -9, 1973.
12. Swinbank, W.C., Long Wave Radiation from Clear Skies, Quart. J. Royal Met. Soc., *89*:339-348, 1963.
13. Swinbank, W.C., The Exponential Wind Profile, Quarterly Journal of the Royal Meteorological Society, Vol. 90, No. 384, 1964, pp. 120-122.
14. Lettau, W.H., Note on Aerodynamic Roughness—Parameter Estimation on the Basis of Roughness Element Description, ECOM Dept. 66-624-F, Univ. of Wisc., April, 1969.

18

AN APPROXIMATE ANALYSIS OF THE MOMENTUM BALANCE FOR THE AIRFLOW IN A PINE STAND

JAMES D. BERGEN

Rocky Mountain Forest and Range Experiment Station,
Fort Collins, Colorado, USA

An approximate model of airflow in a pine canopy is used to estimate the velocity profiles, volume drag coefficient, and effective viscosity from windspeed and foliage distribution measurements made in a pine stand. The results suggest that the live branches are the characteristic drag element and that the effective viscosity has an appreciable dispersive component.

NOTATIONS:

- ε = effective porosity
- ρ_f = mass of foliage per unit volume of canopy space (gm cm^{-3})
- r = characteristic drag element crosswind dimension (cm)
- x_i = horizontal coordinate axes; i = 1,2 (cm)
- x_3 = height (cm)
- U_* = friction velocity above the canopy (cm sec^{-1})
- U_i = airflow velocity along the x_i axis (cm sec^{-1})
- t = time (sec)
- S = airspeed (cm sec^{-1})
- F_i = aerodynamic drag force along the x_i axis (dynes)
- P = static pressure (dynes cm^{-2})
- ρ = air density (gm cm^{-3})
- μ = effective viscosity (gm cm^{-1} sec^{-1})
- T_{ij} = turbulent Reynolds stress (dynes cm^{-2})
- s = S/U_*
- u_i = U_i/U_*
- τ_{ij} = $T_{ij}/\rho U_*^2$
- p = $P/\rho U_*^2$
- d = first drag coefficient eq (3) (dynes gm^{-1} sec cm^2)

c = second drag coefficient eq (3)

L = length (cm) eq (7)

T = air temperature (°C)

K_* = scaled effective viscosity

$(\widetilde{})$ = time average

$(\overline{})$ = horizontal average (eq 7)

$()'$ = local deviation from the horizontal average

ν = summation index

c_o = drag coefficient for the forest floor

g = gravitational constant (cm sec^{-2})

This paper attempts to relate the vertical profile of windspeed observed in a pine stand and at edges in that stand to the structure of the canopy evident from physical measurements. The physical model developed not only is highly approximate, but the approximations made can be justified largely only by the measurements from the stand studied.

The windspeed data to be considered were partly reported in an earlier paper [1] with details of method and site which can only be briefly summarized here. The physical structure of the stand will be described more completely in a separate publication [7].

STAND CHARACTERISTICS

The study was carried out in an even-aged stand of lodgepole pine (*Pinus contorta*), located at an altitude of 3,000 m near Foxpark, Wyoming.

The pine stand was 70 years old with a density of 14 stems per 100 m^2, an average diameter at breast height of 10.4 cm, and an average height of 10.0 m.

Tree crowns averaged 257,820 needles, with a total surface area of 43 m^2 and total dry needle weight of 45 kg. The live canopy began at a height of about 2.6 m and there was a foliage density maximum near a height of about 7 m (fig. 1).

A typical tree crown from the stand is shown in vertical section in the photograph of figure 2. The structure is open, with a projected silhouette diameter of about 10-30 cm. The optical porosity of such a crown in terms of the envelope defined by connecting the branch tip with a smooth curve falls near 70% as compared with about 20% for the branch silhouettes consisting of the needles on the main branch and on the attached twigs.

THE AVERAGE WINDSPEED PROFILE

A principal result of the windspeed measurements is the composite scaled speed profile plotted as a dashed line in figure 3 [1]. The

An Approximate Analysis of the Momentum Balance for the Airflow in a Pine Stand

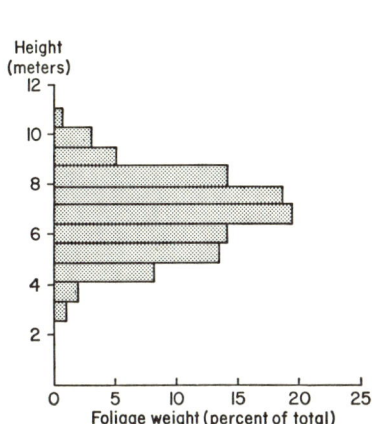

Fig. 1. *Relative distribution of foliage weight with height in the experimental stand.*

Fig. 2. *Horizontal photograph of typical tree crown from the experimental stand.*

local 7-minute average speed was observed to be a constant fraction of (U_*) for the measurements. The average is taken for profiles of (S/U_*) at 34 locations in the stand to form the composite. The profile then represents an average index of flow for the stand as a whole, in contrast to the measurements at separate locations which showed considerable variation between locations for the same values of (U_*) and the same wind direction above the canopy. The profile locations were at the center of the local inter-tree spaces; i.e., the quadrilateral formed by the nearest four stems.

POROSITY

After the windspeed measurements were made, a narrow opening was cut in the stand. The cutting was a 10-m by 50-m rectangle with the long axis perpendicular to the southwesterly winds which prevailed during the previous measurements. Windspeed profiles were measured during southwesterly flow at seven locations equally spaced along the clearing axis in two rows located at 1 m from the upwind edge and at 2 m from the upwind edge, respectively. Windspeed profiles were also measured at some of the original positions upwind from the clearing. When the vertical velocity at the windward clearing edge, as estimated by a simple continuity calculation on the two sets of profiles, is applied as a correction to the apparent flow, it is possible to estimate the horizontal speed at the windward edge. A profile of this estimated average speed scaled by (U_*) is shown as the solid curve on figure 3.

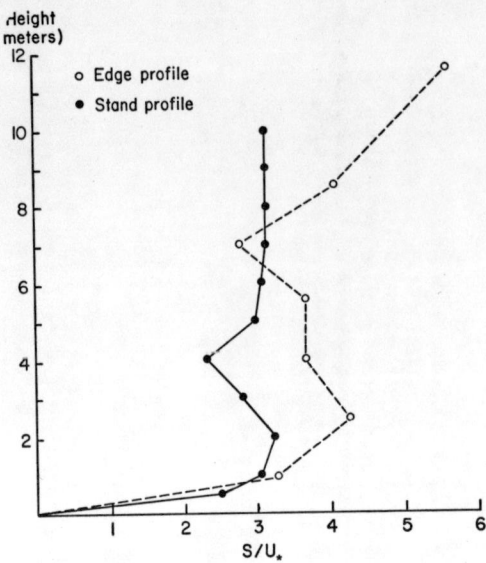

Fig. 3. Vertical profiles of scaled speed in the stand and at the clearing edge.

The windspeed profiles upwind from the clearing showed no significant change, indicating that the clearing induced no large-scale upstream divergence. Therefore the horizontal mass flux through the clearing edge should be equal to that in the upwind canopy flow over the same height interval. When the average scaled profiles of figure 3 are compared, we find that the average windspeed below 11 m at the clearing edge is about 70% of the average speed below 11 m in the upwind canopy. The subcanopy region offers little obstruction to the flow, and thus such an average implies much lower values in the midcanopy region. Since the fraction of the canopy space occupied by foliage material is less than 2%, a considerable portion of the canopy must be occupied by separated flows.

We may define an "effective porosity" (ε) as the fraction of any horizontal plane which is occupied by flow which contributes to the large-scale mass flux. If the rest of the plane is occupied by foliage elements and zones of separated flow created by these elements of some characteristic crosswind dimension (r) and if these elements are widely spaced compared to (r), then it is plausible to assume that

$$\varepsilon = 1 - \beta\rho_f \qquad (1)$$

where (ρ_f) is the volume density of foliage and (β) is a constant which depends on (r) and the porosity of the characteristic foliage element, assumed to be the same at all levels in the canopy. If the average value of (ρ_f) is computed from figure 1, and the other foliage data, then the previous porosity estimate implies that (β) is about 4.2 $m^3 gm^{-1}$. The calculated profile for (ε) using this value in relation (1) is shown in figure 4.

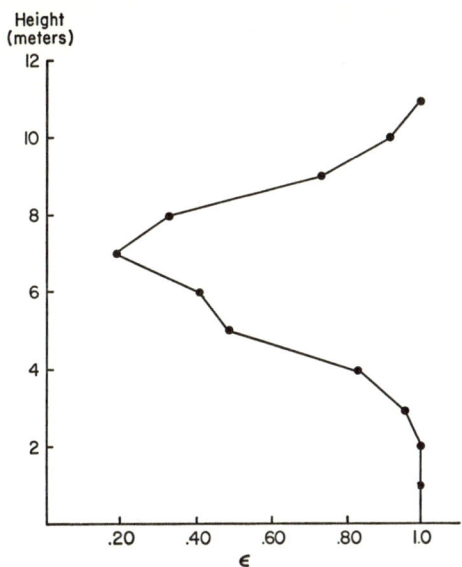

Fig. 4. Estimated variation of effective porosity through the canopy.

MODEL ANALYSIS

The region below the treetops will be considered as composed of two parts: a throughflow phase which is simply connected as far as the streamline pattern is concerned occupying a fraction (ε) of the local volume, and a remaining region consisting of clumps of foliage material and volumes of separated flow to their lee. If we assume that this composite medium is locally homogeneous and isotropic on a scale which is small, compared to the total height of the canopy, where (ε) will be defined as the effective porosity of a differential volume of such a size; and that air density variations may be ignored. It may be shown that [6] the continuity condition is:

$$\frac{\partial}{\partial x_\nu}(\varepsilon U_\nu) = -\frac{\partial \varepsilon}{\partial t} \qquad \nu = 1,2,3 \qquad (2)$$

where, as throughout the remainder of the text, the summation convention is applied to the Greek letter subscripts: x_1, x_2 referring to the horizontal coordinates and x_3 to the height. In this expression U_1, U_2, U_3 are the velocity components in the throughflow phase of the volume. The foliage materials and separated flows will be assumed to be rigid; i.e., the righthand side of equation (2) vanishes.

The force exerted on the stationary phase by the "through phase" can be supposed to be of the form

$$F_i = d(1-\varepsilon)U_i + c(1-\varepsilon)SU_i \qquad i = 1,2,3 \qquad (3)$$

where the coefficients (d) and (c) are independent of (S) and the local flow direction.

The momentum balance for the throughflow phase in a differential control volume may thus be written as

$$\frac{\rho \varepsilon d U_i}{dt} = -\varepsilon \frac{\partial P}{\partial X_i} + F_i + \delta_{3i} g \rho \varepsilon + \theta_i \qquad i = 1,2,3 \qquad (4)$$

where
- P = the pressure
- g = the gravitational constant (980 cm sec^{-2})
- δ_{3i} = the Kronecker delta which equals unity for (i) equal to 3 and vanishes for other values of (i)

and where [6], θ_i represents the viscous effects

$$\theta_i = \frac{\partial}{\partial X_\nu}\left(\mu \frac{\partial U_i}{\partial X_\nu}\right) - \frac{2}{3}\left[\frac{\partial}{\partial X_i}\left(\mu U_\nu \frac{\partial \varepsilon}{\partial X_\nu}\right)\right] \qquad \begin{array}{l} i = 1,2,3 \\ \nu = 1,2,3 \end{array} \qquad (5)$$

The effective viscosity (μ) may exceed the values appropriate to molecular diffusion in the fluid by a component due to the net transport or dispersion by random motions which are comparable to the dimensions of the foliage drag elements [10]. However, if we may assume that most of the turbulent mixing in the medium is by motions with a scale much larger than the characteristic foliage dimension (r), then the dispersive flux on the scale of the differential volume can be treated as gradient diffusion with an effective diffusion constant which is independent of the local speed.

To form the time average momentum balance, from equation (4) we follow the conventional procedure [9] of resolving the velocities into a steady state labeled by the serrated overbar and fluctuating component after subsequent averaging

$$\varepsilon \rho \widetilde{U}_\nu \frac{\partial \widetilde{U}_i}{\partial x_\nu} + \frac{\partial}{\partial x_\nu}(\varepsilon \rho T_{\nu i}) = -\varepsilon \frac{\partial \widetilde{P}}{\partial x_i} + \widetilde{F}_i + \delta_{3i} g \rho + \widetilde{\theta}_i \qquad i = 1,2,3 \qquad (6)$$

where (T_{ij}) is the turbulent Reynolds stress as conventionally indicated [9]. The windspeed observations indicate that the advective terms are comparable to the foliage drag term [1]. If the local invariance of (S/U$_*$) may be taken to imply a similar scaling for (U_i/U_*) then the former could be observed only if both ($\widetilde{\theta}_i$) and the linear term of ($\widetilde{F}_i$) are negligible and if the remaining terms are proportional to (U_*^2) for (i) equal to (1) and (2).

Relation (6) refers to the differential volume of the forest canopy layer. To derive the horizontal average at a particular height, we must carry out a process analogous to that for the time interval average.

The space average velocity to be noted with an overbar may be defined as

$$\overline{u}_i = \lim_{L \to \infty}\left[\frac{1}{4L^2}\int_{-L}^{+L}\int_{-L}^{+L} \widetilde{u}_i \, dx_1 \, dx_2\right] \qquad (7)$$

where the assumption that such a limit exists is equivalent to the steady state assumption implied in the time average. The averages for p, s, and ε are similarly defined.

The only horizontal variation to be considered in the horizontal average variables will be the pressure; i.e., $(\partial P/\partial x_i)$ which will be assumed to correspond to the synoptic scale pressure gradient.

The variations of (ρ) associated with the point-to-point deviations of the air temperature in the horizontal plane and with the vertical variation of the average temperature may be neglected for the horizontal components of the plane-averaged momentum balance, which may be written as

$$\frac{\partial}{\partial x_3}(\overline{\varepsilon\,\overline{U_3^\bullet}\,\overline{U_i^\bullet}} + \overline{\varepsilon^\bullet \overline{U_3^\bullet}\,\overline{U_i}}) + \frac{\partial}{\partial x_3}(\overline{\varepsilon\,\overline{T}_{i3}} + \overline{\varepsilon^\bullet T_{i3}}) = -\overline{\varepsilon}\frac{\partial \overline{p}}{\partial x_i}$$

$$- c(1-\overline{\varepsilon})\overline{s}\,\overline{u}_i - c(\overline{\varepsilon^\bullet s^\bullet u_i^\bullet} + \overline{u}_i\,\overline{\varepsilon^\bullet s^\bullet} + \overline{s}\,\overline{u_i^\bullet \varepsilon^\bullet}) - \overline{\varepsilon^\bullet \frac{\partial p^\bullet}{\partial x_i}} \quad i = 1,2,3 \quad (8)$$

where the time-average variables have been scaled by (U_*) or (U_*^2) as noted by the corresponding lower case symbol and where the dot superscript indicates the local deviation from the plane average.

The continuity equation yields:

$$\frac{\partial}{\partial x_3}(\overline{\varepsilon^\bullet U_3^\bullet}) = 0 \quad (9)$$

The averaged equation for the vertical component reduces to the normal form of the hydrostatic relation with the foregoing assumptions.

Physically, the first term of equation (8) represents the divergence of a momentum flux due to a random distribution of steady vertical motions in the canopy. Since the differential volume of equation (6) is large compared to the individual branch silhouettes, these motions must be of the scale of the tree crowns or the spaces between crowns.

In general, the distribution of the steady-state vertical velocities in the canopy may depend on the temperature distribution as well as the geometry of the canopy. Thermal circulations consistent with the assumptions made for the model could consist of subsidence in relatively cool inter-tree regions and ascent where the temperature is above the average for the stand. Appreciable momentum transfer in this manner would cause a negative correlation between the point-to-point variations in air temperature and those for airspeed in the upper canopy, and the reverse situation below the foliage maximum. Significant correlations are not found for measurements made at the inter-tree space centers [2]; however, the correlation for the entire horizontal plane cannot be estimated from that for such measurements unless we postulate some form for the local variation of (s) and the air temperature. For high solar altitudes, when such circulations might be expected to be most energetic, it is not unreasonable to suppose that the horizontal variation of (s) and a suitably scaled air temperature are similar [5].

$$\frac{s - s_a}{T - T_a} = \frac{s_a - \overline{s}}{T_a - \overline{T}} \quad (10)$$

where the subscript (a) marks the value at the center of the
inter-tree space. Analyses of the available measurements indicate
that neither (s_a) or (T_a) is significantly correlated with the size of
the inter-tree space [2,4], so that the average indicated by
equation (7) may be replaced by the average over the ensemble of
inter-tree spaces. On the further assumption that (s') and (T') are
small and may be approximated by the differential, we find that $\overline{s'T'}$
must be proportional to $\overline{s_a^{\bullet} T_a^{\bullet}}$ and thus close to zero.

Since the differential volume of equation (6) is large compared
to individual live branch silhouettes, the variations in effective
porosity (ε') and the associated variations in (U_3') and the other
variables must be of the scale of the individual tree crowns or the
space between adjacent crowns.

An order of magnitude estimate may be made for the terms involving
(ε') in equation (8) if the similarity of the point-to-point deviations
in s, U_i, and ε may be accepted; i.e.

$$\frac{s - s_a}{s_a - \overline{s}} = \frac{\varepsilon - \varepsilon_a}{\varepsilon_a - \overline{\varepsilon}} \tag{11}$$

and the analogous relation for U_i. Physically, such an assumption
would imply that these random and steady circulations represent a
balance between local deviations in the advective terms, the
horizontal-shear stress and the aerodynamic foliage drag of equation
(8).

To derive an upper bound for $\overline{\varepsilon' s'}$ from equation (11) we assume as
above that the deviations $\varepsilon_a^{\bullet}$, ε', $s_a^{\bullet}$ may be approximated by the
differentials. The lack of correlation between s_a and the size of the
inter-tree space found in the previous study [4] together with the
relative uniformity of the tree crown weights appear to imply that
$\overline{\varepsilon_a^{\bullet} s'}$, $\overline{s_a^{\bullet} \varepsilon'}$ must vanish. The porosity near the center of the inter-
tree space must approximate unity while the average speed over the
entire plane in the canopy must be less than $\overline{s}$. With these
considerations the product $|\overline{\varepsilon' s'}|$ must be less than $\overline{s}(\varepsilon')^2/\overline{\varepsilon}$ at all
levels in the canopy. Since $\overline{(\varepsilon')^2}/\overline{\varepsilon}$ as computed from the foliage
survey data and equation (1) above is less than 10 percent at all
levels, it is not implausible to neglect the fifth term in equation (8)
compared to the fourth.

Entirely analogous estimates follow for $(\overline{U_i' \varepsilon'})$ and $(\overline{T_{i3}' \varepsilon'})$ if we
make the conventional mixing length assumption for the local turbulent
exchange coefficient.

We find that the first five terms involving (ε') of equation (8)
may be dropped by comparison with the corresponding mean value
products with an error of less than 10%.

We may apply the random walk model presented by [10] to represent
the remaining dispersive flux terms in equation (8) by a gradient
process

$$\overline{U_3' U_i'} \sim \frac{\partial \overline{U_i}}{\partial x_3} \tag{12}$$

where the constant of proportionality varies directly with $(\overline{S})$ since the scale of turbulent motions responsible for local mixing is comparable to the scale of the irregularities associated with ε'.

If as above we make the conventional mixing length assumption for $(\overline{T}_{i3})$

$$\overline{T}_{i3} \sim \frac{\partial \overline{U}_i}{\partial x_3} \qquad (13)$$

where the constant of proportionality is a "turbulent viscosity," which also varies directly with (S).

The equations (12) and (13) are of the same form in (S) and $(\overline{U}_i)$ and we may define an "effective viscosity" (K_*) as the sum of a turbulent mixing (K) and a dispersive component.

The remaining term corresponds physically to a concentration of the pressure gradient in the spaces between trees. The basic assumptions of the model rule out separation on the scale of the crowns, clearly implying that the total pressure drop from windward to the lee canopy boundaries is small relative to the foliage drag terms and the horizontal shear stress.

We are now in a position to eliminate or parameterize the terms in the momentum balance corresponding to the point-to-point variations.

The approximate equations are

$$\frac{\partial}{\partial x_3}\left[\overline{\varepsilon} K_* \frac{\partial \overline{u}_i}{\partial x_3}\right] = - \overline{\varepsilon} \frac{\partial \overline{p}}{\partial x_i} - c(1 - \overline{\varepsilon})\overline{s}\, \overline{u}_i \qquad i = 1,2 \qquad (14)$$

ESTIMATE OF THE PRESSURE GRADIENT

The value of $(\partial \overline{p}/\partial x_i)\frac{1}{\rho U_*^2}$ used in the calculation was estimated from the surface isobar pattern over the Wyoming, Montana, and Utah area as indicated by interpolation between the 4-hour surface weather maps for each wind profile measurement used to derive the scaled speed profiles of figure 2. The average of the pressure gradients for the (x_1) direction taken as southwesterly, was 3.1×10^{-5}; the result for the other component was 2.5×10^{-5}.

THE CALCULATION

For calculation purposes, the region below 12 m in the stand can be divided into three regions defined by the magnitude of the foliage drag terms as compared to the viscous terms in equation (14). In the region of high foliage density near mid-crown levels, the flow tends to be a balance between the foliage drag terms and the pressure gradient. In the upper crown, the flow verges on a constant shear stress condition, while in the subcanopy space this condition must be approached near the floor of the stand.

For regions of low foliage density, equation (14) may be approximated by the integral form

$$\overline{\varepsilon} K_* \frac{\partial \overline{u}_i}{\partial x_3} = 1 + \int_z^\infty \overline{\varepsilon}\, \frac{\partial \overline{p}}{\partial x_i} dx_3 - c\int_z^\infty (1 - \overline{\varepsilon})\overline{s}\overline{u}_i\, dx_3 \qquad i = 1,2 \qquad (15)$$

where the integrals were evaluated by the simple trapezoidal rule and the derivative as a centered difference. One of the constraints imposed on the solution is that neither velocity component change sign throughout the layer. K_*, $\bar{u}_2$, $\bar{u}_1$ are evaluated from the integrals of the previous estimate in the iteration.

For the high foliage density levels, the first order finite difference form of (14) was used directly to evaluate $\bar{u}_2$, $\bar{u}_1$, and K_* in that order.

Ten equally spaced grid points were used in the calculation with interpolated values for the $\bar{s}$ and $\bar{\epsilon}$ arrays.

The upper boundary conditions on $\bar{U}_1$ and $\bar{U}_2$ were taken as $\bar{S}$ and zero, respectively; i.e., the (x_1) axis was taken along the direction of the wind above the canopy.

The upper boundary condition for (K_*) was taken as that associated with the displaced law of the wall, with a displacement height of 7 m and roughness length of 50 cm.

At the lower boundary, the shear stresses $(\bar{T}_{3i})$ at the level of the subcanopy speed maximum were assumed to vary directly with the local speed and velocity, allowing an apparent drag coefficient to be defined as

$$\bar{T}_{3i} = c_o \bar{S} \bar{U}_i$$

Since, as noted above, the pressure gradient must have a constant ratio to $(U_*)^2$ and thus to $\bar{s}$ and $\bar{u}_i$ at any point, this assumption does not imply a constant shear stress layer below the speed maximum.

No condition was placed on the sign of (K_*).

CALCULATION RESULTS

The immediate results of the calculation are shown in figures 5 and 6. The calculated values of (c) and (c_o) are 5×10^{-3} and 10^{-9}, respectively. As was expected from the results of the pressure gradient calculation, the cross flow $(\bar{U}_2)$ was small at all levels. It reaches a maximum of 1.3 U_* at 8 m height, just above the level of minimum windspeed, apparently reflecting the minimum in the effective volume drag coefficient, then decreases sharply with increasing depth into the canopy. The maximum directional shear relative to the above-canopy flow is about 20°.

The vertical distribution of the apparent viscosity shows a maximum just above the level of minimum speed, with an increase from 10 U_* to 120 U_* from the base of the live canopy to the 8-m level.

The estimate of the synoptic pressure gradient term in equation (15) is highly approximate; as a check on the effect of error in this term, the calculation was repeated with an estimate based on the Montgomery-Rossby model [11] and the measured aerodynamic roughness of the canopy; i.e., 50 cm [1]. For the same magnitude of the surface shear stress, the model predicts an angular departure of 70° for the above-canopy wind from the direction of the pressure gradient. The results for K_* are shown as unconnected points on figure 6. The most drastic change in the results occurs in U_1, U_2 in the subcanopy region. The effects on the (K_*) distribution are relatively slight. The value of (c) remains at approximately 0.5×10^{-2}.

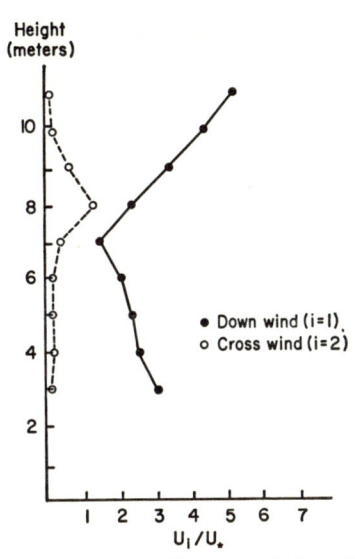
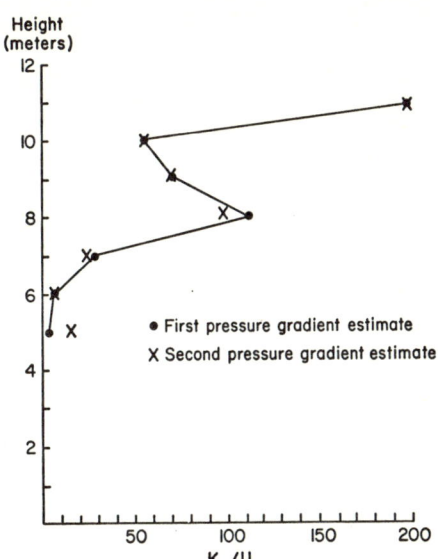

Fig. 5. *Vertical profile of scaled velocity components calculated for the stand.*

Fig. 6. *Vertical profile of effective scaled viscosity calculated for the stand.*

The volume drag constant (c) may be used to estimate the characteristic dimension of the foliage drag elements (r).

If we assume that such elements resemble a random array of cylinders, in regard to wake geometry and drag coefficient [8], then (r) should approximately equal (1/8c) or about 50 cm, which is in reasonable agreement with the observed crosswind dimension of the live branch silhouettes of figure 2.

This value for (r) has definite implications as to the components of K_*. Even with superposition of a large number of wakes, the apparent viscosity to be expected from cylinders of this dimension for the profile of figure 3 is less than 10 U_* [12]. Thus, a major part of K_* in the upper canopy must be due either to turbulent large-scale motions induced by vertical shear, or to dispersion. By analogy with wake flow, however, we would expect dispersion to increase steadily with height above the speed minimum. It seems likely that the local maximum near 8 m must be due to the variation of the dispersivity with $\bar{\varepsilon}$, implying that dispersivity is a major component of K_*, at least in this region.

CONCLUSIONS

There do not appear to be serious discrepancies between the model and the available observations, although the observations cannot furnish a definite confirmation of the model assumptions.

A less speculative treatment of the problem would appear to require extensive sampling of the velocity components and the air temperature in space, particularly through the tree crowns. The analysis of such data in terms of a random process in time and space by conventional spectral techniques should establish the scale and intermittency of the local flow separation. Similar sampling of the local static pressure fluctuations are also needed to shed some light on the pressure term neglected in equation (8).

The characteristic dimension calculated from the computed bulk foliage drag coefficient is in fair agreement with the average crosswind silhouette dimension of the individual live branches with their associated twigs and needle clusters.

The variation of the velocity components with height found for the observations through the canopy cannot be taken as generally typical of such stands, since the observation sequence was dominated by conditions where the synoptic scale flow is strongly ageostrophic. For stronger winds and more level terrain, we might expect the crosswind component to show a maximum in the subcanopy space as well as near the speed minimum. The variation of the scaled viscosity is, however, of some general consequence insofar it suggests a large dispersive component which could be expected to have radically different properties from the turbulent viscosity commonly supposed to account for momentum exchange in the canopy.

REFERENCES

[1] Bergen, J. D. 1971. Vertical windspeed profiles in a pine stand. For. Sci. 17:314-321.
[2] Bergen, J. D. 1974. The independence of the point-to-point variations in windspeed and temperature in a pine stand. USDA For. Serv. Res. Note RM-258, 2 p. Rocky Mt. For. and Range Exp. Stn., Fort Collins, Colo., USA.
[3] Bergen, J. D. 1974. The variation of air temperature with canopy cover within a pine stand. USDA For. Serv. Res. Note RM-253, 3 p. Rocky Mt. For. and Range Exp. Stn., Fort Collins, Colo., USA.
[4] Bergen, J. D. 1974. The variation of windspeed with canopy cover within a pine stand. USDA For. Serv. Res. Note RM-252, 4 p. Rocky Mt. For. and Range Exp. Stn., Fort Collins, Colo., USA.
[5] Bergen, J. D. In press. Vertical air temperature profiles in a pine stand: Spatial variation and scaling problems. For. Sci.
[6] Faizullaev, D. F. 1966. Laminar motion of multiphase media in conduits. Translation (1969) Consultants Bureau. Plenum Publ. Corp.: New York. 144 p.
[7] Gary, H. 1974. Crown structure and the vertical distribution of biomass within a lodgepole pine canopy. (To be published as a Research Paper, Rocky Mt. For. and Range Exp. Stn., Fort Collins, Colo., USA)
[8] Goldstein, S. (ed). 1938. Modern developments in fluid dynamics. Dover, N. Y. 702 p.
[9] Hinze, J. O. 1959. Turbulence. McGraw-Hill: N. Y., 586 p.
[10] Scheidigger, A. E. 1960. The physics of flow through porous media. Univ. Toronto Press. 313 p.
[11] Sutton, E. G. 1953. Micrometeorology. McGraw-Hill: N. Y. 333 p.
[12] Yano, M. 1966. Turbulent diffusion in a simulated vegetative cover. Tech. Rept. FDDL. College of Engineering, Colo. State Univ. 194 p.

A FIELD STUDY OF ATMOSPHERIC EXCHANGE PROCESSES WITHIN A VEGETATIVE CANOPY

GERRIT den HARTOG* and ROGER H. SHAW†

*Atmospheric Environment Service, Department of Environment, Toronto, Canada.
†Purdue University, West Lafayette, Indiana, U.S.A.

ABSTRACT

Measurements within a mature canopy of corn (Zea mays L.) included mean temperature and wind speed profiles, the eddy fluxes of sensible heat and horizontal momentum, and leaf area density. These measurements permitted, for the first time, direct evaluation of leaf drag coefficients and the eddy transport coefficients for heat and momentum within a natural corn canopy. The leaf drag coefficient was found to be nearly constant with height and with wind speed. Measured transport coefficients for heat and for momentum were within 30% of each other and decreased approximately exponentially with depth in the canopy in a manner similar to that of the mean wind speed profile.

LIST OF SYMBOLS

Symbol	Description	Units
$A(z)$	Plant area density	$m^2\ m^{-3}$
C_d	Bulk drag coefficient	
C_d'	Leaf drag coefficient	
C_p	Specific heat of air at constant pressure	$J\ Kg^{-1}\ K^{-1}$
$H(z)$	Eddy flux of sensible heat	$W\ m^{-2}$
$K_h(z)$	Turbulent transport coefficient for sensible heat	$m^2\ s^{-1}$
$K_m(z)$	Turbulent transport coefficient for momentum	$m^2\ s^{-1}$
$T(z)$	Temperature	K
$V(z)$	Total horizontal velocity	$m\ s^{-1}$
a	Canopy flow index	
h	Height of the top of the canopy	m
u	Horizontal velocity component in the direction of the mean wind	$m\ s^{-1}$
v	Horizontal velocity component perpendicular to the direction of the mean wind	$m\ s^{-1}$
w	Vertical velocity component	$m\ s^{-1}$
z	Height above ground	m
ρ	Air density	$Kg\ m^{-3}$
$\tau(z)$	Eddy flux of horizontal momentum	$N\ m^{-2}$

An overbar denotes a time average and a prime the deviation from the average.

INTRODUCTION

A knowledge of turbulent transport processes within plant layers is important because these processes directly affect the exchange of heat, mass, and momentum between the air and both plant parts and the soil surface. These processes thus determine to a large extent the microclimate of such regions. Present knowledge concerning turbulent transfer within plant communities is, however, insufficient to provide a quantitative description of this exchange.

A central problem is the relationship of the drag characteristics and the density and vertical distribution of plant material to the aerodynamic drag. Information on this relationship is needed to model exchange processes involving energy, mass and momentum and thus provide an understanding of an important aspect of the interaction between plants and their environment.

Interest in the microenvironment of plant communities over the past decade or so, has led to the development of a number of mathematical models which attempt to describe wind flow and the vertical fluxes of heat, water vapor, momentum and carbon dioxide within and above plant canopies. Examples of such models can be found in Brown and Covey [1] and Uchijima and Wright [2]. These models usually depend upon an assumption of one dimensionality and rely on parameters such as drag coefficients, eddy transport coefficients and canopy flow indices.

The most common description of the wind profile within relatively dense vegetation has been an exponential relationship of the form

$$V(z) = V(h) \exp[a(z/h - 1)] \qquad [1]$$

where "a" is referred to as an index of canopy flow. Although the above relationship is emperical, some physical meaning has been attributed to the index "a". A list of canopy indices for various types of vegetation, quoted by Shinn and Cionco [3], indicates that factors such as canopy density and element flexibility affect the value of "a". According to these authors, canopy index values range from 0.44 for citrus orchards to 4.40 for gum-maple forest canopies. They suggest that values less than about 2.5 define "ideal" canopies, in which the atmospheric boundary layer shear is still coupled to the flow within the canopy. The value they list for the canopy flow index of mature corn was 1.97.

The eddy transport coefficients of heat and momentum are defined by

$$H(z) = -\rho C_p K_h(z) \frac{dT(z)}{dz} \qquad [2]$$

$$\tau(z) = \rho K_m(z) \frac{dV(z)}{dz} \qquad [3]$$

Various forms of the coefficients have been suggested including a linear increase with height by Tan and Ling (in [4]). An exponential height dependence has been reported by a number of authors (e.g., Uchijima [5] and Cionco [6]). The application of a flux profile relationship to define the transport coefficient, however, can be seriously questioned when distributed sources and sinks are contained in the volume considered. An estimate of the transport coefficient of momentum is often obtained and equality to the other transport coefficients (heat, water vapor, and carbon dioxide) is assumed, in the evaluation of the fluxes involved. Again this method can be questioned.

The bulk drag coefficient C_d is defined by

$$\tau(z) = \rho C_d V^2(z) \qquad [4]$$

and in the air layers above the vegetation is known to be height dependent since τ is approximately constant in the surface boundary layer while the wind speed increases with height. For one particular height, however, C_d is fairly constant as long as measurements are made reasonably close to the surface [7].

The drag force per unit volume of the plant air layer can be expressed as

$$\frac{d\tau(z)}{dz} = \rho\, C_d'\, A(z)\, V^2(z) \qquad [5]$$

The leaf drag coefficient C_d' so defined, has been used by a number of authors in the modelling of the canopy environment (e. g. Uchijima and Wright [2] and Wright and Brown [8]). Uchijima and Wright found, for an immature corn canopy in an indirect determination, that the leaf drag coefficient ranged in value from 0.047 to 0.54 with a mean value of 0.17. The coefficient appeared to decrease slightly with wind speed. A decrease with height and an increase with velocity of the leaf drag coefficient was reported by Wright and Brown.

In this study we present data taken in September and October 1971 at Elora, Canada within and above a mature canopy of corn (Zea mays L.). The measurements included vertical profiles of mean temperature and wind speed, and determinations of the eddy fluxes of sensible heat and horizontal momentum within the canopy. Canopy structure was also determined and included evaluation of the vertical distribution of plant area $A(z)$. Using these measurements, a direct evaluation of the parameters defined above was made possible for the first time.

THE MEASUREMENTS

Measurements were made in corn planted in rows 0.76 m wide at a density of approximately 54,000 plants per hectare. Upwind fetch during the experiments exceeded 115 m at all times and generally was greater than 200 m. The terrain at the site was quite flat and there were no large irregularities in the corn crop with respect to dry or wet areas or holes in the canopy in the immediate vicinity of the measurements.

Differential temperature measurements were made within and above the canopy between seven different levels giving a continuous temperature profile between 0.40 and 4.20 m. Thermocouples of 0.125 mm teflon-coated copper and constantan were butt welded to form a junction not much larger than the diameter of the wire itself. The sensors were painted with white reflectance paint to minimize radiation errors. The output from each thermopile was electronically integrated to yield mean profiles.

Wind speed measurements were made midway between the rows of corn plants at six levels within the canopy using small cup anemometers (Rimco, CSIRO). Above the canopy wind speed measurements were made with Thornthwaite cup anemometers.

Additional wind speed measurements, turbulence and covariance data were available from four fast response, three-dimensional hot-film anemometers [9]. These instruments are cylindrical hot-film anemometers which derive their directional sensitivity from the variation in local heat-transfer coefficient around their circumferences. Each film is split along its length to create two electrically isolated sensitive elements which together detect the angle of incidence of the instantaneous wind velocity. The anemometers were servo controlled to keep them pointing into the mean wind. Turbulence spectra and cospectra of heat and momentum transport within the corn canopy are presented elsewhere [10].

Each anemometer was operated in conjunction with a fast response fine-wire resistance thermometer and analysed on-line by a digital computer. Each of the four parameters (u, v, w, T) was calculated at each sampling point together with its square and all cross products. At the end of each 15 minute sampling period, all means, variances and covariances were computed. The system thus yielded values of sensible heat flux $H(z)$ and momentum flux $\tau(z)$ at up to three levels within, and one level above the corn canopy.

Canopy structure was determined by taking plants from the field and superimposing them on a large grid drawn on sheets of paper. The grid was subdivided into eleven 0.10 m intervals in the horizontal plane and twenty 0.15 m intervals in the vertical. Plant shapes were then drawn on the grid noting the width of plant parts at each intersection. Leaf length and elevation angle were then measured for each subcompartment and a computer program used to calculate total plant area and leaf elevation angles as functions of height.

RESULTS AND DISCUSSION

Canopy structure

Canopy structure was measured on September 22. The plant area index, including stalk, ear and tassel was 3.0 with a standard deviation of 0.30 for the eight plants analyzed. The plants were, as a fraction of total area, 82% leaves, 13% stalk, 3% ear and between

A Field Study of Atmospheric Exchange Processes within a Vegetative Canopy

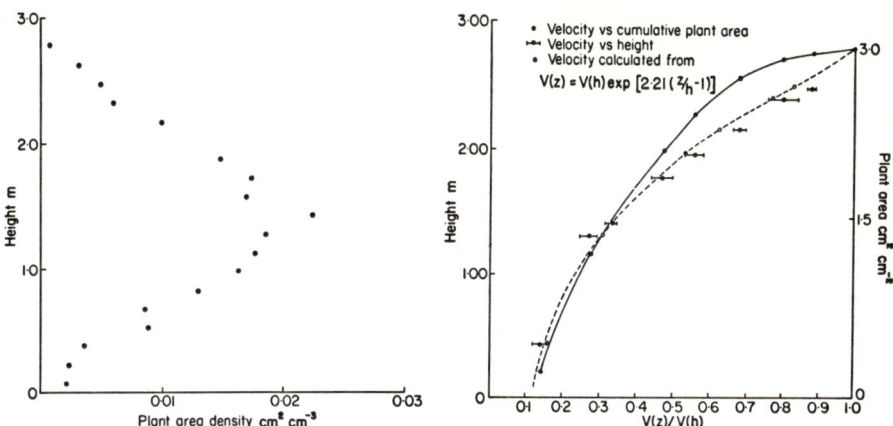

Fig. 1. Plant area density in corn.

Fig. 2. Normalized velocity profile in corn.

1 and 2% tassel. Area for the stalk, ear and tassel was taken as the area of a bisecting plane and all areas represent one side only. The average height of the canopy was 2.80 m with a standard deviation of 0.11 m for 21 plants measured. Plant area density is shown in Fig. 1 as a function of height, and forms a rather symmetrical pattern about the mean canopy height. Changes in the canopy structure due to senescence during the period of observation are discussed later.

Mean velocity profiles

An example of the mean velocity profile, normalized with respect to the top of the canopy and as a function of both height and of cumulative plant area, is shown in Fig. 2. No reversal of the velocity profile was noted down to the lowest height of 0.43 m although additional profiles observed between the 0.10 m and 0.70 m level indicate virtually no shear in this region.

The velocity profile within the canopy can be expressed in terms of an exponential relationship, as has been mentioned, according to equation [1]. For the velocity profile of Fig. 2, the canopy flow index "a" was 2.21 and the correlation coefficient of the regression analysis was 0.99 for the ten 15 minute periods between 1400 and 1630 EST on October 5. The index did not change appreciably during the course of the experiments even though quite drastic changes occurred in the canopy structure toward the end of the period.

A comparison between cup anemometer velocity and the total horizontal velocity as measured with the hot-film sensor indicated an overestimate by the cup anemometers of close to 10%. This overestimate did not appear to change significantly with height in the canopy.

Table 1. Bulk drag coefficients above and within the corn canopy.

Height m	Bulk drag coefficient	Correlation coefficient
1.40	0.20	0.90
1.95	0.14	0.98
2.40	0.095	0.95
2.80	0.064	
3.55	0.043	0.90

Drag coefficients within the corn canopy

Bulk drag coefficients, as defined in equation [4], and calculated from data recorded on October 5 are presented in Table 1. Values are shown for each height at which shear stress was measured, together with correlation coefficients resulting from linear regression analyses of measured shear stress with the square of the total horizontal wind velocity at the same height. An additional value at 2.8 m, the canopy top, was calculated from interpolated values of shear stress and wind speed.

Figs. 3 and 4 show examples of the shear stress-velocity squared relationship at 1.95 m (within the canopy) and at 3.55 m, respectively. The bulk drag coefficient showed no dependence on velocity at any of the measuring heights. Its value increased with depth into the canopy in a manner that can be approximated by an exponential relationship of the form

$$C_d(z) = C_d(h) \exp[2.34(1 - z/h)] \qquad [6]$$

Because shear stress measured at each height was a linear function of the square of the wind speed at that level, a similar relationship holds for the momentum lost within any increment of height within the canopy. We can state, therefore, that the leaf drag coefficient C_d^i, as defined in equation [5], was independent of wind speed over the range of velocities experienced during our measurements.

Shear stress measurements with the hot-film anemometer yield the total drag in the direction of the mean wind. The appropriate velocity term in equation [5] is, therefore, the mean value of the square of the instantaneous total wind resolved in the mean wind direction and defined by

$$\overline{V^2(z)} = \overline{(u^2 + v^2 + w^2) \cos \theta} \qquad [7]$$

where the overbar indicates a time average and the angle θ is measured between the instantaneous wind vector and the mean wind direction. It can be shown [11] that this expression is approximated by

$$\overline{V^2(z)} = \overline{u}^2 + \overline{u'^2} + \overline{v'^2}/2 + \overline{w'^2}/2 \qquad [8]$$

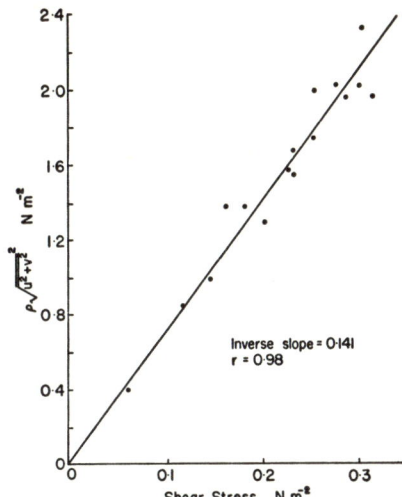
Fig. 3. Shear stress versus velocity squared at 1.95 m.

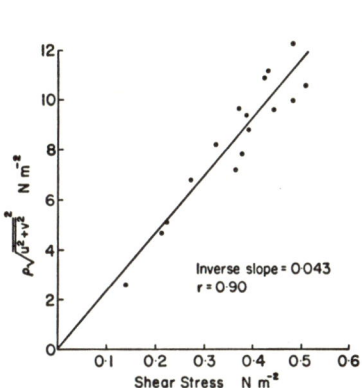
Fig. 4. Shear stress versus velocity squared at 3.55 m.

Each of the terms in equation [8] was evaluated on-line. The ratio of the velocity defined in this manner to the total horizontal velocity $\sqrt{u^2 + v^2}$, also evaluated on-line, did not change significantly with height in the canopy or with time during the experimental period. The mean value of the ratio was 1.15. Use of $V^2(z)$ calculated from equation [8] leads to values of C_d' 0.75 times those obtained if the total horizontal velocity (true cup velocity) had been used.

Leaf drag coefficients calculated from equation [5] using equation [8], for September 30 and October 5, are shown in Table 2.

Analysis and interpretation of these data are limited due to the lack of canopy structure information throughout the period. Canopy measurements were obtained on September 22. From visual observations,

Table 2. Leaf drag coefficients in the corn canopy.

Date	Height interval m	Leaf drag coefficient	Standard deviation	Number of runs
Sept. 30	1.67-2.03	0.119	0.030	26
	2.03-2.80	0.163	0.046	
	1.67-2.80	0.156	0.029	
Oct. 5	1.40-1.95	0.076	0.020	16
	1.95-2.40	0.083	0.029	
	2.40-2.80	0.069	0.040	
	1.40-2.80	0.077	0.012	

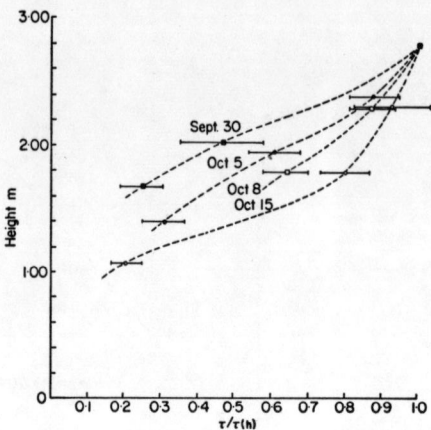

Fig. 5. Normalized momentum flux profiles in corn.

very little structural change occurred during September but leaves started to turn brown and tassels began to break early in October. The most rapid deterioration of the canopy occurred between October 8 and October 15 during cold wet weather. By October 15 nearly all leaves were dead and many tassels and stalks were broken resulting in deeper penetration of the external flow. This is illustrated most clearly by the change with time of the normalized shear stress profiles in Fig. 5. Hortizontal bars indicate the magnitude of the standard deviation of the normalized shear stress at each height. Normalized profiles of heat flux inside the canopy [10] are not presented here but were of similar form and were displaced in the vertical by less than 0.1 m from equivalent shear stress profiles in all but the lowest portions of the curves. This implies that the relative source and sink strengths for heat and momentum were similarly distributed within this canopy at this time.

For the reasons given, the leaf drag coefficients measured on September 30 are believed to be the more representative values for a corn canopy. The mean leaf drag coefficient measured in the upper half of the canopy on this occasion was 0.16, very close to the mean value determined indirectly by Uchijima and Wright [2].

The lower values reported for October 5 arise because no correction was made for the decrease in leaf area that occurred between the two dates. These data indicate that the change of C_d' with height was small but this can be taken to be true only if we can correctly assume that the change in the canopy structure occurred uniformly throughout all levels. This, of course, is open to question. However, it does appear that large variations of the leaf drag coefficient do not exist. This result, plus the observed constancy of C_d' with wind speed, over the range experienced, suggests that the momentum balance procedure for evaluating drag distribution within the canopy [8] is an acceptable approach.

Table 3. Ratio of the transport coefficient for heat to that for momentum.

Height m	1.40	1.95	2.40	2.80	3.55
K_h/K_m	0.76	0.70	1.21	1.02	0.91
Standard deviation	0.4	0.5	0.2	0.2	0.2

Wind tunnel measurements of drag on single leaves, reported elsewhere [11], produced drag coefficients ranging between 0.13 and 0.30 depending primarily on angle of attack and, to a smaller degree, on wind velocity and leaf size. The fact that the field measurement of C_d^l lies at the lower end of the range measured in the wind tunnel is probably a result of mutual shading of leaves. Shading tends to decrease the overall drag coefficient [12].

Eddy transport coefficients

Transport coefficients were calculated from equations [2] and [3] using a graphical evaluation of the slope of the temperature and velocity profiles. Neither the fluxes of heat and momentum nor the gradients of temperature and wind speed reversed sign, at least down to the lowest point of measurement.

Ratios of the mean value of the transport coefficient for heat to that for momentum are given in Table 3. Considering the relatively large standard deviation of the individual observations as well as the possibility for errors in the evaluation of the gradients, no evidence is presented that suggests that the ratios differ significantly from the value of 1. Similarly, no significance should be applied to the increase, indicated by the data, in the ratio between 1.95 and 2.40 m.

Spectral analyses of the heat and momentum fluxes, both above and within the canopy [10], indicate that, for this canopy, eddies carrying heat and momentum were of similar frequency scale. It is recognized that the structure of the corn, particularly at the end of the experimental period, was not typical of a healthy crop and equality of the transport coefficients and similarity in the scales of motion associated with the transport may not be representative. This might be especially true if gradients of the mean quantities reverse sign within the canopy.

Both of the transport coefficients decreased in magnitude with depth in the canopy. An exponential relationship, suggested by Uchijima [5] and used by a number of authors including Brown and Covey [1], appears to fit our experimental data reasonably well.

SUMMARY

Data on vegetation structure, wind speed and temperature profiles and the vertical distribution of heat and momentum fluxes were used to investigate, by direct measurement in the field, the behavior of drag coefficients and transport coefficients within a mature corn canopy.

The profile of mean wind speed within the canopy was quite accurately represented by the exponential relationship of equation [1], the canopy flow index remaining approximately constant during a period in which the canopy showed considerable senescence.

The bulk drag coefficient increased with decreasing height in the canopy at least to the 1.40 m level. It was not dependent on velocity. There appeared to be little change in the leaf drag coefficient with respect to height or velocity within the canopy. This result tends to validate the momentum balance procedure for evaluating the drag distribution within the canopy (e.g. Wright and Brown [8]).

The velocity term, used to define the leaf drag coefficient, was the mean value of the square of the total instantaneous velocity, resolved in the direction of the mean wind. The ratio of this velocity term to the square of the total horizontal velocity (true cup velocity) was found to be 1.32. Use of true cup velocity therefore would lead to an overestimate in the leaf drag coefficient by a factor equal to this ratio. Large errors can thus result in determining the momentum flux profile within the corn canopy if the correct velocity term is not used.

Transport coefficients within the canopy appeared to be constant proportions of the value at the top of the canopy and decreased approximately exponentially with height to the 1.40 m level. The ratio of the transport coefficient for heat to that for momentum ranged from 0.70 to 1.21 within the canopy.

The results of this research indicate that canopy flow models, based on the measurements of mean values at the top of the canopy are possible at least for crops such as corn. The corn canopy presents a relatively homogeneous environment in terms of plant area distribution where preferred regions of upward or downward motion of air do not appear to be present [10]. More measurements are needed in the lower regions of the canopy where leaf area diminishes and velocity gradients become small.

REFERENCES

[1] Brown, K.W., and W. Covey., 1966: The energy budget evaluation of the micrometeorological transfer processes with a cornfield, Agr. Meteorol. 3, 73-96.

[2] Uchijima, Z., and J.L. Wright., 1964: An experimental study of air flow in a corn plant air layer. The Bulletin of the National Institute of Agricultural Sciences (Japan). Series A., 11, 19-66.

[3] Shinn, J. H., and R. M. Cionco, 1973: A note on observations of turbulence and mean flow in vegetation canopies. Presented at the 11th National Conference on Agricultural and Forest Meteorology (AMS), Duke University, Durham, North Carolina, 8-10, January 1973.

[4] Lemon, E.R., 1963: The energy budget at the earth's surface. Part II. Production research report No. 72. Agricultural Res. Service U.S. Dept. of Agriculture. 49 pp.
[5] Uchijima, Z., 1962: Studies on the microclimate within plant communities. I. On the turbulent transfer coefficient within plant layers. J. Agr. Meteorology (Tokyo), 18, 1-9.
[6] Cionco, R.M., 1965: A mathematical model for air flow in a vegetative canopy. Journal of Applied Meteorology., 4, 517-522.
[7] Bradley, E.F., 1972: The influence of thermal stability on a drag coefficient measured close to the ground. Agricultural Meteorology., 9, 183-190.
[8] Wright, J.L., and K.W. Brown, 1967: Comparison of momentum and energy balance methods of computing vertical transfer in a crop. Agronomy Journal, 59, 427-432.
[9] Shaw, R.H., G. Kidd, and G. W. Thurtell., 1973: A miniature three dimensional anemometer for use within and above plant canopies. Boundary Layer Meteorology, 3, 359-380.
[10] Shaw, R. H., R. H. Silversides and G. W. Thurtell., 1974: Some observations of turbulence and turbulent transport within and above plant canopies. Boundary Layer Meteorology. In press.
[11] den Hartog, G., 1973: A field study of the turbulent transfer of momentum between the atmosphere and a vegetative canopy. Ph.D. Dissertation. University of Guelph, Ontario. Canada. 88 pp.
[12] Hoerner, S. E., 1965: Fluid dynamic drag. S. F. Hoerner, Midland Park, N.J., U.S.A., 452 pp.

20

ENERGY AND MASS EXCHANGE OF A NATIVE GRASSLAND IN SASKATCHEWAN

EARLE RIPLEY and BERNARD SAUGIER

*University of Saskatchewan, Saskatoon, Canada,
Centre National de la Recherche Scientifique, Montpellier, France*

Fluxes of sensible heat, water vapour and carbon dioxide were calculated above a natural grassland, during most of a growing season, using measured profiles of windspeed, temperature, humidity and CO_2 concentration, as well as radiation and soil heat flux measurements. These micrometeorological measurements have been related to soil, plant and atmospheric factors and a model of water flow and CO_2 fixation has been constructed.

LIST OF SYMBOLS

E – evaporation (g m^{-2} s^{-1})
F – atmospheric CO_2 flux (g m^{-2} s^{-1})
G – global radiation (W m^{-2})
H – sensible heat flux (W m^{-2})
L – latent heat of evaporation (J g^{-1})
LAI – green leaf area index
P – CO_2 chloroplast flux (g m^{-2} s^{-1})
R_i – Richardson number
R_n – net radiation (W m^{-2})
S – soil heat flux (W m^{-2})
T, T' – dry- and wet-bulb temps. (°C)
W – leaf respiration (g m^{-2} s^{-1})

e – water vapour pressure (mb)
k – root conductivity (cm s^{-1} bar^{-1})
v – water extraction by roots (cm s^{-1})
z – height (m)
α – initial slope of photosynthesis/radiation curve (g CO_2 J^{-1})
β – Bowen ratio (H/LE)
γ – psychrometric constant (mb °C^{-1})
Δ – slope of saturation vapour curve with temperature (mb °C^{-1})
χ_a – absolute humidity (g m^{-3})
χ_s – saturated abs. humidity (g m^{-3})
ψ – water potential (bars)

C_a, C_c, C_i – atmos., chloroplast and intercellular CO_2 concentrations (g m^{-3})

$r_b, r_s, r_{cut}, r_{can}$ – boundary layer, stomatal, cuticular, and canopy resistances for water vapour (s m^{-1} or s cm^{-1})

r_b^c, r_s^c, r_m, r_c – boundary layer, stomatal, mesophyll and carboxylation resistances for CO_2 (s m^{-1} or s cm^{-1})

Note: Energy fluxes in graphs are ly min.$^{-1}$ (1 ly min^{-1} = 698 W m^{-2})
CO_2 fluxes in graphs are g m^{-2} h^{-1}. (1 g m^{-2} h^{-1} = 1/3600 g m^{-2} s^{-1}).

1. INTRODUCTION

The research reported in the present paper formed part of a total ecosystem study of a natural grassland in the mixed-grass prairie region of western Canada (1). This study, the Matador Project (2), was instituted under the International Biological Programme and was aimed at analyzing and synthesizing the flows of energy and mass (water, carbon and nutrients) through the system. The ultimate objective has been the production of a computer model of the complete ecosystem, capable of explaining the observed year-to-year variation in production and of predicting the effects of man-induced stresses (such as fire, irrigation, grazing, addition of nutrients) on the system. The use of such a model would provide a sound basis for the utilization and management of the natural, and adjacent agricultural, areas.

The micrometeorological studies have been directed towards analysis of the microclimate in the biotic zone and to the exchanges of energy and mass between the major components of the system. To these ends, pertinent field measurements were gathered over several seasons and have been analyzed in relation to plant and soil factors. Models are being constructed that will simulate the microclimate and energy and mass exchanges in relation to the macrometeorological driving variables, and the vegetation and soil properties.

The present paper describes briefly the micrometeorological measurements and the derivation of the fluxes. It then discusses the variation of the fluxes, both seasonally and diurnally, and gives a tentative explanation of the observed variations by means of a model based on the interactions between evaporation, carbon dioxide flux, and soil and plant water status.

2. METHODS

2.1 Site and instrumentation

The project site was located at latitude 50°42'N, longitude 107° 43'W, at an elevation of 684 m above sea level in the mixed-grass region of the northern Great Plains of North America. The micrometeorology study area (3) was situated on a plain that had been the bed of a former glacial lake. The land is quite flat, rising gradually to the north to a range of hills of height 120 m above the plain at a distance of 8 km. Two to three kilometers to the south there is an abrupt drop of about 120 m to the valley of the South Saskatchewan River.

The vegetation has been classified as the Agropyron-Koeleria faciation of the Stipa-Bouteloua association (1). On the heavy clay soil of the study area, the vegetation is fairly uniform and composed principally (75 percent of above ground biomass) of the two grass species, Agropyron dasystachyum and A. smithii (4).

The micrometeorological site (5,6) had a minimum fetch of 500 m over ungrazed prairie and there was a distance of 2 to 3 km to the nearest major change in surface roughness. The above-canopy profile measurements consisted of dry- and wet-bulb temperatures, windspeed, and CO_2 concentration (Table 1) with a top sampling height of 6.5 m.

Table 1. Measurement Summary.

Height (cm)	Dry-Bulb Temp.	Wet-Bulb Temp.	Windspeed	CO_2 Conc.
650	X		X	X
330	X		X	X
250	X			
170	X	X	X	X
90	X	X	X	X
50	X	X	X	X
30	X	X	X	X
20	X	X	X	

Radiation (global, reflected, net, total) at a height of 2 m.
Canopy air temperatures at heights of 2, 5, 7.5, 10, 15 cm.
Soil temperatures at depths of 1, 2, 5, 10, 15, 20, 50, 100 cm.
Soil heat flux, leaf temperatures, wind direction, atmospheric pressure, rainfall.

The sensors (6) were spaced at approximately logarithmic intervals above the zero plane (about 10 cm). Measurements were taken at three-minute intervals over most of the growing season using a digital datalogger recording on magnetic tape (6). Platinum resistance thermometers were used for the measurement of air temperature; miniature cup anemometers for windspeed (7,8); and an infrared gas analysis system (6, 9,10) for carbon dioxide concentration. Conventional thermopile-type radiation sensors and soil heat flux plates were used, and leaf and soil temperatures were measured with thermocouples. All data were averaged for 30-minute periods before further analysis.

2.2 Calculations of Diffusivities and Fluxes

Both the energy-balance and aerodynamic methods were used to calculate the fluxes (of heat - H, water vapour - LE, and carbon dioxide) from the measured data (11).

In the energy-balance case, the Bowen ratio was calculated using an orthogonal regression between the five dry- and wet-bulb temperatures:

$$\beta = \frac{H}{LE} = \gamma \frac{dT/dz}{de/dz} = \frac{\gamma}{(\Delta+\gamma)dT'/dT - \gamma} \qquad (1)$$

Combining this relationship with the one-dimensional energy conservation equation for the ground surface:

$$H + LE = R_n - S \qquad (2)$$

yields values of the fluxes H and LE.

For the aerodynamic case, wind, temperature and humidity profiles were used to derive a stability parameter for each particular run. The integrated flux profile relationships then allowed calculation of the fluxes from theoretical profiles fitted to the measurements. The log-linear profile (12) was used under stable conditions, while flux-profile relationships of Dyer and Hicks (13) as integrated by Paulson (14) were used under unstable conditions.

2.3. Comparison of Energy-Balance and Aerodynamic Methods

As an assessment of the methods the sum of the sensible plus latent heat fluxes calculated by the aerodynamic method $(H+LE)_{aero}$ was compared in Fig. 1 with net radiation minus soil heat flux (R_n-S in equation 2). Gradient Richardson numbers (at a height of 1 m) during the period of comparison ranged from -0.10 to +0.16. Although more stable conditions did occur, the latter value was considered to be near the upper limit for fully turbulent flow and no aerodynamic calculations were made when $R_i > 0.16$. $(H+LE)_{aero}$ may be shown to be directly proportional to the wet-bulb temperature gradient and only indirectly (through the stability correction) dependent on the dry-bulb gradient. Although there is a considerable amount of scatter of the points in Fig. 1, the general agreement is good except at the two extremes. In stable conditions, the negative aerodynamic values are

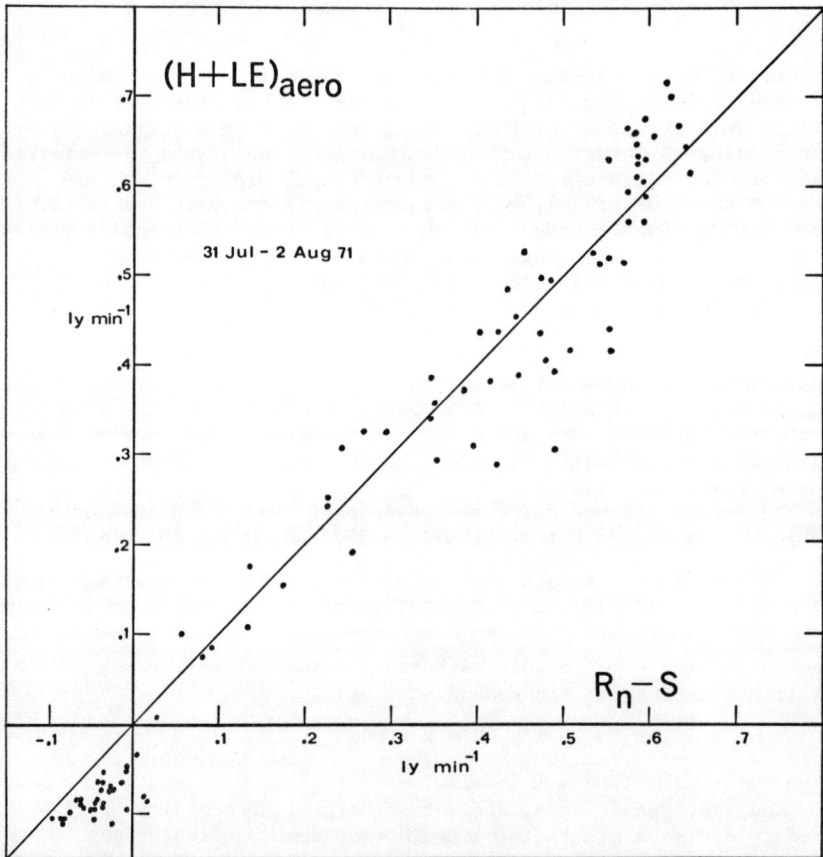

Fig. 1. Comparison of $(H + LE)_{aero}$, the sum of sensible and latent heat fluxes computed using the aerodynamic method, and the measured term $R_n - S$, net radiation minus soil heat flux.

consistently greater than the energy-balance estimates. This might indicate that the log-linear profile is not the best choice for stable conditions (15), although measurement error in $R_n - S$ is important at night. The discrepancy at high values of $R_n - S$ may reflect the difficulty in the measurement of soil heat flux, computed from changes in soil temperatures.

3. SEASONAL AND DIURNAL PATTERNS OF THE FLUXES

3.1. Seasonal Patterns

While the seasonal variation of the net incoming energy is mainly dependent on earth-sun geometry and local cloud patterns, the partitioning of this energy into sensible and latent atmospheric fluxes and soil heat flux depends mainly on plant and soil properties. Some of these are summarized in Fig. 2 while a more detailed description may be found in (16) and (17). The green leaf area index (4) rose from 0.2 in early May, 1971, to a peak of 1.0 in early July, then dropping to 0.6 and remaining there through August and September. The 0-30 cm soil moisture content was high after the snowmelt recharge in early April, dropping during a dry May, and rising again during a relatively rainy period in June and early July. Almost no rain fell during the remainder of the season, resulting in a complete lack of water available to the vegetation in this layer during August and September.

The pattern of net radiation shows a somewhat erratic rise to a peak of over 300 ly d^{-1} in late July, followed by a steady decline as the days grew shorter in early September. About 8 percent of the net radiant energy went into the soil during early and mid-summer with the remainder being divided between the sensible and latent atmospheric fluxes. The small amount of energy (roughly 1 percent of R_n) used in photosynthesis has been neglected. In mid-May the greater part of the atmospheric flux was in the form of sensible heat, but with the increase in green leaf area the latent heat flux soon dominated, utilizing almost 80 percent of the net radiation in mid-July. With the drying-out of the soil and decrease in green leaf area, the latent heat flux declined during August and September.

The net atmospheric carbon dioxide flux has been separated into day and night-time components. The daytime flux represents the CO_2 assimilated by the vegetation minus the CO_2 respired by the soil while the night-time flux is the sum of soil (root plus micro-organism) respiration and shoot respiration. Some measurements inside the vegetation canopy have permitted an estimate to be made of the CO_2 flux from the soil (18). Over the season, the soil supplied as much CO_2 to the vegetation during the day as did the atmosphere. The relative daytime CO_2 flux has been plotted in Figure 2a for comparison with the green leaf area index. The two curves closely parallel each other until August when the CO_2 flux declines rapidly in spite of a continuing fairly high green leaf area. This difference will be discussed later. The daytime CO_2 flux reached its peak in early July, more or less in phase with green leaf area index and soil moisture content. The night-time CO_2 flux, however, did not reach its peak until several weeks later when soil temperatures were at their highest.

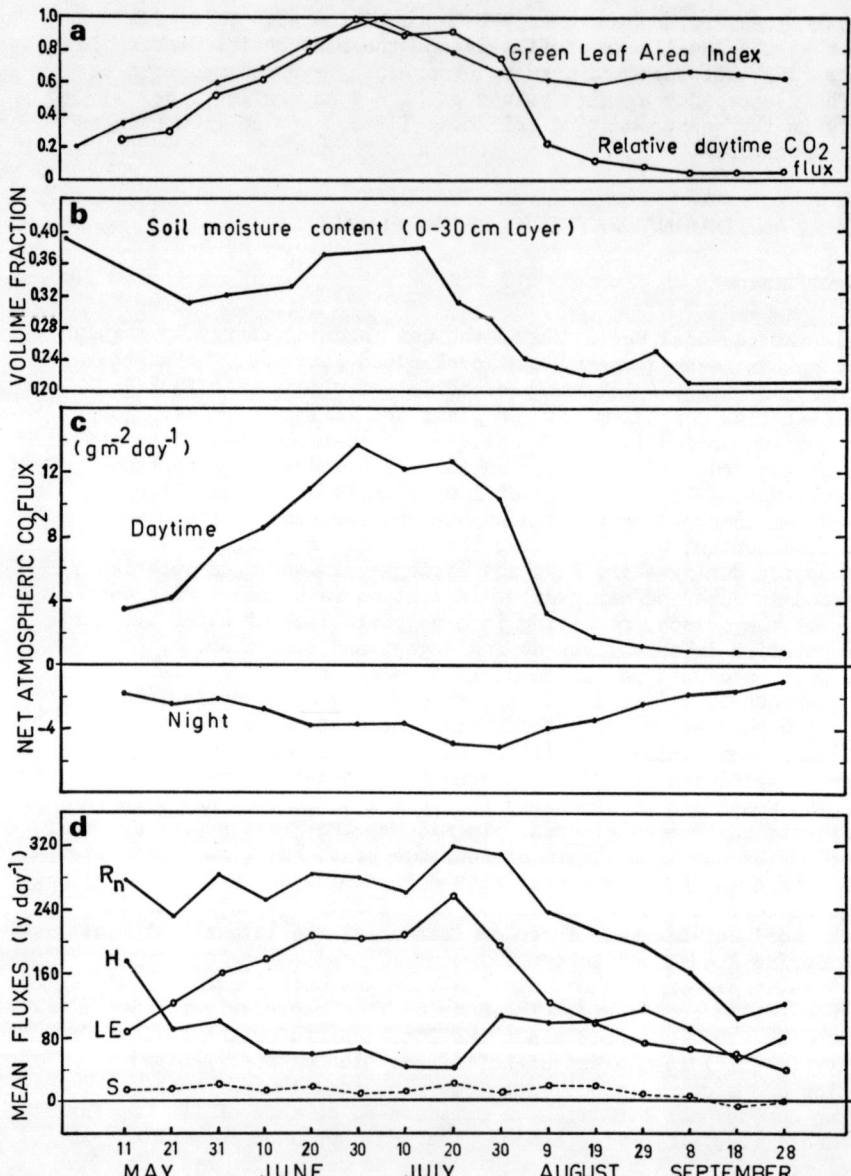

Fig. 2. Variation during 1971 growing season at Matador of: a. Green leaf area index and daytime atmospheric CO_2 flux (10-day means relative to maximum -13.7 g m^{-2} for 30 June); b. Soil moisture content of the 0-30 cm layer; c. 10-day means, centered on indicated date, of day- and night-time atmospheric CO_2 fluxes (taken as positive downwards); d. 10-day means of the surface energy fluxes; H is the aerodynamic estimate and LE is the residual after substituting Haero in the energy-balance equation (2).

3.2. Diurnal Patterns

Ten-day means of the water vapour and carbon dioxide fluxes for a number of periods during the season have been plotted as functions of time of day in Fig. 3. In addition to the seasonal trends noted earlier, these plots reveal some other interesting features.

The water vapour fluxes are always positive, showing that evaporation usually continues through the night. The peak latent heat fluxes occur one to two hours after solar noon. This may be contrasted with the CO_2 fluxes which usually reach a peak in mid or late morning. Although both fluxes encounter the same two major resistances (stomatal and boundary-layer) on their paths which would tend to keep them in phase, the driving forces are quite different. In the case of water vapour the driving force is related to the difference between the leaf temperature that usually reaches a peak at mid-afternoon and the air dew-point, that decreases during the afternoon. In the case of carbon dioxide, the driving force is the difference between the ambient air CO_2 concentration and that at the leaf chloroplasts. The former has only a slight daytime variation while the latter is highly dependent on photosynthetic rate, i.e. on incident radiation, leaf water status and leaf temperature.

The earliest 10-day period (21 May) in Fig. 3 shows a very broad mid-day peak in the CO_2 flux, as does that for 20 June. By mid-July there is evidence of an afternoon depression. This effect is most obvious on the 19 August curve as a pronounced mid-day depression showing some recovery in the late afternoon as leaf temperatures drop. By mid-September, in spite of similar soil moisture conditions, the mid-day depression has disappeared, presumably because of the much reduced driving force and leaf temperature.

4. MODELLING

It is clear from the preceding results that the available soil water affects in a decisive manner the fluxes of water vapour and CO_2, and consequently the growth of the grassland. Thus in the modelling approach, emphasis is put on plant-water relations rather than on light interception, treating the canopy as a single leaf. This may be justified on the basis of the low green leaf area index of the canopy.

4.1. Water Model

Fig. 4 represents the main aspects of the water model, separated for convenience into three sections. Section A describes the input of water to the plants, v, that is computed as:

$$v = \sum_{i=1}^{6} v_i$$

where v_i is the water uptake by the roots of the layer i of soil, taken to be proportional to the difference of water potential between the soil ψ_{si} and the plant ψ_p when this difference is positive:

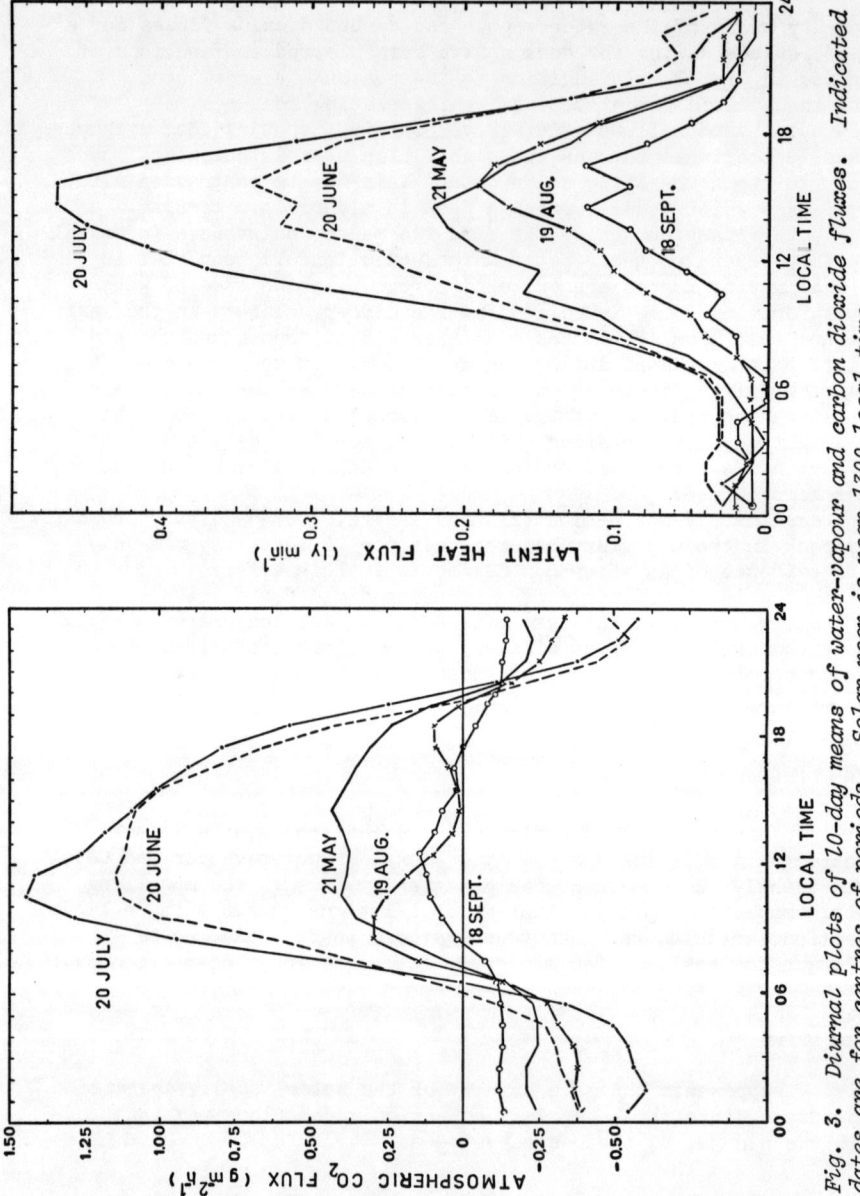

Fig. 3. Diurnal plots of 10-day means of water-vapour and carbon dioxide fluxes. Indicated dates are for centres of periods. Solar noon is near 1300 local time.

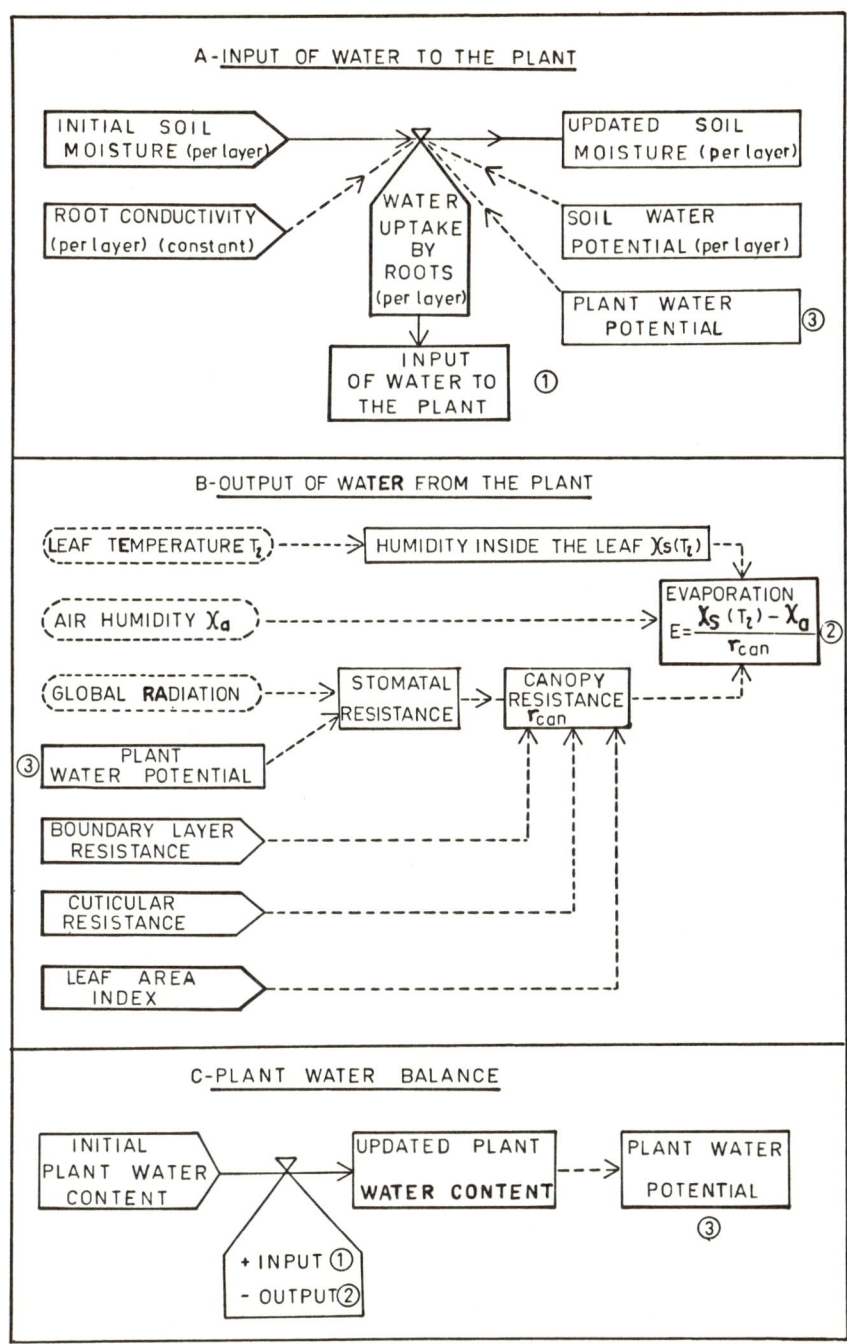

Fig. 4. Model of water movement in the soil-plant-atmosphere system.

$$v_i = k_i (\psi_{si} - \psi_p) \quad \text{if } \psi_{si} > \psi_p$$

$$v_i = 0 \quad \text{if } \psi_{si} < \psi_p$$

where the bulk root conductivity of layer i, k_i, has been taken equal to the total root conductivity times the relative biomass. Soil resistance of the root zone to water flow has been neglected, since it is usually much smaller than the resistance across the root (19).

Fig. 4B summarizes the movement of water from the plant by evaporation through the stomata and the cuticle and then through the boundary layer of the leaf. The canopy resistance r_{can} is computed as:

$$r_{can} = [r_s r_{cut}/(r_s + r_{cut}) + r_b]/LAI$$

with r_s, r_{cut} and r_b respectively stomatal, cuticular and boundary layer resistances of the two sides of the leaf and LAI leaf area index. r_{cut} is taken as constant (40 s cm^{-1}), as is r_b (0.1 s cm^{-1}) which is always very small on account of the narrow width of the leaves (0.3 cm) and high wind speed.

The stomatal resistance is taken as a function of plant water potential ψ_p and global radiation, G:

$$r_s = 1.2/G^{(0.20-0.02\psi_p)} \quad \text{if } \psi_p > -15 \text{ bars.}$$

$$r_s = 1.2 \exp[-0.14(\psi_p+15)]/G^{(0.20-0.02\psi_p)} \quad \text{if } \psi_p < -15 \text{ bars.}$$

In these formulae, G is in ly min^{-1}, ψ_p in bars and r_s in s cm^{-1}. r_s increases quite quickly when ψ_p drops below the threshold value of -15 bars; the inclusion of ψ_p in the radiation term allows for a more gradual stomatal response to light under water stress. These formulae have been derived from field measurements.

Fig. 4C shows how the plant water potential is computed from the changes in water content of the plants. The relation between water content and water potential was obtained from field data.

4.2. CO_2 Model

The photosynthesis model is adapted from Chartier (20) for plants with photorespiration. The photorespiration W is supposed to join the main CO_2 path at the intercellular level (C_i) as shown in the diagram, where C_a is the air CO_2 concentration, C_c the CO_2 concentration at the chloroplasts, r_b^c and r_s^c the boundary layer resistance and the stomatal resistance of the leaf to CO_2, r_m the mesophyll resistance and r_c the carboxylation resistance. From the diagram one may write:

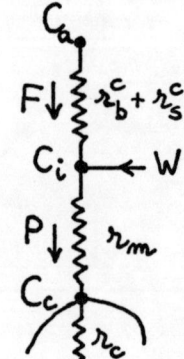

$$C_c = C_a - F(r_b^c + r_s^c + r_m) - W r_m.$$

A second relation combines the effects of radiation and CO_2 concentration on chloroplast activity (21, 20):

$$P = F + W = \alpha G/(1 + \alpha G r_c/C_c)$$

In these relations P is the CO_2 flux into the chloroplasts, F the net CO_2 flux entering the leaf, G the global radiation and α a constant related to the maximum quantum yield of photosynthesis.

Elimination of C_c between the two relations allows calculation of F from G and C_a. The mesophyll resistance is taken as a constant (3 s cm^{-1}). Temperature influence is introduced through the carboxylation resistance that varies from 1.4 s cm^{-1} at 20°C to 50 s cm^{-1} at 45°C and 21 s cm^{-1} at 0°C. These values are derived from laboratory studies of photosynthetic response to temperature of individual leaves. Water potential influences stomatal resistance as discussed in section 4.2; r_s^c is taken as equal to r_s, the resistance to water vapour, times 1.64, the ratio of molecular diffusivities of water vapour and of CO_2 in air. The leaf respiration in the light is taken as a constant fraction (0.2) of the CO_2 flux into the chloroplasts, which is a reasonable assumption at atmospheric CO_2 concentrations (22), while the dark respiration is taken as 0.025 x $2^{T_1/10}$. Finally, the soil respiration has to be subtracted from the net photosynthesis to get the atmospheric CO_2 flux; it was estimated from the values measured during the preceding nights. A diurnal variation of soil respiration based on soil temperature at 5 cm has been introduced for 15 July but for 20 August a constant value was taken because microorganism activity was thought to be then confined to the deeper, moist layers of soil where temperature changes are very small.

4.3. Model Output Compared to Measurements

Fig. 5 presents results obtained with the model for two clear days contrasting with respect to initial soil moisture conditions as shown in the following table, adapted from (23), where water potential has been computed from moisture retention curves given in (24).

	Depth (cm)	0-2	2-5	5-15	15-30	30-45	45-90
15 July	Moisture (vol %)	33.0	34.0	36.5	38.5	35.9	35.8
1971	Water potential (bars)	-3.9	-2.9	-1.5	-0.9	-1.8	-1.8
20 Aug.	Moisture (vol %)	16.0	17.5	20.5	25.3	28.2	30.8
1971	Water potential (bars)	-960	-590	-220	-76	-25.7	-9.9
	Root conductivity (in 10^{-7} cm s^{-1} bar^{-1})	1.36	1.12	2.50	2.0	1.36	2.9

On 15 July, roots were able to extract water from all layers while on 20 August, extraction was confined to the two bottom layers and most of the time, to only the bottom layer. As a result, plant water potential reached a minimum of only -16.2 bars on the July day, which caused practically no stomatal closure, and a minimum of -28.3 bars on the August day, which led to a significant increase in stomatal resistance and lowered the evaporation rate to less than half of what it was in July, in spite of higher leaf temperatures. The dot above the curve of water potential (Fig. 5) represents one measurement of leaf water potential taken from (25).

The crosses in the evaporation curves represent rates derived from micrometeorological measurements. There is a good overall agreement, although evaporation from the model reaches its maximum earlier in the afternoon. Direct soil evaporation, that is not taken into account by the model, may be responsible for the difference.

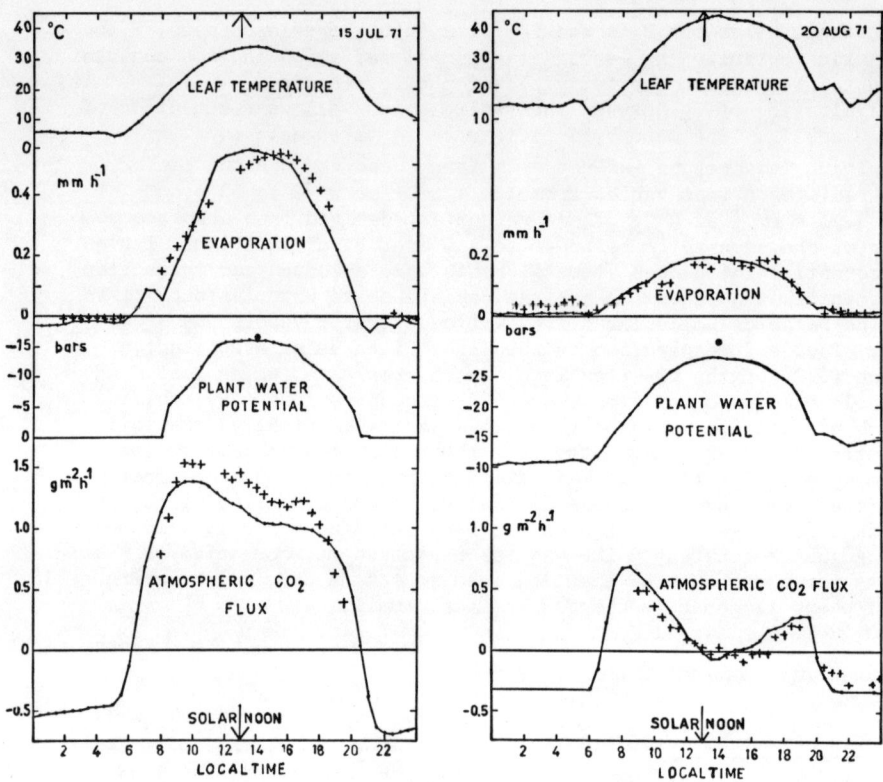

Fig. 5. Comparison of model output with measured water vapour and CO_2 flux data. The lines represent the model output and the crosses the measured values. For reference are also included measured leaf temperatures and modelled plant water potential. The single dots on each of the latter curves are measured plant water potentials.

The main features of atmospheric CO_2 flux are also adequately simulated, as seen from comparison between the solid line (model) and the crosses, that represent calculations of the fluxes using the aerodynamic method. The decrease in CO_2 flux between 1000 and 1600 on 15 July arises from two factors: a decrease of photosynthesis caused by high leaf temperature and an increase in soil respiration. The midday depression observed on 20 August is caused mainly by excessive leaf temperatures (max. 43.5°C) and only to a lesser extent by stomatal closure. While the canopy resistance to CO_2 increases from 7.5 s cm^{-1} at 0800 to 16.5 s cm^{-1} at 1400, the carboxylation resistance per unit area of soil surface increases from 2.35 s cm^{-1} to 50 s cm^{-1}.

4.4. Future Developments

The model presented here requires leaf temperatures as an input. This could easily be replaced by air temperature, using an energy balance equation of the leaf as done in the model of Penning de Vries (26). Treating the canopy as a single leaf leads to an oversimplification of the photosynthesis/light relation leading to the discrepancies between model and measurements observed at moderate radiation levels on 15 July. This could be improved by using a simple model of radiation penetration in the canopy.

5. CONCLUSIONS

One of the major goals in this study was to understand how weather affects the growth of the grassland, directly, and through its effects on soil moisture. Fig. 2a shows daytime CO_2 flux into the soil + plant system as closely related to leaf area index until the end of July, after which it drops quite quickly, following the depletion in soil moisture (Fig. 2b). Evaporation (Fig. 2d) follows a similar pattern, although the decrease in August is not so drastic. The reason for the rapid decrease in CO_2 flux in August is made clear by comparing diurnal curves of CO_2 flux for the periods centered on 20 July (LAI = 0.8) and on 19 August (0.6) (Fig. 3): the photosynthetic rate is obviously severely reduced between 0800 and 1800 because of plant water stress and high leaf and soil temperatures.

A dynamic model based on soil-plant water relationships adequately simulates the diurnal course of evaporation, plant water potential, and CO_2 flux. The decrease in photosynthetic activity in mid-August has been attributed mainly to high leaf temperatures, and only to a lesser extent to stomatal closure.

One might wonder why only cool-season, C_3 grasses are found in this type of grassland, when C_4 grasses, that have higher temperature optima for photosynthesis and greater water use efficiency, would be favoured during the dry, hot periods of August and early September. Apparently C_3 grasses grow very well during the cooler months of May, June and July and thus are too strong competitors for C_4 plants to survive during that period. Also not all years have as hot and dry August weather as 1971. It would require only a small increase in mean temperature to permit the appearance of C_4 species, which may be found in the Matador area on most southern facing slopes.

ACKNOWLEDGMENTS

The authors are grateful for the funding provided for this study by the National Research Council of Canada through its Canadian Committee for the International Biological Programme as part of the Matador Project.

REFERENCES

(1) Coupland, R.T. 1961. A reconsideration of grassland classification in the northern Great Plains of North America. J. Ecol. 49, 135-167.
(2) Coupland, R.T. 1972. Technical Report No. 1. Operational Phase (1967-72): A Summary of Progress. Matador Project, University of Saskatchewan, Saskatoon, 31 pp.

(3) Ripley, E.A. 1972. Technical Report No. 2. Meteorology and Climatology; I. Description of sensors and measurement programme, Matador Project, University of Saskatchewan, Saskatoon, 51 pp.
(4) Coupland, R.T., E.A. Ripley, and P.C. Robins. 1973. Technical Report No. 11. Description of Site: I. Floristic composition and canopy architecture of the vegetative cover,
(4) ... Matador Project, University of Saskatchewan, Saskatoon, 54 pp.
(5) Ripley, E.A. 1972. Man, Matador and Meteorology, Atmosphere $\underline{10}$, 113-127.
(6) Ripley, E.A. and B. Saugier. 1972. Technical Report No. 4. Micrometeorology: I. Description of sensors and measurement programme, Matador Project, University of Saskatchewan, Saskatoon, 68 pp.
(7) Bradley, E.F. 1969. A small, sensitive anemometer system for agricultural meteorology, Agric. Meteor. $\underline{6}$, 185-193.
(8) Ripley, E.A. and O. Olm. 1972. A readout circuit for the Bradley anemometer, Agric. Meteor. $\underline{10}$, 461-466.
(9) Saugier, B. and E.A. Ripley. 1974. A sensitive device for recording vertical gradients of CO_2 concentration, J. Appl. Ecol. $\underline{11}$, in press.
(10) Ripley, E.A., B. Saugier and O. Olm. 1973. A control and readout system for the measurement of atmospheric CO_2 profiles with a digital data logger, J. Phys. E. (Scientific Instruments) $\underline{6}$, 183-185.
(11) Ripley, E.A. and B. Saugier. 1973. Technical Report No. 25. Micrometeorology: VI. Fluxes of radiation, matter and energy, 1971, Matador Project, University of Saskatchewan, Saskatoon, 162 pp.
(12) Webb, E.K. 1970. Profile relationships: the log-linear range and extension to strong stability, Quart. J. R. Met. Soc. $\underline{96}$, 67-90.
(13) Dyer, A.J. and B.B. Hicks. 1970. Flux-gradient relationships in the constant flux layer. Quart. J. R. Met. Soc. $\underline{96}$, 715-721.
(14) Paulson, C.A. 1970. The mathematical representation of windspeed and temperature profiles in the unstable atmospheric surface layer. J. Appl. Met. $\underline{9}$, 857-861.
(15) Pruitt, W.O., D.L. Morgan and F.J. Lawrence. 1973. Momentum and mass transfers in the surface boundary layer, Quart. J. R. Met. Soc. $\underline{99}$, 370-386.
(16) Ripley, E.A. 1974. Technical Report No. 45. Micrometeorology: VIII. The seasonal and diurnal variation of evapotranspiration, 1970-71, Matador Project, University of Saskatchewan, Saskatoon, 29 pp.
(17) Ripley, E.A. 1974. Technical Report No. 50. Micrometeorology: IX. Seasonal and diurnal variation of net atmospheric CO_2 flux 1971, Matador Project, University of Saskatchewan, Saskatoon, 16 pp.
(18) Ripley, E.A. 1974. Technical Report No. 38. Micrometeorology: VII. Canopy CO_2 concentrations and fluxes, Matador Project University of Saskatchewan, Saskatoon, 46 pp.

(19) Newman, E.I. 1973. Permeability to water of the roots of five herbaceous species, New Phytol. 72, 547-556.
(20) Chartier, P. 1970. A model of CO_2 assimilation in the leaf, in: Prediction and Measurement of Photosynthetic Productivity: Proceedings of the Trebon Symposium, PUDOC, Wageningen, Netherlands.
(21) Monteith, J.L. 1963. Gas exchange in plant communities, in: (L.T. Evans, ed.) Environmental Control of Plant Growth, Academic Press, New York, pp. 95-111.
(22) Jackson, W.A. and R.J. Volk. 1970 Photorespiration, Ann. Rev. of Plant Physiology 21, 385-426.
(23) De Jong, E. 1973. Technical Report No. 22. Soil Physics: II. Soil water, Matador Project, University of Saskatchewan, Saskatoon, 42 pp.
(24) De Jong, E. and J.W.B. Stewart. 1973. Technical Report No. 36. Description of Site: III. Soil characterization, Matador Project, University of Saskatchewan, Saskatoon, 37 pp.
(25) Redmann, R.E. 1973. Technical Report No. 29. Plant water relationships, Matador Project, University of Saskatchewan, Saskatoon, 84 pp.
(26) Penning de Vries, F.W.T. 1972. A model for simulating transpiration of leaves with special attention to stomatal functioning, J. Appl. Ecol. 9, 57-71.

RADIATION EXCHANGE IN PLANT CANOPIES

J. ROSS and T. NILSON

*Institute of Astrophysics and Atmospheric
Physics, Tartu, Estonian S.S.R., USSR*

Basic problems in the radiative transfer theory for plant canopies are considered. The radiative transfer equation is derived and some methods for its approximate solution in different regions of the spectrum are pointed out.

1. THE RADIATIVE TRANSFER EQUATION

The radiative transfer theory formulated in astrophysics and atmospheric physics /1/ holds for turbid media. Several difficulties arise when one tries to extend this theory for the description of the radiation exchange in plant canopies.

Consider a plant canopy homogeneous (in the statistical sense) in the horizontal. The plant canopy is illuminated by the direct solar radiation and the diffuse sky radiation. Let us denote by S_0 the irradiation density $w \cdot m^{-2}$ of direct solar radiation on a plane perpendicular to the direction of sunrays $r_0 = (\vartheta_0, \varphi_0)$ and by $d_0(r)$ the brightness ($w \cdot m^{-2}$ sterrad^{-1}) of the sky in the direction $r = (\vartheta, \varphi)$, where ϑ is the zenith angle and φ the azimuth of the direction r, ϑ_0 – the zenith angle and φ_0 – the azimuth of the sun, respectively. The first difficulty we meet is due to the statistical nature of the radiation regime within the canopy caused by large dimensions of phytoelements (leaves, stems, flowers or their parts) that absorb and scatter radiation. So it is natural first to try to formulate the theory for the values of radiation characteristics averaged over the horizontal. It seems reasonable that the real plant canopy would be modelled by a medium consisting of randomly placed small thin plates with a given vertical distribution of their area per unit volume $u(z)$ and with a given statistical distribution $g(r_L)$ of their inclination (ϑ_L) and orientation (φ_L) angles. Consider the height z and the direction $r=(\vartheta,\varphi)$ within a model stand. Derivation of the transfer equation along a similar way may be performed as in atmospheric physics /1/. Thus we have to consider the so-called elementary volume. This elementary volume must be large enough to contain a sufficient amount of phytoelements so that the functions u and g may be realised with sufficient accuracy within the volume. Moreover, the statistical average of the brightness of radiation $i_\lambda(z,r)$ incident upon the elementary volume for any direction r must be equal to the mean value of the radiation brightness for the given height z. Hence the dimensions of the elementary volume must be considerable in the horizontal.

On the other hand, the dimensions of the elementary volume must be small enough to guarantee that there is no mutual shading of different phytoelements within the volume when viewed from any direction r. It is difficult to choose such a form of the elementary volume that these limitations would be met. However, one possible form of the elementary volume is a long horizontal rectangular parallelepiped with one face perpendicular to the direction r under consideration. Then the above limitations will be satisfied, except a horizontal direction, for which mutual shading cannot be avoided.

Likewise, we cannot proceed with derivation without making any assumptions on canopy structure. To avoid further complications a random dispersion of foliage must be assumed. In other words, the phytoelements must be placed independently of each other, according to the Poisson distribution. This assumption is needed to ensure that the statistical character of the distribution of foliage elements within the elementary volume would be independent of that of the radiation incident upon the volume. But since the statistical character of radiation is due to the displacement of adjacent phytoelements, the mutual independence of the positions of phytoelements follows from this fact.

Now writing the balance conditions for the photons of the wavelength interval $d\lambda$ incident upon the elementary volume through its surface area $d\sigma$ within the solid angle $d\Omega$ in the direction r within the time interval dt, and emerging from the volume in that direction, we are led to the following transfer equation /2,3/

$$\frac{1}{u(z)} \cdot \frac{\partial i_\lambda}{\partial z} \cos \vartheta = -G(r) i_\lambda(z, r) + \frac{1}{\pi} \eta_\lambda(z, r), \qquad (1)$$

where $u(z)$ is the foliage area per unit volume at the height z and $G(r)$ — projection of unit foliage area in the direction r onto a plane perpendicular to the direction r. The term $\eta_\lambda(z, r)$ characterizes the radiation sources and is different for shortwave and longwave radiation. In the shortwave region this term describes only the redistribution of radiation entering the elementary volume from all directions r' and scattered in the direction r within the volume:

$$\eta_\lambda(z, r) = \int_4 \int_\pi \Gamma_\lambda(r', r) i_\lambda(z, r') d\Omega' + \int_2 \int_\pi \Gamma_\lambda(r', r) q_\lambda(z, r') d\Omega', \qquad (2)$$

where $\Gamma_\lambda(r', r)$ is the monochromatic scattering phase function for the elementary volume, r' being the direction of incident radiation, r — the direction of observation; $q_\lambda(z, r)$ — the mean brightness of radiation penetrating the canopy up to height z without being intercepted by phytoelements; $d\Omega' = \sin\vartheta' d\varphi'$ — an elementary solid angle; $4\pi, 2\pi$ — spherical and hemispherical regions of integration.

The first term on the right-hand side holds for multiple scattering of the radiation within the stand, while the second term is valid for the first-order scattering of radiation.

At the same time, in the longwave region scattering may be neglected and the emission of radiation by phytoelements according to the Stephan-Boltzmann law must be considered

$$\eta = G\epsilon\sigma T^4, \qquad (3)$$

where ϵ — the emissivity of phytoelements, σ — the Stephan-Boltzmann constant, T — the absolute temperature of phytoelements.

The radiative transfer equation (1, 2) holds for monochromatic brightness of radiation and cannot correctly be used for wide spectral intervals where the optical properties of phytoelements undergo considerable changes. For instance, the absorption of radiation has significant differences in the visible and the near infrared regions. When calculating the radiation regime for the whole spectral region of shortwave radiation at least two different equations must be considered, one for the visible region and the other for the near infrared.

In contrast to the transfer equations in atmospheric physics the coefficient of extinction $G(r)$ is only a geometric characteristic of the stand and is independent of wavelength. This is due to the significant optical thickness of phytoelements and so the probability of a photon penetrating a phytoelement without being absorbed or scattered is zero. An optical anistropy of plant stands must be taken into account, since the extinction coefficient $G(r)$ depends on the direction r, the scattering phase-function being a function of both directions r

and r', and not only on the scattering angle $\widehat{rr'}$, as ordinarily in turbid media. Only a special case of the spherical orientation of phytoelements is isotropic /3/. The scattering phase function Γ is related to that of a single phytoelement γ by means of the formula

$$\Gamma(r', r) = \frac{1}{2\pi} \int_2 \int_\pi j(r'_i, r, r_L) g(r_L) |\cos \widehat{rr_L} \cos \widehat{r_L r'}| d\Omega_L, \qquad (4)$$

where $g(r_L)$ is the distribution function of foliage inclination and orientation, $\widehat{rr_L}, \widehat{r_L r'}$ are the angles between directions r and r_L or r_L and r' respectively. Similarly

$$G(r) = \frac{1}{2\pi} \int_2 \int_\pi g(r_L) |\cos \widehat{rr_L}| d\Omega_L. \qquad (5)$$

The transfer equation (1, 2) must be solved on the following boundary conditions, which describe the radiation conditions on upper $(z = z_0)$ and lower $(z = 0)$ boundaries of the stand (we use subscript 1 for downward radiation and 2 for upward radiation respectively):

$$i_1(z_0, r) = 0, \quad \text{if} \quad 0 \leq \vartheta < \frac{\pi}{2}, \qquad (6)$$

$$i_2(0, r) = A_g \int_2 \int_\pi [i_1(0, r') + q(0, r')] \cos \vartheta' d\Omega', \quad \text{if} \quad \frac{\pi}{2} < \vartheta \leq \pi \qquad (7)$$

for the shortwave region and

$$i_2(0, r) = \sigma \epsilon_g \frac{T_g^4}{\pi}, \quad \text{if} \quad \frac{\pi}{2} < \vartheta \leq \pi, \qquad (7a)$$

in the longwave region. Here A_g is the albedo, ϵ_g — the emissivity and T_g — the temperature of the ground surface.

According to the treatment given the transfer equation describes only the part of shortwave radiation, which is at least once intercepted. Radiation penetrating without interception is considered separatly.

2. METHODS OF SOLVING THE TRANSFER EQUATION

The integro-differential equation (1, 2) is rather difficult to solve particularly due to the anisotropy of plant stands as optical media. However, a precise solution may be given in a very special case when the phytoelements are horizontal and dull /4/. The general case can be solved only by approximate methods. One possible method for solving the transfer equation is a simple iterative method. A first approximation is obtained when only the second term in the right-hand side of formula (2) is considered, i.e. only first-order scattering is taken into account. For instance, the brightness of the first-order scattering of direct solar radiation within the plant canopy is as follows:

$$i_1 = \frac{S_0 \cos \vartheta_0}{\pi} \ell(r_0, r)(e^{-K_0 L} - e^{-KL}), \quad \text{if} \quad 0 \leqslant \vartheta < \frac{\pi}{2},$$

$$i_2 = \frac{S_0 \cos \vartheta_0}{\pi} \left\{ \ell(r_0, r) e^{-K_0 L} + [A_g - \ell(r_0, r)] e^{-(K_0 - K)L_0 - KL} \right\}, \quad (8)$$

$$\text{if} \quad \frac{\pi}{2} < \vartheta \leqslant \pi,$$

where $L = \int_z^{z_0} u\, dz$ is the downward cumulative foliage area index at the height z, $K = G(r)/\cos \vartheta$, $K_0 = G(r_0)/\cos \vartheta_0$, $G_0 = G(r_0)$ and $\ell(r_0, r) = \Gamma(r_0, r)/(G \cos \vartheta_0 - G_0 \cos \vartheta)$ is the brightness coefficient of a very dense canopy for the first-order scattering; $L_0 = L(0)$.

It appears that multiple scattering may be neglected in the greater part of the visible region of the spectrum (except the green region) /3/ and formula (8) serves as a good approximation in that region, but in the near infrared multiple scattering is important. This results from the fact that in the visible region the phytoelements absorb 70 to 95 per cent of the radiation while in the near infrared - only 15 to 35 per cent of it. In the near infrared the iteration procedure must be continued, adding at each step of iteration one order of scattering, until the desired accuracy is achieved.

In the majority of practical problems the downward and upward irradiation densities F_1 and F_2 are needed instead of radiation brightness. They may be calculated as integrals below:

$$F_1 = \int_{2\pi} \int i_1 \cos \vartheta \, d\Omega, \quad F_2 = \int_{2\pi} \int i_2 |\cos \vartheta| \, d\Omega. \qquad (9)$$

Another way of evaluating the irradiation densities F_1 and F_2 is to solve the transfer problem (1, 2, 6, 7) by means of an approximate method similar to the Schwarzschild approximation known in atmospheric physics. The method makes use of dividing the radiation field into downward and upward irradiation densities only. For instance, the albedo of a very dense $(L(0) \to \infty)$ plant stand (for direct radiation) obtained by the method cited above is as follows:

$$A = \frac{\omega}{2} K_0 \left(\frac{1 - K_0}{1 - \omega - K_0^2} - \frac{1 + K_0}{1 - \omega - K_0^2} \cdot \frac{1 - \sqrt{1 - \omega}}{1 + \sqrt{1 - \omega}} \right), \qquad (10)$$

where $K_0 = K(r_0)$, and ω is the scattering coefficient for phytoelements. When considering the longwave radiation field the solution of the problem (1, 2, 6, 7a) is quite trivial provided that the temperature of the foliage elements is given /5/. But calculating the temperature distribution over the foliage elements is a serious problem, which needs information about all the components of the energy balance of the foliage elements, including the transpiration of leaves.

3. TOTAL RADIATION FIELD

There is another important problem that needs further specification. The transfer equation in the form given above does not include the sun and sky radiation penetrating the canopy without being intercepted. The total field of downward radiation is the sum of $i_s(z,r) + q(z,r)$ or in terms of irradiation densities $F_1(z) + Q(z)$. We have

$$q(z,r) = d_0(r)a(z,r) + S_0 a(z,r_0)\delta(r - r_0) \quad \text{and}$$
$$Q(z) = S_0 a(z,r_0) \cos\vartheta_0 + \int_{2\pi}\int_{\pi} d_0(r)a(z,r) \cos\vartheta \, d\Omega, \qquad (11)$$

where $a(z,r)$ is the proportion of gaps in the foliage wieving in the direction r at the height z, δ is the Dirac function. In the longwave region $S_0 = 0$.

It appears that radiation penetrating the canopy is the most important part in the visible region of the spectrum. Also $g(z,r)$ must be known to solve the transfer equation, since it is included in the source function η.

The problem of describing the penetration of direct and diffuse radiation is essentially a problem of stand architecture, since only the shading effects must be investigated. This is the problem of estimating the free path of a photon within the plant canopy. The calculation of the probability of gaps within the foliage, if the stand architecture characteristics are given, is one of the most important problems in modern phytoactinometry. Most of the calculation models assume that phytoelements have a random dispersion of foliage, which may be described by the Poisson distribution. This leads to a well-known exponential formula for the gap proportion:

$$a(z,r) = e^{-K(r)L(z)} \qquad (12)$$

4. DISCUSSION

The transfer of shortwave radiation within plant canopies is determined by the following factors, i.e. by the characteristics which enter into the transfer equation:

(i) incident solar radiation S_0 and diffuse radiation from the sky d_0, their spectral composition and angular dependence;

(ii) optical properties of phytoelements (phase function γ) and those of the ground surface (albedo A_g), their spectral and angular dependence;

(iii) stand architecture, i.e. the foliage area density $u(z)$ and the distribution of foliage inclination and orientation $g(r_L)$. Investigations into the radiation regime within plant canopies have revealed an exceptionally important effect of stand architecture on irradiation densities in the shortwave region as well as in the longwave one. It seems that now the major problem is how to account for the deviations from the Poisson distribution of phytoelements that exist in real plant stands and cause discrepencies between theory and experiments. This problem is first to be solved for the direct and diffuse radiation penetrating the canopy, and then for the scattering problems. Evidently more detailed geometrical characteristics of the plant stand, such as the distribution of plants in the horizontal, the habitus of plants, the size of leaves etc. must be taken into account.

Finally, a good statistical treatment of the transfer problems is needed. Here one has to keep in mind the spatial statistics of radiation within plant stands as well as the temporal one.

REFERENCES

/1/ Chandrasekhar, S.: "Radiative Transfer". Oxford (1950).

/2/ Ross, J. K.: "The Mathematical Modelling of the Field of Photosynthetically Active Radiation (PhAR) within Plant Canopies". In: "Actinometry and Atmospheric Optics", Moscow (Russian) (1964).

/3/ Ross, J. K. and Nilson, T. A.: "A Mathematical Model of the Radiation Regime of the Plant Cover". In: "Actinometry and Atmospheric Optics", Tallinn (Russian) (1968).

/4/ Ross, J. and Nilson, T.: "The Radiation Regime within Plant Canopies with Horizontal Leaves". In: "Phytoactinometrical Investigations of Plant Canopies", Tallinn (Russian) (1967).

/5/ Budagovsky, A. I., Ross, J. K. and Tooming, H. G.: Distribution of the Longwave Fluxes and Radiation Balance in Plant Cover". In: "Actinometry and Atmospheric Optics", Tallinn (Russian) (1968).

22

AN EDDY CORRELATION METHOD FOR THE DETERMINATION OF MOMENTUM, HEAT AND MASS TRANSFER, USING HOT–WIRE ANEMOMETRY

ALAIN BAILLE and JEAN-PIERRE CHIAPALE

*Station de Bioclimatologie d'Avignon—Montfavet,
INRA, 84140 Montfavet – France*

ABSTRACT

An eddy correlation method for the determination of momentum, heat and mass transfers in atmospheric conditions has been developed, using hot-wire anemometry and associated analogic circuitry for signal analysis. The basis of this eddy correlation technique, the experimental set-up and the operational aspects are presented. Advantages and shortcomings of the method are outlined. Some preliminary results, essentially spectra and autocorrelation functions, are presented which seem in good agreement with previous results.

LIST OF SYMBOLS

The overbar over a symbol represents the mean value of the quantity. The sign represents the fluctuation of the quantity.

$u, \bar{u}$: instantaneous and mean horizontal velocity
e', e'_1, e'_2 : voltage fluctuations
u', w' : horizontal and vertical velocity fluctuations
t' : temperature fluctuation
ϕ' : vertical angular fluctuation
α, β : sensibility coefficients of the hot-wires to respectively horizontal and vertical velocity
γ : sensibility coefficient of the hot-wires to the temperature
Φ_M, Φ_H, Φ_E : momentum flux, heat flux and water-vapor flux
n : frequency
ζ : time delay
$R(\zeta)$: autocorrelation function

INTRODUCTION

The determination of momentum, heat and mass transfers by eddy correlation technique is an interesting method for agronomic studies, especially when it is desired to estimate the energy-balance near the ground in heterogeneous sites where the classical methods are not suitable (/̄1 /̄, /̄2 /̄, /̄3 /̄), such as West-France bocage, or windbreak networks /̄ 4 /̄. In the past decade, different versions of this method have been carried out for field-use, some of them based on hot-wire anemometry (/̄5 /̄, /̄6 /̄, /̄7 /̄, /̄8 /̄, /̄9 /̄), others based on sonic anemometry (/̄ 10 /̄ /̄ 11 /̄). The main limitation of the eddy correlation technique was the high cost of the apparatus and experimental set-up. However, the recent development of hot-wire anemometry and the progress realized in the field of signal analysis in the last years /̄ 12 /̄ seem to afford a new interest in this method which can be now developed at a reasonable price. It is the reason why we have been trying to carry out an eddy correlation technique of flux measurements by mean of hot-wire anemometry. The aim of this paper is to present the principles of the method, its operational aspects and some preliminary results.

PRINCIPLES OF THE METHOD

The method is based on the linearized equations of hot-wire anemometry, which can be used only in moderately turbulent flows. For a single inclined wire situated in the xOz plane -Ox beeing the mean flow direction (see Figure 1)- and consequently, insensitive to lateral angular fluctuations, we have

$$e' = \alpha . u' + \beta_\phi . \phi' + \gamma . t' \quad (1)$$

where
  e' = fluctuation of wire voltage
  u', ϕ', t' = fluctuations of horizontal velocity, vertical angle and temperature
  α, β_ϕ, γ = sensibility coefficients of the wire to respectively horizontal velocity, vertical angular deviation and temperature

If the vertical angular fluctuations ϕ' and the horizontal velocity fluctuations u' are not too large, we can write

$$w' = (\overline{u} + u') . tg\,\phi' \sim \overline{u} . \phi' \quad (2)$$

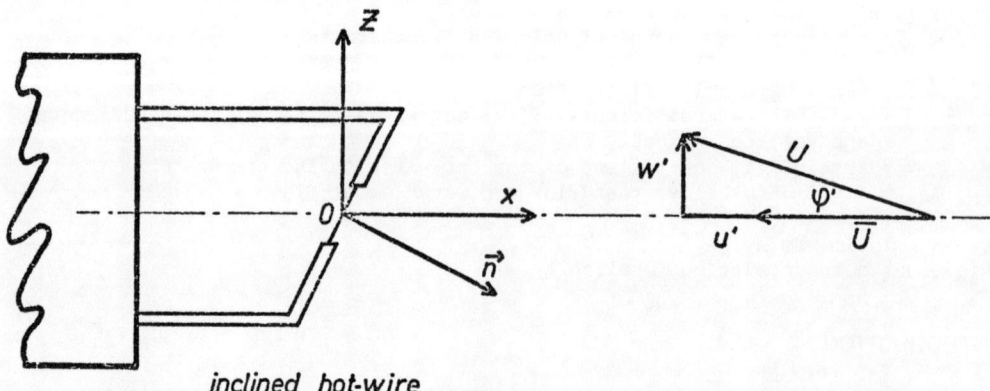

Figure 1 : Schematic representation of an inclined hot wire in the flow w' beeing the vertical velocity fluctuation.

Then, equation (1) becomes

$$e' = \alpha \cdot u' + \beta \cdot w' + \gamma \cdot t' \quad (3)$$

with

$$\beta = \frac{\beta \phi}{\overline{u}} \quad (4)$$

The determination of the momentum and heat flux is possible if we can extract the kinematic fluctuations u' and w' from the temperature fluctuations t', i.e. if we have three hot-wires giving a system of three equations similar to (3).

For the sake of simplicity, we have choosen the combination of a X-probe constituted by a pair of platinum wires of diameter 5 μ, arranged at approximately ± 45° to the flow direction, and by a low-heated platinum wire of diameter 2 μ, sensitive only to the temperature fluctuations. The system of equations can be written :

$$e'_1 = \alpha_1 \cdot u' + \beta_1 \cdot w' + \gamma_1 \cdot t'$$
$$e'_2 = \alpha_2 \cdot u' + \beta_2 \cdot w' + \gamma_2 \cdot t' \quad (5)$$
$$e'_3 = \gamma_3 \cdot t'$$

(Suscribs 1 and 2 refer to the X probe wires, and suscrib 3 to the temperature-wire).

The separation of the kinematic and thermal variables is realized by adding to the voltages e'_1 and e'_2 respectively the quantities $-\gamma_1 \cdot t'$ and $-\gamma_2 \cdot t'$ obtained by amplifying the e'_3 signal.

Thus (5) reduces to

$$e''_1 = \alpha_1 \cdot u' + \beta_1 w'$$
$$e''_2 = \alpha_2 \cdot u' + \beta_2 \cdot w' \quad (6)$$

From these two equations, we can easily extract the u' and w' fluctuations. Then by multiplication and integration, we obtain the momentum flux

$$\Phi_M = -\rho \cdot \overline{u'w'} \quad (7)$$

and the heat flux

$$\Phi_H = -\rho\, C_p \cdot \overline{w't'} \quad (8)$$

The water-vapor flux $\quad \Phi_E = -\rho\, L \cdot \overline{w'q'} \quad (9)$

can also be obtained if the specific humidity fluctuations q' can be measured. However, the measurement of rapid fluctuations of this quantity is still very difficult, and its study, which is presently undertaken in our laboratory, is beyond the scope of this paper.

EXPERIMENTAL SET-UP

The experimental set-up, which is based on the primitive one used by SCHON and BAILLE /13/ is divided in two parts. The schema of the first part, which is destined to field-use is presented in the figure 2.

The X probe wires are supplied by constant-temperature anemometers suitable for field use, with a overheating coefficient (1) of approximately 0,8.

The temperature wire, supplied by a 15 volts battery is operated at constant current, the intensity of the current beeing about 1 mA. A build-up of operational amplifiers with an adjustable gain between 0 and 1000 allows to amplify the temperature wire signal. The output signals e'_1, e'_2 and e'_3 are simultaneously recorded on a magnetic tape recorder, and analyzed ulteriously at the laboratory with the second part of the electronic set-up, which is presented on the figure 3. This part is constituted by two groups of operational amplifiers which realize the separation of the kinematic and thermal variables, and give output signals proportional to u', w' and t'. A multiplier modul at the end of the analogic chain give either of the correlations $\overline{u'w'}$, $\overline{u't'}$ or $\overline{w't'}$. Spectra and autocorrelation functions are calculated by mean of adequate numerical programs on a PDP 8-E mini-computer.

OPERATIONAL ASPECTS

Once the hot-wires calibrated in a wind-tunnel regulated in velocity and temperature (see /11/ for the process of calibration), they are placed on a wind vane which is itself placed on a moving arm along a mast of 6 meters high. Each run lasts 10 minutes and during this time are recorded the three signals e'_1, e'_2, e'_3, the mean velocity and the mean temperature.

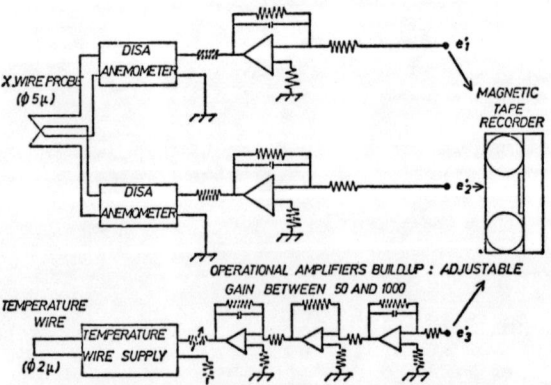

Figure 2 : Schema of the field apparatus

---

(1) The overheating coefficient of a hot-wire is defined by the ratio $(R_W-R_0)/R_0$, R_W beeing the wire resistance at the operating temperature, and R_0 the wire resistance at a reference temperature.

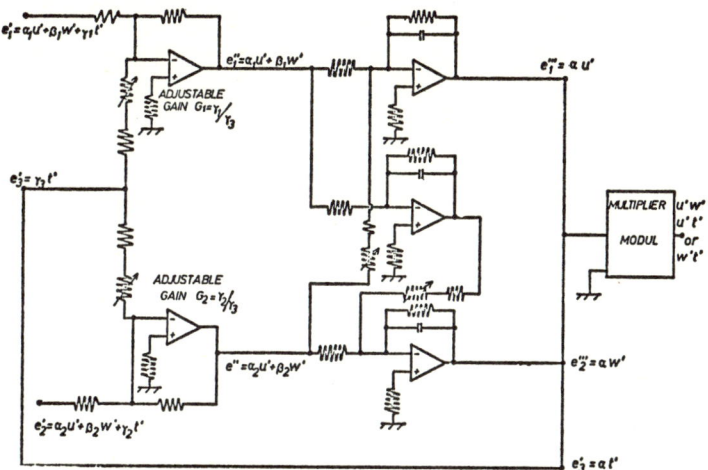

Figure 3 : Schema of the analogic chain used for the separation of u', w' and t'

ADVANTAGES AND SHORTCOMINGS

This eddy correlation technique, based on the utilization of very thin wires, has the main advantage to give all the turbulent characteristics of the flow (fluxes, but also turbulence intensities, spectra and co-spectra, correlation functions), and this, for frequencies until 1000 Hz and more. Thus, a very large frequency range can be investigated. The main shortcoming is constituted by the fact that this method cannot be used in highly turbulent flows. Other shortcomings are the risk of breaking of the wires, especially in the case of strong bursts of wind, and the contamination by dirt and dust, which produces a change in wire calibration. Thus, the cleaning of the probes in a solvent bath is required to regular intervals, as well as the check-up of the calibration curves after a long time exposure in natural conditions.

PRELIMINARY RESULTS

We present here some examples of spectra and autocorrelation functions of the variables u', w' and t', measured at 1 m high above a tall fescue cover. Figure 4 shows the spectra of the longitudinal and vertical fluctuations in the range 0,01 Hz - 100 Hz. We can observe that they are in good agreement with previous results ($/\!\_9\_/$, $/\!\_14\_/$, $/\!\_15\_/$, $/\!\_16\_/$) : the spectra suit well the classical - 5/3 power law at high frequencies, and their maxima are situated at about 0,1-0,2 Hz. The same remarks can be made about the temperature spectrum (Figure 5)

The figure 6 shows the autocorrelation functions $R_u(\zeta)$ and $R_T(\zeta)$ of the longitudinal velocity fluctuations and of the temperature fluctuations. They can be adjusted by the following functions :

$$R_u(\zeta) = 1 - (\frac{\zeta}{T_1})^{m_1} \qquad (10)$$

and

$$R_T(\zeta) = 1 - (\frac{\zeta}{T_2})^{m_2} \qquad (11)$$

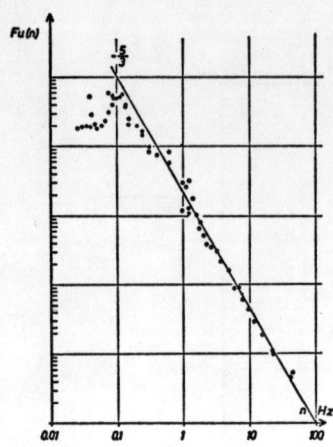

Figure 4 : u' and w' spectra at 1 m. high

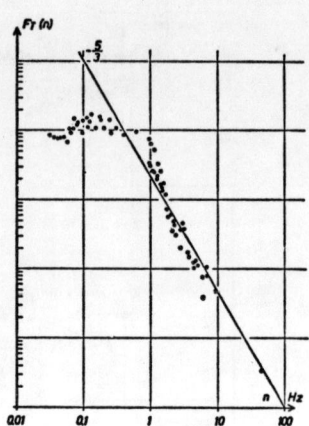

Figure 5 : t' spectra

T_1 and T_2 are characteristic times of the autocorrelation function. We find

$m_1 \sim 0,84$ $T_1 \sim 0,35$ sec

$m_2 \sim 0,78$ $T_2 \sim 1,35$ sec

The values of the exponents m_1 and m_2 are not too far from the theoretical value 2/3 suggested by some authors ([17], [18]).

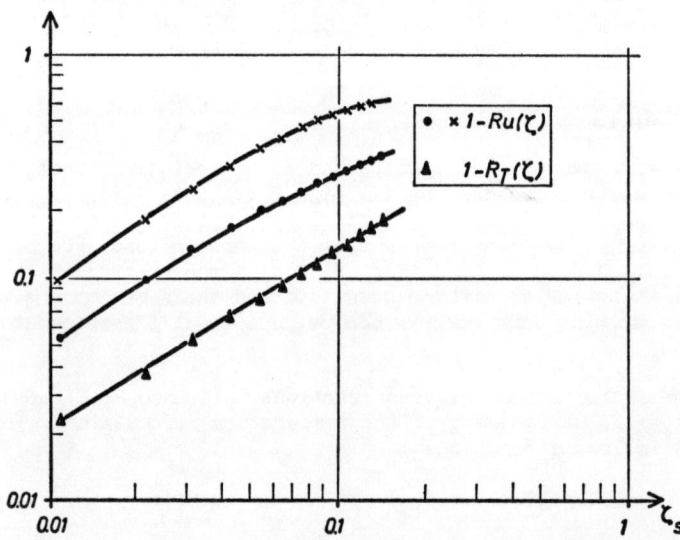

Figure 6 : Autocorrelation functions $R_u(\zeta)$ and $R_T(\zeta)$

CONCLUSION

These first results, though they do not prove with certainty the fiability of the method for flux measurements, seem to demonstrate its suitability for high frequencies turbulence studies in and above canopies. In the immediate future, our aim is to check the values of the flux obtained by this eddy correlation technique against values measured in the same time by other methods (aerodynamical or energy-balance methods). This is the work presently undertaken at the Station de Bioclimatologie d'Avignon.

ACKNOWLEDGMENTS

This research is supported partly by the Direction Générale à la Recherche Scientifique et Technique. Thanks are due to Dr SCHON (Ecole Centrale de Lyon) for his helpful remarks.

BIBLIOGRAPHY

/ 1 / - DYER A.J. (1963). The adjustment of profiles and eddy fluxes. Quart. J. Roy. Meteo. Soc., Vol. 89, pp. 276-281

/ 2 / - RIDER N.E., PHILIP J.R., BRADLEY E.F. (1963).The horizontal transport of heat and moisture. Quart. J. Roy. Meteo. Soc., Vol. 89, pp. 507-521

/ 3 / - DE VRIES D.A. (1959). The influence of irrigation on energy balance and climate near the ground. J. Meteor., Vol. 16, pp. 256-270

/ 4 / - GUYOT G., SEGUIN B., VERBRUGGHE M. (1974). Modification of land roughness and resulting microclimatic effects : a field study in Brittany (France). (A paraître)

/ 5 / - CRAMER H.E., RECORD F.A. (1953). The variation with height of the vertical flux and momentum. J. Meteor., Vol. 10, pp. 219-226

/ 6 / - DYER A.J., MAHER F.J. (1965). Automatic eddy-flux measurement with the evapotron. J. Appl. Meteo., Vol. 4, pp. 622-630

/ 7 / - BUSINGER J.A., MIYAKE M., DYER A.J., BRADLEY E.F. (1967). On the direct determination of the turbulent heat flux near the ground. J. Appl. Meteo., Vol. 6, pp. 1025-1032

/ 8 / - MIYAKE M., DONELAN M., Mc BEAN G., PAULSON C., BADGLEY F., LEAVITT E. (1970). Comparison of turbulent fluxes over water determined by profile and eddy-correlation techniques. Quart. J. Roy. Meteo. Soc., Vol. 96, pp. 132-137.

/ 9 / - SHAW R.H., SILVERSIDES R.H., THURTELL G.W. (1974). Some observations of turbulence and turbulent transport within and above plant canopies. Bound. Lay. Meteo., Vol. 5, pp. 429-450

/‾10_/ - WYNGAARD J.C., COTE O.R. (1971). The budgets of turbulent kinetic energy and temperature variance in the atmospheric surface-layer. J. Atm. Sci., Vol. 28, pp. 190-204

/‾11_/ - POND S., PHELPS G.T., PAQUIN J.E., BEAN G.M., STEWART R.W. (1971). Measurements of the turbulent fluxes of momentum, moisture and sensible heat over the ocean. J. Atm. Sci., Vol. 28, pp. 901-907

/‾12_/ - BRADSHAW P. (1971). An introduction to turbulence and its measurements. PERGAMON PRESS

/‾13_/ - SCHON J.P., BAILLE A. (1972). Méthode d'isolement des fluctuations turbulentes cinématiques et thermiques au moyen d'une sonde anémométrique à trois fils. C.R. Acad. Sci. Paris, Vol. 274, Ser. A, pp. 116-119.

/‾14_/ - LUMLEY J.L., PANOFSKY H.A. (1964). The structure of atmospheric turbulence. Interscience Publishers.

/‾15_/ - VAN ATTA C.W., CHEN W.Y. (1970). Structure functions of turbulence in the atmospheric layer over the ocean. J. Fl. Mech., Vol. 44, pp. 145-159

/‾16_/ - KAIMAL J.C., WYNGAARD J.C., IZUMI Y., COTE O.R. (1972). Spectral characteristics of the surface layer turbulence. Quart. J. Roy. Meteo. Soc., Vol. 98, pp. 563-590

/‾17_/ - INOUE E., TANI N., IMAI K. (1955). Measurements of wind turbulence over cultivated fields. Bull. Nat. Inst. Agr. Sc., Ser. A, n° 4, pp. 1-37

/‾18_/ - TANI N. (1963). The wind over cultivated fields. Bull. Nat. Inst. Agr. Sc., Ser. A, n° 10, pp. 1-99.

23

MEASUREMENT OF ATMOSPHERIC INFRARED RADIANT FLUX AND TESTING OF SOME EMPIRICAL FORMULAE FOR ESTIMATING THIS FLUX

C. L. PALLAND

*Research Department, N. V. Heidemaatschappij Beheer,
Arnhem, The Netherlands*

ABSTRACT

The paper deals with the practical problems met with in the continuous measuring of atmospheric infrared radiation. Using extensive data, the empirical constants in the formulae of Brunt and Swinbank have been determined for longitude 05°53'E, latitude 52°04'N. Both formulae and some related formulations have been compared mutually to arrive at a conclusion as to which one of them describes the phenomenon most accurately.

1. INTRODUCTION

The net radiation of the evaporating surface plays an important part in solving evaporation problems. Formulae for the calculation of the evaporation, describing the physical phenomenon with reasonable accuracy, will necessarily have the net radiation as a starting-point.
Data about net radiation are not or only seldom collected on a routine basis. These data should also refer to the specific properties of the evaporating surface to which the data are to be applied. Likewise, the absorption and reflection properties for radiation have to correspond with those of the evaporating surface.
In practise, these properties will mostly vary widely because of the large number of factors that influence the absorption and transmission of radiation such as:
— in the case of watersurfaces: wind, waterdepth, colour of bottom, clearness of water
— in the case of crops: species, stage of growth, moisture supply and moisture condition of the leaves.
Consequently, series of measurements of electro-magnetic energy as received from the hemisphere over the evaporating surface are required. If these data are availavle, the net radiation can be computed by means of assumpitons as regards the properties of the surface for each specific case.
This paper discusses the practical problems involved in the continuous measuring of atmospheric infrared radiation. With the help of extensive data, the emperical constants in the formulae of Brunt and Swinbank have been defined for longitude 05°53'E, latitude 52°04'N. Both formulae and some related formulations have been compared mutually to arrive at a conclusion as to which one of them describes the phenomenon most accurately.

2. LIST OF SYMBOLS

A	surface of a thermopile	m^2
R	reciprocal sensitivity	$W.m^{-2}.mV^{-1}$
a	empirical constant	–
b	empirical constant	$mb^{-½}$
e	partial vapour pressure in air	mb
h	coefficient of heat transfer	$W.m^{-2}.K^{-1}$
q	radiant flux density	$W.m^{-2}$
θ	temperature	K
ϕ	rate of heat gain or loss	W
α	absorption factor	–
α	empirical constant	K^{-2}
β	empirical constant	$W.m^{-2}$
ϵ	residual term	$W.m^{-2}$
ρ	reflection factor	–
σ	Stefan – Boltzmann constant	$W.m^{-2}.K^{-4}$
τ	transmission factor	–

Sub- and superscripts:

ai	air
gl	glass
i	instrument
k	observation number
lo	long-wave radiation (wave-length $> 3.10^{-6}$m)
p	thermopile
pl	polyethylene
rs	reflected on soil surface
s	soil surface
sh	short-wave radiation (wave-length $< 3.10^{-6}$m)

3. MEASURING INSTRUMENTS

3.1 To measure the long-wave radiant flux from the atmosphere on a horizontal surface (q_{lo}^{ai}), a fully weatherproof radiometer is required. The receiving surface must be protected against dust, moisture and cooling by wind. Only polyethylene has been found to be suitable for this purpose under all weather conditions and at low temperatures. This material has a high transmission for both infrared radiation and solar radiation (q_{sh}). A disadvantage is that q_{sh} is also registrated by such a pyrradiometer. The pyrradiometer used is manufactured by Dr. B.Lange - Berling - Zehlendorf according to a design of Professor Schulze. This type of radiometer is available on the market as a net pyrradiometer, the upper and lower part of which can be used separately.

Radiometers measure the exchange of radiation energy between the receiving surface and its immediate surroundings. The rate of heat gain can be expressed by [12]:

$$\phi \approx (\alpha_{sh}^{p} \tau_{sh}^{pl} q_{sh} + \alpha_{lo}^{p} \tau_{lo}^{pl} q_{lo}^{ai} + \rho_{lo}^{pl} \alpha_{lo}^{p} \sigma \theta_{p}^{4} + \alpha_{lo}^{pl} \alpha_{lo}^{p} \sigma \theta_{pl}^{4}) A \qquad (1)$$

The heat loss by:

$$\phi \approx (\alpha_{lo}^{p} \tau \theta_{p}^{4} + h_{p} \Delta\theta) A, \text{ with } \Delta\theta = \theta_{p} - \theta_{i} \qquad (2)$$

Under steady state conditions with $\theta_i = \theta_{pl}$ and $\theta_p^4 = (\theta_i + \Delta\theta)^4 \approx \dot{\theta}_i^4 + 4\theta_i^3 \Delta\theta$, and further omitting geometrical effects and multiple reflections one gets:

$$\Delta\theta = \frac{\alpha_{sh}^{p} \tau_{sh}^{pl} q_{sh} + \alpha_{lo}^{p} \tau_{lo}^{pl} (q_{lo}^{ai} - \sigma\theta_i^4)}{4\alpha_{lo}^{p} (1 - \rho_{lo}^{pl}) \sigma\theta_i^3 + h_p} \qquad (3)$$

See list of symbols.

Atmospheric Infrared Radiant Flux and Testing of Some Empirical Formulae

From (3) follows that the electrical signal is generated by q_{lo}^{ai}, q_{sh} and $\sigma \Theta_i^4$, the so called own radiation of the instrument.

To registrate q_{lo}^{ai} continuously (also during daytime) it is necessary that:
- q_{sh} is measured separately by a pyranometer;
- Θ_i, the temperature of the mass of the instrument, is recorded;
- the instrument is calibrated in both wave-length regions 0.3 – 3 micron and 3 – 100 micron, because

$$\alpha_{lo}^p \tau_{lo}^{pl} \neq \alpha_{sh}^p \tau_{sh}^{pl}$$

Furthermore it appears that the sensivity of the radiometer in these spectral regions (R_{sh} and R_{lo}) depends on $\Theta_i = \Theta_{pl}$, in fact the airtemperature (Θ_{ai}). Hence, to calculate q_{lo}^{ai} correctly from the recorded signal, the condition $\Theta_i = \Theta_{pl} = \Theta_{ai}$ must be fullfilled. For this, the instrument should be artificially ventilated with embient air, both the mass of the instrument and the protecting polyethylene dome.

q_{sh} has been measured by a Kipp Solarimeter, a pyranometer manufactured by Kipp en Zonen, Delft - Holland. This radiometer resembles a pyrradiometer, but it has a glass dome (an outer and inner dome). The same relations (1), (2) and (3) are valid. For glass $\tau_{lo}^{gl} \approx 0$, so only the first term in the nummerator is left in (3). This means that the electrical signal is only generated by q_{sh} and is independently of q_{lo}^{ai}.

$\alpha_{lo}^p \tau_{lo}^{pl} \neq \alpha_{sh}^p \tau_{sh}^{pl}$ as mentioned previously. The receiving surface of the Schulze radiometer if blackened with Eppley/Parsons' optical matt black lacquer of which $\alpha_{lo} = \alpha_{sh} = 0.985$ [7.9]. $\tau_{lo}^{pl} \neq \tau_{sh}^{pl}$; both transmission-factors depending on the thickness of the material. Lupolen – H is the trade mark of the polyethylene used for the Schulze radiometer. It has a standard thickness of 0.1 millimetre. But the seperate Lupolen – H domes differ widely in thickness. In literature τ_{sh}^{pl}-values are given from 0.90 up to 0.96 [1.8] and for new Lupolen – H, τ_{lo}^{pl}-values from 0.75 up to 0.83 have been determined [12].

The transmission of polyethylene decreases owing to incoming ultraviolet radiation.

In spectographic measurements, a reduction of 1 to 2 per cent a month, depending on the amount of solar radiation has been determined [12]. In view of the foregoing, both instruments have been calibrated every three months. The Schulze-radiometer has been calibrated with the old dome as well as the new one.

3.2. Calibration of the instruments

3.2.1. Determination of R_{sh}

The best and most direct way to determine R_{sh} is to shade off the termopile of the radiometer against the solar-radiation by means of a disk for a few minutes. The decrease in deflection of the signal on the recorder during, before and after shading off equals the direct solar radiation measured by a pyrheliometer and converted to the flux on a horizontal plane. The diameter of the disk and the distance from the disk to the centre of the pile should be in agreement with the aperture angle of the standard instrument.

R_{sh} depends on the sun's elevation and azimuth. Next to a slight deviation from the cosinus law it is principally caused by refractions in the dome. The size of the dome and the shape of the thermopile play a great part in it. Moreover, the Lupolen – H domes often deviate from the ideal spherical shape. All these factors give rise to a course in R_{sh} during day-time and throughout the seasons.

To determine R_{sh} as accurately as possible, more calibration points should be taken during day-time at regular time-intervals. From these momentaneous R_{sh}-values, a weighed daily mean can be calculated, using q_{sh} at the moment of calibration as a factor of weight [9].

In figure 1, the single relative values of R_{sh} are given in relation to the sun's zenith angle. A cross indicates the weighed daily means at the smallest zenith angle on the day concerned.

Schulze upper part

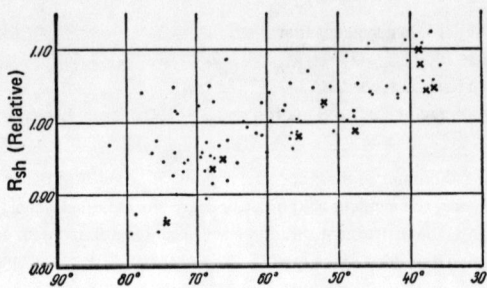

Kipp Solarimeter

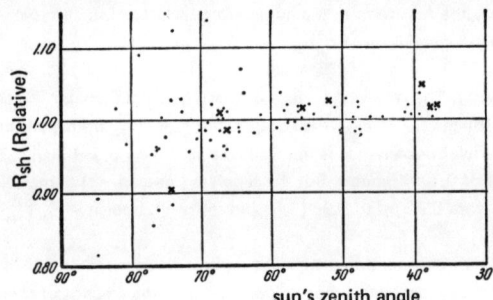

sun's zenith angle

Figure 1. Relative reciprocal sensitivity in relation to the sun's zenith angle.

The course of R_{sh} during the day is clearly manifested by the spread of the momentaneous relative R_{sh}-values in the graph, considering that the inaccuracy of the calibration is less than two per cent. The spread of the points of the Solarimeter is less, possibly because a decreasing transmittance of the dome and a deviation from the ideal spherical shape do not play a part here. In this respect it may be pointed out that radiometers should be always exposed and calibrated at the same direction of orientation. From considerations of symmetry it is advisable that the length or broadwise axis of the pile coincides with the south direction.

3.2.2. Determination of R_{lo}

R_{lo} has been calculated from

$$R_{lo} = \frac{\alpha_{sh}^{p} \, \tau_{sh}^{pl}}{\alpha_{lo}^{p} \, \tau_{lo}^{pl}} \, R_{sh}$$

The absorptionfactors for Eppley/Parsons' optical matt black lacquer $\alpha_{sh}^{pl} = \alpha_{lo}^{p} = 0.985$.
For new Lupolen – H, $\tau_{sh}^{pl} = 0.94$ as a mean value. This value is used as a standard to determine τ_{sh}^{pl} of a new dome. As the spectograph gives too low τ_{sh}^{pl}-values, these transmissionfactors have been merely used as a relative measure to fix τ_{sh}^{pl} of the Lupolen – H dome concerned, before and after use. τ_{lo}^{pl} can be determined directly by spectrographical measurements. A linear change in transmission with time has been assumed [9.12].

3.3. Ventilation

As stated earlier, relation (3) is only valid if the condition $\Theta_{pl} = \Theta_i$ is fulfilled. Much attention should be paid to this during both, the continuous recordings and the calibration procedure. This means that artificial ventilation of both the mass of the instrument and the polyethylene dome is necessary.

The Schulze radiometer is provided with a ventilation system by the factory. But its capacity has proved inadequate and had to be changed. Three ventilation openings have been made in the screen plate asymmetrically with respect to the thermopile. A strong ventilation was realised with an airstream velocity of about 5 m.s^{-1} near the dome, in the axis of the opening. Via another opening the mass of the instrument was ventilated as well. The ventilationair was drawn form the open air at the height of the instrument. This ventilation system prevented dew and hoar frost formation, it kept the dome clear from snow and evaporated rapidly attached waterdrops. The Kipp Solarimeter was supplied too with such a ventilation system, which reduced the so-called "zero-depression" to 25 per cent [2].

4. RESULTS OF THE MEASUREMENTS

q_{lo}^{ai} has been measured at the Deelen airfield (longitude 05°53'E, latitude 52°04'N). The recordings, covering November 1964 till January 1966 have been fully worked out. The monthly incoming long-wave radiation energy from the atmosphere is given in table 1.

Table 1. Monthly incoming long-wave irradiation from the atmosphere

	q_{lo}^{ai} (M J.m^{-2} month^{-1})		q_{lo}^{ai} (M J.m^{-2} month^{-1})
Nov. 1964	644	July 1965	975
Dec. "	877	Aug. "	937
Jan. 1965	837	Sept. "	871
Febr. "	729	Oct. "	862
March "	799	Nov. "	782
April "	817	Dec. "	869
May "	883		
June "	911	Year 1965	10272

5. THE TESTING OF SOME FORMULAE

5.1. The problem

The formulae of Brunt and Swinbank were examined, as well as the latter formula, extended by a constant term. The Brunt-formula assumes the clear sky to be a grey radiator with an emission factor depending upon the water vapour pressure [3]:

$$q_{lo}^{ai} = a\sigma\Theta_{ai}^4 + b\sqrt{e}\sigma\Theta_{ai}^4 \qquad (4)$$

Swinbank found a lineair relaitonship between q_{lo}^{ai} and $\sigma\Theta_{ai}^6$ [11]:

$$q_{lo}^{ai} = \alpha\sigma\Theta_{ai}^6 \qquad (5)$$

For symbols used, see list of symbols.

Both models were extended with a constant term since it is not impossible that a certain amount of background radiation independently of the air temperature would exist.

The coefficients in the models are expected to depend upon the condition of the atmosphere (temperature, humidity, dust, etc.) and consequently upon the season and the type of circulation. Therefore the test was made by grouping of the data according to both month and the following wind direction categories.

category 1: $355° - 55°$
category 2: $55° - 115°$
category 3: $115° - 175°$
category 4: $175° - 235°$
category 5: $235° - 295°$
category 6: $295° - 355°$

The observations at variable wind and those, at which the wind direction at ground surface differed considerably from that at the 900 mb level, were grouped in a separate category [13]

5.2. The data

From the continuous recordings of q_{Io}^{ai} at the Deelen airfield, 231 momentaneous observations in 1956 and 1966 were selected. All of them were taken at darkness and a cloudless sky. This was done on purpose, because day-time measurement of q_{Io}^{ai} yields data with extra inaccuracies, since q_{sh} is recorded as well (see 3.1 and 3.2). Moreover, the differing temperature profile of the lower atmosphere during day-time would call for a correction on the models [10].

In practice, the formulae are predominantly used to calculate q_{Io}^{ai} as a mean period value. With a view to the testing of the models, momentaneous observations have been taken as a basis in order to avoid the risk of mutual differences between the models becoming mitigated as a result of taking averages.

Processing of data requires mutual independence of the observations on the one hand and a sizeable sample on the other hand (see 5.3). A compromise has been found by limiting the number of observations to 2 per night at a time interval of at least 4 hours.

The numbers of observations per month and per wind category are shown in table 2.

In view of the low accuracy of R_{Io} in the months of December and January, no observations made in these months have been considered.

5.3. Testing of the models

The coefficients in the models have been determined with the aid of simple regression in the form as given in (4) and (5) and with a constant term added [6]. The residual variance of the models has been mutually compared for significant differences.

It has been assumed that the residuals are mutually independent and normally distributed, it has also been assumed that Θ_{ai} and e have been measured correctly. Hence, if we write for the kth observation:

$$\left. q_{Io}^{ai} \right]_k = f(\Theta, e) + \epsilon_{k'}$$

the residual ϵ_k is the only error that arises because q_{Io}^{ai} as measured, differs from q_{Io}^{ai} as computed by the model concerned.

The method using concomittant variables of Kendall and Stuart, without the need for the assumptions as mentioned above, yielded approximately the same coefficients [5]. The empirical constants are given in table 2.

The preference for a grouping according to months or to wind direction has been established by a test as regards internal consistence and a simple Wald-Wolfowitz runtest.

5.4. Conclusions

Obviously the manner of grouping has a great impact. Yet the preference for grouping according to month or according to wind direction is not clear. For the application it is important to know more about the reasons underlying the differences which will allow to make a choice of grouping.

The model of Swinkbank with a constant term added to it describes the phenomenon better thans that without a constant term. The model of Brunt (without a constant) is found to suit the purpose better in contrast to results obtained elswhere [4.10.11.13]. Ofcourse, the statements are only valid for computing momentaneous values of q_{lo}^{ai} at a cloudless sky during the night, on the basis of this sample which is representative for Western-European conditions at moderate latitude. Data of the model of Swinbank with constant term and the model of Brunt are given in table 2.

Table 2. Empirical constants in Brunt's formula and Swinbank's formula with a constant term.

Month	Number of observations	Brunt $q_{lo}^{ai} = (a+b\sqrt{e})\sigma \Theta_{ai}^4$		Swinbank+constant term $q_{lo}^{ai} = \alpha\sigma\Theta_{ai}^6 + \beta$	
		a (−)	b(mb$^{-1/2}$)	$\alpha \cdot 10^4$(K^{-2})	β(W.m^{-2})
February	19	0,663	0,046	0,087	36,8
March	25	0,661	0,045	0,082	47,2
April	25	0,619	0,056	0,078	57,0
May	20	0,600	0,057	0,076	61,2
June	20	0,751	0,017	0,069	95,5
July	10	0,601	0,064	0,098	10,3
August	30	0,653	0,049	0,078	73,2
September	33	0,731	0,030	0,083	60,6
October	34	0,739	0,029	0,085	56,4
November	15	0,747	0,023	0,083	57,3
Year	231	0,678	0,041	0,088	39,5
Wind direction category					
1	43	0,669	0,039	0,087	35,0
2	33	0,765	0,012	0,075	74,4
3	24	0,688	0,038	0,072	80,9
4	39	0,682	0,045	0,089	42,7
5	52	0,650	0,053	0,091	35,4
6	32	0,663	0,044	0,092	26,0
7	8	0,703	0,032	0,084	48,6

REFERENCES

1. Ambach, W., E. Beschorner und H. Hoinkes: Über die Eichung des Strahlungsbilanzmessers nach R. Schulze (Lupolengerät). Arch. Met. Geoph. Biokl., B, 13, 76 (1963).

2. Bener, P.: Untersuchung über die Wirkungsweise des Solarigraphen Moll-Gorczynski. Beiträge zur Strahlungsmeszmethodik III, Arch. Met. Geoph. Biokl. B. 2, 188 (1951)

3. Brunt, D.: Physical and Dynamical Meteorology, P. 136. Cambridge University Press, 1944.

4. Deacon, E.L.: The Derivation of Swinbank's Long-Wave Radiation Formula. Quart. J. Roy. Met. Soc. 96, 313 – 319 (1970).

5. Kendall, M., and A. Stuart: Advanced Statistics II. 406, London: Ch. Griffin & Co. Ltd. 1967.

6. Mandel, J.: Statistical Analysis of Experimental Data. P. 175 – 181. New York: Interscience Publishers, 1964.

7. Marchgraber, R., and R.M. Armstrong: Solar Radiation Measurement Instrumentation USA SRDL. Technical report 2251, U. S. Army Signal Research & Development Laboratory, Fort Monmouth, N.J. (Personal communication).

8. Schulze, R.: Über die Verwendung von Polyäthylen für Strahlungsmessungen. Arch. Met. Geoph. Biokl., B. 11, 211 (1961).

9. Palland, C.L., and L. Wartena: Investigations on the Calibration Factor of the Schulze and the Funk Radiation Balance Meters and Comparison of Some Measured Results. Arch. Met. Geoph. Biokl., B, 16, 95 – 104 (1968).

10. Paltridge, G.W.: Day-Time Long-Wave Radiation from the Sky. Quart. J. Roy. Met. Soc. 96, 645 – 653 (1970).

11. Swinbank, W.C.: Long-Wave Radiation from Clear Skies. Quart. J. Roy. Met. Soc.: 98, 339 – 348 (1963).

12. Wartena, L., C.L. Palland, and. A. Koetsier: Some Experiences on the Measuring of Long-Wave Radiation Fluxes, Arch. Met. Geoph. Biokl., B. 14, 189 – 205 (1966).

13. Wartena, L., C.L. Palland, and G.H. v.d. Vossen: Checking of Some Formulae for the Calculation of Long-Wave Radiation from Clear Skies, Arch. Met. Geoph. Biokl., B 21, 335 – 348 (1973).

24

LASER-DOPPLER ANEMOMETRY AND ITS APPLICATION TO FLOW INVESTIGATIONS IN THE ENVIRONMENT OF VEGETATION

F. DURST*, G. WIGLEY† and M. ZARE*

Sonderforschungsbereich 80, Universität Karlsruhe, Germany
Atomic Energy Research Establishment, Harwell, England

1. Introduction

The continuing and still growing interest in detailed velocity measurements in the environment of vegetation is a direct result of an increasing need for quantitative understanding of flow fields in natural systems and of their influence on heat and mass transfer from leaves, plants, and complete fields of plants. Up to date, most studies of transport phenomena in the environment of vegetation have been fashioned so as to obtain integral information, e.g. see [1], [2], and mean properties like temperature, humidity of the surrounding air, or air content of the water, etc. have been the main variables of the investigations. More recently, however, the experimental emphasis has been shifted to local investigations, e.g. [3], [4], in order to yield a closer understanding of local heat and mass transfer phenomena occuring from plant leaves or the plants as a whole.

The unstrumental requirements for measuring local flow properties in the environment of vegetation strongly depend on the ultimate application and on the flow properties to be measured. In spite of the large spectrum of instruments available, only a few can be recommended as suitable devices to locally measure turbulence properties. It is shown in reference [5] that laser-Doppler anemometers are such suitable instruments.

The application of a laser-Doppler anemometer to investigations of the flow field near plant leaves are described in section 2. Results of these investigations are provided and their implication for heat and mass transfer calculations from plant leaves are discussed.

Section 3 provides a summary of investigations of a water flow around a two-dimensional obstacle in order to demonstrate the need for direction-sensitive laser-Doppler anemometers for measurements in complex flow fields [6] as they occur in the environment of vegetation. The results presented in section 3 represent preliminary measurements obtained to understand the growth of sea-weeds' influence on the structure of the flow.

An overall summary of the present paper is given in section 4, and the importance of laser-Doppler anemometry to flow investigations in the environment of vegetation is stressed.

2. APPLICATION OF LASER-DOPPLER ANEMOMETERS TO FLOW INVESTIGATIONS NEAR PLANT LEAVES

Local heat and mass transfer studies from plant leaves are essential to obtain a closer understanding of the transport phenomena that occur near leaf surfaces. These phenomena are influenced by the propoerties of the flow field around individual leaves, and the parameters of this flow should be measures to allow quantitative transport parameters to be obtained. The flow fields are, however, influenced be several factors like the roughness of the leaf surface, the turbulence of the oncoming flow, flutters, the direction of the leaf with respect to the wind velocity and many others. Hence, the phenomena of interest are complex and laborious studies are needed to investigate the influence of the different parameters.

The present study concentrates on the measurements of mean velocity and turbulence intensity near plant leaves using a laser-Doppler anemometer for the measurements. The slope of the measured mean-velocity profile near the wall is employed to calculate the local surface shear stress as a basis for heat and mass transfer calculation according to Reynolds' Analogy.

2.1. The Experimental Equipment

In order to carry out the boundary-layer studies on plant leaves, an open-jet wind tunnel was designed and built with a nozzle diameter of 300 mm and a velocity range up to approximately 3.00 m/sec. The main constructional details are schematically shown in Figure 1a. Air blown into the pressurized section of the setting chamber escaped through the nozzle after being passed through an arrangement of honeycombs and wire meshes to straighten and smoothen the flow. A parabolic contraction section provided a uniform velocity profile over a large part of the measuring section. This was checked with hot-wire anemometers prior to the measurements together with an assessment of the overall performance of the wind tunnel. Wire meshes attached to the end of the nozzle section allowed turbulence-intensity variation between 0.5 and 5.5%.

The laser-Doppler anemometer employed for the velocity measurements inside the boundary layer of the leaf surface is schematically shown on Figure 1 and its major components can be seen from the photograph on Figure 1d. It consisted of a 5 mW He-Ne-laser (Spectra Physics 120), an integrated optical unit, a light-collecting and imaging lens, and a photomultiplier (EMI 9558) to convert the optical signal into an electrical one for further processing. The resulting signals were processed by two trackers manufactured by DISA and TPD-TNO. The optical and electronic systems of the laser-Doppler anemometer were carefully tested prior to the measurements on the leaf.

Figure 1b and c show samples of the beam leaves employed in the present experiments. These leaves were planted in pots and positioned in the test section as indicated on Figure 1d.

Laser-Doppler Anemometry

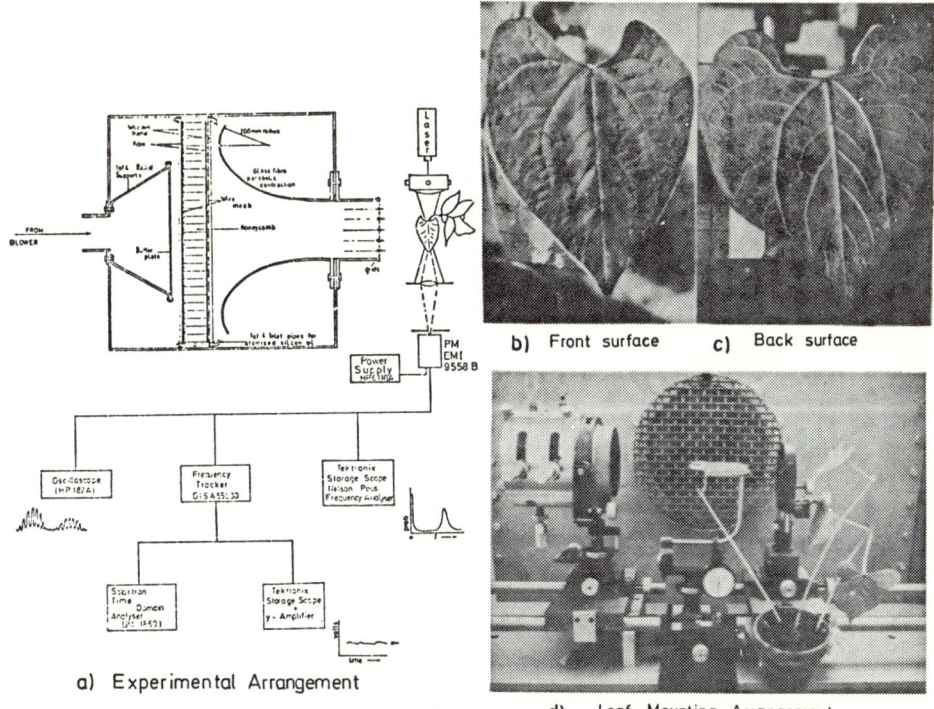

Figure 1 - Experimental Set up for Bean Leaf Measurements

2.2. Laser-Doppler Measurements

Velocity measurements were first carried out with the metal-leaf model mounted in the wind-tunnel measuring section. To obtain accurate velocity profiles, the flat plate was traversed vertically by a precision screw mechanism and its vertical position was measured relative to the optical bench with a resolution better than 0.005 mm by a dial gauge. The flat plate was positioned as far below the scattering volume as the screw mechanism would allow, and the dial gauge was fixed in this position. The flat plate was then raised in increments of either 0.25, 0.20, or 0.1 mm depending on the velocity profile shape, until the flat plate intercepted the laser beams. Then the frequency trackers would begin to loose track too often for meaningful results to be taken. For horizontal traverse, the whole of the flat plate support and vertical traverse unit was moved manually along a rigid straight rail, rigidly mounted on the optical bench and graduated in millimeters.

In this way velocity measurements were made in horizontal traverses at one centimeter intervals across the flat plate and vertically within 6.5 mm of the plate surface, to a high degree of positional accuracy. As many as 45 points velocity measurements were made in some velocity profiles.

Results of the measurements are shown in Figure 2 for horizontal distances from the leading edge of 1 cm to 9 cm. These are plotted in normalized coordinates and agree well with the analytical solution for laminar boundary-layer flows. Further away from the leading edge, the profiles deviated from the laminar solution as it is indicated by the friction factor values shown in Figure 3. The deviation is possibly caused by the streamlines divergence.

The remainder of the work was performed on primary leaves of Phaseolus vulgaris. The effect of the natural surface roughness on the boundary-layer velocity profiles was investigated in parallel flow and in one case of the leaf inclined at 10° to the flow.

The vertical height of the plant leaf relative to the optical bench was measured using the flat-plate arrangement. The plant leaf was stuck over the flat plate with the leading edge of the leaf projecting over the leading edge of the flat plate by approximately 2.0 mm. This is shown on photographs in Figures 1b and 1d. Examples of the investigated upper and lower surfaces of leaves that were used in the experiments are shown in Figures 1b and 1c.

Earlier experiments showed that if wilting occured the leading edge of the plant would sag and increase the boundary-layer thickness over the leaf. To prevent this sagging the leading edge was supported with plastic foil. A thin foil was employed to avoid changes of the effective leading-edge thickness.

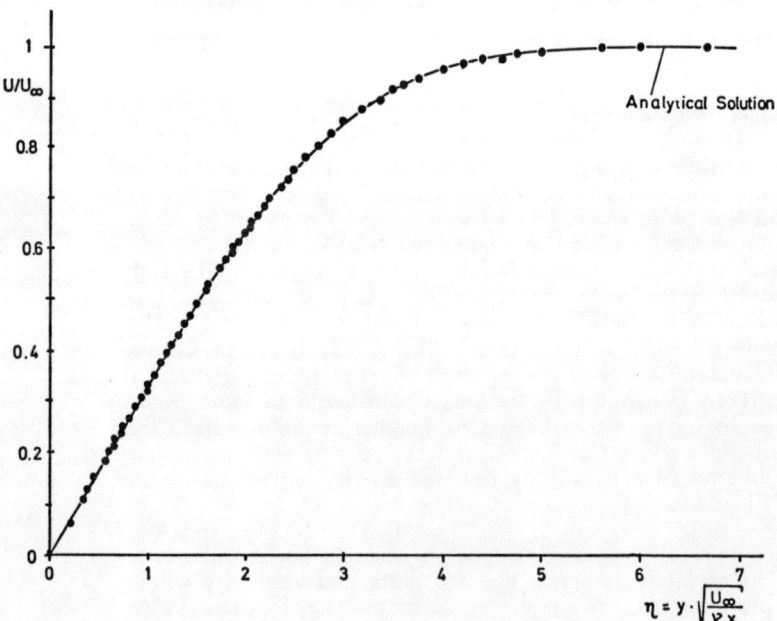

Figure 2 - Normalized Velocity Profiles for Leaf Model

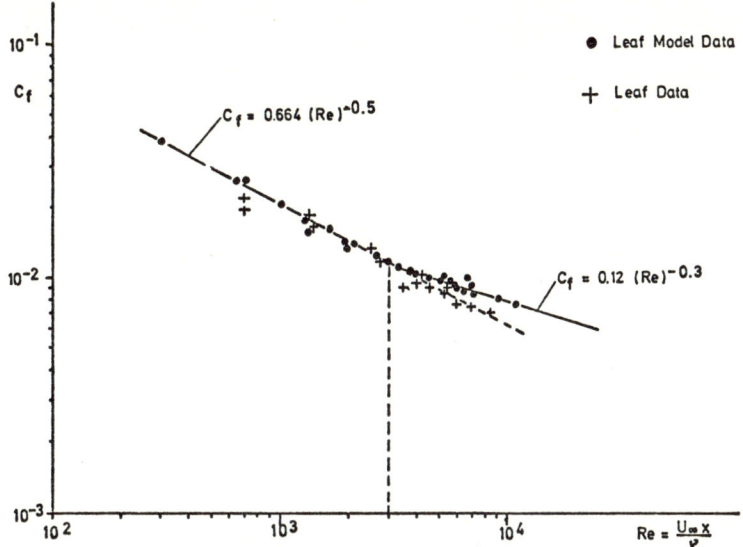

Figure 3 - Friction Factors for Leaf and Leaf Model

Studies above the upper surface of live leaves were straightforward but difficulties were encountered with lower surface due to problems of mounting. These were minimized by excision of the stem before mounting onto the flat plate.

Similar to the experiments with the metal model mean velocity measurements were carried out and friction factors obtained by measuring the velocity gradient near the wall :

$$\tau_w = \mu \left(\frac{\partial U}{\partial y}\right)_{y=0} \quad \text{and} \quad c_f/2 = \frac{\tau_w}{\rho U_\infty^2} \qquad (1)$$

Velocity profiles for the leaf are shown in Figure 4 and the corresponding friction factors are plotted in Figure 3. The latter are well described by the relationship for laminar flow :

$$c_f/2 = \frac{0.332}{\sqrt{Re_x}} \quad \text{with} \quad Re_x = \frac{U_\infty \cdot x}{\nu} \qquad (2)$$

2.3. CONCLUSIONS

The above measurements provided information that can be utilized to obtain information on heat and mass transfer on plant leaves using Reynolds' analogy. This analogy yields the following relationship :

$$\frac{Nu_x}{Re_x \, Pr^{1/3}} = \frac{c_f}{2} \qquad (3)$$

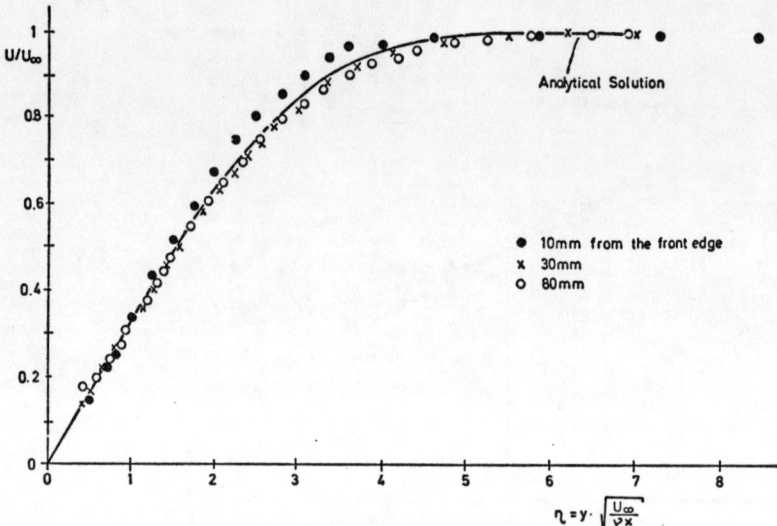

Figure 4 - Normalized Velocity Profiles for Leaf

Hence, the local heat transfer expressed by the Nusselt number reads for the above leaves:

$$Nu_x = 0.332 \cdot \sqrt{Re_x} \sqrt[3]{Pr} \qquad (4)$$

The averaged Nusselt number can be shown to be:

$$Nu_m = 0.664 \sqrt{Re_x} \sqrt[3]{Pr} \qquad (5)$$

A similar relationship can be obtained for the Sherwood number.

The above results suggests that engineering data on boundary-layer flows can be used to predict heat and mass transfer from plant leaves. For cases where there is no seperation, the total heat transfer is obtained by integrating over the leaf area yielding a shape factor to account for the special shape of the leaf and the surface area.

Relationship (5) is in some disagreement with some published data and this can presently not be explained satisfactorily. The authors are convinced, however, that proportionality factors in excess of 0.664, given in equation (5), must either result from experimental conditions different from those imposed in the present experiments or have been deduced from erroneous measurements.

3. APPLICATIONS OF LASER-DOPPLER ANEMOMETERS TO RECIRCULATING FLOWS IN WATER

In the present section the application of a laser-Doppler anemometer is described which was developed with the aim to provide a reliable instrument for measurements in recirculating flows. This anemometer was applied to a flow near an obstacle to study the characteristics of a recirculating flow, modeling flows in streams and rivers. The experiments described are preliminary in nature and will be extended to explore those factors that influence the different growth rates of sea-weeds observed in different parts of river beds. Up to date, reliable investigations of such recirculating flows have been prevented due to lack of suitable measuring techniques for high turbulence levels. Laser-Doppler anemometry has, however, opened up new possibilities for investigations of such flows.

3.1. The Experimental Set-up

The measurements near the flow obstacle were carried out in an open water channel of 300 x 300 mm cross-section with glass side walls over the whole channel length, see Figure 5. The water was pumped to a head tank from which it passed a contraction section and flow straighteners positioned ahead of the developing flow region of 3.00 m length before the obstacle. A gate valve at the pump outlet and a weir at the end of the channel allowed the flow rate and water height to be adjusted and kept constant during the experiments. The channel was set up for subcritical flow with an upstream Froude number of

$$Fr = \frac{U_m}{\sqrt{gh}} = 0.3$$

and a Reynolds number of

$$Re = \frac{U_m \cdot h \cdot w}{\nu (2h + w)} = 23\,600$$

The optical anemometer designed and built for the present investigations consisted of a 5 mW He-Ne laser (Spectra Physics 120), optical components to guide and focus the laser beam on a rotating diffraction grating (TPD-TNO, ≈100 lines/mm). A lens of f = 150 mm focused the two first-order beams inside the measuring control volume to provide approximately

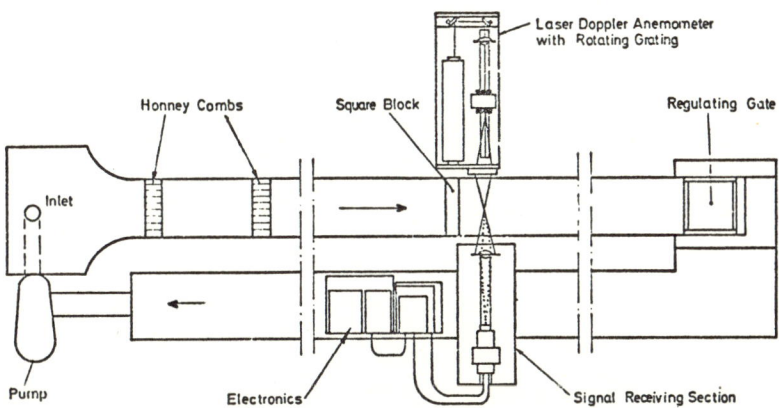

Figure 5 - Schematic Diagram of the Open Water Channel

50 fringes inside the crossing region of the two beams. The scattered light was collected in the forward direction by a light collecting lens set-up in a manner described in ref. [7] and focussed on to a pinhole of 500 μm diameter in front of a photomultiplier (EMI 9658R). The signal from the photomultiplier was fed to the electronic signal-processing equipment, see Figure 6a. The two parts of the anemometer optics were positioned on two separate tables which could be traversed to obtain measurements at different points in the flow.

Although there were sufficient scattering particles naturally present in the water to obtain laser-Doppler signals, titanium dioxide particles of approximately 1 μm mean diameter were added to increase the data rate. All measurements were carried out with a TSI frequency tracker demodulator (Model 1090) and the output signal was passed through a low pass filter of 10 kHz (TSI model 1057) prior to being processed by a digital voltmeter (DISA 55D30) and a rms-voltmeter (DISA 55D35). Figure 6c shows the experimental set-up, together with the block diagram of the electronics applied in the measurements. A photograph of the electronic equipment is given in Figure 6c.

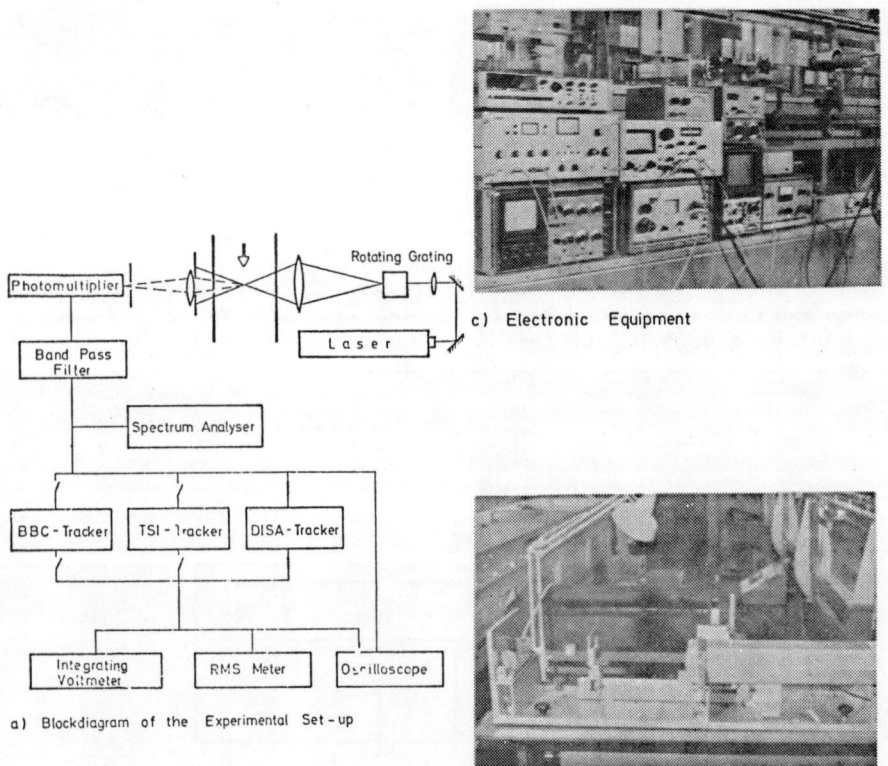

Figure 6 - Experimental Arrangement for Vector Velocity Measurements

3.2. Experimental Results

Measurements upstream and downstream of the 50 × 50 mm square flow obstacle were carried out to obtain mean velocity and fluctuating flow properties in the different flow regimes and to demonstrate the applicability of the developed laser-Doppler anemometer to complex flow fields.

At the beginning of the experiments, the appropriate rotational speed of the diffraction grating was chosen by means of a spectrum analyzer to avoid interference of the spectra at zero and at the signal frequency. Figure 7 shows the unshifted and upshifted signal spectra and indicates the interference with the spectrum at zero for the unshifted case.

Laser-Doppler anemometer measurements of the axial mean velocity component are provided in Figure 8a. The measurements indicate that the shear layer separating from the front top corner does not reattach and therefore the downstream separation zone extends over the top of the square block.

The measured fluctuations of the axial velocity components are shown in Figure 8b and have been normalized with the channel free-stream velocity. The results show that the shear layer that develops from the front block edge produces turbulent fluctuations and a region of high axial velocity fluctuation extends downstream. Diffusion of turbulence causes this region to spread as the measurements clearly show. The high mixing rate in the recirculation zone causes even distributions of rms-velocity fluctuations.

It is worthwhile to mention that the measurements described above cannot be obtained with conventional measuring techniques. Even laser-Doppler anemometers without frequency preshifting devices are subject to serious errors.

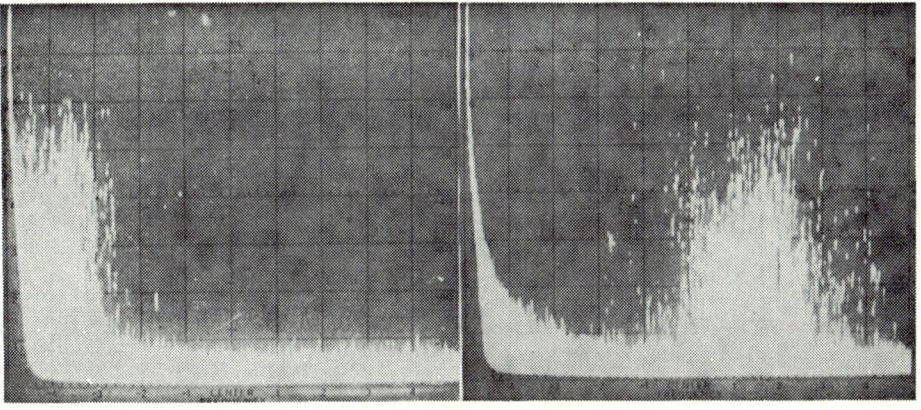

a) unshifted b) upshifted

Figure 7 - Spectrum of Laser-Doppler Signals

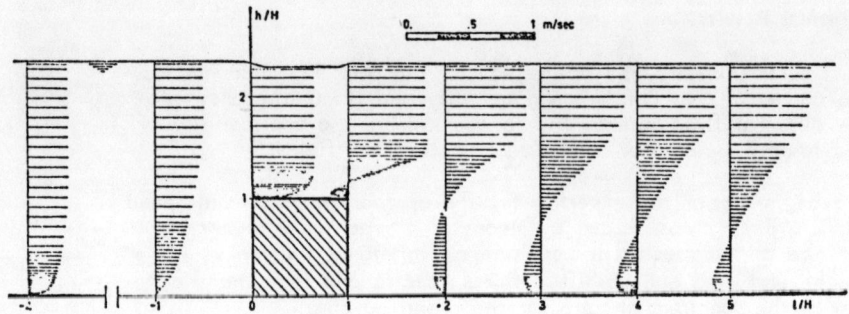

a) Mean Velocity Profiles

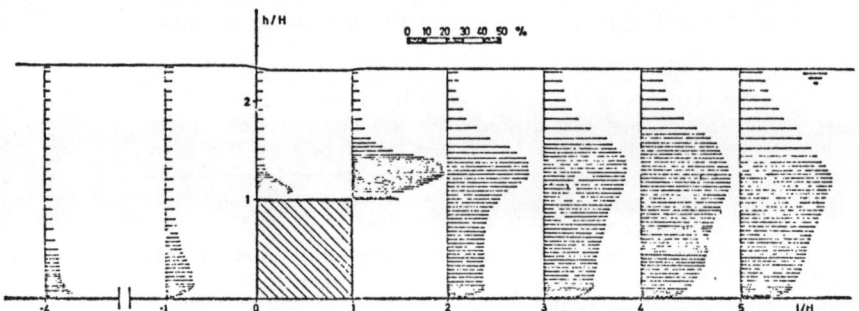

b) Turbulence Fluctuations Profiles

Figure 8 - Subcritical Flow over an Obstacle in an Open Water Channel

To demonstrate the aforementioned errors, the above measurements were repeated with the frequency preshift switched off. Several problems were experienced when the anemometer was operated in this way:

a. There is a lower limit to the frequency detectability of the processing electronics. In regions of low velocity no measurements could be carried out or erroneous data were recorded, see Figure 9a.

b. In regions of high turbulence intensity the tracking range of any available electronic system is exceeded and not all velocity components are recorded correctly. In addition, changes in the sign of the velocity fluctuations are not taken into account, yielding erroneous measurements. This can clearly be seen from Figure 9b, which shows that in regions of low turbulence intensity ($\sqrt{\overline{u'^2}}/U$) results of the two measurements are idential- In regions of high turbulence intensity, like the recirculating zones, large differences occur between the data obtained with and without optical preshift.

Laser-Doppler Anemometry

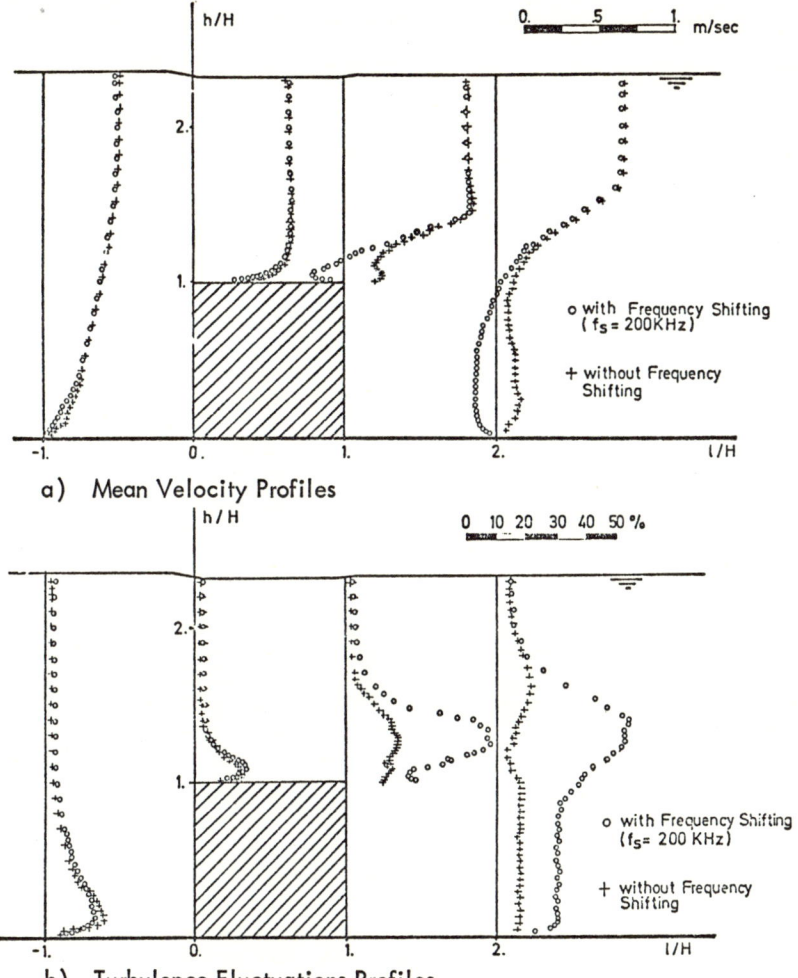

a) Mean Velocity Profiles

b) Turbulence Fluctuations Profiles

Figure 9 - Comparison between Measurements with and without Frequency Shifting

3.3. CONCLUSIONS

The above investigations showed that measurements can be attempted in flow regions of low turbulence level by conventional laser-Doppler anemometers. In regions of low velocity and large velocity fluctuations, the application of direction-sensitive laser-Doppler anemometers is essential to obtain correct measurements.

The investigations of the flow around a square obstacle revealed the presence of strong velocity gradients in some flow regimes. In regions of this kind, one usually finds short sea-weeds and the present investigations suggest that this is due to the sea-weeds being torn short in the region of high shear. Although the growth rate in these regions may be large, the occurence of sea-weed growth is prevented by the action of the high shear stresses.

The fact that the growth rate of sea-weeds is apparently high in the recirculating flow regions near obstacles placed under water, is due to the high turbulence transport of mass in these flow regimes and to the low shear stresses present there.

FINAL REMARKS

The present investigations revealed the great potential of laser-Doppler anemometers to flow investigations in the environment of vegetation. Instruments developed by the authors were applied to measurements near plant leaves and the results obtained were utilized to calculate heat and mass transfer data on the basis of Reynolds analogy. It is worth mentioning that earlier attempts failed to obtain the required velocity information by means of a hot-wire anemometer since natural hairs on leaves produced erroneous data near the surface. For distances closer than 1.0 to 1.5 mm the wire touched the natural hairs, yielding higher heat transfer from the wire and, hence higher velocity readings.

A direction-sensitive laser-Doppler anemometer employing optical frequency preshift was applied to the highly turbulent flow regions near a square obstacle positioned perpendicular to the flow. The results allowed the growth rate differences for sea-weeds observed in rivers to be quantitatively explained. Comparisons of measured mean velocity and rms-velocity fluctuations with those obtained with a conventional laser-Doppler anemometer clearly showed the need for optical preshifting to obtain accurate measurements in highly turbulent flows.

ACKNOWLEDGEMENT

The authors are very thankful to Professor Dr.-Ing. E. Naudascher, Director of the Institute of Hydromechanics of the University of Karlsruhe, for his continuous interest in the work. Sincere thanks are also due to Professor J.L. Monteith and Dr. J.A. Clark from the University of Nottingham for permitting G. Wigley to spend three and a half months at the University of Karlsruhe. Part of the work described was submitted by G. Wigley in partial fulfillment for the degree of Ph.D. at the University of Nottingham. Mrs. A. List, Miss G. Bartman and Mrs. M. Cherdron were very helpful in preparing the final version of the manuscript. Professor E.O. Macagno, visiting Professor at the Institute of Hydromechanics, was very helpful proof reading the final version of the manuscript.

The authors thankfully acknowledge the final support of this work by the "Deutsche Forschungsgemeinschaft" and partial support through a NATO contract (No. 659) which the senior author holds. Personal support to one of the authors (G.Wigley) in the form of a DAAD scholarship is also gratefully acknowledged.

REFERENCES

[1] Kumar, A., Barthakur, N., "Convective Heat Transfer Measurements of Plants in a Wind Tunnel". Boundary Layer Meteorology, 2, pp. 218, 1971.

[2] Thom, A.S., "The Exchange of Momentum, Mass and Heat between an Artificial Leaf and the Airflow in a Wind Tunnel" Quart. J. Roy. Met. Soc., 94, p. 44, 1968.

[3] Wigley, G., "Transport in the Boundary Layers of Real and Model Leaves", University of Nottingham, Ph.D. Thesis, May 1974.

[4] Perrier, E.R., Robertson, J.M., Millington, R.J., and Peters, D.B., "Spatial and Temporal Variation of Wind Above and Within a Soybean Crop", Agricultural Meteorology, 10, p. 421, 1972.

[5] Durst, F. and Whitelaw, J.H., "Integrated Optical Units for Laser Anemometer". Journal of Physics E: Scientific Instruments, Vol. 4, 1971.

[6] Durst, F. and Zaré, M., "Removal of Pedestals and Directional Ambiguity of Optical Anemometer Signals". To be published in Applied Optics, November 1974.

[7] Durst, F. and Whitelaw, J.H., "Light Source and Geometrical Requirements for the Optimization of Optical Anemometer Signals" Opto-Electronics, 5, pp. 137-151, 1973.

SECTION 3
PLANTS

WATER TRANSFER IN PLANTS

PAUL G. JARVIS*

Coniferous Forest Biome, University of Washington, Seattle, WA 98195, U.S.A.

ABSTRACT

The plant is considered as a largely hydraulic system in which water moves along gradients of hydrostatic pressure and water potential along pathways which have frictional resistance to flow and storage capacity associated with them. The origin of the driving force and of water potentials and water deficits are discussed. The relative sizes of resistances in the pathway are reviewed and the storage capacity of leaves, roots and stems evaluated. The concepts are extended to water transfer in forest stands.

1. INTRODUCTION

Water moves through the plant as a liquid by viscous flow in the xylem and by diffusion across cell membranes. The driving force for these transfers is the fall in water potential in the leaf as a result of the loss of water by evaporation. Water moves to the sites of evaporation from sources within the plant and from the soil along the gradients of water potential (or relevant component potential) which result. In this essay I shall mainly consider the properties of the transfer system and the sources within the plant of stored water which exchanges in significant quantities with the transpiration stream. My approach will be largely qualitative and descriptive and will make use of only simple one-dimensional, one-parameter models. In a useful, critical treatment of the soil-plant-atmosphere continuum (SPAC), Philip [1] argued for more adequate models of water transport and others (e.g. [2]) have advocated the simultaneous treatment of water and solute transfer. However, I consider that at the present time we have insufficient knowledge to justify more complex models. Consequently, my intention in writing this essay is to collect together some ideas which may stimulate further investigation and possibly form the basis of a future water transfer model.

2. THE PATHWAY

Location of the Major Axial Pathway

Water enters the plant from the soil through the root hairs and epidermis of young growing roots and through breaks in the periderm of older suberised

---

*Permanent address: Department of Botany, University of Aberdeen, St. Machar Drive, Aberdeen, AB9 2UD, U. K.

roots (3, 6). It traverses the root cortex and endodermis to reach the root xylem. It moves in the xylem to the leaves where it passes from the xylem endings by way of two or three mesophyll cells to the sites of evaporation (4). There are several points of uncertainty regarding the details of this pathway (5,6).

In the root cortex and leaf mesophyll there are at least two alternative pathways available for water movement. Water can pass from cell to cell by osmosis across the cell membranes and by diffusion and cyclosis through both the symplasm and cell vacuoles along gradients of water potential, or it can move through the intermicellar spaces of the cell walls along gradients of hydrostatic pressure or matric potential. Movement of water into cells undoubtedly occurs in growing and expanding tissues, and movement in and out of cells is a regular consequence of the diurnal change in pressure in the xylem (see Section V). However, which pathway conducts the major part of the normal transpiration stream will depend upon the relative size of the resistances to flow along the alternative routes. In a recent review, Weatherley (5) concluded that the resistance in the cell to cell pathway is so large as to require unreasonable differences in water potential across both the leaf mesophyll and root cortex to support the known flow rates. Additionally, cell to cell transfer is not in accord with observations of tracers and kinetic experiments. He concludes: (p. 184) "Taken together the lines of evidence discussed above point to the cell walls as being the main transpirational pathway from the xylem to the evaporating surface.", and (p. 186) "The balance of evidence favors the view that as a working hypothesis movement of water occurs through the cortical and stelar cell walls with passage through the vacuoles largely restricted to the endodermis.". However, in a detailed discussion of transport across the root, Newman (6) inclines towards the symplasm pathway.

The cell walls can be considered as a hygroscopic matrix of cellulose micelles with intermicellar pores of about 1 to 10 nm in radius. Hydrostatic pressure differences of 30 to 300 bar are required to drain pores of these sizes and consequently they normally remain water filled. Thus, transfer of water in the cell wall may be likened to the movement of water in a saturated soil and Darcy's law applied. I can find no direct measurements of the hydraulic conductivity of cell walls and values for artificial cellulose membranes are very variable and probably too low (5), but indirect estimates vary from 10^{-8} to 10^{-11} m s^{-1} bar^{-1} depending on the location and water stress (7, 33).

Transfer of water in the xylem has recently been reviewed by Zimmermann (8). Unquestionably, water moves by mass flow through the xylem vessels and tracheids along gradients of hydrostatic pressure. The lumen of differentiated vessels is continuous over distances of a few centimeters up to meters so that water need only pass through pits infrequently, whereas the lumen in tracheids is a few mm in length (Table 1). Flow of water in the lumen of vessels is laminar rather than turbulent (9) and obeys the Hagen-Poiseuille law for laminar flow in a bundle of capillaries (10). Thus the volume flow rate of water under a pressure difference ΔP through a length ℓ of vascular tissue is:

$$q = \frac{\pi N r^4}{8 \eta \ell} \Delta P \qquad (1)$$

where N is the number of conducting vessels of mean equivalent hydraulic radius r (alternatively, $Nr^4 = \sum_{i}^{m}(n_i r_i^4)$) and η is the viscosity of the sap ($\approx$ that of water (10)). The frictional resistance to the flow of water is $R = \Delta P/q$ so the bracketed term on the right is the conductance. After a careful study of the vascular anatomy of the tomato, Dimond (10) obtained reasonable agreement between the pressure drop across the vascular bundles in the petioles and internodes, necessary to maintain flow at the transpiration rate, calculated from equation (1), and determinations from measurements of the conductivity of segments of the stem and the petioles. Thus his results are consistent with the hypothesis that in vessels the main resistance to flow is in the lumen and that the pits between the vessels offer little resistance. However, fungal toxins and other large molecules can plug the pit membranes so that they become a limiting flow resistance (11).

In the tracheids of conifers a major part of the total resistance normally lies in the margo of the bordered pits. Petty and Preston (12) show that flow through the pores in the margo is laminar (Re<10) and that a modified Hagen-Poiseuille equation may be applied. The total resistance of tracheid lumen and pit pairs in a length ℓ of xylem is then:

$$R = \frac{\Delta P}{q} = \frac{8\eta\ell}{\pi N r^4} + \frac{8\eta(t+1.15a)}{\beta\pi n a^4} \quad (2)$$

where the second term on the right is the contribution to the total resistance of n pores in the margo of depth t and mean equivalent radius a. The addition of 1.15a to the length of the pit membrane pores is the Couette correction, which becomes important when the radius of the capillary is of the same order as its length; β (= 0.75 ℓ_t/ℓ, where ℓ_t is mean tracheid length and 0.75 allows for tracheid overlap) takes account of the sap passing through a number of bordered pits in series. Using (see Table 1) experimentally determined values of the variables in equation (2), Petty and Puritch (13) showed that about 60% of the total axial flow resistance was in the bordered pits and 40% in the lumens of solvent-dried wood of _Abies grandis_. Numerous earlier observations that most of the resistance resides in the pits probably result from pit aspiration during sample preparation. Since the pit resistance is inversely proportional to (πa^4), it is very sensitive to small changes in the size and number of the pores in the margo. Thus the pit resistance would be expected to decline as the pits mature and the margo becomes more perforate (14) and to increase as incrustations develop on the margo fibers with age. For similar reasons, the resistance to flow is much higher in heartwood than in sapwood (15,18), in latewood than in early wood (16,17), and in diseased wood than in healthy wood (17, 18, 19).

The Plant as a Hydraulic System

Numerous experiments on small herbaceous plants rooted in solution culture have shown that the addition of metabolic inhibitors or osmotica to the culture

Table 1. Properties of the tracheids and bordered pits in wood of _Abies grandis_ (13, 19).

			air dried	solvent dried	infested*
Tracheid	ℓ_t	(mm)	2.6	2.6	2.5
	N	(mm^{-2})	320	830	670
	r	(μm)	9.9	14.0	12.6
Pit	t	(μm)	0.1	0.1	0.1
	n	(tracheid^{-1})	600	27 000	10 500
	a	(μm)	0.12	0.09	0.10

*tree infested by Balsam Woolly Aphid, _Adelges piceae_

solution, or the excision of the roots, result in rapid changes in transpiration rate (20, 21, 22). The changes are usually apparent within a minute, their rate of appearance being limited by the response time of the measurement system. Similarly, application of mechanical pressure sufficient to compress the vessels in the stem causes a rapid change in transpiration rate (23). In an elegant series of experiments, Raschke (24, 25) showed that the stomata in detached maize leaves responded to a change in the hydrostatic pressure in the water supply to the leaf base within 0.1 s as measured with a mass flow porometer, i.e. instantaneously, and in all parts of the leaf simultaneously. Thus he proved that the leaf is a hydraulic system to which the stomatal apparatus is linked. We may conclude with Raschke (24) that: "Any change in xylem pressure will be transmitted to other parts of the plant; however, the amplitude will decrease with distance from the site of the pressure change. Friction in the water path and the elasticity of the tissues attenuate the pressure wave as it travels through the plant." In addition, attenuation and phase shifts resulting from the exchanges of water between the xylem and the surrounding tissues further obscure the hydraulic nature of the system and phase lags in the propagation of changes in water potential or flow rate are of the order of minutes (26, 27) or hours (28, 29) (Fig. 1) depending on the bulk of the tissue.

The frictional resistances to the flow of water in the roots and stems of plants, together with the gravitational potential in tall trees, are expected to lead to substantial reductions in pressure in the xylem of -10 bar or lower in rapidly transpiring plants, if water is pulled up the tree from above according to the cohesion theory. The main objection to the cohesion theory has always been that such tensions could not occur since the water would be metastable and bubbles would appear and break the columns (8,30). To a considerable extent, the problem of whether such tensions can and do exist in the xylem has been answered by measurements with the pressure chamber (31). In addition, tension or vapor pressure transducers inserted into the conducting sapwood of trees show the negative hydrostatic pressure gradients expected on the basis of the cohesion theory (Fig. 1).

3. THE DRIVING FORCE

Evaporation Rate

The driving force to move water against the gravitational potential and the frictional resistances in the pathway is derived from the evaporation of water in the leaf. The environmental driving variables for evaporation are conveniently illustrated by the Penman-Monteith combination equation (32):

$$E = \frac{sA + c_p \rho \delta e k_a}{\lambda (s + \gamma(1 + \frac{k_a}{k_s}))} \tag{3}$$

The rate of evaporation per unit stand area (E) is a strong function of the available radiant energy (A) and the vapor pressure deficit (δe), and a weaker function of windspeed, through its effect on boundary layer conductance (k_a), and of air temperature, through its effect on s (the slope of the relation between saturation vapor pressure and temperature). λ, c_p, ρ and γ are the latent heat of vaporisation of water, the specific heat of air at constant pressure, the density of air and the psychrometric constant, respectively:

they are all weak functions of temperature. E also strongly depends on the stomatal conductance (k_s) and consequently is affected by all the environmental and physiological variables which change k_s (Section VI).

In a stand forming a complete cover, the rate of water transfer through an individual stem (q) depends upon the number of stems (n) supporting the canopy: the more stems there are, the smaller is the flow through each one, i.e. q = E/n (Table 2).

Water Potential and Water Deficit

As a result of the loss of water by evaporation, the water potential of the cell wall matrix adjacent to the liquid-air interface falls. Because of the continuous hydraulic nature of the pathway, liquid water moves towards the interface from sources of higher water potential which are distributed both throughout the plant and in the soil. As a consequence of the water movement through the plant along a series of frictional resistances, gradients of water potential result, with the largest drops in potential where the flow rates and resistances are largest. The lowered water potential in the transpiration pathway provides the driving force for the movement of water out of adjacent tissues such as the leaf mesophyll, cortex and phloem. The consequent loss of water from tissues in the root, stem and leaf constitutes the water deficit (33). Thus the water deficit is the inevitable consequence of the flow of water along a pathway in which frictional resistance or gravitational potential have to be overcome. Water deficits do not occur because the loss of water from the leaves in evaporation exceeds the supply from the roots, as is so often stated. In a completely rigid flow-path, for example, a potential difference is necessary to drive steady state flow against the frictional resistance when input and output flow rates are equal (Fig. 2A). The water deficit arises because the water potential of cells adjacent to the pathway tends to equilibrate with that in the pathway by means of movement of water to the pathway (Fig. 2B). The size of the water deficit depends on the moisture characteristic of the cells (i.e. the relation between relative water content and water potential), the resistance of the axial pathway between the location of the tissue and the major source of water, and the flow rate of water through the plant (Fig. 2C). In general, lower potentials and hence larger deficits occur higher up the plant as shown in Fig. 1.

Table 2. The flow rate of water in a single stem (q, mg s^{-1}) in stands of different densities of individuals but with a complete foliage cover. Absorbed radiant energy = 500 W m^{-2}.

Crop	No. of Stems m^{-1}	Bowen Ratio		
		0	1	4
Forest	0.01	2×10^4	10^4	4×10^3
	0.1	2×10^3	10^3	4×10^2
	0.4	500	250	100
Horticultural	1	2×10^2	10^2	4×10
	10	2×10	10	4
Field	100	2	1	4×10^{-1}
	1000	2×10^{-1}	10^{-1}	4×10^{-2}

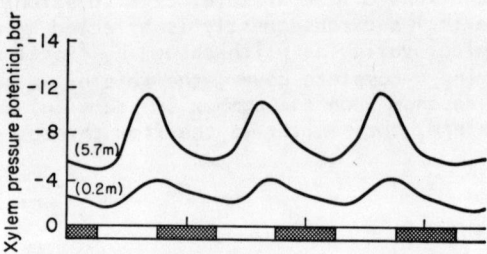

Fig. 1 The diurnal course of xylem pressure potential in the stem sapwood of a 25-year old Sitka spruce at two heights on sunny days in August. From (38).

Phase lags

Because of the hydraulic nature of the pathway, water will tend to be withdrawn simultaneously from all sources when transpiration begins. However, it is to be expected that initially most water will move to the liquid-air interface from nearby available sources of water such as the leaf mesophyll, petiolar and branch tissues, rather than from tissues further away in the plant or the soil, because the pathway is short and the difference in potential large. Since the cells are finite sources, their capacity to supply water decreases as it is withdrawn and their water potential tends towards that of the xylem. Consequently, a front of water potentials and water deficits spreads down through the plant appearing at lower levels somewhat attenuated and with a phase shift. In small herbaceous plants with small storage capacity, both the attenuation and the phase shift may be negligible but in the stems of trees they are considerable (Fig. 1). Both attenuation and phase shift in the propagation of water

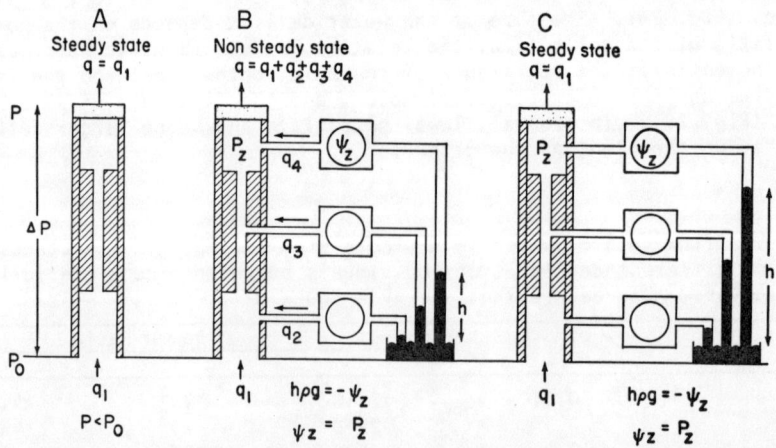

Fig. 2 Diagrammatic representation of a transpiring plant with tensiometers plugged into the cells adjacent to the xylem at three levels. q is the loss of water in transpiration, q_1 is the uptake of water by the roots and q_2--q_4 are the flows of water out of storage in the cells. P_z is the xylem pressure potential and Ψ_z is the water potential in the adjacent cells at height z.

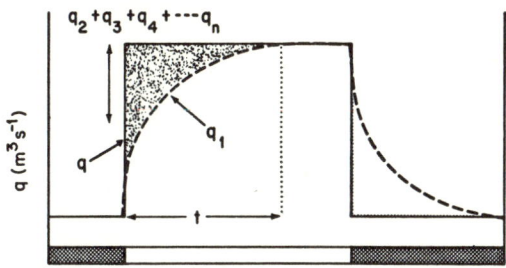

Fig. 3 A diagram of the flow of water in the plant when the transpiration rate is changed in a step by irradiation. The solid line is the loss of water from the leaves in transpiration, q; q_1 and q_2--q_n are the uptake by the roots and the flows out of storage, respectively, as in Fig. 2. Adapted from (5).

potential result from the storage capacity of the conducting pathway and associated tissues (Section V). Pathway resistance alone cannot cause lags. This is illustrated in Fig. 3. The solid line shows both transpiration and uptake from the soil in a rigid, inelastic 'plant' without storage capacity as in Fig. 2A. In a plant with appreciable storage capacity, uptake from the soil also starts immediately, but at a much slower rate, and follows the dashed line. The vertical distance between the two lines represents the flow of water out of storage during the day, the stippled area being the accumulated water deficit. The shaded area shows uptake of water continuing after transpiration has ceased until the deficit is filled. The length of period t depends upon the storage capacity of the plant or tissue and is of the order of minutes in small plants and hours in trees (Section V) (27, 87)

Development of Leaf Water Potential

From the foregoing, it is clear that leaf water potential (ψ_ℓ) depends upon the rate at which water flows through the plant. Attempts to correlate ψ_ℓ with particular environmental variables such as irradiance or vapor pressure deficit generally show a marked hysteresis because transpiration from the plant is not linearly dependent upon any one environmental variable (equation 3). A linear dependence of ψ_ℓ upon the flow rate of water through the plant is formalized by the one-dimensional SPAC hypothesis (34):

$$\psi_\ell = \psi_s - \sum_{i=s}^{i=\ell} q_i R_i - h\rho g \qquad (4)$$

where the subscripts ℓ and s indicate leaf and soil, respectively, and q_i and R_i are the partial, series-linked flows and resistances in the pathway from soil to leaf. Clearly this equation is difficult to apply in practice if the pathway resistances are not constant but are a function of flow rate (Section IV) and if appreciable quantities of water move to the leaf from parallel sources in the plant (Section V). In a number of herbaceous dicotyledons ψ_ℓ has been found to be largely independent of q over a wide range with the implication that a limiting resistance decreases with increasing q (5) (see

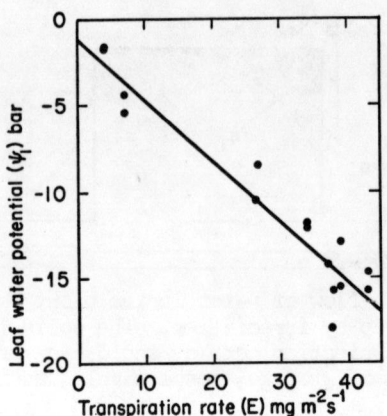

Fig. 4 The diurnal variation in leaf water potential as a junction of transpiration rate in unstressed apple trees. From (36).

Section IV). However, in some herbaceous (35) and woody plants (36), ψ_ℓ is a linear function of E (Fig. 4).

The Gradient of Water Potential

Under static conditions of no water flow, an axial gradient of water potential in the plant of 0.1 bar m^{-1} height is to be expected because of gravity. Under the dynamic conditions of rapid water flow through the SPAC, much larger gradients are to be expected because of the frictional resistances encountered by the flowing water. Recently interest has centered on the size of the axial gradient in normally transpiring plants because of two reports claiming that 0.1 bar m^{-1} was not exceeded in the stems of tall trees, even at midday (31, 37). In contrast, much steeper gradients of up to 2 bar m^{-1} have been found in a number of trees (Figs. 1 and 6) and even steeper gradients can be inferred in some herbaceous plants (38). Richter (39) points out that this discrepancy probably arises from estimating the gradient in the stem from measurements of water potential in leaves or twigs attached to the plant at different heights. The leaf samples are not on the same flow path and this can lead to the mistaken inference of zero gradient (Fig. 5). Dimond's careful study of the pressure and flow relations in the vascular bundles of tomato (10) showed that the pressure difference required to move water to the leaves was up to 10x that required to overcome gravity, and Heine (40) predicted gradients in conifers of over 1 bar m^{-1} from measurements of conductivity and flow rate.

Fig. 5 emphasizes the problems in determining gradients in plants with current techniques. Nearly all measurements of potential are made at the extremities of the plant: very few potentials have been measured at points along the pathway. In a few cases, tension or vapor pressure transducers (29, 41, 42) have been inserted into the stem xylem of trees at different heights, as in Fig. 1. Another alternative, initially used on tobacco (43), which we have used on spruce trees (38), is to enclose a branch so that it does not transpire and then to regard it as a tensiometer plugged into the stem xylem. The xylem pressure potential of twigs on the branch should then be equal to that at the

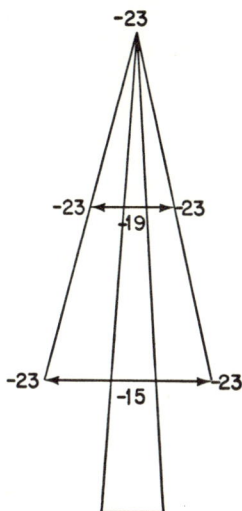

Fig. 5 A diagram of leaf and stem water potentials in a tree to show the apparent absence of a vertical gradient of water potential in the leaves. From (39).

junction of the branch and stem xylem. Fig. 6 shows the axial gradient in xylem pressure potential in the stem obtained in this way. The difference in potential between enclosed and transpiring branches with adjacent insertion at the same node is the drop in potential along the 1-2 m long transpiring branch. Gradients along the branches are at least as large as the gradients in the stem.

Water potentials vary from point to point in leaves and gradients along leaves of tobacco of about 20 bar m^{-1} have been recorded (44): in roots, axial gradients of up to 3 bar m^{-1} have been found in spruce and tobacco (38, 45). However, I know of no measurements of radial gradients in either roots or stems.

Control of Leaf Water Potential

It is a curious fact that leaf water potentials in the majority of un-droughted vascular plants fall within the same range of about -5 to -25 bar, irrespective of the size or leaf area of the plant: much the same range in ψ_ℓ is found in small seedlings and in large trees, and in canopies of small or large leaf area index. However, it might be expected that larger reductions in ψ_ℓ would be found in larger plants to support the transport over a longer pathway of larger volumes of water, as the result of the absorption of energy by the larger leaf area.

In the short term, a lower limit to ψ_ℓ may be set by stomatal closure with consequent reduction in q. In many plants, stomatal closure occurs over a narrow range of ψ_ℓ as guard cell turgor falls to a critical level (46) (Section VI). However, if this were the only control, the stomata would be closed for much longer periods in large plants than in small ones, and this does not

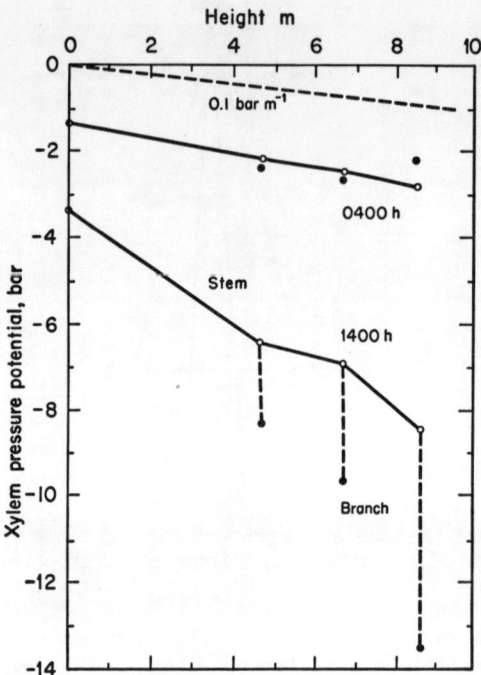

Fig. 6 Xylem pressure potentials at different heights in the stem of a 25-year old spruce tree (open circles) and in shoots at the ends of branches (closed circles). The dashed line above shows the gradient to be expected from the effect of gravity. The length of the dashed lines joining the open and closed circles gives the drop in potential along the branches in the afternoon. Redrawn from (38).

happen. A variable plant resistance, inversely proportional to the flow rate, would also help to keep ψ_ℓ within narrow limits during the course of a day (see Section IV). Similarly, a narrow range ψ_ℓ among plants of different sizes most probably results from relative constancy in (qR) (equation 4) as plants grow. Thus morphological homeostasis probably reduces the resistance to flow in proportion to the increase in the volume flowing.

In the branch and root systems, this is achieved by branching of the vascular system into additional parallel-linked conduits and by the secondary growth of additional vascular tissue so that A_ℓ/A_b (Fig. 7) remains constant. Thus the supply resistance to a leaf increases only slightly as the branches grow in length and the potential drop across a branch remains about the same. Local variations in anatomy alter this picture. For example, Huber (8) found the conductivity of shoots to be higher nearer the top of a fir tree than lower down, and the conductance of unbranched vascular bundles in tomato varies among internodes, depending upon the variation in size and number of vessels in a bundle (10).

In the stem the cross-sectional area of the conducting tissue must increase in proportion to the flow if the pressure drop is not to increase, i.e. A_ℓ/A_s must remain constant. Consequently, it is of considerable interest that Grier

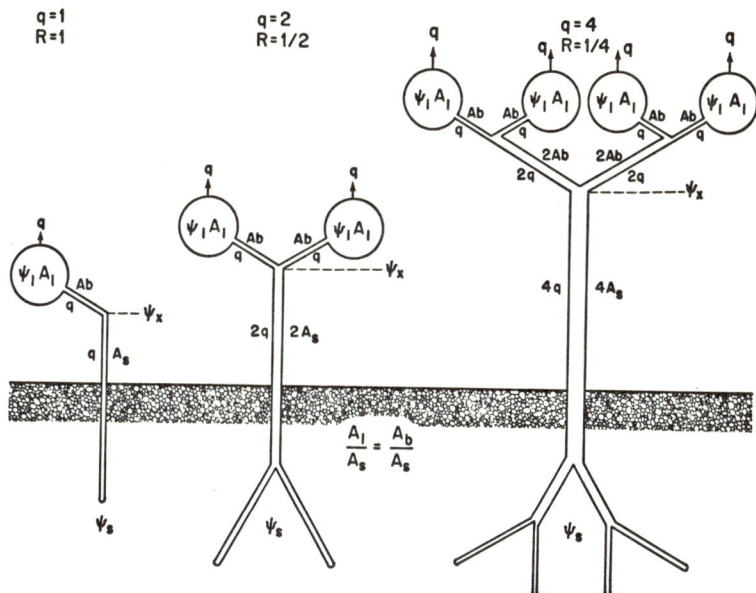

Fig. 7 Diagrammatic representation of a growing plant to illustrate how (qR) and hence ($\Psi_\ell - \Psi_s$) remain approximately constant as a result of the growth of the conducting tissue. A_ℓ is the leaf area, and A_s and A_b are the cross-sectional areas of the conducting tissue in the stem and branch, respectively, of the initial small plant.

and Waring (47) have found highly significant linear correlations between leaf biomass and sapwood area in the stem in Douglas fir, noble fir and ponderosa pine. Clearly in the stems of conifers in which large gradients in hydrostatic pressure occur, the cross-sectional area of the sapwood is a variable of major significance to the development of ψ_ℓ.

The Water Potential in a Canopy

Water is transferred from the soil to the foliage canopy through a number of parallel conduits - the stems. When there are many conduits present, the flow through any one is smaller than when there are fewer present (Table 2), and hence the drop in potential will also be smaller, if the resistance of the individuals is the same. That is, the resistance of the pathway between soil and canopy is $R_c = R_i/n$ where n is the number of stems of resistance R_i per unit area. Thus if there are fewer stems, the resistance of the pathway will increase unless the resistance of each stem also decreases.

In tall perennial crops the density of the stand may be reduced by natural death of some of the plants or by management. The consequences of removing half the trees in a stand at ceiling leaf area index are illustrated in Fig. 8. The immediate initial result is to double R_c and to redistribute the absorbed radiant energy so that a considerable amount is absorbed by the ground, and more is absorbed by an individual crown than previously. Consequently, evaporation from the stand drops ($E' < E$) but the flow of water in each remaining tree increases ($q' > q$) so that ψ_ℓ' falls below ψ_ℓ. This is one of the main symptoms of post-thinning stress. As the canopy fills in to the initial leaf area index, by doubling the leaf area per tree, E' increases again to E and q' rises to $2q$. Thus unless R_t is reduced proportionately, much lower values of ψ_ℓ would result. However, the sapwood cross-sectional area, which is a function of leaf biomass (47), also increases rapidly (indeed this is the main reason for thinning trees) so that R_t falls to a new low value as R_c returns to its original value. The end result, if lower ψ_ℓ is not to occur, is half the number of trees with the original leaf area index and twice the leaf area and sapwood area per tree. Thus the sapwood area index may be a good indication of the capability of a stand to withstand water stress, and a constant value a desirable management goal.

4. PATHWAY RESISTANCE

Definition

Liquid water moves in the xylem by mass flow along gradients of hydrostatic pressure (P) in the absence of membranes, so that xylem pressure potential is the appropriate driving force. Movement into cells adjacent to the pathway is by osmosis across membranes along gradients of water potential (ψ). Since ψ is in dynamic equilibrium with the xylem water potential ($P + \pi$), and since the solute potential (π) of xylem sap is small, the assumption is usually made that $\psi(z) = P(z)$ so that ψ can be regarded as the driving force for water movement through the plant. In that case the flow at any point is:

$$q = \frac{1}{R} \int_{z_2}^{z_1} \frac{d\psi}{dz} \cdot dz = \frac{\psi_1 - \psi_2}{R} \tag{5}$$

where the dimensions of q are $[L^3 \, T^{-1}]$ and R, the resistance of the pathway, has dimensions of $[ML^{-4} \, T^{-1}]$ but is often expressed in pressure units of bar s m^{-3}. The equivalent pathway conductance is $K = 1/R$.

A plethora of often misleading definitions of resistance, conductance, conductivity and permeability, with various combinations of units, currently exists in the literature and some rationalization is very desirable (34, 48). The resistance defined above is a property of the pathway between z_1 and z_2. As in electricity, the intrinsic property of the constituents of unit length (ℓ) and area (A) of pathway can be described by a resistivity (or conductivity, $k = 1/\rho$):

$$\rho = \frac{\Delta P}{q} \frac{A}{\ell} = R \frac{A}{\ell} \tag{6}$$

with dimensions of $[ML^{-3} \, T^{-1}]$ or units of bar s m^{-2}. There is some scope in choosing the appropriate area. For a membrane or cell wall, the total cross-sectional area conducts and is appropriate: for a stem or petiole, the area of vascular tissue, or of vessel and tracheid lumens seems more appropriate. However, I can see little justification for using the leaf area, although this is frequently done, i.e. resistance is often defined as $\Delta\psi$/(transpiration rate per unit leaf area). There is then a tacit, unverified assumption that there is a relationship between the leaf area and the cross sectional area of the

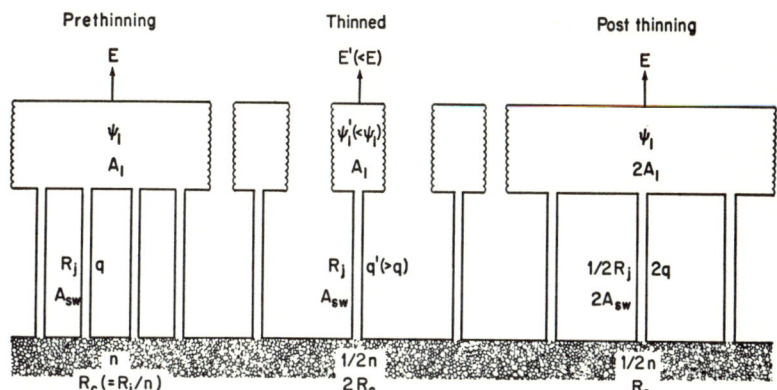

Fig. 8 A diagram to show the expected effects of thinning a stand of trees on the leaf water potential and cross-sectional area of the sapwood. q is the flow of water through an individual tree, with a cross-sectional area of stem sapwood, A_{sw}, a flow resistance, R_j, and a leaf area, A_ℓ, prior to thinning. The primes indicate the corresponding values after the number of trees has been reduced from n to $\frac{1}{2}$n. R_c is the resistance of the liquid flow pathway between soil and canopy.

conducting pathway, and that this relationship is constant and independent of species and physiological condition of the plant (8, 47).

In the above definitions, the viscosity of water ($\eta = 10^{-3}$ N s m^{-2}) is implicitly included in the resistance and resistivity (48). Since viscosity is a function of temperature, pressure and solute concentration, it may be better isolated, thus leading to a parallel set of definitions (Table 3).

The size of the resistance in plants and parts of plants is poorly known because we do not have adequate techniques for its measurement. The resistance of whole plants growing in nutrient solution can readily be obtained from steady-state measurements of transpiration rate, average ψ_ℓ and solution solute potential (e.g. 49). Similar estimates can be made in the field if q can be measured, and the soil water potential is known. Such estimates include a resistance to transfer across the rhizosphere, but this is small in relation to that in the plant, at least when the soil is wet (6, 50). In absolute terms, such estimates of whole plant resistance must depend as much on the plant

Table 3. Definitions of flow resistance and conductivity (see also 48)

Name	Symbol	Definition	Dimensions	Units	Reciprocal	Symbol
resistance	R	$\Delta\psi/q$	$ML^{-4}T^{-1}$	bar s m^{-3}	conductance	K, G
resistivity	ρ	$(\Delta\psi A)/(q\ell)$	$ML^{-3}T^{-1}$	bar s m^{-2}	conductivity	σ
relative resistance	R_r	$(\Delta\psi A)/(q\ell\eta)$	L^{-2}	m^{-2}†	relative conductivity* permeability constant	k

*not specific conductivity; in SI specific means divided by mass.
†$\eta = 10^{-3}$ N s m^{-2} for water.

architecture as on transfer properties of the pathway, but changes in response to treatment and environment can be informative (49).

Quantitative estimates of the relative size of the resistance in different parts of the pathway are almost always obtained by inference. I know of no instances in which accurate measurements have been made <u>in situ</u> of both the flow through a segment of pathway and the drop in potential across it. Most measurements of ψ are made in the leaves with psychrometers or in shoots with pressure chambers. We are unable to measure ψ at intermediate positions in the pathway, and current measurements of sap flux are not accurate in terms of the volume flowing. When estimates of ψ in the plant are available, errors of interpretation may result from oversimplifying assumptions about the phyllotactic arrangement of the vascular tissues (51), the definition of which can be very exacting (10). In trees, accurate measurement of pressure potential in roots, stems and branches is somewhat easier (Fig. 1 and 6)(29, 38) and errors arising from the arrangement of vascular bundles are less likely, but accurate estimation of q_t is still a problem which sap flux meters and micrometeorological techniques only solve approximately.

More usually, the conductivity of stem cores and stem, branch and petiolar segments is measured in the laboratory in steady state flow conditions (15, 16, 18). However, application of such determinations to the transfer of water in intact plants is again limited by lack of knowledge of q_t or ψ at appropriate points in the pathway. Thus estimates of the relative size of resistances in different parts of the plant are nearly all obtained by inference from potential differences or experiments involving mutilation of the plant. Consequently there is still controversy as to where the largest resistances lie.

Relative Size of the Resistance in Roots, Stems and Leaves

In a number of experiments with small herbaceous plants (e.g. 52, 53, 54) and with large trees of Scots pine (55) removal of the roots has been found to result in a substantial rise in ψ_ℓ, or a more rapid flow of water through the plant, or both. In those experiments where the stem was removed as well, there was little or no additional effect. From such experiments, it has been deduced that the root system is the major source of resistance to water flow in the plant and that the resistance of the xylem in the stem is comparatively small. Boyer (56, 57) enlarged upon such experiments by measuring the half-time for the rise in ψ_ℓ, when transpiration was prevented, in intact plants and after excision of roots and stem. He found the following relative distribution of resistances in the root system, stem and leaves: 2:1:1 in sunflower; 2:1:1.5 in bean and 10:1:2.5 in soybean. Steady state measurements of flow through the plant before and after excision of parts have led to similar estimates of the relative size of the resistances: tomato 2.5:1:1 (53); sunflower 2:1:1 (53); safflower 2.5:1:10 (54); Scots pine 4:1:- (55). Thus in some cases the root system is the largest resistance and in others the foliage, but the stem resistance is always comparatively small.

In contrast, in 20 year old Sitka spruce the stem resistance may be proportionately much larger. At times of high evaporation, the xylem pressure potential in the roots or at the base of the stem varies between about -2 to -4 bar (38)(Figs. 1 and 6), whereas 6 m higher up in the stem in the middle of the canopy, values of ψ of -10 to -15 bar are found. The implication is that the resistance in the stem is about 3x that in the roots. However, the volume flowing in the roots may be less than that in the stem because of transfer of water out of storage in the stem into the transpiration stream. Consequently, such a conclusion, based on potential differences alone, is uncertain. At the ends of the primary branches, xylem pressure potentials are 2 to 3 bar lower than in the stem, and they fall another 2 bar in the secondary branches, thus dropping up to 5 bar between trunk and shoot (38). Although the flow is not known, one must conclude that a major part of the overall flow resistance is in the branches. Furthermore, Richter (34) found in Taxus baccata that the potential in non-transpiring twigs was the same as that in transpiring twigs throughout a day (-7 to -13 bar), thus indicating negligible leaf resistance. Clearly, there is scope for more well-designed experimentation on resistances to flow in all kinds of plants, and especially trees.

If equation (4) holds, ψ_ℓ should fall in proportion to increases in transpiration rate when ψ_s is essentially zero in wet soil or solution culture. Experiments carried out by a number of different people on a range of species indicate that this may be so (35, 36, 58, 59, 60), but that ψ_ℓ is often independent of transpiration rate over a wide range (35, 50, 60, 61) (Fig. 9). Some investigators have obtained both kinds of result with different species (35, 60). However, it is not only species differences which are involved since both kinds of results have been obtained with the same species by different investigators, e.g. pepper A (35), B (60); sunflower A (58), B (61); maize A (58), B (35). The only valid generalization at the moment seems to be that in all the woody plants so far studied (apple (36), pear, citrus (60) and Gossypium barbadense (35), ψ_ℓ decreases with increasing transpiration rate. Presumably differences in physiological condition or variety account for the different responses within the same species.

Because the root resistance is large in relation to stem and leaf resistance in most plants, it has generally been assumed that a type B response indicates a fall in root resistance with increasing transpiration rate (5). Evidence in support of this comes from experiments on cotton (61) in which the B response was converted to an A response by dipping the roots in boiling water (line C, Fig. 9). Since the root resistance may be only 25% of the total plant resistance (57), the decline in root resistance must be proportionately larger than the increase in transpiration rate to maintain (qR) and hence ψ_ℓ constant. Possibly, there is a type A response rather than a type B, when the root resist-

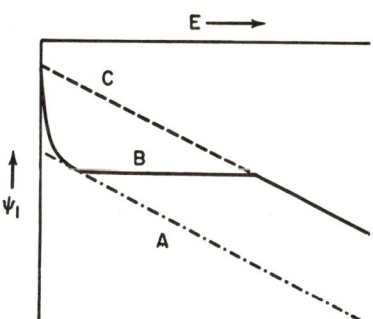

Fig. 9 Alternative relationships between leaf water potential and transpiration rate, E, which have been found experimentally. Curve A includes a constant plant resistance. Curve B indicates a plant resistance which falls over part of the range of increasing transpiration rate.

ance is not a large proportion of the total resistance so that reductions in root resistance are inadequate to maintain (qR) constant as q rises.

A number of possible explanations of the mechanism by which the root resistance could change with transpiration rate are discussed by Newman (6). All are speculative and none are completely convincing, largely because of uncertainty about the pathway of water movement through the root. Whatever the mechanism, clearly it is of considerable ecological and physiological advantage to the plant to be able to maintain a high ψ_ℓ at high transpiration rates, thereby keeping the stomata open and avoiding physiological trauma.

In young conifer stems, the use of dyes has shown that the smaller diameter tracheids in an annual ring only become effective in conducting water axially when the transpiration rate is high (62). This suggests that the resistance in the stem should also decrease as the rate of flow of water increases, and preliminary estimates of diurnal variation in stem resistance in Scots pine support this (55). Resistance of the xylem may be expected to vary as the result of the regular cavitation and refilling of the conducting xylem elements, daily and seasonally, but there is little quantitative information about the influence of cavitation on water flow through the xylem (8).

Soil Resistance

In plants rooted in soil, ψ_ℓ falls with increasing transpiration rate (curve A, Fig. 9) unless the soil is saturated in the rooting zone (50, 64, 65). Such observations have been taken to show that in unsaturated soil the zone of soil around the roots is a major resistance (the perirhizal resistance) to the flow of water in the SPAC, so that changes in root resistance have little influence on ψ_ℓ (5, 52). Initially, this interpretation received support from calculations which showed appreciable gradients of water potential and hydraulic conductivity close to the root (66, 67). However, the use of more realistic larger values of root length in the equations have led to the conclusion that gradients of conductivity in the vicinity of the roots are small and unlikely seriously to impede water uptake (6). As a result, Newman (68) suggested that the main resistance causing reduction in ψ_ℓ was the transfer of water to the rooting zone from sources further away. However, using compartmentalized root systems, Faiz (65) has shown that the rise in ψ_ℓ which follows a reduction in transpiration rate is accompanied by redistribution of water within the rooting zone and movement across the perirhizal resistance and does not require long distance movement into the root zone. Thus there was the paradox that experiments indicated the presence of a perirhizal resistance which could not be explained by a reduction in hydraulic conductivity, but which might perhaps lie at the root-soil interface. The roots of transpiring cotton plants shrink considerably (69). Faiz measured in situ water potentials of -13 bar, shrinkage in diameter of 25% and relative water contents of 0.7. He showed that repeated squeezing of the soil around the roots or regular vibration of the pots for 1 minute every 1 to 2 hours resulted in 6 bar higher values of ψ_ℓ. The most likely interpretation is that these treatments prevented the formation of a vapor gap and reduction in the area of liquid contact between the roots and the soil as root shrinkage occured.

5. CAPACITANCE

Changes in Water Content

The water content of all tissues changes seasonally and diurnally; these changes are usually most evident as changes in tissue volume or linear dimension, although they are extremely small in woody tissues (70). In the introduction to his review, Kozlowski (70) wrote: "Diurnal contraction and expansion of plant tissues are related to higher transpiration than absorption of water during the day, and the reverse at night. During the day absorption of water through the roots lags behind transpiration because of resistance to water movement through the plant." It will be clear from the foregoing sections that this interpretation could be misleading. The changes in water content are caused by the movement of water from the regions of high water potential in the plant tissues to the region of lower pressure potential in the xylem. Therefore, the water potentials of tissues in the plant follow the xylem pressure potential, usually with some phase shift which is dependent upon the storage capacity of the tissue and the resistance associated with the pathway: lags do not result from the occurrence of resistance alone, but occur only if there is some storage capacity.

The movement of water out of storage may make a substantial contribution to the amount of water transpired. The tissues in the plant can be regarded as a number of alternative sources of water linked in parallel with each other and with the soil. Thus the total flux from the plant in evaporation (E) is made up of a number of partial flows within the plant (see Fig. 2):

$$E = q_1 + q_2 + q_3 + \cdots + q_n \tag{7}$$

where q_1 is the flow from the soil and q_{2-n} are flows out of storage in the plant. The flow from a particular store is:

$$q_i = (\psi_i - \psi_{xylem})/r_i \tag{8}$$

where:

$$\psi_i = f(R_i) \tag{9}$$

The volume of water which can come out of storage is:

$$\Delta V_i = V_i \cdot \Delta R_i = \int_{t_1}^{t_2} q_i \cdot dt \tag{10}$$

where V_i is the volume of water in the turgid tissue of relative water content R_i.

As Weatherley has pointed out (5), the relative sizes of the flows out of storage at any one time and their relative phasing depends upon the resistance between the store and xylem, r_i, the capacity of the store, V_i, and the relationship between ψ_i and R_i (the tissue moisture characteristic). Thus it would be expected that water would move out of storage initially from the tissues closest to the sites of evaporation, so that the largest dimensional changes would initially occur in or near the leaves. However, as the sources are of finite size, their capacity to supply water falls as it is removed so that the main sources of supply are found to occur progressively lower down the plant (71). If the sources within the plant are large enough and the pathway to the xylem of low enough resistance, sufficient water may be withdrawn from tissues to satisfy the requirements of evaporation for considerable periods. Consequently the assumption that uptake from the soil equals evaporation from the leaves is almost certainly incorrect at any instant. In trees, diurnal uptake of water and tissue rehydration can continue most of the night and, when periods of dull weather follow bright weather, may continue for several days (29, 71, 72, 73). Conversely, under drought conditions tissue shrinkage may continue for three weeks.

We know very little about the resistances to water movement between the stores and the xylem. The moisture characteristic of tissues is relatively well known for leaves (4) and wood (74) but can only be guessed at for other tissues such as cambium, cortex and phloem, or roots. Probably in these extensible tissues a large proportion of the water can be withdrawn for small changes in water potential, as in leaves of tomato (4). In the following sections the main consideration is the capacity of different parts of the plant to exchange water, since this can be estimated approximately.

Storage in the Canopy

The relative water content of leaves (R) may fall during the day in conjunction with the decline in ψ_ℓ, from about 0.98 at dawn to below 0.8. In the extensible leaves of corn and cotton, for example, such changes are accompanied by reductions in leaf thickness of up to 20%, but more rigid leaves, such as holly, shrink little in external dimension, the cell volume falling and the intercellular space volume rising within the same confines (25, 75). The volume of turgid leaves in a canopy is L.d where L is leaf area index and d is the thickness of the turgid leaves. If α is the fraction of the leaf volume which is water-filled, and n the number of stems per unit ground area, the volume of water in the turgid leaves on a stem is:

$$V = (Ld\alpha)/n \quad (11)$$

The change in mass of water in the leaves during the day is:

$$\Delta M = V\rho \cdot \Delta R \quad (12)$$

where ρ (10^9 mg m^{-3}) is the density of water. The change in mass of water stored may be compared with the rates of transpiration listed in Table 2. This approach is applied to a herbaceous crop and a dense young (20 year) conifer stand in Table 4. The final line gives the period of time for which the total amount of water removed from storage in the leaves could supply the water

Table 4. A comparison between the change in the amount of water stored in the foliage during the day and the rate of transpiration

	herbaceous crop	conifer stand
L	4	10
d (m)	200×10^{-6}	500×10^{-6}
α	0.6	0.8
n (stem/m^2)	100	0.4
V (m^3/stem)	5×10^{-6}	10^{-2}
ΔR	0.2	0.1
ΔM (mg/stem)	10^3	10^6
transpiration (mg s^{-1}/stem)*	1	250
Δt (hour)†	0.28	1.1

*from Table 2, $\beta = 1$; †$\Delta t = \Delta M/$(transpiration)

transpired. Thus the normal diurnal changes in storage of water in the leaves can provide for only about 1/4 of an hours transpiration in a herbaceous crop, but for over an hour in a dense young conifer stand. The main differences between the two examples which are responsible are the leaf thickness and leaf area index: the large difference in number of stems per unit area compensates for the difference in the rate of transpiration per stem.

Storage in the Roots

Diurnal shrinkage of the roots is presumably largely the result of water loss from the living cells of the cortex. Huck, Klepper and Taylor (69) found that on a sunny dry day the roots of cotton shrank to as little as 60% of their turgid diameter. If the volume fraction of the root that is water-filled remains constant, this represents a decline in relative water content to below 40% ($R=(V_m\alpha)/(V_t\alpha) = r_m^2/r_t^2$, where r is radius and the subscripts t and m indicate the turgid and minimum condition). This is an enormous decrease in water content and about twice that observed by Faiz in cotton (65). We need considerably more information about the generality of these phenomena.

With such information we may estimate the contribution of stored water in the roots to transpiration. The change in mass of water in the root system during a day is given by equation (12). An approximate estimate of the volume of water in the turgid root system of a plant may be obtained from:

$$V = (\pi r_t^2 \ell \alpha)/n \qquad (13)$$

where r_t is an appropriately weighted mean turgid root radius, and ℓ is the length of the root system per unit area of ground. Newman (6) gives the range of ℓ as 10^3 to 4×10^5 m of root per m^2 of ground area. The higher values are for grasses and the lower ones for woody species, with other herbaceous plants intermediate. Estimates of the volume of water in two root systems are given in Table 5. Conservatively assuming the same changes in R as in leaves leads to estimates of much larger changes in water content because of the larger volume of water stored in the roots than in the leaves. Consequently it seems that the water stored in the roots could supply the transpiration requirement for substantial periods of time. Increasing ΔR to the size suggested by the data of Huek et.al. (69) or Faiz (65) would increase ΔM and Δt proportionately

Table 5. A comparison between the change in the amount of water stored in the root system during the day and the rate of transpiration.

	herbaceous crop	conifer stand
ℓ (mm^{-2})	5×10^4	5×10^3
r_t (m)	0.2×10^{-3}	2×10^{-3}
α	0.8	0.8
n (stem/m^2)	100	0.4
V (m^3/stem)	5×10^{-5}	0.125
ΔR	0.2	0.1
ΔM (mg/stem)	10^4	12.5×10^6
transpiration (mg s^{-1}/stem)*	1	250
Δt (hour)†	2.8	14

*from Table 2, $\beta = 1$; † $\Delta t = \Delta M/$(transpiration)

by large amounts. The estimates of ΔM and Δt are also very sensitive to the values of ℓ and, especially, r_t, which could be in considerable error.

Storage in the Stem

It has been known for a long time that the trunks of trees swell and shrink diurnally (70 - 73) and more recently the same phenomenon has been described in cotton (77, 78). It has also been known for many years that the water contents of the stems of herbaceous plants and trees change both diurnally and seasonally (3, 78, 91, 92). Clearly, the changes in water content and dimension are related, but we must show some discrimination here because the water content of both extensible, living tissues and inextensible, woody tissues changes while only the living tissues would be expected to change in dimension.

- In Living Tissues

Concurrent measurements of diurnal change in dimension of the xylem cylinder and the entire stem (excluding the corky bark in trees) in cotton and in conifers, have shown that 92% or more of the change in dimension occurs in the peripheral living cells, including the cambial initials and derivatives and the phloem (29, 71, 79) (Fig. 10). These cells are mostly thin-walled and extensible and change by up to x2 in water content during the course of a day (80).

Diurnal changes in stem diameter (Δd) have been found to vary from a few μm up to about 1600 μm: typical values are about 300 μm (81). To some extent, Δd is larger in larger stems because the phloem is thicker, so that the percentage change in diameter is largest in small stems, where it may reach several percent, and it increases markedly with height up a tree (29, 81).

The volume of water exchanged is:

$$\Delta V = \Delta A \ell \alpha = \pi \bar{d} \frac{\Delta d}{2} \ell \alpha \qquad (14)$$

where ΔA is the change in cross-sectional area of the stem. Thus the volume of exchangeable water increases linearly with stem diameter. While Δd may be somewhat dependent upon $\bar{d}$, it is not unreasonable to assume a fairly constant number of living cells across the cambium and phloem except perhaps in very small stems. Dormant cambial cells are about 5 μm across but the derivatives

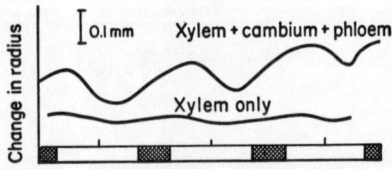

Fig. 10 Diurnal changes in stem radius in a 25-year old Sitka spruce tree on sunny days in June. The upper curve was obtained with the transducer shaft on shaved down bark just above the living phellogen. The lower curve was obtained with the transducer shaft on the sapwood surface where the bark peels readily. From (38).

expand to about 40 μm. The cambium and phloem may be about 4 mm thick, so that about 100 cells may be involved in the deformation. If 25% of these are inextensible fibers and Δd = 300 μm, each cell would change in size by about 4 μm or 10% of its turgid volume. This would correspond to a water potential of between -10 and -20 bar depending on the tissue moisture characteristic. The mass of water coming out of storage is $\Delta M = \Delta V \cdot \rho$. This is compared with the transpiration rate for two stands in Table 6 to show that the deforming tissues of the stem can supply, by moderate contraction, a substantial transpiration demand for over an hour. The larger volume of exchangeable water in a tree is counteracted by the much larger transpiration rate resulting from the smaller number of stems per unit area.

The water content of the deforming tissue changes rapidly in response to changes in xylem pressure potential. This is readily seen in the field under conditions of variable cloudiness. Sharp changes in irradiance, or a shower of rain, cause initial responses in stem diameter and xylem pressure potential (measured directly with "Aquapots" or inferred from leaf water potentials) within one or two minutes (29, 82). The half-time for the exchange of water between xylem and phloem has been shown to be 20-30 minutes at temperatures above 20°C in cotton, but much longer at lower temperatures (83). It would seem that the main barrier to radial movement is the cambial initials since water must largely move through them in the symplasm, whereas in the differentiating xylem and phloem it may also move in the free space and vacuoles. Recently it has been shown that the passive diffusion equation by Philip (1) is adequate to describe the changing distribution with time of water potential and water content in the phloem (84, 85). Consequences of the application of this equation are : (1) that the water does move radially from the xylem to the phloem; (2) that the movement of water and the propagation of ψ are linear diffusion processes; (3) the radial diffusivity of water potential is about 1.6×10^{-6} cm^2 s^{-1}; and (4) the pattern of swelling and shrinkage can be accurately predicted from changes in xylem or leaf water potential.

– In Non-Living Tissues

The water content of the heartwood of trees remains at a fairly constant, usually low level throughout the year, whereas the water content of the sapwood in ring-pored, diffuse-pored and non-pored species declines during the summer and autumn to a winter minimum (78). Expressed as a percentage of the dry weight, the water content may change by x2, or by a ΔR of over 0.4 (88). In

Table 6. A comparison between the change in the amount of water stored in the cambium and phloem and the rate of transpiration.

	herbaceous crop	conifer stand
$\bar{d}$ (m)	0.01	0.2
ℓ (m)	1	10
Δd (m)	300×10^{-6}	300×10^{-6}
α	1	1
ΔA (m^2)	4.7×10^{-6}	9.4×10^{-5}
ΔM (mg/stem)	4.7×10^3	9.4×10^5
transpiration (mg s^{-1}/stem)*	1	250
Δt (hour)†	1.3	1

*from Table 2, β = 1; † Δt = ΔM/transpiration

a perceptive review, Stewart (86) wrote "the water in the sapwood, especially in the outer sapwood which contains the transpiration stream, can be regarded as a supplementary reservoir for the tree, particularly during diurnal peak conditions. It is probable that the transpiration stream withdraws water from adjacent fibrous elements because these cells have lost the capacity to control their water content." In view of the large changes in water content and the rigidity of the wood, most of the water which is removed must be replaced by air. In the late winter the wood is refilled over a few weeks (78), presumably as the result of sap pressure in deciduous hardwoods, but it is less clear how it refills in Gymnosperms. Thus the sapwood is an important, renewable source of water within the tree. Whether the internal tissues of the stem in herbaceous plants are also an important reserve of exchangeable water is unknown.

The amount of sapwood increases linearly with leaf mass (47) and asymtotically with diameter of the tree (89), and decreases with height within a tree (90). There are large specific differences in sapwood thickness. For example, in 50 cm diameter trees, the sapwood thickness was 13, 8, 5, 5 and 2.5 cm in ponderosa pine, lodgepole pine, Douglas fir, Engelmann spruce and western red cedar, respectively (89). The volume of water stored in a ring of sapwood of thickness x is:

$$V = A \cdot \ell \cdot x = \pi \bar{d} \times \ell \propto \qquad (15)$$

Thus the volume of water increases linearly both with the diameter of the tree and the thickness of the sapwood. The change in mass of water stored in the sapwood is given by equation (12). In Table 7 a comparison is made between the amount of exchangeable water stored in a young dense stand, and an older thinned stand of larger trees. Clearly the amount of exchangeable water stored in the sapwood can supply the transpiration requirement of a large tree for several days. However we do not know whether all of this water is normally part of the transpiration stream or how readily it moves into it.

Evidence from heat pulse techniques and dyes indicates that maximum axial flow in the sapwood occurs several rings in from the cambium (62, 81, 93) because the margo in the bordered pits nearer to the cambium is imperforate (93), but that most of the sapwood conducts water to a lesser extent. Furthermore, in any one annual ring most of the axial flow is in the earlywood, but

Table 7. A comparison between the change in the amount of water stored in conifer sapwood and the rate of transpiration.

	young dense stand	older thinned stand
n (stem/m^2)	0.4	0.1
$\bar{d}$ (m)	0.2	0.5
ℓ (m)	10	40
x_* (m)	0.05	0.07
α ‡	0.5	0.5
V (m^3/stem)	0.15	2.2
ΔR	0.3	0.3
ΔM	4.6×10^7	6.5×10^8
transpiration (mg s^{-1}/stem)*	250	10^3
Δt (hour)+	50	180

‡ tracheid lumen area/area of wood ≈ 0.7 in noble fir and 0.3 in Sitka spruce (13)
* from Table 2, β = 1; + Δt = ΔM/transpiration.

more of the ring conducts at high transpiration rates (62). Thus the pathlength for movement of water out of storage in the latewood of a ring is relatively short. Nonetheless, lateral movement of water from tracheid to tracheid in the xylem is along a high resistance pathway. In the majority of conifers most bordered pits are found on the tapering tangential end walls of the tracheids with few on the radial walls, although this varies among species (94). Consequently, radial movement must either be in the cell walls or in the ray tracheids and parenchyma. To reach the rays, water may move tangentially from tracheid to tracheid through a number of bordered pits in series. Comstock (95) has reviewed ratios of axial to tangential permeability in sapwood ranging from 820 to 81 600 and of axial to radial permeability ranging from 190 to 35 500. This anisotropy will make modeling changes in water content and water flow in the sapwood difficult. At a first approximation, some success can probably be achieved by the application of Darcy's law and the conservation of mass to derive a diffusion equation as applied in soil water movement.

6. CONTROL OF WATER TRANSFER

Steady State Stomatal Behavior

Ultimate control of water movement through the plant rests with the stomata, since stomatal conductance (k_s) controls the rate of evaporation from the foliage (equation 3). Steady state stomatal conductance is controlled by several environmental variables and by ψ_ℓ through their effects on guard cell turgor pressure. Thus, change in k_s may be in direct response to environmental variables and to their effects on water transfer through the plant by negative feedback through ψ_ℓ.

Photon flux, leaf temperature, vapor pressure deficit and intercellular space CO_2 concentration all influence k_s independently of ψ_ℓ as shown in Fig. 11a-d, but are themselves influenced by the plant to only a limited extent. On the other hand, k is also a function of ψ_ℓ (Fig. 11e), which depends upon the properties of the water transfer pathway and the rate of flow of water in the plant. Curve A is typical of observations made in the field on a number of species; curve B seems to be more frequent in laboratory studies on plants grown in artificial environments and may reflect lack of stress during development. It is note worthy that the A response allows ψ_ℓ to fluctuate over its entire normal diurnal range with virtually no effect on k_s. This is an adaptation of considerable advantage to CO_2 uptake by the plant, but at the same time, the solute potential and hydrostatic pressure in the cells also fluctuate during the day over ranges which in most species curtail cell division and inhibit some cellular processes.

Thus in steady state conditions (no stomatal oscillations), ψ_ℓ controls k_s only near its lower limit. Environmental conditions which lead to high evaporation rates and low ψ_ℓ, particularly irradiation and vapor pressure deficit, can cause stomatal closure but this does not usually happen because limiting values of ψ_ℓ are not usually reached, even at high transpiration rates, unless there is also a shortage of water at the roots. In addition, osmotic adaptation as the result of exposure to low ψ_ℓ results in further depression of the inflexion point in curve A. Increasingly, diurnal closure of stomata is being found to be a response to temperature and vapor pressure deficit, independent of effects on ψ_ℓ.

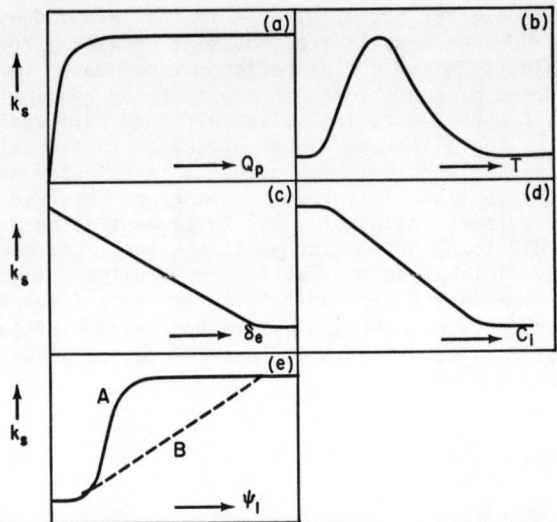

Fig. 11 Idealised relationships between stomatal conductance, k_s, and (a) photon flux density, Q_p (b) temperature, T (c) vapour pressure deficit, δe (d) intercellular space CO_2 concentration, C_i and (e) leaf water potential, Ψ_ℓ.

REFERENCES

(1) Philip, J.R. 1966. Ann. Rev. Plant Physiol. 17, 245-268.
(2) Cowan, I., & F.L. Milthorpe. 1968. IN: Water Deficits and Plant Growth, Vol. 1. ed. T.T. Kozlowski. 137-192. Academic Press, New York.
(3) Kramer, P.J. 1969. Plant and Soil Water Relationships: A Modern Synthesis McGraw-Hill, New York.
(4) Slatyer, R.O. 1967. Plant-Water Relationships. Academic Press, New York.
(5) Weatherley, P.E. 1970. Adv. in Botanical Res. 3, 171-206.
(6) Newman, E.I. 1973. IN: The Plant Root and Its Environment, ed. E.W. Carson. 363-440. Univ. Virginia Press, Blacksburg.
(7) Jarvis, P.G. & R.O. Slatyer. 1970. Planta. 90, 303-322.
(8) Zimmermann, M.H. & C.L. Brown. 1971. Trees, Structure and Function, Springer, Berlin-Heidelberg-New York.
(9) Waggoner, P.E. & A.E. Dimond. 1954. Am. J. Bot. 41, 637-640
(10) Dimond, A.E. 1966. Plant Physiol. 41, 119-131.
(11) Van Alfen, N.K. & N.C. Turner. 1974. Plant Physiol. suppl. P.56.
(12) Petty, J.A. & R.D. Preston. 1969. Proc. Ray. Soc. B 172, 137-151.
(13) Petty, J.A. & G.S. Puritch. 1970. Wood Science & Technology. 4, 140-154.
(14) Mark, W.R. & D.L. Crews. 1973. For. Sci. 19, 291-296.
(15) Lin, R.T., E.P. Lancaster, R.L. Krahmer. 1973. Wood & Fiber. 4, 278-289.
(16) Banks, W.B. 1968. J. Institute of Wood Sci. 20, 35-41.
(17) Puritch, G.S. & R.P.C. Johnson. 1971. J. exp. Bot. 22, 953-958.
(18) Puritch, G.S. 1971. J. exp. Bot. 22, 936-945.
(19) Puritch, G.S. & J.A. Petty. 1971. J. exp. Bot. 22, 946-952.
(20) Anderson, N.E. C.H. Hertz, & H. Rufelt. 1954. Physiol. Plant. 7, 753-767.
(21) Allerup, S. 1960. Physiol. Plant. 13, 112-119.

(22) Falk, S. 1966. Physiol. Plant. 19, 602-617.
(23) Brogårdh, T. & A. Johnsson. 1973. Physiol. Plant. 28, 241-245.
(24) Raschke, K. 1970. Plant Physiol. 45, 415-423.
(25) Raschke, K. 1970. Science. 167, 189-191.
(26) Cowan, I. 1972. Planta. 106, 185-219.
(27) Lang, A.R.G., B. Klepper, & M.J. Cumming. Plant Physiol. 44, 826-830.
(28) Sheriff, D.W. 1973. J. exp. Bot. 24, 796-803.
(29) Richards, G.P. 1973. Ph.D. Thesis, Univ. Aberdeen.
(30) Oertli, J.J. 1971. Z.
(31) Scholander, P.F., M.T. Hammel, E.D. Bradstreet, & E.A. Hemmingsen. 1965. Science. 148, 339-346.
(32) Monteith, J.L. 1965. SEB Symposium. 19, 205-234.
(33) Pospíšilová, J. 1969. Biol. Plant. 11, 130-138.
(34) Richter, H. 1973. J. exp. Bot. 24, 983-994.
(35) Barrs, M.D. 1970. IN: Proc. Symp. Plant Response to Climatic Factors. UNESCO. Uppsala, Sweden. 249-258.
(36) Landsberg, J.L. et al. 1974. J. appl. Ecol. (in press)
(37) Tobiessen, P., P.W. Rundel & R.E. Stecker. Plant Physiol. 48, 303-304.
(38) Hellkvist, J., G.P. Richards & P.G. Jarvis. 1974. J. appl. Ecol. (in press).
(39) Richter, H. 1972. Ber. deutsch. Bot. Ges. 85, 341-351.
(40) Heine, R.W. 1970. Ann. Bot. 34, 1019-24.
(41) Spomer, G.G. 1968. Science. 161, 484-485.
(42) Wiebe, M.H., R.W. Brown, T.W. Daniel & E. Campbell. 1972. BioSci. 20, 225-226.
(43) Begg, J.E. & N.C. Turner. 1970. Plant Physiol. 46, 343-346.
(44) Hoffman, G.J. & W.E. Splinter. 1968. Agron. J. 408-413.
(45) DeRoo, M.C. 1969. Agron. J. 61, 511-515.
(46) Turner, N.C. 1974. IN: Mechanisms of Regulation of Plant Growth. Roy. Soc. of New Zealand (in press).
(47) Grier, C.C. & R.H. Waring. 1975. Forest Sci. (in press).
(48) Heine, R.W. 1971. J. exp. Bot. 22, 503-511.
(49) Janes, B.E. 1970. Plant Physiol. 45, 95-103
(50) Tinklin, R. & P.E. Weatherley. 1968. New Phytol. 67, 605-615.
(51) Fiscus, E.L., L.R. Parsons & R.S. Alberte. 1973. Planta. 112, 285-292.
(52) Tinklin, R. & P.E. Weatherley. 1966. New Phytol. 65, 509-517.
(53) Jensen, R.D., S.A. Taylor & M.M. Wiebe. 1961. Plant Physiol. 36, 633-638
(54) Duniway, J.M. 1974. In press.
(55) Roberts, J. 1974. personal communication.
(56) Boyer, J.S. 1969. Science. 163, 1219-1220.
(57) Boyer, J.S. 1971. Crop. Sci. 11, 403-407.
(58) Neumann, H.H., Thurtell, G.W. & Stevenson, K.R. 1974. Can. J. Plant Sci. 54, 175-184.
(59) Hailey, J.L. E.A. Hiler, W.R. Jordan & C.H.M. VanBavel. 1973. Crop. Sci. 13, 264-267.
(60) Camacho-B, S.E., A.E. Hall & M.R. Kaufmann. 1974. In press.
(61) Stoker, R. & P.E. Weatherley. 1971. New Phytol. 70, 547-554.
(62) Kozlowski, T.T., J.F. Hughes, & L. Leyton. 1966. Biorheology. 3, 77-85.
(63) Milburn, J. 1966. Planta. 69, 34-42.
(64) Macklon, A.E.S. & P.E. Weatherley. 1965. New Phytol. 64, 414-427.
(65) Faiz, S.N.A. 1973. Ph.D. Thesis, Univ. of Aberdeen.
(66) Gardner, W. R. 1960. Sact. Sci. 89, 63-73.
(67) Cowan, I.R. 1965. J. appl. Ecol. 2, 221-239.

(68) Newman, E.I. 1969. J. appl. Ecol. 6, 261-272.
(69) Huck, M.G., B. Klepper & H.M. Taylor. 1970. Plant Physiol. 45, 529-530.
(70) Kozlowski, T.T. 1972. IN: Water Deficits and Plant Growth. ed. T.T. Kozlowski, 1-64. Academic Press, New York and London.
(71) Dobbs, R.C. & D.R.M. Scott. 1971. Can. J. For. Res. 1, 80-83.
(72) Waggoner, P.E. & N.C. Turner. 1971. Bull. Conn. Expt. Stn. 726, 1-87.
(73) Lassoie, J.P. 1973. For. Sci. 19, 251-255.
(74) Stamm, A.J. 1964. Wood and Cellulose Science. The Ronald Press Co., New York.
(75) Jarvis, P.G. 1971. IN: A Manual of Photosynthetic Methods. ed. Z. Sestak, J. Catsky & P.G. Jarvis. 566-631. Junk, the Hague.
(76) Namken, L.N., J.F. Bartholic & J.R. Runkles. 1971. Agron. J. 63, 623-627.
(77) Klepper, B., V.D. Browning & M.M. Taylor. 1971. Plant Physiol. 48, 683-685.
(78) Gibbs, R.D. 1958. IN: The Physiology of Forest Trees. ed. K.V. Thimann. The Ronald Press Co., New York.
(79) Molz, F.J. & B. Klepper. 1973. Agron. J. 65, 304-306.
(80) Stewart, C.M., S.H. Tham & D.L. Rolfe. 1973. 242, 479-480.
(81) Lassoie, J.P. 1974. Ph.D. Thesis, Univ. of Washington.
(82) Stansell, J.R., B. Klepper, V.D. Browning & H.M. Taylor. 1973. Agron. J. 65. 677-678.
(83) Klepper, B. F.J. Molz & C.M. Peterson. 1973. Plant Physiol. 52, 565-568.
(84) Molz, F.J., & B. Klepper. 1972. Agron. J. 64, 469-473.
(85) Molz, F.J., B. Klepper & V.D. Browning. 1973. Agron. J. 65, 219-222.
(86) Stewart, C.M. 1967. Nature. 214, 138-140.
(87) Doley, D. 1967. J. Ecol. 55, 597-617.
(88) Chalk, L., & J.M. Bigg. 1956. Forestry. 29, 5-21.
(89) Lassen, L.E., & E.A. Okkonen. 1969. USDA Forest Res. Paper. FPL 124.
(90) Smith, J.H.G., J. Walters & R.W. Wellwood. 1966. For. Sci. 1, 97-103.
(91) Klemm, W. 1966. Flora. 156, 232-235.
(92) Klemm, W. 1966. Biol. Zentralblatt. 6, 781-783.
(93) Mark, W.R., & D.L. Crews. 1973. For. Sci. 19, 291-296.
(94) Jane, F.W., K. Wilson & D.J.B. White. 1970. The Structure of Wood. Adam & Charles Black, London.
(95) Comstock, G.L. 1970. Wood and Fiber. 1, 283-289.

WATER TRANSPORT IN WHEAT

O.T. DENMEAD and B.D. MILLAR

*CSIRO Division of Environmental
Mechanics, Canberra, Australia.*

ABSTRACT

A field method for determining resistances to water flow in intact plants is described. In studies with wheat, large resistances were found in the roots and stems. These may often be more important in determining the plant's water status that the transmission resistance of the soil. Finally, some observations on water transport to the ears of wheat during grain drying are presented.

SYMBOLS AND UNITS

ℓ	distance along leaf from node of attachment to centre of leaf blade	m
Q	total flux of water through roots	$m^3 s^{-1}$
q_i	flux of water to leaf i	$m^3 s^{-1}$
q_ℓ	flux of water to leaf	$m^3 s^{-1}$
q_x	flux of water in stem	$m^3 s^{-1}$
R	effective resistance to water flow of whole root system	$s\, m^{-2}$
r_ℓ	net resistance to water flow in leaf per unit leaf length	$s\, m^{-3}$
r_x	net resistance of xylem elements per unit stem length	$s\, m^{-3}$
S	effective resistance of soil in root-zone to water movement to roots	$s\, m^{-2}$
s_i	length up stem to node of attachment of leaf i	m
ψ_i	average water potential in leaf i	m
ψ_ℓ	average water potential in leaf	m
ψ_o	water potential at base of stem	m
ψ_r	water potential at surface of root	m
ψ_s	average soil water potential in root zone	m
ψ_{x_i}	water potential in xylem elements at distance x_i along stem	m

INTRODUCTION

Models of water transport in the soil-plant-atmosphere continuum were first developed several years ago (1,2,3). These analyses were necessarily simplified, but they have proved adequate to explain the combined influences of soil water supply and atmospheric evaporation demand on the development of leaf water potentials, on the course of transpiration and on the onset of wilting. One aspect which they left open, however, was the influence of resistances to water flow within the plant. The importance of these, relative to the resistance to water flow in the soil, has been the subject of some speculation, e.g. Newman (4), but few reliable measurements of plant resistance are available. This paper describes a method for determining flow resistances in intact plants and deals specifically with the resistances to flow through the roots, stems, leaves, and ears of wheat. Full details of the work will be published elsewhere. Below, we set out the basic assumptions and procedures.

WATER TRANSPORT THROUGH SOIL AND PLANT

The flow of water is initiated by evaporation from the leaves, which in turn: lowers leaf water contents, decreases leaf water potentials, and generates a water potential gradient which extends progressively through the plant water system and into the soil. Examples of the distributions of evaporation sites and water potentials in a wheat crop are given in Fig. 1. It is obvious that there is no single value of leaf evaporation rate, nor is there a unique leaf water potential. The problem in analysing such a system is to identify segments of the water pathway through which the flow is conserved or to which an effective resistance can be ascribed. First, we examine flow in the various segments.

<u>In the soil</u>. Physical analyses of water movement in the root-zone, e.g. Philip (1), Gardner (2) and Cowan (3) lead to a relation between the total flux of water to the roots, Q, and the transmission resistance of the soil, S, of the form.

$$Q = (\psi_s - \psi_r)/S. \qquad [1]$$

S is a function of the soil moisture conductivity and diffusivity, and the root geometry. In wet soils, at water potentials of say > -10 m, S is small, but it increases rapidly as the soil dries, and at water potentials of about - 150 m it becomes a large component of the total resistance to water flow.

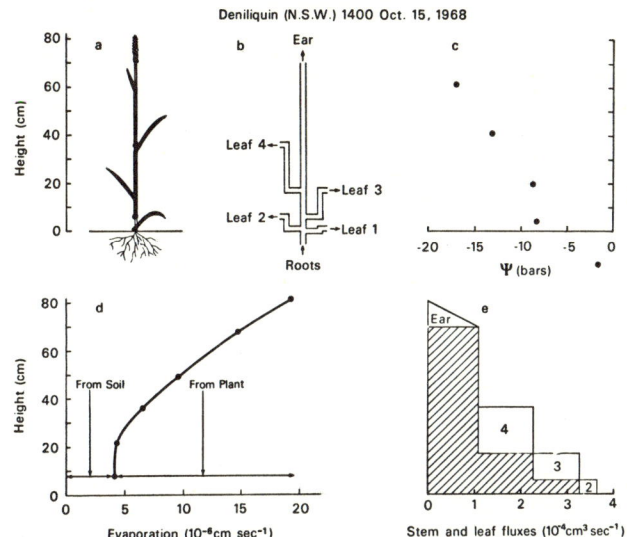

Fig. 1. (a) A wheat tiller to scale. (b) Model of the water conducting system of the stem to same scale. (c) Water potentials in root zone and in each leaf. (d) Flux density of water vapour in canopy. (e) Stem and leaf water fluxes calculated from (b) and (d).

<u>In roots</u>. We extend Cowan and Milthorpe's (5) analysis of water uptake by a single root to uptake by a branching net-work of roots. In this case it is possible to derive a relationship between Q and the effective resistance of the whole root system, R:

$$Q = (\psi_r - \psi_o)/R, \qquad [2]$$

with R being a function of the conductances of the root cortex and xylem, and the lengths of root segments.

<u>In stems</u>. A model of the water conducting system of the wheat stem, based on anatomical studies (6,7,8) is shown in Fig. 1. Water is carried in sets of conducting vessels, the xylem elements, which divide at the nodes. The elements are long, cylindrical, thick-walled cells continuous in the internodal segments. There are about 200 elements in an internode with internal radii typically between 3 and 16 µm. The set carrying water to a particular leaf separate from the set carrying the main flow at the node <u>below</u> the one at which the leaf is attached. The sets are about equal in number and size and we have assumed that the resistance per unit stem length of each set of xylem elements, r_x, is constant. In the internodal segments where the flow is conserved, we have that

$$q_x = (\psi_{x_1} - \psi_{x_2}) / r_x(s_2 - s_1). \qquad [3]$$

In leaves. Because water is reticulated along the leaves through a net-work of veins, and evaporation occurs along the whole length of the leaf sheath and blade, it is not possible to identify segments in the leaf through which the flow is conserved. We have used the approximation,

$$q_\ell = (\psi_x - \psi_\ell) / \ell r_\ell. \qquad [4]$$

In the whole pathway. Equations [1], [2], [3] and [4] can be combined to yield the relationships:

$$\psi_1 = \psi_s - Q(S + R) - q_1 s_1 r_x - q_1 \ell_1 r_\ell \qquad [5]$$

and, for i > 1,

$$\psi_i = \psi_s - Q(S + R) - \{\sum_{j=1}^{i-1}[(Q - \sum_{k=1}^{j} q_k)(s_j - s_{j-1})] + q_i(s_i - s_{i-1})\}r_x - q_i \ell_i r_\ell. \qquad [5]$$

Equations [5] and [6] give the leaf water potentials as linear functions of the soil water potential, the water fluxes through the various segments of the plant, and the segment lengths. The coefficients are the flow resistances. If the variables can be measured in a series of experiments, and if the resistances remain constant, the equations provide a means for calculating the resistances through multiple regression analysis. We have used this approach in studies of water transport in wheat.

RESISTANCES TO WATER FLOW IN THE WHEAT PLANT

Basic measurements. Micrometeorological methods based on the energy balance were used to measure the flux densities of water vapour at different heights in the canopies of wheat crops - see (9) for a description of the methodology.

From these data and supplementary measurements of crop structure it was possible to calculate the fluxes of liquid water and the appropriate segment lengths. An example of one such set of measurements is given in Fig. 1.

Soil water potentials were measured in situ by means of osmotic tensiometers (10) installed throughout the root zone. Water potentials in the leaves were measured (after detachment) with thermocouple psychrometers (11). The distribution of water potentials at one sampling occasion is shown in Fig. 1.

In all experiments, soil water potentials throughout the root zone were high, generally > -10 m. Sample calculations indicated that in these circumstances S would be negligible. Accordingly later calculations were made on the assumption: S = 0. Also, because of the small variation in soil water potential it was possible to ascribe a value to ψ_s with little error.

The resistances. Forty sets of measurements were used for estimating the resistances through equations [5] and [6]. The resistance in the leaves was found to be negligibly small, but large resistances were found in the roots and the stem. The estimate of the resistance of the root systems of individual tillers was $(1.52 \pm 0.69) \times 10^{11}$ s m^{-2} and that of the resistance of a set of xylem elements per unit stem length was $(1.81 \pm 0.39) \times 10^{12}$ s m^{-3}.

The small resistance in the leaves is not surprising when one considers the extent of vein reticulation (7). The root resistance is similar to that measured for wheat seedlings by E.F. Cox (reported in reference 5) and is comparable with root resistances for tobacco (12), corn (13) and onions (14) when account is taken of the ground area occupied by each plant. That is, the same crop evaporation rate requires much the same drop in water potential across the roots.

No such comparable data are available for stem resistance. We have, however, made two other independent estimates for wheat plants, one by measuring water flow when a pressure difference was applied across sections of detached stems, and another by calculating flow resistance by Poiseuille's equation from measurements of the xylem elements using the ensemble averaging technique suggested by Cowan and Milthorpe (5). These estimates were respectively 0.44×10^{12} and 0.32×10^{12} s m^{-3} - the same order of magnitude as the field measurements but three to four times less. The difference is somewhat larger than we expect from experimental error or from differences in technique, and it is interesting to speculate on the cause. One possibility is that in the field, not all the xylem elements are functional in water transport due perhaps to loss of liquid continuity in some of the larger elements through water stress. Another is that the xylem elements of field grown plants may be smaller than those of the plants used in the laboratory measurements, which were raised in growth cabinets. If, for instance, the radii of the elements were only one-third less than those measured in the laboratory plants, the Poiseuille flow estimate would be 1.6×10^{12} s m^{-3}.

We conclude this section by noting that these within-plant resistances are large and may often be more important in determining leaf water potentials than the transmission resistance of the soil. For instance, in the example of Fig. 1 a moderate transpiration rate of 15×10^{-8} m s^{-1} (15×10^{-6} cm s^{-1}) resulted in a drop in water potential of 70 m (7 bars) across the root system and a further drop of 90 m (9 bars) between bottom and top leaves.

WATER TRANSPORT TO THE EAR

In this final section, we present some qualitative observations of the water relations of the ears of wheat, which arise from recent work on the processes of grain drying in intact plants. Our first studies have been concerned with the hydraulic connection between the ear and the rest of the plant during senesence. We have been particularly interested in the proposition that the ears may still retain sufficient connection with the soil for changes in soil moisture content to have a significant influence on the rate of grain drying.

To examine this, we measured changes in plant water potential with miniature, thermocouple psychrometers (15) attached at the bottoms and near the tops of stems, and on the ears. (The leaves at this time were completely dead.) Other measurements were made of the water potential of the soil in the root zone and of the water content of the ears. Figure 2 shows the rate of drying of the ears, and Fig. 3 shows plant and soil water potentials measured on the four occasions indicated by arrows in Fig. 2.

We can make the following observations:

(a) The differences in water potential between soil and stem, and their hourly and day-by-day changes, suggest that a hydraulic connection existed from the soil through the roots and the stem at all stages; that some upward transport always occurred ; and that changes in the resistance of roots and the stem are relatively slow.

(b) On the first day shown in Fig. 3, Dec. 4, some hydraulic connection between the ear and the rest of the plant was apparent - the water potentials of ear and stem reached comparable, high values overnight and underwent similar changes throughout the day.

(c) It is evident that as drying progressed, the connection between ear and stem became more tenuous. Within ten days, the difference in water potential between the ear and the top of the stem increased from a maximum of 50 m (5 bars) to one of 1000 m (100 bars), suggesting a very large change in the resistance to water transport from stem to ear. It is probable that

Water Transport in Wheat

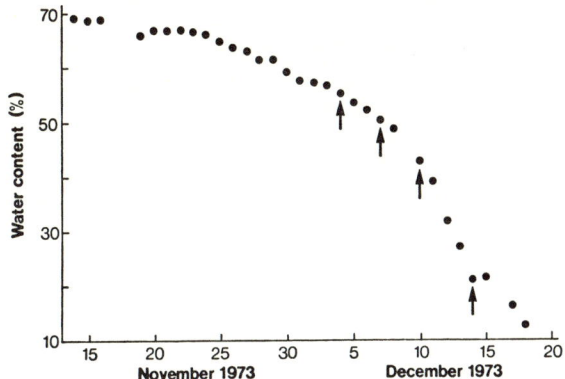

Fig. 2. Drying rate of ears of wheat at Bungendore field site.

the water status of the ear at this very late stage of senesence depended more on the weather than soil moisture.

More detailed examinations of the influences of both soil moisture and weather on the water balance of the ears are proceeding.

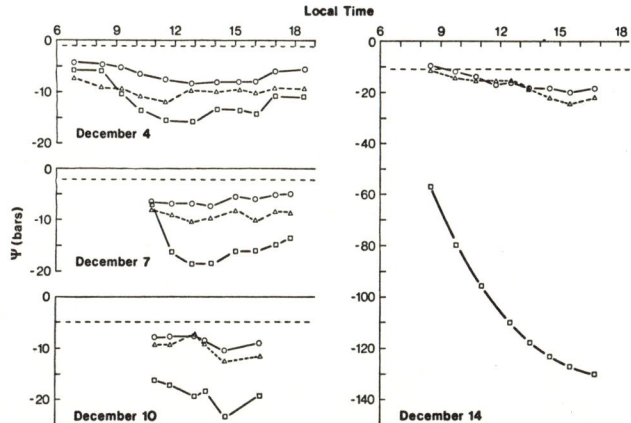

Fig. 3. Changes in water potentials in stem at 0.02 m (O) and 0.68 m (Δ) above ground, and in ears (□) on the days marked by arrows in Fig. 2.

REFERENCES

(1) Philip, J.R. 1957. Proc. Congr. Int. Comm. Irrig. Drain. 3rd (San Francisco, 1957). pp. 8. 125-154.

(2) Gardner, W.R. 1960. Soil Sci. 89, 63-73.

(3) Cowan, I.R. 1965. J. appl. Ecol. 2; 221-239.

(4) Newman, E.I. 1969. J. appl. Ecol. 6, 261-272.

(5) Cowan, I.R., and Milthorpe, F.L. 1968. In 'Water Deficits and Plant Growth'.(Ed. T.T. Kozlowski.) Vol. 1, pp. 137-193. (Academic Press: New York.)

(6) Percival, J. 1921. 'The Wheat Plant: a Monograph'. (Duckworth: London.)

(7) Esau, K. 1965. 'Plant Anatomy'. 2nd Ed. (John Wiley and Sons, Inc.: New York.)

(8) Patrick, J.W. 1972. Aust. J. Bot. 20, 49-63.

(9) Denmead, O.T., and McIlroy, I.C. 1970. In 'Plant Photosynthetic Production/Manual of Methods'. (Eds. Z. Sestak, J. Catsky, and P.G. Jarvis.) pp. 467-516. (Dr W. Junk N.V.: The Hague.)

(10) Peck, A.J., and Rabbidge, R.M. 1969. Soil Sci. Soc. Am. Proc. 33, 196-202.

(11) Millar, B.D. 1971. J. exp. Bot. 22: 875-890.

(12) Begg, J.E., and Turner, N.C. 1970. Plant Physiol. 46, 343-346.

(13) Shinn, J.H., and Lemon, E.R. 1968. Agron. J. 60, 337-343.

(14) Miller, A.A., Gardner, W.R., and Goltz, S.M. 1971. Agron. J. 63, 779-784.

(15) Hoffman, G.J., and Rawlins, S.L. 1972. Science 177, 802-804.

WATER VAPOUR DIFFUSION POROMETRY FOR LEAF EPIDERMAL RESISTANCE MEASUREMENTS IN THE FIELD

C.J. STIGTER

Agricultural University, Department of Physics and Meteorology,
Wageningen, The Netherlands

ABSTRACT

After a short indication of the problems where measurement of leaf epidermal resistance may contribute to a solution, the closed diffusion porometer is introduced as the most suitable device for field measurements. After dealing with the theory of the apparatus we describe problems which troubled the users up till now. On this basis we report on improvements in design of such an instrument, in our calibration and measuring strategy and in our understanding.

SYMBOLS USED

I , water vapour flux density ($kg/m^2 s$)
t , time (s)
e_l, absolute humidity (vapour concentration) at the evaporating surface within the leaf or at an artificial wet surface (kg/m^3)
e_p, absolute humidity within the porometer (kg/m^3)
R_l, (dummy) leaf diffusional resistance for water vapour (s/m)
R_p, porometer diffusional resistance for water vapour (s/m)
S_p, surface of the porometer entrance opening (m^2)
V_p, air volume of the porometer during measurements (m^3)
V_t, calibration volume of the porometer (m^3)
K , slope of lines in the resistance/transient time diagrams (m)
T , temperature (°C)

INTRODUCTION

An important opposition to water flow in the Soil-Plant-Atmosphere chain is the water vapour diffusional resistance of the epidermis of the leaves. Moreover this resistance provides information on part of the resistance to CO_2-flow, which is, together with the existing concentration gradients, determining potential uptake in photosynthesis.

Still many scientific controversies do exist in relation to epidermal resistance (the cuticular resistance normally held to be of only minor importance, this can be understood as stomatal resistance). Much contrasting information exists on influence of epidermal resistance on potential canopy transpiration (e.g. discussions around the paper of Lee [1], Comp. [2] and [3] and even on the relative role of its components in leaf transpiration [4]. Also the problems related to the dependence of epidermal resistance on environmental parameters (e.g. [5],[6],[7]) and to the determination of the true mechanisms responsible for stomatal movements (Comp. [3],[5],[6])) are far from solved. Moreover the role of epidermal resistance in determining canopy and leaf microclimate, as it is for example of influence on the life of insects and parasites, is still mainly known qualitatively.

One of the more recent tools that may be able to contribute to the solutions of these problems containing many interacting variables is the use of simulation models (e.g. [8],[9],[10]).). Our preliminary results [3] were in agreement with the statement of Monteith [11] that the main limitation of microclimatic simulation models is ignorance about spatial changes of stomatal resistance within a canopy. These latter models may also give more definite answers to the potential crop transpiration problem. For checking output of, or as an input into, these and other models reliable measurements of stomatal resistance are needed [3]. The same holds true for progress in plant physiological research, dealing among other things with the above mentioned problems of stomatal behaviour, where epidermal gas flows have to be known [5],[6]. Also irrigation timing is still interested in cross-checking of different methods. Considering these tasks ahead there was ample space for improvement on measuring reliability although a wide variety of more or less quantitative methods existed.

Among these the quantitatively most promising methods make use of detection of the direct vapour stream as it diffuses from within the leaf through the epidermis. These methods form part of a class of diffusion porometers. Some recent authors [12] still trust to viscous flow porometers for amphistomatal leaves, if calibrated anyway against a diffusion porometer. It has to be admitted that in solving the important problem of canopy sampling [11], [13],[14]) the possibility of a relatively high number of measurements in a unit of time with such methods would be an advantage. We feel, however, that the well documented drawbacks of introduced uncertainties and bias (Comp. e.g. [15],[16],[17]) prohibit obtainment of reliable results on diffusional resistance, even with thorough additional knowledge on leaf epidermal and mesophyll structure. Most of these objections do also apply to diffusion methods making use of diffusion of gases from one side of the leaf to the other (Comp. [3]). Therefore diffusion porometry detecting separately the vapour coming from the two leaf sides is the most accurate.

Although several detection mechanisms can be used in these porometers, electrical humidity elements have advantages, especially in portable equipment. An other choice within this class of methods is that between a closed, semi-open or open circuit method [18]. Open methods are not recommendable because of gradients within the porometer house. In the closed diffusion porometer the electrical humidity sensor is used in a dynamical (non-stationary) method [19]. This has originally been a drawback when compared with stationary use in a semi-open method (Comp. [19] for relevant discussion and literature). However, only the closed method is not in need of a constant gas supply over the leaf, which makes it much more attractable for field use.

THE DIFFUSION POROMETER: THEORY AND PRACTICE

In regard to the above we have tried to improve the closed water vapour diffusion porometer as originally proposed by Wallihan and Van Bavel (Comp. [3], [14] and [19] for a review of the relevant literature and more details regarding the subsequent parts of this paper).

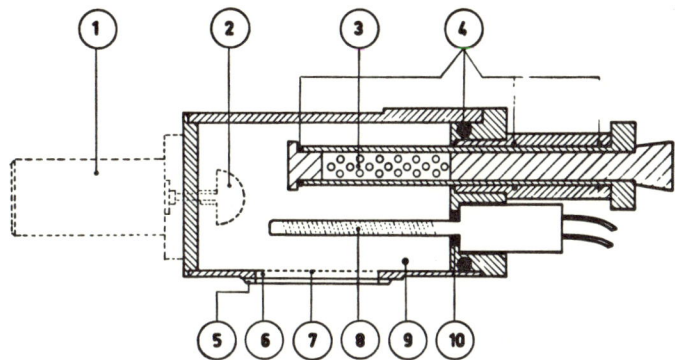

Fig. 1. The porometer as designed. Cross-section through the central longitudinal axis of the cylindrical cup ($V_p = 39.9 \pm 0.2$ cm^3). Material: polypropylene. One may discern:
1. the motor (Mauthe GmbH, Type 16-35-12). 2. the fan (four mutually perpendicular flat blades). 3. silica gel holder for drying pellets (dust-poor "Kali-Chemie AG Trockenperlen"). The holder may be moved into and out of the cup.
4. O-rings for sealing. 5. rubber sealing fringe around the sensor cup opening. (Saba silicone sealant, Nr.25, grey). 6. sensor cup opening (2.03 ± 0.01 cm^2).
7. perforated membrane for suppression of turbulent exchange between cup air and ambient air (VECO-125K, calculated diffusion resistance: 0.14 s/cm (25°C)).
8. sensor (Hygrodynamics Inc., TH 7 15-1284; cylinder of 4 cm length and 3 mm diameter). 9. thermistor (YSI-precision-thermistor, 1 MΩ (25°), Nr.44015, held by its own streched wires, perpendicular to the given cross-section).
10. luting material to fix the sensor (Bucarid).

The principle of the method as used by us is simple. The porometer consists of a small cup containing the humidity sensor (Plate I, Details in Fig. 1). When clamped onto a leaf the initially low water vapour concentration, which is made spatially constant by permanently using a small fan within the cup, increases. The transpiration speed of the leaf determines the rate of increase of absolute humidity and by the way the rate of decrease of electrical resistance of the sensor. The time needed for a fixed decrease of sensor resistance, between two fixed points of the log-linear part of its calibration diagram of electrical resistance against relative humidity, is now taken as a measure for leaf diffusion resistance. Because of temperature dependence of the calibration curves the sensor (cup air) temperature has to be measured. Temperature of the evaporating surface is needed also, because it determines the vapour concentration potential inside the leaf. After each measurement the cup air is dried to its original starting point, which is a good deal higher in resistance (drier) than the two fixed values used in the measurement, to cancel out starting up effects.

The evaporation (vapour flux density) from a leaf surface of constant temperature into the porometer volume, where the water vapour concentration is time-dependent, may be expressed, using the model of Ohm's law, as:

$$I(t) = \frac{e_1 - e_p(t)}{R_1 + R_p} \qquad (1)$$

This brings the equation expressing the course of porometer concentration with time to:

$$\frac{d e_p(t)}{dt} = \frac{S_p}{V_p} \frac{e_1 - e_p(t)}{R_1 + R_p} \qquad (2)$$

Integrating (2) between the two times at which the fixed electrical resistance values are passed (t_i and t_f) gives, apart from dynamical use and other sensor properties involved:

$$\Delta t = \frac{V_p}{S_p} (R_1 + R_p) \ln \frac{e_1 - e_p(t_i)}{e_1 - e_p(t_f)} = K (R_1 + R_p) \qquad (3)$$

It may be seen from (3) that indeed in theory, if the moisture absorbed by the sensor for indication and the effect of time lag both could be disregarded, the ratio between total resistance and transient times measured would be constant. Measurements over one or more dummy epidermes placed over a saturated surface of known temperature would be sufficient to obtain R_p. Even if the sensor was not temperature dependent different constants would be obtained at different surface temperatures, because e_1 is temperature dependent. This is apart from temperature influences on R_1 and R_p which are known in principle as they depend only on the relation of the diffusion coefficient of water vapour in air with temperature.

Many workers found that several existing varieties of the method as described were not very reliable in practice and moreover quite cumbersome in calibration and use. We found during our investigations that this must have been due to (different) combinations of the following facts:

a) Insufficient awareness of the properties of the electrical humidity sensors and their influence on calibrations and measurements.
b) Use of some incorrect calibration methods.
c) Use of shadow over sunlit leaf parts in the field before measurement.
d) Use of wrong initial conditions for subsequent measurements.
e) Use of a porometer wall material with an absorption capacity for water vapour.
f) Incorrect measurement of leaf temperature.
g) Impossibility to check any influence of the measuring device on the actual opening situation of the stomata.

THE IMPROVED POROMETER.

We designed an improved instrument, using an electronical self-timing circuit to measure the transient times, trying to meet (and to explain) the above mentioned problems as follows:

Ad a) Doing measurements over a series of dummy epidermes, imitating correctly (from the point of diffusion theory) leaf epidermal resistance (Plate II), we obtained straight lines such as ① shown in Fig. 2. Calculating V_p

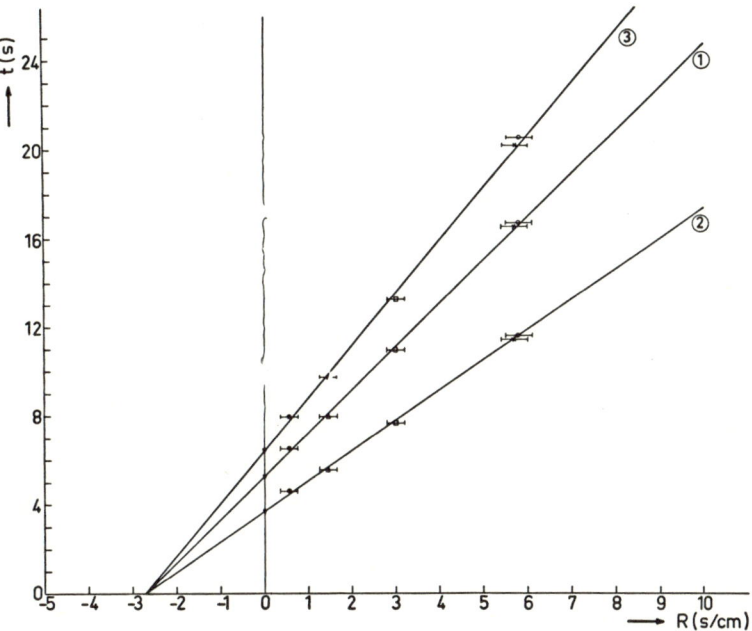

Fig. 2. Line ① represents a calibration, after six weeks of intensive sensor use, with cup and surface at about 27°C, yielding a V_t of 213 cm³. Line ② represents a calibration made at the same day as ①, with the same cup temperature but with the surface now at about 31°C, yielding a V_t of 212 cm³. With cup and surface both at 27°C lines with a slope such as ② were found when this sensor was still hardly used before, with a V_t in the neighbourhood of 140. The calibration marked ③, giving V_t = 272, under the same temperature conditions as for ①, was made after five months in which periods of intensive use were succeeded by periods of storage under dry conditions. See [19] for more details.

from the slopes of such lines, using K (3) as a check, we found values much higher than expected. Moreover the volumes measured in this way became higher in the course of time under the same measuring conditions (Fig. 2, 3). Changing the cup temperature yielded relations between volumes and temperature as given in Fig. 3.

Making use of what was found by scanning literature on our Li-Cl sensors and doing complementary investigations such a behaviour could be explained from the following sensor properties: 1. a change in calibration with time (first ageing effect); 2. a time lag which has to be taken into consideration; 3. a change of this response time in the course of time (second ageing effect); 4. temperature dependence of the response time; 5. insensitivity of this response time to the speed of transpiration into the cup; 6. absorption of water vapour which is not neglectible in comparison with V_p; 7. mere dependence of the response time on this amount of absorbed water per unit of (relative) humidity change in the cup.

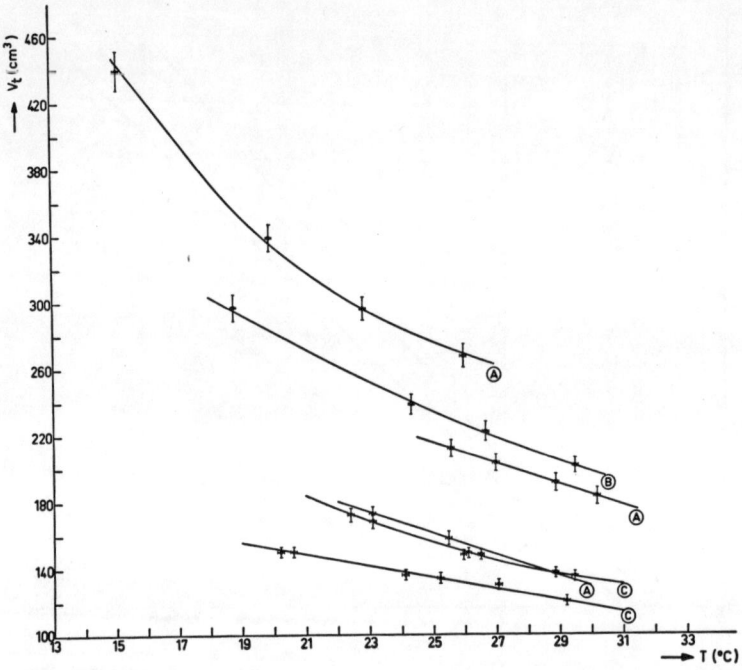

Fig. 3. The relation of V_t with temperature in the course of time for three sensors. Sensor Ⓐ: lowest straight line in the first week of its use, second straight line one month (of intensive use) later, upper curve after five months of alternate intensive use and storage. Sensor Ⓑ: only the curve after half a year of use in preliminary investigations and try out of other calibration methods succeeded by half a year of dry storage. Sensor Ⓒ: lowest straight line in the first week of its use, upper curve after one month of intensive use. Curves such as the latter one and Ⓑ may of course also be approximated by two straight lines. [Note: at the point of intersection of the lowest Ⓐ-line and the upper Ⓒ-curve we have accidentally three measuring points, two equal ones on Ⓐ and one on Ⓒ.] The figure illustrates clearly the individual character of the sensors. This does however nowhere invalidate the calibration method.

The latter property provides the theoretical explanation for the unexpected fact that, sensor properties included, the lines in Fig. 2 remained straight throughout. The values for R_p, measured from these lines (Fig. 2, $t = 0$), were confirmed by calculations from turbulent transport theory to be about 1.9 s/cm. Taking the above into consideration a calibration volume V_t could be defined using (3). Following in the course of time the behaviour of V_t, representing all sensor changes, could now be excellently used for obtaining very accurate field measurements.

Ad b)   Calibration methods not taking into account that all sensor properties as mentioned above are different for individual sensors, such as the two sensors method, or calibration methods imitating incorrectly the diffusion pattern through leaf epidermes, such as the use of pipes of different lengths or injection of saturated water vapour, are bound to fail. The dummy resistances preferably consist of metal multipore membranes from which the resistances can be calculated from pore number and dimensions and/or determined independently from comparative evaporation measurements. Such measurements made by us did agree very well with calculations, for our nickel VECO-multipore membranes with cone-shaped holes, using an early formula by Penman and Schofield.

Ad c)   Using values of the slope K (3) for calibration, without further calculations from this slope, has been a common practice. Comparable to what has been shown to be valid in theory, each combination of sensor (cup) temperature and surface temperature actually yields its own calibration line. Moreover their slope changes in the course of time. This makes thorough calibration in this way impossible. Having leaf and cup at nearly the same temperature would at least reduce the number of calibration measurements. Therefore shadowing the leaf before attaching the cup has been practised. This however enhances the danger of a change in stomatal opening before the end of the measurement.

In relation to our arguing above, V_t being calculated from a line including different transpiration speeds, it must be concluded that V_t is indeed independent from this speed. But in that case surface temperature must be also of no influence on V_t. This was confirmed by a series of measurements such as shown in Fig. 2, line ① and ②, resulting in the same V_t. So our calibration method makes it possible to measure sunlit leaf parts directly.

Ad d)   Different drying speeds to bring the porometer humidity back to the starting point after each measurement result in variable influences of hysteresis and porometer wall absorption and releasing of water vapour. These effects have also influenced measurements. For obtaining reliable results we found it absolutely necessary to have the water distribution within the sensor stabilized, for about two minutes, to be sure that at a (necessarily fixed) starting point indication this distribution was the same prior to each measurement. Only in this case behaviour during a new absorption cycle was consistent.

Ad e)   To prevent sensible sensor drift phenomena to occur, as experienced by others, we used polypropylene as wall material in stead of perspex applied by most users. We investigated separately this material and teflon. They have indeed preferable properties in relation to short time absorption and adsorption.

Ad f)   The importance of accurate e_l-measurements brings about the need for accurate estimation of the internal leaf temperature. We found temperature measurement at the opposite side of the leaf as used for resistance measurement most suitable. By isolation from the porometer clamp and special contact construction for pressing a thermistor to a leaf the time constant of the system could be held relatively low (4 s). Temperatures measured immediately after the shortest transient times in the field ($\simeq$ 5 s) and after 20 s from passing the first fixed resistance point differed 0.3 °C at maximum. The influence on R_l was in this case 5 %.

Ad g) To be able to check such an influence of the time constant of the temperature system we registrated automatically two transient times at each measurement, a short one and a long one. The long transient time was measured over the complete log-linear part of the sensor calibration diagram ($\simeq 20 \rightarrow$ $\rightarrow 27$ % R.H.), the short transient time only over a smaller first part. Comparison of calculations with short and long transient times revealed that V_t was almost constant over diagram parts used.

A second important advantage of knowledge of the ratio of long and short times is the check it provides on eventual influence on stomatal resistance by application of the measuring device (shadow!). From field measurements on Indian corn (Zea mays) we found that, with of course a high degree of variation, in the morning at long transient times of about 1.5 to 2 minutes the risks became quite real of influencing the stomatal opening with the apparatus. In the afternoon this time was even from 1 to 1.5 minute.

Short time	Ratio	Leaf side	R_1	Short time	Ratio	Leaf side	R_1
5.34 s	3.60	U	2.00 s/cm	6.22 s	3.60	U	2.70 s/cm
5.64	3.57	L	2.15	5.46	3.65	L	2.65
5.90	3.64	U	2.65	5.15	3.55	U	1.95
5.70	3.64	L	2.45	4.80	3.59	L	1.85
5.61	3.61	U	2.20	4.57	3.56	U	1.65
6.64	3.65	L	3.15	4.47	3.60	L	1.65
6.35	3.60	U	2.75	5.20	3.62	U	2.25
5.08	3.59	L	1.90	5.04	3.63	L	2.15
4.93	3.61	L	1.65	4.69	3.60	U	1.70
7.57	3.60	U	3.50	5.10	3.64	L	2.10

Table 1. For the 20 subsequent measurements concerned, with a mean resistance of 2.25 s/cm, the mean ratio between long and short transient times was found to be 3.61 ± 0.02 at $27^o.5$ C mean cup temperature. From calibrations the day before and the day after the measuring day we expected at this resistance and this temperature 3.64 ± 0.04.

In table 1 as an example a series of subsequent measurements on upper and lower sides of fully sunlit corn leaves gives an idea on the constancy of the ratio of long and short transient times and the variability of the resistance of identical places of different leaves. For this purpose measuring places 10 cm from the top of big leaves, near the leaf central vein, in a 1 m high crop were used. Mean values for upper and lower sides are 2.3 and 2.2 s/cm respectively. This symmetry was found to hold for our corn under a variety of circumstances. This will be elaborated on elsewhere.

ACKNOWLEDGMENTS

I am most thankful to my colleagues Ir.J.Birnie and Ing.B.Lammers for designing the electronical equipment and other fine cooperation and to Mr.A.E.Jansen and his people of our mechanical workshop for skillful assistance in design and construction of our porometer. I am indebted to Prof.Dr.Ir.J.Schenk for reading and improving this text.

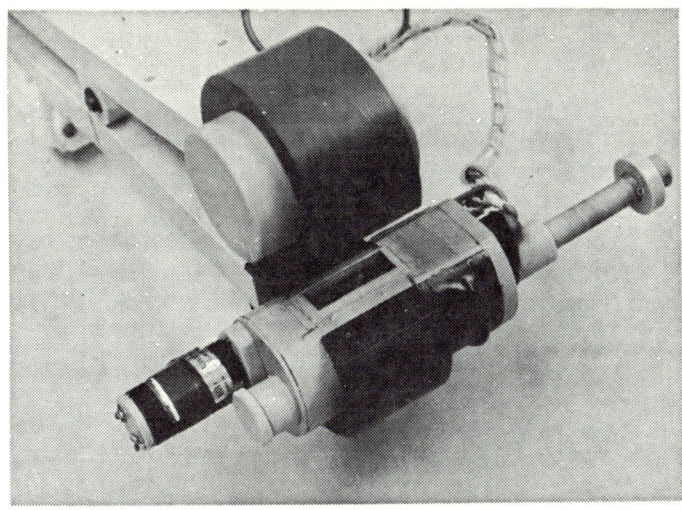

Plate I. The Wageningen Laboratory of Physics and Meteorology diffusion porometer (prototype). One discerns from left to right: motor head, opening with rubber sealing fringe and anti-convection membrane, sensor head with wires, silica gel container (pulled out).

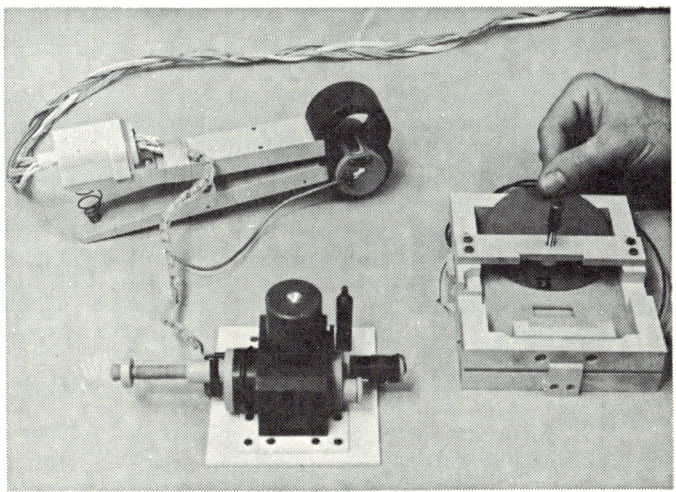

Plate II. A high resistance dummy epidermis is placed over a 2 cm^2 opening above saturated filter paper. The porometer, in its calibration clamp, will be brought over it, in a fixed position, for a measurement.

REFERENCES

1) Lee, R., 1967. The hydrologic importance of transpiration control by stomata. Water Res.Res. 3:737-752.
2) Shepherd, W., 1972. Some evidence of stomatal restriction of evaporation from well-watered plant canopies. Water Res.Res. 8:1092-1095.
3) Stigter, C.J., 1972. Leaf diffusion resistance to water vapour and its direct measurement. I. Introduction and review concerning relevant factors and methods. Meded.Landb.Hogesch. (Comm.Agric.Univ.), Wageningen, 72-3:1-47.
4) Tanton, T.W. and Crowdy, S.H., 1972. Water pathways in higher plants. III. The transpiration stream within leaves. J.exp.Botany 23:619-624.
5) Meidner, H. and Mansfield, T.A., 1968. Physiology of stomata. McGraw Hill, London, 179 pp.
6) Lange, O.L., 1972. Wasserumsatz und Stoffbewegungen. Fortschr.Botanik 34:93-112.
7) Akita, S. and Moss, D.N., 1972. Differential stomatal response between C_3 and C_4 species to atmospheric CO_2-concentration and light. Crop Sc. 12:789-793.
8) Waggoner, P.E., 1969. Environmental manipulation for higher yields. In: J.D.Eastin et al. (editors): Physiological aspects of crop yield. Am.Soc.Agron., Madison, 396 pp.
9) Goudriaan, J. and Waggoner, P.E., 1972. Simulating both aerial microclimate and soil temperature from observations above the foliar canopy. Neth.J.agric.Sc. 20:104-124.
10) Penning de Vries, F.W.T., 1972. A model for simulating transpiration of leaves with special attention to stomatal functioning. J.appl.Ecol. 9(1):57-77.
11) Monteith, J.L., 1973. Principles of environmental physics. Edward Arnold, London, 241 pp.
12) Downey, L.A., Anlezark, R.N. and Muirhead, W., 1972. Construction, calibration and field use of a rapid-reading viscous flow porometer. J.appl.Ecol. 9 (2):431-437.
13) Brun, L.J., Kanemasu, E.T. and Powers, W.L., 1973. Estimating transpiration resistance. Agron.J. 65:326-328.
14) Stigter, C.J. and Lammers, B. Leaf diffusion resistance to water vapour and its direct measurement. III. Results with the improved diffusion porometer in the growth room and in fields of Indian corn (Zea mays). To appear as Meded.Landb.Hogesch. (Comm.Agric.Univ.) Wageningen, 1974.
15) Gale, J. and Poljakoff-Mayber, A., 1967. Resistance to gas flow through the leaf and its significance to measurements made with viscous flow and diffusion porometers. Isr.J.Bot. 16:205-211.
16) Turner, N.C., 1970. Response of adaxial and abaxial stomata to light. New Phytol. 69:647-653.
17) Domes, W., 1971. Unterschiedliche CO_2-Abhängigkeit des Gas-austausches beider Blattseiten von Zea mays. Planta (Berl.) 98:186-189.
18) Hand, D.W., 1973. Techniques for measuring CO_2-assimilation in controlled-environments enclosures. Symp. on Greenhouse Climate; evaluation of research methods. ISHS Techn.Comm. 32:133-147.
19) Stigter, C.J., Birnie, J. and Lammers, B., 1973. Leaf diffusion resistance to water vapour and its direct measurement. II. Design, calibration and pertinent theory of an improved leaf diffusion resistance meter. Meded.Landb.Hogesch. (Comm.Agric.Univ.), Wageningen, 73-15:1-55.

28

HEAT AND MASS TRANSFER FROM REAL AND MODEL LEAVES

J.A. CLARK* and G. WIGLEY†

*University of Nottingham School of Agriculture,
Sutton Bonigton, Loughborough, England.
† Chemical Engineering Division, AERE, Harwell, England.

ABSTRACT

This paper discusses the validity of formulae for heat transfer from isothermal surfaces, derived from engineering sources, used to estimate heat loss from leaves. Temperature distributions obtained by measurements with a thermal imaging camera and by calculation illustrate differences between isothermal and constant flux surfaces and suggest that related distributions of stomatal resistance may exist in real leaves.

INTRODUCTION

The study of heat transfer between plants and their environments has received considerable attention since the review by Raschke (1), but has made few major advances. The Polhausen equation for transfer in laminar flow or its simple multiple derivatives have generally been used in the analysis of measurements on leaves and leaf models both to estimate the resistance to convective heat transfer and for comparison with measured values (2,3,4). Enhanced rates of heat transfer from leaves in natural flows have been described by multiplying the value predicted by the Polhausen equation by a constant factor β. Another factor has been the widespread use of metal isothermal models of leaves in studies of transfer which give only mean transfer coefficients, a notable exception being in the paper by Parlange et al (5).

The appropriate form of the Polhausen equation for local heat transfer from a constant flux thin flat plate parallel to a laminar flow (6) is:

$$Nu_d = 0.453 \ Re_d^{\frac{1}{2}} \ Pr_d^{\frac{1}{3}} \tag{1}$$

where Nu_d is the local Nusselt number

$$Nu_d = \frac{Hd}{k(T_d - T_a)} \tag{2}$$

and Re_d is the local Reynolds number

$$Re_d = \frac{Ud}{\nu} \tag{3}$$

and Pr is the Prandtl number for the fluid, equal to 0.71 in air.

Here H_d is the local heat flux (W m^{-2}), d is the distance downwind from the leading edge, k is the thermal conductivity of air, $(T_d - T_a)$ is the local temperature difference between the surface (T_d) and the airflow (T_a), U is the velocity of the free stream flow and ν is the kinematic viscosity of air.

The equivalent local resistance for convective heat transfer r_H, in units of seconds per metre in the SI system, is related to Nu_d by the equation

$$r_H = \frac{\rho c_p (T_d - T_a)}{Hd} = \frac{\rho c_p d}{Nu_d k} \tag{4}$$

where ρ and c_p are the density and specific heat of air at some chosen arbitrary temperature. Wigley and Clark (7) have recently reported deviations from the form depicted by Eqn (1) in measurements of heat transfer from realistically shaped leaf models of 110 mm width in both laminar and turbulent parallel flows. Similar results were obtained by Chamberlain (8) who measured the local rates of deposition of a radioactive tracer to models of shape identical to those of Wigley and Clark, but in the flow above a plant canopy in a wind tunnel. The local heat transfer coefficients for leaf models in parallel turbulent flows (7), estimated from measurements of the steady state energy balance of electrically heated models, showed a relationship between Nu_d and Re_d consistent with transfer in a fully turbulent boundary layer, i.e. the exponent of Re_d in the equation

$$Nu_d = \text{const } Re^n Pr^{0.33} \tag{5}$$

was approximately 0.8.

The empirical equation for dimensionless heat transfer fitted to these published measurements, was for parallel turbulent flow

$$Nu = 0.045 \, Re^{0.84} \, Pr^{0.33} \tag{6}$$

shown as line 3 in Fig. 1 (reproduced from Boundary Layer Meteorology). In parallel laminar flow at low Reynolds numbers the measurements were indistinguishable from the above, but for Reynolds numbers greater than 10^3 the empirical fitted equation was

$$Nu = 0.22 \, Re^{0.6} \, Pr^{0.33} \tag{7}$$

shown as line 2 in Fig. 1. In both cases the greatest deviations from the predictions of Eqn 1 (Fig. 1, line 1) occur close to the leading and trailing edges of the model. The measured values of Nu are below line 1 at small Reynolds numbers, implying less heat transfer per unit temperature gradient than that predicted, while the converse applies near the trailing edge.

Local surface temperatures on leaves may therefore be expected to differ considerably both from those predicted by application of the Polhausen equation, and from values measured on isothermal metal models. However, the <u>mean</u> Nusselt numbers for the total heat transfer from the surface will fall within a smaller range of values.

Philip (8) and Budagorsky (9) suggested that the boundary layer over real leaves in the field may be turbulent in character, while measurements of <u>mean</u>

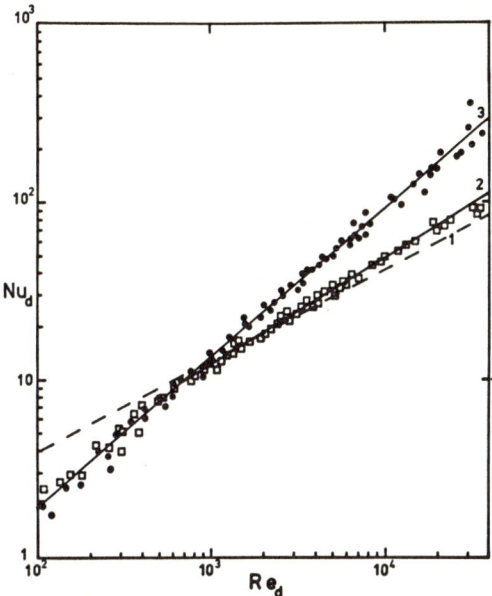

Fig. 1. Comparison of heat transfer for a constant flux surface predicted by theory (line 1) and measured in parallel laminar and turbulent flows (lines 2 and 3 respectively).

transfer from leaf models have also shown deviations from the Polhausen equation (4,10). However, alternative mechanisms of boundary layer transfer do not appear to have been considered in detail by previous workers.

The present paper reports estimates of the local vapour transfer resistance from real leaves in the wind tunnel and in the field. These are based on partitioning of the energy balance by the application of local heat transfer coefficients measured for a model of similar shape and dimensions in the same flow conditions.

THEORY AND METHODS

The model used in these measurements was made from a 1 mm thick perspex sheet and was geometrically similar to a primary leaf of Phaseolus vulgaris. It was heated uniformly by a fine wire grid moulded into the perspex so that it could be considered to approximate to a constant flux surface. The radiation balance was simplified by painting the surface with Parsons optical black. In laboratory measurements the model was supported in the vertical plane within the horizontal air flow from an open jet laminar wind tunnel as shown in Fig. 2. The turbulent wake flow was generated by a 25 mm x 20 mm cros-section bluff body placed vertically in the flow 150 mm upstream from the model. Laminar flow velocities were measured using an orifice plate integral with the wind tunnel. Turbulent flow velocities in the range 0.7 m s^{-1} to 5.0 m s^{-1} were used and the turbulent intensity of each flow was measured

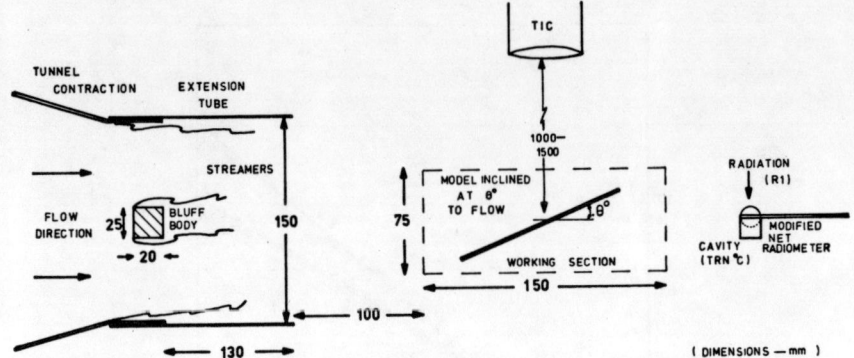

Fig. 2. Plan view of wind tunnel orifice showing turbulence generator and working area.

using a DISA 55A01 hot wire anemometer. The turbulent intensity of the laminar flow was approximately 1% and in the turbulent flow greater than 35%.

The net radiation absorbed by the model was calculated from the difference between the thermal radiation emitted by the surface and the measured radiation, R, incident on a miniature net radiometer held in the same plane as the model. A black cavity of known temperature T_c was used as a reference and the radiation flux corrected to the local model surface temperature using the Stefan-Boltzmann Law.

The local temperatures, T_d, and the temperature distribution over the model surface were measured with an AGA thermal imaging camera (hereafter referred to as TIC). Although a TIC measures a relative temperature, absolute temperature measurement is possible by including in the field of view a black body at a known temperature. Chromel-alumel thermocouples were used to measure the temperature of the black body reference for the TIC and the air temperature (T_a).

The local energy balance, in units of $W\,m^{-2}$, for the <u>surface</u> of the dry model leaf may be written as

$$e + R = \frac{\rho c_p}{r_H} (T_d - T_a) + \sigma (T_d^4 - T_c^4) \tag{8}$$

where e is the input of electrical power, $\rho c_p (T_d - T_a)/r_H$ is the <u>local</u> rate of convective heat loss and σ is the Stefan-Boltzmann constant. All temperatures are in °C unless used in conjunction with σ when they are in K.

The experimental technique was the same for the real leaf except that solar heating was simulated using a quartz iodine photoflood lamp, and the net radiation absorbed by the leaf (Rn) was measured using the unmodified net radiometer. Primary leaves of <u>Phaseolus vulgaris</u> were used, selected for regular shape and similarity to the model dimensions. The plants were grown in pots in a greenhouse and secondary growth excised. The pot and stem were enclosed in a sealed polyethylene bag during measurements and a check was kept on the evaporative loss by periodic weighing. Measurements were, however, made only at windspeeds of $0.7\,m\,s^{-1}$ and $1\,m\,s^{-1}$ since at higher windspeed fluttering prevented estimation of the local temperature by thermography. The local energy balance for the real leaf can be written

$$Rn + \sigma (T_m^4 - T_d^4) = \lambda \frac{c_p}{r_H} (T_d - T_a) + \lambda \rho \frac{(x^* - x)}{(r_v + r_s)} \tag{9}$$

Averaging of the temperature distribution across the leaf gave a mean leaf temperature, T_M. A correction for the local value of the net radiation was then calculated for a local leaf temperature of T_d by using the Stefan-Boltzmann Law. The last term on the right hand side of Eqn (6) is the evaporative heat flux where λ is the latent heat of vaporisation of water and $(x^* - x)$ is the calculated difference in the humidity mixing ratio between a water saturated surface at temperature T_d and the air at temperature T_a. Work on a wet model in the turbulent wake flow (Wigley and Clark, in preparation) showed almost equal convective transfer from the two surfaces at angles of incidence up to 20°. In similar conditions, the convective transfer processes from the model and the real leaf should be sensibly identical.

An estimate of the evaporative flux was made from the energy balance and the Lewis relationship employed to determine the boundary layer resistance to water vapour transfer r_b, assumed equal to 2 r_H since the real leaf was considered to be hypostomatous. The physiological control of transpiration by the leaf, expressed by the stomatal resistance r_s, was then estimated as a function of the distance downwind from the leading edge d. The local energy balances for the model and real leaf were also determined in the field and used as the basis for similar calculation. Measurements in the field closely followed those in the laboratory. The model and real leaf were placed vertically in a barley crop at the same height as the flag leaves. The windspeed was measured at the same height with a Hastings omnidirectional thermistor anemometer. The wind direction was recorded by a wind vane 5 metres above the ground.

RESULTS

Examples of the temperature distributions over the model and real leaf in parallel turbulent flow are shown by Fig. 3. The wind flow is from right to

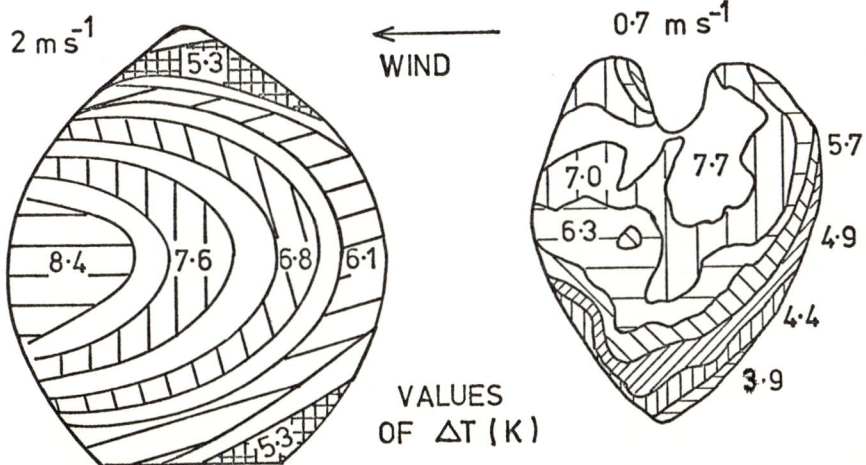

Fig. 3. Tracings from original thermographs showing temperature distributions over the model and a real leaf in the wind tunnel in a turbulent airflow. Temperatures are shown as differences from air temperature.

left in both cases. The relative temperatures are indicated on the figures, traced from the original thermograph. Measured local surface temperatures for the leaf are shown in Table 1a, which presents the local energy balance for the real leaf in tabular form. The external resistance to water vapour transfer, r_v, is not shown in the table since it is given by 2 r_H. The variation in stomatal resistance, r_s, with the distance downwind from the leading edge, shown in Fig. 4, is the most important result of this analysis. The method employed here allows estimation of r_s over areas generally precluded from diffusion porometer measurements, i.e. near the leading and trailing edges of the leaf and in areas with raised veins. The estimated value of r_s is lower

Table 1. Experimental conditions and local energy partition for a leaf in the wind tunnel (a) and in the field (b)

(a) Windspeed: 0.7 m s^{-1} (turbulent, parallel)   Air temperature: 25.58°
    Relative humidity: 29%   Measured evaporative flux: 70 W m^{-2}
    Measured leaf resistances:
        Top of leaf   — 1300 m s^{-1}
        Middle of leaf — Leading regions 1300 – 1000 s m^{-1}
                        Trailing regions 400 s m^{-1}

Distance from leading edge d mm	Surface temperature °C	Net radiation W m^{-2}	Convection W m^{-2}	Evaporation W m^{-2}	s m^{-1} Vapour resistance	s.m^{-1} Leaf resistance
Transect 20.5 mm below petiole						
2.0	30.5	163	−113	−51	93	1210
5.0	31.3	159	−113	−46	108	1420
9.0	31.9	155	−114	−41	119	1680
16.2	32.6	150	−115	−34	130	2110
34.4	33.3	145	−113	−33	147	2320
51.2	32.6	150	−96	−54	157	1270
64.8	32.6	150	−93	−57	163	1180
72.4	31.9	155	−82	−73	165	840
82.0	32.6	150	−89	−61	169	1100
85.6	31.9	155	−79	−75	170	300

(b) Estimated windspeed: 1.2 m s^{-1} Time: 09.22 hours GMT
    Air temperature: 18.5°C   Relative humidity: 46%

Distance from leading edge d mm	Surface temperature °C	Net radiation W m^{-2}	Convection W m^{-2}	Evaporation W m^{-2}	s m^{-1} Vapour resistance	s.m^{-1} Leaf resistance
8.9	18.5	147	−1	−146	77	150
29.9	20.7	134	−46	−88	104	320
42.0	22.3	125	−71	−54	113	600
53.0	20.7	134	−40	−94	119	300
63.9	22.3	125	−64	−61	125	530
75.4	20.7	134	−36	−98	130	290
88.1	22.3	125	−59	−66	135	490

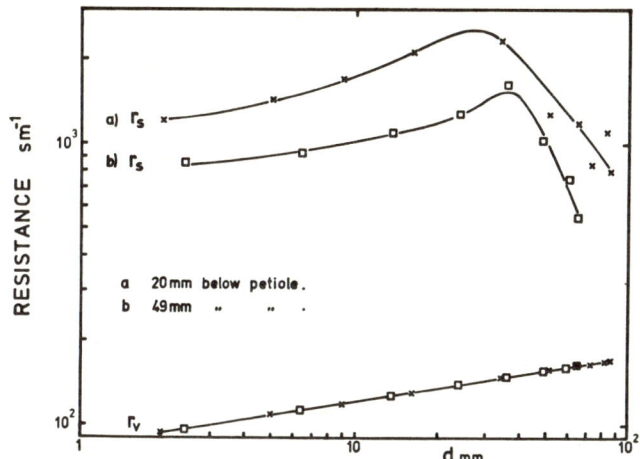

Fig. 4. Resistance to water vapour transfer v distance downwind on a bean leaf. r_s = stomatal resistance; r_v = boundary layer resistance.

near edges than over the rest of the leaf, with the lowest values recorded at the trailing edge. The highest values of r_s are recorded in the central area near the veins.

The measurements of convective heat transfer from the model in the field are presented in Fig. 5 as the local dimensionless heat flux, Nu_d, plotted logarithmically against the local Reynolds number, Re_d. Classification of the results by windspeed yields lines with a slope of 0.75, i.e. intermediate between the values of 0.6 and 0.84 observed for laminar and turbulent flows respectively in the wind tunnel (7). The constant term, in a fitted relationship assumed to follow Eqn 5, varied with windspeed, the value at 1 m s^{-1} being 0.1. Measurements of the temperature and temperature distribution over the surface of the real leaf in similar conditions are presented in Table 1b, together with values of the calculated local energy balance. These exhibit a realistic partition of energy between convective and evaporative fluxes, lending confidence to estimates of the stomatal resistance also presented in the table.

DISCUSSION

The temperature distributions presented, measured by thermography, emphasise that leaves are not isothermal surfaces. This point was recognised by Raschke (11) as early as 1956, but has often been overlooked in the search for simple models of the convective transfer between vegetation and its environment, based on available formulae from engineering sources.

Measurement of the local transfer illustrates deviations from the predictions of the Polhausen equation in heat transfer from realistically shaped leaf models. These have frequently been overlooked when isothermal models have been employed. Such local deviations may be explained by measurements of the boundary layer velocity field over leaves and models by optical anemometry reported elsewhere in these proceedings (12).

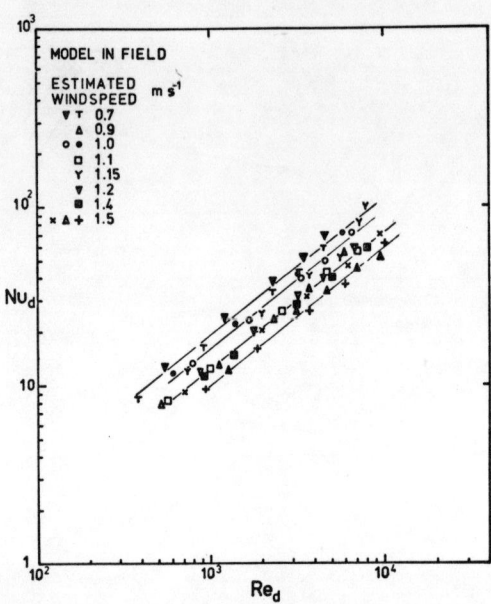

Fig. 5. Logarithmic plot of Nusselt number against Reynolds number for a dry model leaf in field conditions.

The predictions of theory and the results embodied in the empirical equations presented here and in (7) are most easily illustrated and compared by tabulating values of surface temperature over a leaf. The simplest model of a dry flat plate will suffice. Dissipation of the energy input by the alternative paths of convection and radiation produces a difference (ΔT) between the surface temperature (T_L) and that of the air (T_a) which changes with the rate of transfer, such that

$$\Delta T = R/\{\sigma\,(T_L^4 - T_a^4) + Nu\,k/L\} \tag{10}$$

where L is the appropriate characteristic dimension. For a leaf R is the 'isothermal net radiation', the net radiation relative to a surface at temperature T_a, expressed per unit of surface area. If for simplicity we adopt a constant radiative conductance of say $5.6\ W\ m^{-2}\ K^{-1}$, Eqn 10 becomes

$$\Delta T = R/\{5.6 + 0.025\,Nu/L\} \tag{11}$$

in SI units.

Table 3 presents surface temperatures calculated using this equation. A large value of R of $250\ W\ m^{-2}$ has been assumed equivalent to bright sunshine, a flow velocity of $1\ m\ s^{-1}$ and a maximum surface length in the wind direction of 0.1 m. Values of Nu and L are appropriate to:

(a) heat transfer from isothermal surfaces of the width 0.1 m, Nu being obtained from the equation

$$Nu_D = 1.08\ Re_D^{\frac{1}{2}}\ Pr^{0.33} \tag{12}$$

a simple multiple of the Polhausen equation fitted to their measurements of heat transfer from metal discs (3);

Table 3. Comparison of temperatures at point L downwind on a dry flat surface in a 1 m s^{-1} airflow. For conditions of (a), (b) and (c) see text.

	Temperature differences from air temperature (K)		
	(a)	(b)	(c)
L (m)	Isothermal model, empirical	Constant flux, theoretical	Constant flux model, empirical
0.005	9.7	5.3	6.1
0.01	9.7	7.2	7.1
0.02	9.7	9.5	8.1
0.03	9.7	11.1	8.8
0.04	9.7	12.4	9.4
0.05	9.7	13.4	9.8
0.06	9.7	14.3	10.1
0.07	9.7	15.0	10.4
0.08	9.7	15.7	10.7
0.09	9.7	16.3	10.9
0.10	9.7	16.8	11.2
Mean	9.7	12.5	9.3

(b) local heat transfer from a constant flux surface according to theory (Eqn 1) at points d downwind on the surface;

(c) local heat transfer measured from a constant flux surface in the field and reported in this paper, i.e. using the empirical equation

$$Nu_d = 0.10\ Re^{0.75}\ Pr^{0.33} \qquad (13)$$

The substantial differences between the local temperatures in the constant flux case and the temperature predicted for an isothermal surface are close to the maximum values which might occur outdoors. In practice the variations in local transfer rates in the boundary layers of real leaves are likely to result in significant temperature gradients but of much lower magnitude except in extremely dry conditions. The effects of such temperature gradients should, however, receive attention from plant physiologists both because of possible influences on respiration and translocation and because of implications for the water relations of leaves. For instance, thermal factors, as well as proximity of water in the xylem may be concerned in the differences in wilting often observed between the centres and edges of large leaves such as sugar beet.

The field measurements on the model support the conclusions drawn from wind tunnel work (7), that convective transfer in turbulent or natural conditions cannot be predicted by simple multiples of the Polhausen value. Subsequent application of these results to the real leaf yielded realistic estimates for the convective and evaporative transfers.

Leading and trailing edge effects on laminae of finite thickness appear to be responsible for the local deviations from the Polhausen equation. On a large surface these effects will be minimal compared with the overall convective transfer. However, in the case of grass and cereal leaves the transfer will be dominated by edge effects. The results for the model in parallel laminar and turbulent flows (Fig. 1) show identical transfer at Reynolds numbers below 10^3. If these results can be extrapolated to grass-like leaves then it is possible that convective transfer may sometimes be independent of the flow character.

ACKNOWLEDGEMENTS

G.W. thanks the Agricultural Research Council of Great Britain for the support of a studentship. The Science Research Council provided a grant to JAC for the hire of the Thermal ImagingCamera employed in the measurements. We also wish to acknowledge the continued interest and advice of Professor J.L. Monteith.

REFERENCES

1. Raschke, K. (1960) Ann. Rev. Plant Physiol. $\underline{11}$, 111-126.
2. Gates, D.M. and Papian, L.E. (1971) 'Atlas of Energy Budgets of Plant Leaves', Academic Press, London and New York.
3. Pearman, G.I., Weaver, H.L. and Tanner, C.B. (1972) Agricultural Meteorol. $\underline{10}$, 83-92.
4. Thom, A.S. (1968) Quart. J. Roy. Met. Soc. $\underline{94}$, 44-55.
5. Parlange, J.Y., Waggoner, P.E. and Heichel, G.H. (1971) Plant Physiol. $\underline{48}$, 437-442.
6. Ede, A.J. (1967) 'An Introduction to Heat Transfer Principles', Pergamon Press, Oxford.
7. Wigley, G. and Clark, J.A. (1974) Heat transport coefficients for constant energy flux models of broad leaves. Boundary Layer Meteorology, in press.
8. Chamberlain, A.C. (1974) Mass transfer to bean leaves. Boundary Layer Meteorology, in press.
9. Budagovsky, A.I. (1964) 'Ispareni Po chvennoi Vagi', Nauk, Moscow.
10. Parkhurst, D.F., Duncan, P.R., Gates, D.M. and Kreith, F. (1968) J. Heat Transfer (Trans ASME), $\underline{90}$, 17-76.
11. Raschke, K. (1956) Arch. Meteorol. Geoph. Biokl. B. $\underline{7}$, 240-268.
12. Durst, F., Wigley, G. and Zare, M. 'Laser-Doppler anemometry and its application for flow investigations in the environment of vegetation. These Proceedings.

PART II
APPLICATION

SECTION 1
PHYTO-ENGINEERING

ENERGY AND MASS TRANSFER IN PLANT COMMUNITIES

A.A. NICHIPOROVICH

*Institute of Plant Physiology,
USSR Academy of Sciences, Moscow*

The plant cover (PC) together with the soil medium (S) comprise a living system (PCS) in which energy and mass transfer radically differ from those of dead bodies.

Imagine a soil column in a lysimeter, the column being isolated from the surrounding soil medium and the soil being sterile; there are no microorganisms in the soil and its top layer is bare of plant life; assume, moreover, that the soil is dried to a state of equilibrium with the water potential of the air and is protected from rainfall. Exchange of energy with the environment in this case can be expressed by a simple equation which, however, must take into account the fact that the air-dry moisture of the soil varies reversibly with variation of water conditions of the air medium. Moreover, with variation of temperature a certain amount of gas will physically be evolved or absorbed in a reversible manner. Transfer of energy and mass will be balanced during a 24 hour day and the mean amount of heat and mass of the soil will remain constant.

The situation will be entirely different if liquid water from the environment is permitted to enter such a sterile and dead soil. In addition to the phenomena mentioned above transfer of water in the soil and its evaporation from soil surface will occur and the heat balance will be altered. In absence of filtration and removal of matter to the subsoil material balance of the soil will remain constant, neither its dry nor total weight being altered.

Interaction with the environment will be more complex if the soil is inhabited by microorganisms and it contains a sufficient amount of organic substances available to them. In this case a new factor will be involved in the heat balance of the soil, namely, heat evolved by the microorganisms during their life activity and decomposition of organic compounds. However this contribution to the soil heat balance is ordinarily negligible.

Many chemical transformations of substances in the soil may occur and mass exchange with the environment should now include such components as evolution of CO_2 as a product of soil respiration, absorption of oxygen required for the life activity of the microorganisms; in some cases molecular nitrogen may enter the soil if nitrogen fixers are present or may be evolved in the presence of denitrificaters. Other gases may also be evolved if rotting occurs. All these processes are strictly depended on the composition of the microflora and microfauna, on the humidity, physico-chemical properties of the soil and also on the water and thermal conditions.

In these casses there will be an imbalance of energy and mass exchange processes. For exemple, from 10 to 20 kg of carbon contained in 30-80 kg of CO_2 may be lost by an hectare of soil as a result of respiration. An additional 100-200 g of nitrogen may be incorporated or lost as a result of fixation from the atmosphere or as a result of denitrification.

The presence of a plant cover (PC) of higher plants will alter this picture. Plants intimately interact with the soil medium and with respect to energy and mass transfer the "plant cover-soil" system may be regarded as a single system.

Green plants, which are able to photosynthesize, significantly affect the energy and mass transfer processes in the complex system: carbon from atmospheric carbon dioxide gas and water and mineral nutrient elements (nitrogen, phosphorus, potassium, magnesium etc.) from the soil are rapidly taken up and incorporated into the photosynthetic organs. These elements, and also hydrogen from photochemically decomposed water, combine together to form organic molecules under the action of sunlight energy and simultaneously oxygen is liberated into the atmosphere.

Formally this process is partially reversible: about 25-30% of the organic substances produced during photosynthesis are oxidized as a result of respiration, suitable amounts of O_2 being absorbed and CO_2 being evolved, a corresponding amount of energy stored during photosynthesis being released as heat. However photosynthesis and respiration are spatially and temporally separated from each other. Thus photosynthesis occurs only in the presence of light and in special photosynthesizing organs, whereas respiration proceeds in all living cells, and is not reversed photosynthesis.

A remarkable feature of photosynthesis is that under the action of sunlight quanta it proceeds in a direction which is opposite to that of the thermodynamic potential gradient. An "organic substance + oxygen" (OS + O_2) system is formed from mineralized and, as a rule, completely oxidized substances such as CO_2 and H_2O as well as from mineral salt ions (NO_3^-, $SO_4^=$, $PO_4^\equiv$); an energy gradient of 1,4 ev must be overcome in this case. The initial substrates of photosynthesis are subjected to complex step-like oxidationreduction chemical transformations involving the formation of active oxidizers and active reductants.

The oxidants and reductants formed at various stages of the process, including the final stage, are located at separate structures of the photosynthetic apparatus and thus protected from back reactions. The final result is the formation of photosynthetic products which remains in the plant and of free oxygen (O_2) which is evolved into the atmosphere. The energy of sunlight transformed into chemical bond energy is thus firmly stored by the system.

In the "Plant cover-soil" system photosynthesis proceeds with an active balance: its total mass increases, as does the amount of bound energy.

The contribution of the plant cover and primarily of plant photosynthesis to the energy and mass transfer of the PCS system is enormous and the energy-rich "organic substance + O_2" system is the initial source of all life on Earth, including that of the plants themselves and the existence of the PCS system.

The amount of CO_2 absorbed per hectare of plants during a day of active photosynthesis may reach 800 or even 1000-1200 kg. The amount of oxygen evolved is respectively 640-960 kg, amount of newly formed plant mass reaches 400 or even 500-600 kg per day and the amount of energy stored by the system will then be $1.6-2.4 \cdot 10^6$ (1,2,3).

About 6-10 kg of nitrogen are consumed by the plants from the soil during a 24 hour day and approximately 20-40 kg of other elements of mineral nutrition.

An efficient plant community absorbs large amounts of light and heat and simultaneously evaporates from 40 to 60 tons of water.

Although plants extract mineral nutrition elements and water from the soil they enrich the latter with organic substances as a result of exosmosis and decomposition of roots and other organic remains. This stimulates microbiological processes, increases the humus content of the soil, improves the physico-chemical state of the latter, increases its fertility and permits the plants to enhance their photosynthetic efficiency and hence the crop yields.

High crop yields and their further increase is one of the most important aims of human activity and is certainly the main aim of agriculture and plant-growing.

The essence of this problem is to intensify mass and energy exchange in the PCS system in such a manner which would ensure the formation of maximal amounts of organic substances with highest possible amounts of energy in them; in other words as much carbon dioxide should be absorbed from the air, and sunlight energy stored by the plant during photosynthesis, as is possible. The maximal daily values of photosynthesis of crops, cited above, pertain to optimal conditions when the final biological yields are 15-20 or even 30-35 tons dry mass (fig. 1). This corresponds, for example, to 60-120 centners per hectare of maize grain, 1000-1200 c/ha of maize green organs, 700-1000 c/ha beet roots etc. From 3 to 5% of PAR energy incident during the vegetation period are used in the formation of such crop yields. In practice, however, the average agricultural yields are 2-3, or frequently even 4-5 times lower (1,2,4).

Soil fertility factors are usually among those factors determining the photosynthetic rate and productivity of plants which are at a low level. It is precisely because of this that tillage and fertilization of the soil, stimulation of activity of soil microbiologic flora and improvement of water conditions (including irrigation) are the main means by which high yields are attained in modern plant-growing practice.

However, the dependence of the photosynthetic activity of plants on soil moisture and fertility conditions, and in particular on fertilizer doses, is not linear and can be described by curves which gradually approach a constant value or plateau (fig. 2).

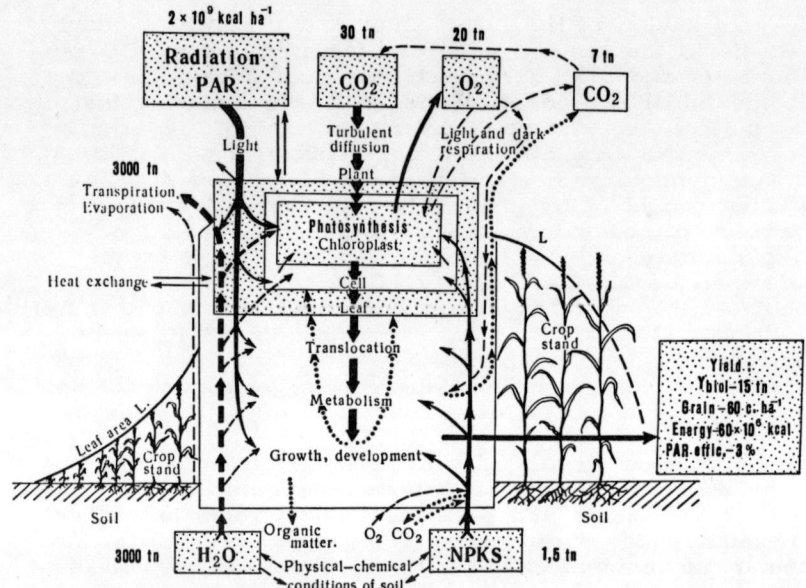

Fig. 1. An example of final results of physiological and photosynthetic activity necessary for obtaining 60 c/h_a of grain.
High yield can be obtained in cases when all processes and factors of photosynthetic activity and productivity are well balanced.

The question arises as to the caused of these limitations. Is the crop yield plateau due to complete exhaustion of the plant potentialities or is the result of such limitations which can be overcome?

One of the causes of the aforementioned response of plants to increasing soil fertility is that along with the positive effect of increasing fertilizer doses and water supply there is also a gro-

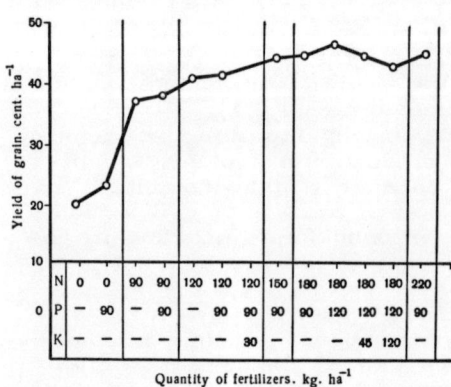

Fig. 2. The dependence of wheat grain on the amount of N, P and K fertilizers under irrigation.

wth of negative side effects. Thus the physical properties of the soil are impaired, ventilation of the soil becomes less satisfactory, the concentration of the soil solution becomes excessive etc.

The number of significant factors of soil fertility is large and only a few of them are optimized by man. As a result, some of them cease to be limiting but then are replaced by others, and new sets of unbalanced factors arise with new restrictive conditions imposed. It is now necessary to study these latter and determined how their restrictive action can be overcome.

It should be mentioned that the reserves of factors of plant productivity as light energy and carbon dioxide in the air are sufficiently great to ensure yields considerably exceeding ordinary or even high presend-day yields. Thus with respect to these factors the farmer's task is not so much to increase these reserves as to use them in a most efficient manner, particularly since carbon and light nutrition of plants are the main functions of green plants.

A situation arises here which resembles that mentioned in the discussion of the efficiency of optimization of soil fertility factors: negative side effects on the photosynthetic activity of plants in phytocenoses gradually increase when the usual methods of raising crop yields are applied.

The ultimate photosynthetic systems are plant communities - phytocenoses, stands (1-5). In order to be highly productive and capable of absorbing with maximal efficiency the incident light and of efficiently utilizing it in photosynthesis and correspondingly of absorbing large amounts of CO_2 from the air, plant communities should possess an optimal spatial and optical density (1-11). The latter condition can be met if the area of the photosynthesizing organs, the leaves, is sufficiently great (L_{max}) but optimal (L_{opt}). This can be attained by choosing optimal plant densities and raising the level of soil fertility.

In the latter case the plant growth rate is higher and hence so is that of the leaf area. Increase of leaf area is accompanied by an increase of the rate of absorption of solar radiation (fig. 3) which is the decisive motive force in transpiration (12-14) and the main component determining the degree of evaporation. In this connection the water requirements of the plants increase.

If the amount of water in the soil available for normal transpiration (W) is below the evaporation tension level (E_v) and the hydrothermal coefficient $H = Wh/E_v$ (where h is the latent heat of water) is much lower than unity, then the leaf area of the crop will be below the optimal value and will strongly depend on the magnitude of the hydrothermal coefficient (fig. 4).

In this case the moisture factor will determine the permissible plant density by regulating the plant growth and restricting its rate and limits. Improvement of water supply and increase of coefficient H increase the growth rate and possible leaf area as well as the duration of functioning of the leaves. The total rate of photosythesis and crop yield usually closely correlate with these characteristics, particularly if better water supply is combined with an improvement of soil fertility factors and especially with an increase of the fertilizer doses.

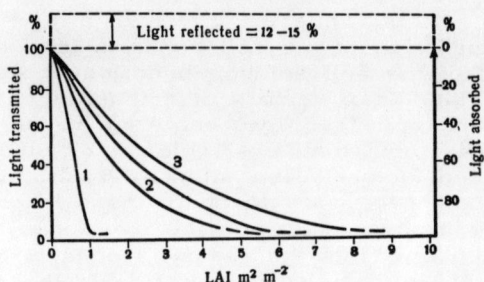

Fig. 3. Light energy absorption by plant communities in dependence on LAI and different structure of foliage;
1) for communities with monolayer of horizontal leaves;
2) for crop stands of majority of agricultural plants;
3) for plant communities with predominantly vertical or with small leaves.

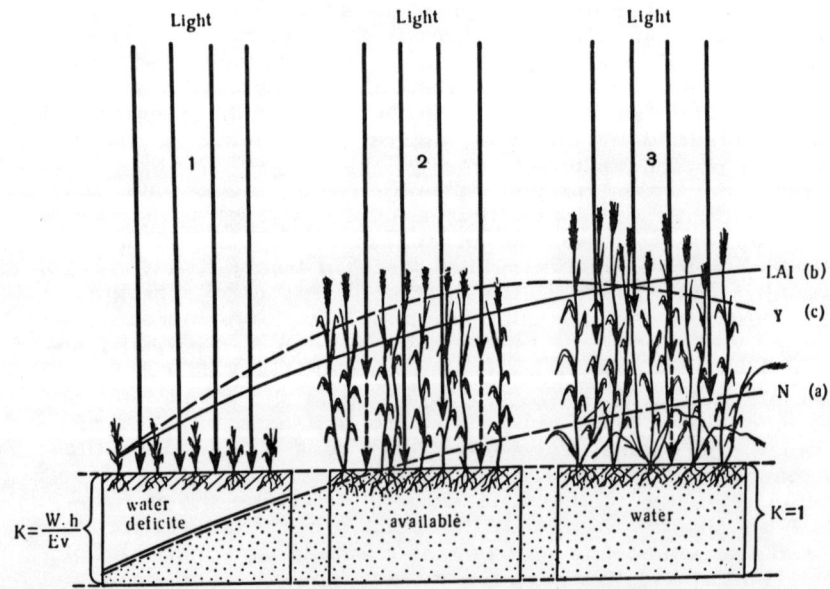

Fig. 4. Plant community structure and density dependent on hydro-thermic coefficient (K) and level of nitrogen supply: 1 - suboptimal; 2 - optimal state; 3 - excessive density;
a) N - nitrogen supply level; b) LAI; c) Y - yield.

However, when the water supply is almost optimal and the physiological hydrothermal coefficient is almost unity (assuming W is measured with respect to the optimal soil moisture level) moisture ceases to be a factor limiting growth processes and this role is now played by the mineral nutrient, and in particular, nitrogen, supply.

When the water supply level approaches a unit hydrothermal coefficient, an optimal leaf area (L_{opt}) is usually formed in the crop. In this case practically all (95-97%) of PAR energy entering the crop can be absorbed by the leaves of about 80-85% of the incident energy, since 12-15% of the PAR energy is reflected.

The highest rates of photosynthesis and organic mass formation (C_{max}) and the yields of crops forming LA_{opt} are usually high at this period of photosynthetic activity and formation of plant community.

However under conditions of good water supply and fertile soil supplied with sufficient amount of fertilizers, the plant growth and increase of leaf area may even exceed the optimal level which signifies that not all potentialities of the plant have been exploited.

Nevertheless the final effect may not be a positive one. One reason is that the leaf area exceeding the optimal value does not absorb more PAR energy than LA optimal. But since plant growth continues and new leaves are formed, the lower ones which are deprived of light and sufficient CO_2 supply, can respire but not photosynthesize; they therefore expend organic substances and soon die out. The active leaf area is shifted to greater heights and this involves greater expenditure of organic matter and energy.

Moreover, when the density is excessive, shading of leaves also becomes excessive, ventilation of the crops and flow of CO_2 to them from the air become impaired and the mean photosynthetic rate strongly drops. As a consequence, relatively larger amounts of nitrogenous substances are formed and correspondingly lower amounts of highly transportable products, which are readily used in secondary syntheses, viz. carbohydrates, are formed. These latter products are important for intense and harmonic growth of all plant organs and in particular for satisfactory growth of reproductive and storage organs.

Excessive growth and self-thickening of plants on fertile soil do not only restrict the plant productivity as a whole but in some cases even lower the economical yields of reproductive and storage organs i.e. of fruits, grain, tubers and root crops and impairs their quality. Moreover, in very dense crops the sickness rate increases and in many cases longing of the plants occurs.

Thus negative side effects in crops may ultimately neutralize the positive effect of optimization of soil fertility factors and also of these factors which lead to rapid plant growth and increase of the size of the photosynthetic apparatus; these effects therefore limit the productivity of the "plant cover-soil" system, although the potential activity of the plants has not been exploited to full extent.

The task of modern plant physiology, agricultural chemistry, micrometeorology and plant breeding is to discover the effect of the restriction as it arises, determine its intensity and find special means of overcoming the restrictions. In this way the productivity of photosynthesizing systems should gradually be raised to those theoretically possible.

Such work should be based on a knowledge of the general laws of photosynthetic activity and plant productivity.

Final biological crop yields (Y_{biol}) are created during photosynthetic activity and are the sum of daily biological mass increments (C): $Y_{biol} = \sum_{o}^{t} C$.

Daily increments which are initially small, gradually increase during the first half of the vegetation period and are maximal (C_{max}) when the photosynthetic apparatus or leaf area is maximal (L_{max}) or optimal (L_{opt}) and when the total growth of biological mass of the crops is most rapid.

This period usually lasts several days and is then followed by the period of formation of reproductive and storage organs and the period of their maturation. At this time the leaf area and photosynthetic activity decrease and tend to zero after termination of vegetation. Correspondingly, the daily increments (C) also decrease.

It has repeatedly been demonstrated that the final magnitude of crop yields of a given plant closely correlates with C_{max} and L_{max} (if they do not exceed L_{opt}).

An important factor determining the magnitude of Y_{biol} is also the duration of the vegetation period T and duration of active operation of the photosynthetic apparatus. Therefore the values of Y_{biol} to a no less degree correlate with the photosynthetic potentials PP which are the sums of mean daily values of L

$$PP = \sum_{o}^{t} L$$

The quantity Y_{biol} can also be expressed by the equation

$$Y_{biol} = C_M \cdot t$$

where C_M is the mean value of the daily increments during the vegetation period which lasts t days. The coefficient $q = C_M/C_{max}$ is usually equal to 0.5-0.6 or on the average 0.55. In some cases it may be higher. Thus Y_{biol} can be characterized by the equation

$$Y_{biol} = t \cdot C_m = t \cdot C_{max} \cdot q = t \cdot C_m / \frac{1}{q}$$

An important quantity, which not only defines the total yield but also its structure and economical value, is the coefficient K_{econ} which is the ratio between dry weight of economically valuable organs and the total biological yield. For cereal crops growing under more or less normal conditions it varies between 0.25 and 0.5. The most common value is 0.4. One of the main tasks of plant growing is enhacement of these coefficients, i.e. increase of the total photosynthetic rate as well as ensuring the best possible distribution of assimilates in growth of, initially, the assimilative organs and then the reproductive and storage organs.

In summary, the crop yield is most closely related to such characteristics of the photosynthetic activity as C_{max}, L_{max}, L_{opt}, NA, K_{econ} and PP. The highest productivity of various plants growing under various conditions can be attained for various combinations of these characteristics.

It is important to assess their real significance for crop formation and to estimate the degree of their variability and their possible maximal levels and hence the maximal productivity levels attainable under optimal conditions.

Let us first consider the concept of an optimal leaf area (L_{opt}). Its main trait is the ability to absorb practically all (95-97%) energy entering the crop. If this were the only requirement the problem could be solved by plant communities with horizontal floating leaves of aquatic plants with $L_{opt} = 1 m^2/m^2$, e.g. Nuphar, Victoria etc. (fig. 4) or multicomponent phytocenoses with tall plants (e.g. forests). This requirement could also be met by plants with vertical leaves ((Typha, partially Phragmites etc.) and finally crops of fodder small-leaf plants including cereals and also clover, lucerne.

In all plant communities of this type L_{max} under good conditions may reach 8-10 m^2/m^2. However these various types of cenoses all possess a number of shortcomings. The reason is that the dependence of photosynthesis on light intensity is described by a hyperbolic curve. As a result at low light intensities, although the rate of photosynthesis may be small its efficiency (E_{PAR}) is high. With increasing light intensity the rate of photosynthesis increases but the efficiency drops (fig. 5).

Consequently, plant communities with a single layer of horizontal leaves are not profitable, even though they are capable of absorbing all incident light, because they utilize the absorbed ener-

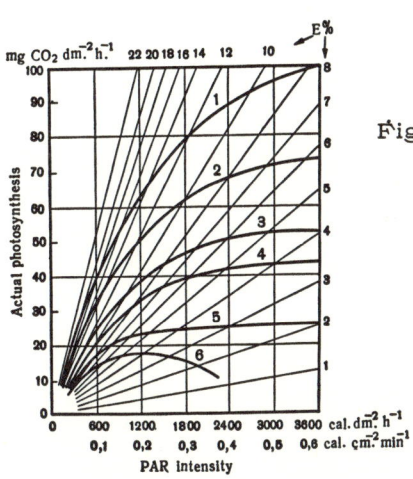

Fig. 5. Light curves of photosynthesis of leaves of different plants: 1, 2 - sorghum, maize, sugar cane; 3, 4 - such cultivated plants as sunflower, sugar beet; 5, 6 - woody and shady plants. Straight lines indicate the levels of PAR utilization efficiency.

gy with a low efficiency as most of the time they are irradiated by high light intensities.

Communities consisting of tall plants are not profitable because of poor air movement which impairs turbulent mixing and hence supply of the leaves with carbon dioxide. Moreover, in woody plants supply of water to the elevated crowns is difficult and very often the leaf structure of such plants protects them from excessive loss of water (stomata located at one side of leaf) but gas supply is then impaired and the photosynthetic rate is lower.

As a result of such contradictory requirements, selection of cultural plants with a high productivity showed that the most productive are those communities for which the mean values of L_{opt} are 4-6 m^2/m^2. However, the value may significantly vary from plant to plant, depending on the illumination level and type of photosynthesis light curve (15). The longer the linear portion of the curve and the greater its slope, the less stringent may be the requirements on the spatial structure of the community and on the value of L_{opt}. The only requirement which should be fulfilled by L_{opt} is complete absorption of PAR. Improvement of the photosynthetic apparatus itself should consist only in an increase of slope of the light "straight lines".

In most present day plants the morphology and activity of the photosynthetic apparatus and also the growth function are far from being perfect. Consequently, these plants do not always form such a L_{opt} which ensures that the potentialities of the photosynthetic apparatus will be used to full extent. Moreover the danger remains of excessive growth and thickening of the phytocenosis under conditions of good water supply and fertilization, or of growth ceasing before all potentialities of the plant have been exploited.

The most effective method overcoming these limitations is to breed plants varieties with highly active photosynthetic apparatus, with modern growth and favorable morphological structure of the plants. In some cases vertical leaves may be favorable. Finally, the response to water and nitrogen supply should also be rational. It is not intense vegetative growth but rather enhancement of photosynthesis and maximal utilization of assimilates for the formation of rapidly growing and capacious, economically valuable reproductive and storage organs which are desirable. The value of L_{opt} for such varieties in dense crops may be greater than that of their predecessor and negative consequences for photosynthesis may be staller. They can therefore respond more favorably to high fertilizer doses and thus produce higher crop yields. (10).

A good example of the effectiveness of work of this kind is the breeding of dwarf Mexican wheat varieties, or the dwarf rice plants breeded by the International Rice Institute in the Philippines.

Of no less importance than the light conditions are the aerodynamic conditions which determine the amount of carbon dioxide carried by the air to photosynthesizing leaves within the cenosis.

In phytocenoses of some plants there may be a deficiency of carbon dioxide during the day, the concentration dropping to

270-250 ppm as compared to the value 320 ppm characteristic of the air above the community (2,6-11). This is due to a discrepancy between the rate of CO_2 assimilation by the plant leaves and the velocity of CO_2 flow from the air. For some plants, and especially for C_3 plants, this reduction of CO_2 concentration may be the cause of a strong reduction of the photosynthetic rate. The C_4 plants in this respect are less sensitive.

Winds of common velocity can carry such quantities of carbon dioxide over crop surface which are several times greater than those quantities which are absorbed by the plants. However below the crop surface the wind velocity and turbulent transfer rate are much lower. The amount of CO_2 in a crop 1 meter high is usually 5-6 kg. An highly active crop with optimal leaf area may assimilate up to 800-1200 kg CO_2 per day. If one assumes that the rate of CO_2 flow to the crop should be sufficient to ensure that the concentration does not drop by more than 10-20%, then from 5 to 10 times more CO_2 should pass through the crop during a day than is assimilated by the plant. This is about 400-1200 kg, or about the amount contained in 800-2000 volumes of air with a height of 1 meter. In other words, the air depleted of CO_2 should be replaced by normal air from 800 to 2000 times per day. At midday when the photosynthetic rate is maximal the exchange rate should be 120-300 per. hour. Expressed in vertical transfer velocity, this is 3.5-5.5 cm.sec^{-1}. Such flow rates can easily occur in crops with low plants (beet, carrot, cabbage). However transfer will be impeded if in crops with tall plants and in cases when the leaf area is large.

The carbon dioxide nutrition problem is solved by various plants in different ways as can be seen from fig. 6. The latter depicts the vertical profiles of the rate of CO_2 absorption by a 2N alkali solution in 2 mm deep flat vessels, the free surface of the solution being 15 cm^2 (16). The vessels were located at various heights within the crop and above it. It can be seen that the carbon dioxide conditions were less satisfactory in the maize crop than in that of beets. This is due, on the one hand, to a higher photosynthetic activity of maize as a C_4 plant as compared to beet (a C_3 plant) and also to a lower transfer rate of CO_2 in the maize crop.

However, even under these difficult conditions maize, being a C_4 plant, could function quite efficiently. The presence of the carboxylating enzyme PEP carboxylase permits this plant to photosynthesize at a high rate even at low CO_2 concentrations (table 1).

As regards the beet plant is should be noted that the low phytocenoses are readily ventilated and hence fransfer of CO_2 from the air is rapid; moreover such plants can utilize CO_2 evolved as a result of soil respiration. In the presence of an extraordinarily active carboxylating enzyme RDP carboxylase, the rate of photosynthesis under sufficiently favorable carbon dioxide may be as high as that of maize.

These data indicate means for further optimization of the photosynthetic productivity of the two plants considered. One would consist in raising the activity of RDP carboxylase in mai-

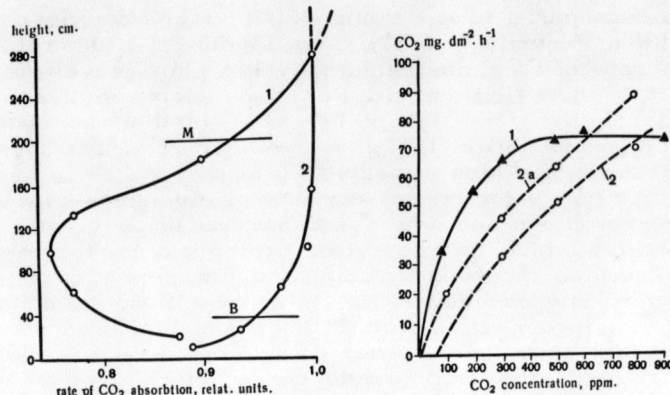

Fig. 6. Photosynthetic activity of plants dependent on CO_2 regime:
  A) intensity of CO_2-regime under and within crop stands of maize (1) and sugar beet (2); M and B-height of plants.
  B) CO_2 curves of photosynthesis of maize (1) and sugar beet leaves at 400×10^3 erg.ct.$^{-2}$sec.$^{-1}$ of linght intensity, 2 with 21% and 2a with 2% O O_2 in the air.

ze, since this activity probably limits the CO_2 saturation rate of photosynthesis. In beet positive results may possibly be attained by reducing the inhibiting effect of oxygen, i.e. by reducing the effect of light respiration via genetic and breeding work (17, 18).

There are therefore several ways of radically intensifying the energy and mass exchange of plant covers, leading ultimately to an increase of plant productivity.

One way consists in improving the morphology and growth function of the plants in such a manner as to permit them to form crops with as large as possible areas L_{opt}, but without impairing the turbulent CO_2 transfer rates. Formally, this problem is similar to that mentioned above in the discussion of utilization of light energy by crops.

However, the mechanisms of assimilation of light and CO_2 and the dependence of photosynthesis intensity on structure of the light beam and on the concentration and motion of CO_2 in air currents are different. The laws of transfer of light energy

Table 1

Activity of carboxylic enzymes in leaves of two Plants

	Activity of enzimes, $\mu m\ CO_2 \cdot dm^{-2}$	
	RDP-carboxylase	PEP-carboxylase
Sugar beet	80 - 127	6,5
Maize	20 - 30	60 - 70

and CO_2 are different in such complex and irregular, heterogeneous and variable media as phytocenoses. Therefore the principles and means of optimization of phytocenosis structures for better utilization of light and CO_2 should also be different although they are closely related components of a single process.

A more universal and important means of radically increasing the photosynthetic productivity of plants would be the improvement of the photosynthetic apparatus itself. This implies an increase of the slopes of the light and carbon dioxide curves and of the plateau levels or, better yet, "straightening out" of the curves; it also implies a lowering of the inhibiting effect of oxygen and a reduction of the light and carbon dioxide compensation point of photosynthesis etc.

Because of the complexity and multi-stage nature of the photosynthetic mechanism, because of the large number of active and regulatory systems determining its activity and finally because of the complex genetic nature of the organization of the photosynthetic apparatus (19), the problem of improving it at a genetic level is not a simple one.

However, there is a great diversity in the activity and nature of the photosynthetic function of higher plants under natural conditions (20). It suffices to mention the existence of C_3 and C_4 plants, photosynthesis of the Crassulaceae and woody plants.

The genetically fixed differences in photosynthesis of economically diferent species of a given family or even distinct differences of various varieties of cultural plants are also great. It may be mentioned that the genetically conditioned differences in the photosynthetic rate may vary between 80 or even 100 mg CO_2 $dm^{-2} \cdot hr^{-1}$ in the C_4 plants and 5-10 mg in woody and shady plants.

At present only in rare cases and only for the most productive plants does the photosynthetic productivity during the period of maximal photosynthesis and growth attain efficiencies of PAR utilization equal to 12-15% and these values are close to those which are the theoretical limit. More commonly the values are 6-10%

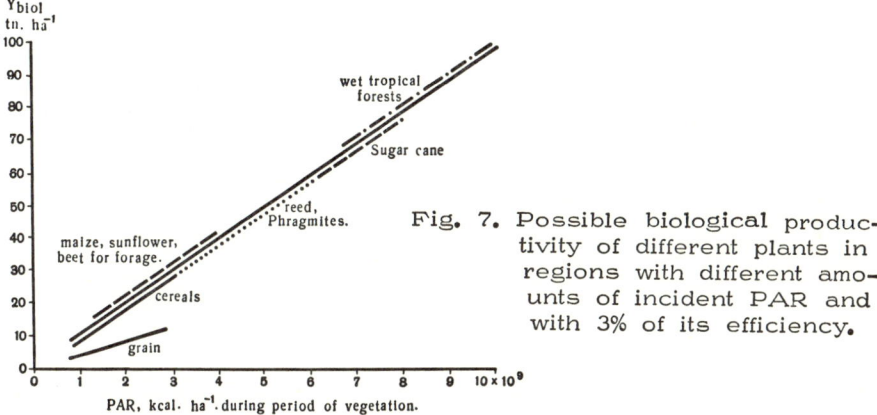

Fig. 7. Possible biological productivity of different plants in regions with different amounts of incident PAR and with 3% of its efficiency.

and the efficiencies for PAR over the vegetation period, at best attain 4-5% and usually are 1-3%.

If a PAR efficiency of 3% could be attained on the average for all cultural plants then the yields shown in the figure 7 would result for various plants growing at various latitudes. This would alter in a radical way the total productivity of agriculture.

Under favorable conditions PAR efficiencies of 4-5% should be feasible at present. Theoretically, efficiencies of 7-8% are possible, or even 10%, in intensive, but of limited size, shelterd ground with maximum control of radiation, thermal, water, mineral and carbon dioxide conditions, it being assumed that special plant varieties breeded for this purpose are cultivated.

REFERENCES

1. Theoretical foundations of the photosynthetic productivity (Russ.). Ed. Nichiporovich A.A. Publ. H. "Nauka", Moscow, (1972).
2. Photosynthesis of productive systems. Ed. Nichiporovich A.A. Israel Progr. Sc. Translation, Jerusalem, (1967).
3. Ničiporovich A.A. Physiological basis of plant productivity. Die Kulturpflanze, Beiheft. 6: 73-109, Gatersleben, DDR, (1970).
4. Photosynthesis and utilisation of solar energy. Level III Experiments, Japan. Nation. Subcomm. PP (IBP), Tokyo, (1968).
5. Solar energy and plant cover productivity (Russ.). Ed. Ross J.K Publ. Eston. Acad. Sc., Tartu, (1972).
6. Functioning of terrestrial ecosystems at the primary production level. Ed. Eckardt F.E., UNESCO, Paris, (1968).
7. Prediction and measurement of photosynthetic productivity. Ed. Šeltlik I., IBP/PP Techn. Meet., Třebon. Publ. PUDOC, Wageningen, (1970).
8. Plant photosynthetic production. Manual and methods. Ed. Šestac S., Čatsky J., Jarvis P., Junk N.V. Publishers, The Hague, (1971).
9. Photosynthesis and matter production of plants. Ed. Togari Y., Yokendo, Tokyo, (1971).
10. Physiological aspects of crop yield. Eds. Eastin J.D., and oth. Publ. Amer. Soc. Agron., Madison, Wisconsin, (1969).
11. Milthorpe F.L., Moorby J. An Introduction to crop physiology. Cambridge University Press., (1974).
12. Budagovsky A.I. Soil water evaporation (Russ.). Publ. H. "Nauka", Moscow, (1964).
13. Slatyer R.O. Plant-water relationships. Academic Press, L., N.Y., (1967).
14. Alpatjev A.M. Water turnover in nature and its transformation (Russ.). Hydrometeoisdat, Leningrad, (1969).
15. Problem of the efficiency of photosynthesis (Russ). Ed. Ross J.K Publ. Eston. Acad. Sc., Tartu, (1969).
16. Nichiporovich A.A., Chmora S.N., Slobodskaya G.A., Avdeeva T.A. Type of photosynthetic gas exchange of leaves of beet and maize in relation to the CO_2 regimes in their crop stands. Plant Science Letters, I, (1973), 399-403.
17. Photosynthesis and photorespiration. Eds. Hatch M.D., Osmond C.B., Slatyer R.O., John Wiley Inc., N.Y., L., Sydney, (1971).

18. Zelitch I. Photosynthesis, photorespiration and plant productivity. Academic Press, L., N.Y., (1971).
19. Net carbon dioxide assimilation in higher plants. Ed. Cl.C. Black. Proceed. Sympos. Univers. South Alabama, (1972).
20. Genetic aspects of photosynthesis, Ed. Nasyrov J.S. Publ. H. "Donish", Dushanbe, (1971).

WATER UPTAKE BY VEGETATION[1]

W. R. GARDNER, W. A. JURY, and J. KNIGHT[2]

University of Wisconsin, Madison, Wisconsin, USA

Current mathematical models of water uptake from soil by a plant root system are discussed. These models suffer from serious deficiencies in that they do not describe accurately such features of the system as the water extraction from non-uniform root systems and the diurnal variation of water potential of the various plant parts. Lack of a verified theory of water movement through the plant is considered to be a major source of error. Until new experiments provide data for developments of better mathematical equations, semi-empirical procedures for relating plant growth and transpiration to soil water will still be necessary.

LIST OF SYMBOLS

P	Precipitation, cm
I	Irrigation, cm
ET	Evapotranspiration, cm
D_R, D_S	Soil water drainage fluxes, cm day^{-1}
PET	Potential evapotranspiration, cm day^{-1}
LAI	Leaf area index (dimensionless)
T	Transpiration rate, cm day^{-1}
E_o	Potential evapotranspiration, cm day^{-1}
R_p	Plant resistance, bar days cm^{-1} or days
L	Effective root length factor, cm^{-2}
r_s	Stomatal resistance, sec cm^{-1}
k	Unsaturated soil conductivity, cm day^{-1}
g	Crop factor (dimensionless)
l_R, l_S	Thickness of soil profile layers, cm
z	Depth, cm
α	Plant conductance factor, cm bar^{-1}day^{-1}

---

[1] Published with the permission of the Director of Research, Wisconsin Agricultural Experiment Station.

[2] On leave from Pye Laboratory, CSIRO, Canberra, Australia.

Ψ_L Leaf water potential, bars or cm
Ψ_R Root water potential, bars or cm
Ψ_s Soil water potential, bars or cm
θ_c Threshold root water content for sub-potential evapotranspiration, $cm^3 cm^{-3}$
θ_R, θ_s Water content of various soil compartments, $cm^3 cm^{-3}$

It has long been known that the state of water in the plant depends upon the energy level of water in the soil. The exact influence of soil water upon transpiration and plant growth became the subject of a vigorous controversy which pervaded the field of soil-plant-water relations for many years. The resolution of the argument lies partly in the plant; particular response to its internal water status or degree of hydration. Since the plant is closely coupled to the soil and only somewhat less coupled to the atmosphere, the processes of transport of water from the soil into the plant and through the plant to the atmosphere must also be understood before the plant response to its environment will be clear. The plant response enigma mainly is for the biologists to decipher. The transport problem will take the best thinking of physical and biological scientists.

THE SOIL-PLANT-ATMOSPHERE CONTINUUM

It is assumed from the outset that the transport of water from the soil pores into the plant roots, through the xylem tissue in the roots, stems, and petioles to the substomatal cavities in the leaves where the change of state from liquid to vapor occurs is along a gradient of ever decreasing potential energy. It is also assumed that the state of water in the plant is best characterized by its degree of hydration or turgor and by the identifiable components of the water potential. We will follow, roughly, the notation of Cowan and Milthorpe (1) who have provided a much more comprehensive review of plant water relations than is possible here.

Conceptually it is deceptively simple to write equations for the flow of water through the various parts of the system:

$$T = f(E_o, r_s, \Psi_L) \qquad \text{atmosphere} \qquad (1)$$

$$= \frac{\Psi_L - \Psi_R}{R_p} \qquad \text{plant} \qquad (2)$$

$$= (\Psi_R - \Psi_s) k\, L \qquad \text{soil} \qquad (3)$$

In equation (1) T is the transpiration rate from the plant, and is assumed to be some function of the potential evaporation E_O and the stomatal resistance r_s. This relationship embodies all canopy factors and meteorological factors pertinent to the transpiration process. The effect of the water potential is supposed to enter directly through the stomatal resistance which, in turn, is somehow related to the leaf water potential Ψ_L. Equation (2) follows a time-honored tradition (van den Honert, 2) and assumes that flow through the plant is proportional to the differences between leaf and root water potential. Equation (3) describes flow of water through the soil to the plant root. This represents the usual equation for the flow of water in unsaturated soil with the added factor L, which takes into account the root length per unit volume of soil and any other appropriate geometrical factors. Finally, Ψ_s is given by the unsaturated flow equation solved for the appropriate initial and boundary conditions. The transpiration function given in equations (1-3) can be thought of as a sink term in this flow equation.

All of the models of the soil-plant-water system represent variations on equations (1-3). Philip (3), who made the initial calculation, and Gardner (4) were first concerned with the soil part of the system and solved equation (3) for flow of water to a single cylindrical root. This had not been done earlier because insufficient data on the unsaturated conductivity were available to make the calculation meaningful. It soon became apparent that, unless the estimates of the extent of the root system were greatly awry, one could not account for the observed drop in water potential between the soil and the plant leaf within the soil alone. This aspect of the problem has been of particular concern to Newman (5,6). This led Gardner and Ehlig (7) to include the plant flow system as described by the plant term in equation (2). In order to extend the equations to non-uniform root systems, which is more often the case in nature, they assumed that the resistance to water flow within the root system itself was negligible (Gardner, 8). They also addressed themselves to the problem of how to average the soil water potential over an entire non-uniform root system. The root distribution parameter L was evaluated from water content distributions and Ψ_R was eliminated from the equation by requiring that the integrated water uptake equal the transpiration rate. Since one could adjust L at will it was not difficult to obtain quite satisfying agreement between theory and experiment. Cowan (9) tidied up the equations by making the conductivity a more rigorous function of soil water potential and by including a diurnal periodicity in the transpiration.

In more recent times a number of workers have modified the equations, seemingly more in the interest of simplification than accuracy. Whisler et al. (10)

examining the steady state upward movement of water into
a root zone, lumped all the geometrical factors into one
parameter leaving only the conductivity and the two soil
potentials Ψ_R and Ψ_s. They appear to have omitted the
gravitational component of the potential in the roots
which led to some rather interesting circulation patterns.
Molz (11) preferred to work with water content exclusively rather than water potential. This can be accomplished
by replacing the potentials in equation (1) by the
appropriate soil water content and the conductivity
by the soil water diffusivity. However, then he
proceeded to omit the water content gradient. It, in
effect, was lumped into an "effective root density."
Nimah and Hanks (12) left all the terms more or less as
written in equations (2) and (3), but included in the
root water potential in equation (2) the osmotic component of the potential. They coped with the problem of
flow resistance within the root system itself by assuming
that a pressure potential drop exceeding the gravitational potential by five per cent was needed to move
water upwards through the roots to the crown. This
eliminated the necessity of specifying the water potential at the root soil interface. The water potential at
the crown was adjusted to give the measured transpiration
rate until it reached a value, such as the permanent
wilting point, below which it was assumed it would not go.
From then on the transpiration rate decreased as the
potential gradient was reduced.

Other models might be noted here. Visser (13) wrote
an explicit relation for the atmospheric term in equation
(1). He assumed that

$$T = gE_o - \alpha \Psi_R \tag{4}$$

He calculated Ψ_R from equation (3) in much the same way
as Philip (3) or Gardner and Ehlig (7). Lambert and
Penning deVries (14) have a very complicated stomatal
model but for equations (1) and (2) root-soil part
differs mainly in that steady state is not assumed.
Goldstein and Mankin (15) describe a similar model
except that an empirical expression for the stomatal
resistance is used, the root resistance is assumed to be
proportional to the soil resistance and inversely proportional to the leaf area index. The latter assumption
allows one to estimate the root extent from the more
easily measured leaf area on the presumption that the
top/root ratio remains constant.

THE TENUOUS ASSUMPTIONS

Remarkably few tests of the equations for the flow
of water from the soil to plant leaves have been made.

The number of adjustable parameters coupled with the almost overwhelming effect of the soil water potential have made it possible to produce reasonable water extraction patterns from computer simulation and to fit theoretical curves to experimental data to the satisfaction of many. The harsh fact is, however, that the agreement between theory and experiment is not good. One cannot seem to find an effective root distribution which gives the correct extraction pattern over a long period of time. Almost invariably one has to assume that the effective root activity decreases with time in the upper part of the root zone and increases in the lower part. If it is necessary to introduce into the system a posteriori an empirical root activity then little is gained by separating out the soil water conductivity or diffusivity, and one might as well abandon the dynamic model altogether and stick to the soil water potential and a static model.

There are some basic reasons why one might expect the simple model described in equations (2) and (3) to fail. It is recognized by most workers that one cannot write an Ohm's law-type equation for the loss of water from the plant leaf (equation 1) unless one is very careful about how he does it since the phase change must somehow be accomodated. Thus, we don't really do this anymore. Instead we look for some rule which will allow us to deduce the stomatal resistance, which controls the vapor transport, if we know the leaf water potential. We are still wrestling with fuzzy empiricisms here, and it is by no means clear how we will finally bridge this chasm.

Less attention has been given to the fact that we also have a serious problem at the other end of the plant. In the soil we take the matric potential gradient as the driving force and we calculate our way up to the boundary between the plant root and the soil. Within the xylem in the root we assume laminar flow due to a pressure gradient. To assume that the pressure within the xylem can be calculated directly from the matric potential outside the root is to pretend that there is no semi-permeable membrane across which the solutes (or at least the cations) must be transported largely by active processes. How much of the potential drop between the soil and the plant leaf occurs at this membrane we do not know, but it may be significant (Dalton, 16). It very probably depends upon metabolic processes. The assumption that it can be represented by a simple single resistance value is not supported by the data.

In addition to membrane effects there may be appreciable longitudinal resistance within the roots, particularly those near the bottom of a root system. This, coupled with nonlinear behavior of the membrane transport, might account for the "moving-sink" type of water extraction patterns which we often observe. Strangely enough, the availability of high speed computers has not encouraged anyone to include these effects in equations (2) and (3). Instead the trend has been towards simplification of the equations.

It seems probable to us that there is little to be gained by trying to fiddle with the equations as they presently exist. It is almost certain that, given enough root parameters, one can fit extraction patterns. This, in itself, may be useful but it does not go far enough. If one is to predict plant response to its environment one must also predict the water potential and the degree of hydration of the various plant parts. Perusal of the literature reveals an almost universal faith that the daytime water potential of the leaf is the all important parameter. We do not know this to be the case. The extent of recovery of turgor at night could be very important and a steady-state model may even be misleading. Furthermore, if one is interested in the production of grain, or tubers, or fruit, the potential of the leaf may not be the correct parameter at all.

A major flaw in the present picture of water transport through the soil-plant-atmosphere continuum may be the assumption that it is a continuum. This tends to obscure the fact that we could understand very well the transport in each part of the system but not yet have them coupled correctly. Much high-powered computing has gone into this problem, but virtually no progress has been made for ten years. This suggests that the experimentalist must come to our aid with imaginative and reproducible experiments which will allow us to join the pieces of the continuum more effectively.

THE WAY OUT FOR NOW

The problems of water and plant growth are far too important to leave until we have worked out our conceptual difficulties. Thus we must do what one always does when he is in theoretical difficulties and go back to the empiricisms. This means that we must forego complete quantitative understanding of the parts of the system and accept it for what it is -- a very black box. This means that one abandons a universal equation which holds for all soil-plant combinations and settle for less general expressions which are specific to a limited number of situations. In our own most recent calculations we have abandoned even such a basic parameter as the soil water potential and related everything to the water content of the plant root zone. We divide the root zone into two regions; an upper zone with a well-developed root system, and a lower zone with sparser root density tapering into the soil zone beneath. These two zones are coupled to the deep soil regime, which may be represented at any

desired level of sophistication by the established models of soil water flow.
These concepts may be represented mathematically by writing a mass balance for the root zone and subroot zone slabs as follows:

ROOT ZONE $-l_R < z < 0$

$$l_R(d\theta_R/dt) = P + I - ET - D_R \qquad (5)$$

SUBROOT ZONE $-(l_R + l_s) < z < -l_R$

$$l_s(d\theta_s/dt) = D_R - D_s \qquad (6)$$

where

P+I = precipitation and/or irrigation rate
ET = evapotranspiration rate
D_R = rate of drainage out of the root zone
D_s = rate of drainage out of the subroot zone
l_R = thickness of root zone
l_s = thickness of subroot zone
θ_R = average water content of the entire root zone
θ_s = average water content of the entire subroot zone

In its present form this system is not soluble. Each of the variables on the right-hand side of the equations must be represented as functions of the average water contents θ_R and θ_s or as input functions of the time in order for the equations to be integrable. In modeling these variables we are guided by two requirements; all possible inputs to the system must be experimentally measurable, and each representation should be more general than the application it is put to. These models will now be discussed in turn.

Drainage

There is considerable evidence that drainage out of a soil layer may be quantitatively related to the water stored in that layer (Black et al., 17; Miller and Aarstad, 18). That this correlation is intimately related to the soil hydraulic conductivity may be seen from the observation that after water has infiltrated to a certain depth below the surface, gradients of moisture potential tend to die out and the resulting downward flow may be considered essentially gravitational (Philip, 19).

Use of a $D(\theta)$ relationship is conceptually preferable to the traditional but nebulous concept of a field capacity, defined as the soil water content where drainage becomes negligible. Furthermore, it has been shown that significant amounts of percolation can occur after the so-called field capacity has been reached, enough to result in a serious error in partitioning the water budget.

The nature of the relationship between drainage and soil water storage is a soil property and should be measured for a given system. In the absence of drainage lysimeters, this could be accomplished by irrigating a plot up to saturation, covering it to prevent evaporation, and monitoring subsequent soil water changes with time.

Evapotranspiration

Soil evaporation and plant transpiration are processes that will be limited either by the amount of external energy available to vaporize the liquid soil and plant water at the sites of evaporation, or by the ability of the soil and plant to transport water to the soil-atmosphere and plant-atmosphere interface.

Energy-limited or potential evapotranspiration is a process that has been modeled with varying degrees of success over the years by Penman (20) and others, and poses considerable difficulties on an hourly or even daily basis, requiring careful aerodynamic measurements as well as radiation and temperature determinations. As the time resolution is decreased, however, the problem becomes more tractable, as diurnal fluctuations and local extremes tend to average out. There is some promise that ET for a fully covered crop or wet surface over several day periods might be represented reasonably well by an empirical correlation that is a function of the incoming net radiation and the mean daily screen temperature, with the same function representing a number of different crops, soils and climates (Priestley and Taylor, 21). Furthermore, since for a given crop and climate the correlation between solar and net radiation is quite high, one may be able to represent potential evapotranspiration in terms of mean daily screen temperature and solar radiation, two relatively accessible measurements.

Non-potential or transport-limited evaporation and transpiration pose more difficulties for a number of reasons. The barrier for soil evaporation is primarily confined to the top few centimeters of dry soil, whereas plant transpiration is ultimately controlled by the stomata, which will begin to close when the integrated water uptake over the entire root zone cannot match the external transpirational demand. Regardless of whether the primary transport barrier between soil and leaf lies in the soil or in the plant the ultimate effect of a decreasing soil water content is to decrease the transpiration rate when the average soil water potential reaches a certain value. We may represent this process abstractly as

$$T = f_1(\text{external meteorological conditions}) \quad \theta_R > \theta_c$$
$$= f_2(\theta_R) \quad \theta_R < \theta_c \quad (7)$$

where θ_c is a threshold average root zone water content below which transport of water to the plant leaves limits the transpiration process. Laboratory experiments under constant external conditions have confirmed this postulate and have also found a reasonably linear decrease of transpiration rate with decreasing water content below the threshold value (Rawlins et al., 22), and found that θ_c is quite distinct and repeatedly obtained. In the field it is more likely to be indistinct and smeared out, but may be approximately determined by examining soil water content in the root zone for a crop entering a stress period, and simultaneously looking at stomatal resistance. Data such as that from Millar et al. (23) in Figure 1 would suffice.
given water, radiation, growth of cover and air temperature inputs as a function of time, and providing we know the drainage-water content relation, the soil moisture potential and hydraulic conductivity, the water table depth, and the average water content at which the sto-

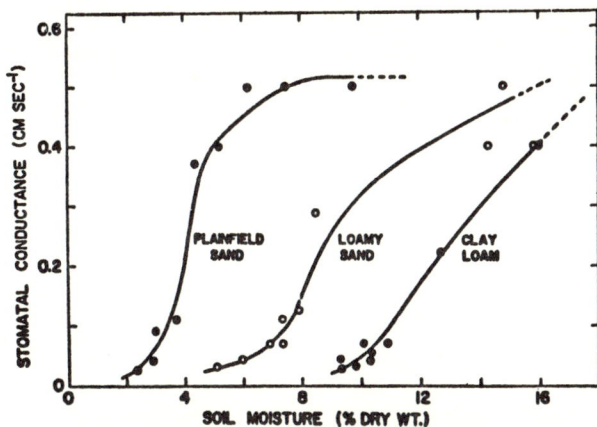

Figure 1. Measured stomatal conductance for seed onions plotted as a function of soil water content. These data were obtained in the growth chamber and the measured growth rates were in good agreement with the stomatal conductances (Millar et al., 23).

Handling soil evaporation under a canopy is more difficult, but for purposes of the zone scheme one might neglect most of the complications and consider it to be part of the total evapotranspiration until soil water begins limiting the transpiration, and negligible below this point. The whole problem of the interaction between soil evaporation and transpiration must be somehow resolved experimentally before this problem can be treated in greater detail. Until that time, we propose consideration of a representation that does not distinguish between them at full crop cover.

During the growth period when the soil surface is not completely shielded this approach will need some modification. Ritchie and Burnett (24), for example, use

$$PET(LAI) = (0.7\sqrt{LAI} - 0.2) PET_{FULL\ COVER} \quad 0.1 < LAI < 2.7 \quad (8)$$

to represent cotton and sorghum ET in Texas when soil evaporation is small.

Having modeled all the inputs to the system of equations we can now generate estimates of soil water content, evapotranspiration, and drainage provided we are mates begin to close. All of this information is readily obtainable experimentally.

It should be re-emphasized that these submodels are not universal. Each climate and soil type will have special problems to negotiate that would not be tractable by an unaltered application of the above model. For example, a finer textured soil with a shallow water table would have need of a representation for capillary rise, which could conceivably be incorporated into the model of the subroot zone. The onus is on the modeler to calibrate his region, consistent with the practical limitations of experimental information.

Plant Growth

The above scheme represents a procedure for calculating the time course of the soil water content in the soil root zone. From this the soil water potential can equally well be calculated. Approximations which are less rigorous than exact solutions of the flow equation are justified on the basis that the plant-water equations are not that good. As insight is gained into such problems of separation of evaporation and transpiration, thermal effects, etc., the calculation can be improved. Prediction or calculation of plant response once the soil water content of potential is known must at this time rest perforce upon a correlation between the growth response in question and the soil water. It turned out for the onions from which the data in Figure 1 were obtained that the rate of dry matter production was directly proportional to stomatal conductance. Figure 1 could thus be used directly for calculating growth.

Figure 2 shows the kind of data which one can now find in the literature relating plant growth to soil water potential. These particular data represent the rate of diameter growth of red pine (<u>Pinus resinosa</u>, <u>Ait</u>) in weeded and unweeded plots as a function of soil water potential (25). Whereas the data in Figure 1 correlate daytime stomatal conductance with average root zone soil water content those in Figure 2 correlate growth rate with soil water potential. In each case these correlations get one across the awkward spots which equation (2) does not come to grips with.

CONCLUSIONS

It is challenging to attempt a deterministic model of the transport of water from soil to plant to atmosphere and to calculate the plant response from environmental stimuli alone. Such a model would be very valuable. As of now our basic understanding of the flow processes involved is not adequate for the task. We need more and better measurements within the plant-soil

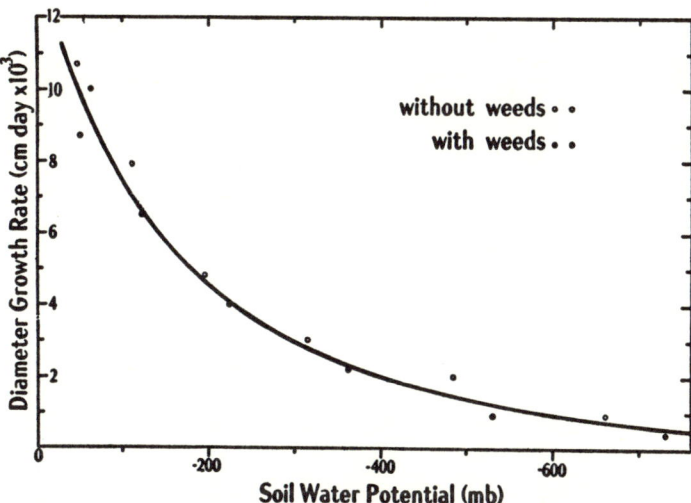

Figure 2. Relationship between diameter growth rate and soil water potential at 30 cm depth beneath a tree on each plot during the dry period in August 1969. From Lambert et al. (25).

system and we need a far better understanding of what these measurements mean.

For the immediate future we see some possibility of describing the soil water environment of the plant in a quantitative way by dealing with gross averages in the root zone. Future experiments and theoretical work should improve these calculations. When we introduce the plant itself into the system our theories fail us, and we must use whatever empiricisms we can obtain. These can assuredly be improved even with present measuring techniques, but we cannot calculate our way into the plant without fresh insight into the flow processes involved.

REFERENCES

(1) Cowan, I. R. and F. I. Milthorpe. 1968. Plant factors influencing the water status of plant tissues. In T. T. Kozlowski [ed.] Water Deficits and Plant Growth, Academic Press, N.Y.

(2) van den Honert, T. H. 1948. Water transport in plants as a catenary process. Disc. Faraday Society 3: 146-153.

(3) Philip, J. R. 1969. The theory of infiltration. Advances in Hydroscience 5: 215-305.

(4) Gardner, W. R. 1960. Some dynamic aspects of water availability to plants. Soil Science 89: 63-67.

(5) Newman, E. I. 1969. Resistance to water flow in soil and plants. I. Soil resistance in relation to amounts of root: Theoretical estimates. Journal of Applied Ecology 6: 1-12.

(6) Newman, E. I. 1969. Resistance to water flow in soil and plants. II. A review of experimental evidence on the rhizosphere resistance. Journal of Applied Ecology 6: 261-272.

(7) Gardner, W. R. and C. F. Ehlig. 1963. The influence of soil water on transpiration by plants. Journal of Geophysical Research 68: 5719.

(8) Gardner, W. R. 1964. Relation of root distribution to water uptake and availability. Agronomy Journal 56: 41-45.

(9) Cowan, I. R. 1965. Transport of water in the soil-plant-atmosphere continuum. Journal of Applied Ecology 2: 221-239.

(10) Whisler, F. D., A. Klute and R. J. Millington. 1968. Analysis of steady state evapotranspiration from a soil column. Soil Science Society of America Proceedings 32: 167-174.

(11) Molz, Fred J. 1971. Interaction of water uptake and root distribution. Agronomy Journal 63: 608-610.

(12) Nimah, M. N. and R. J. Hanks. 1973. Model for estimating soil water, plant and atmospheric interrelations. Soil Science Society of America Proceedings 37: 522-532.

(13) Visser, W. C. 1965. A method of determining evapotranspiration in soil monoliths. In F. E. Eckardt [ed.] Methodology of Plant Ecophysiology, Proceedings of the Montpellier Symposium, UNESCO.

(14) Lambert, J. R. and F. W. T. Penning deVries. 1973. Dynamics of water in the soil-plant-atmosphere continuum: A model named troika. In A. Hadas et al. [eds.] Physical Aspects of Soil Water and Salts in Ecosystems, Springer-Verlag.

(15) Goldstein, R. A. and J. B. Mankin. 1972. PROSPER: A model of atmosphere-soil-plant water flow, pp. 1176-1181. In Proceedings Summer Computer Simulation Conference.

(16) Dalton, F. N. 1972. A Physical-Mathematical Model Describing the Simultaneous Transport of Water and Solutes Across Root Membranes. Unpublished Ph.D. Thesis, University of Wisconsin. 68 p.

(17) Black, T. A., W. R. Gardner, and G. W. Thurtell. 1969. The prediction of evaporation, drainage, and soil water storage for a bare soil. Soil Science Society of America Proceedings 33: 655-660.

(18) Miller, D. E. and J. S. Aarstad. 1971. Available water as related to evapotranspiration rates and deep drainage. Soil Science Society of America Proceedings 35: 131-134.

(19) Philip, J. R. 1957. The physical principles of water movement during the irrigation cycle. Proceedings 3rd International Congress on Irrigation and Drainage 8: 125-128, 154.

(20) Penman, H. L. 1948. Evaporation in Nature. Proc. Phys. 11: 366-388.

(21) Priestley, C. H. B. and R. J. Taylor. 1972. On the assessment of surface heat flux and evaporation using large-scale parameters. Monthly Weather Review 100: 81-92.

(22) Rawlins, S. L., W. R. Gardner and F. N. Dalton. 1968. In situ measurement of soil and plant leaf water potential. Soil Science Society of America Proceedings 32: 468-470.

(23) Millar, A. A., W. R. Gardner and S. M. Goltz. 1971. Internal water status and water transport in seed onion plants. Agronomy Journal 63: 779-784.

(24) Ritchie, J. T. and Earl Burnett. 1971. Dryland evaporative flux in a subhumid climate: II. Plant influences. Agronomy Journal 63: 56-62.

(25) Lambert, J. L., J. R. Boyle and W. R. Gardner. 1971. The growth response of a young pine plantation to weed removal. Canadian Journal of Forest Research 2: 152-159.

31

A NUMERICAL MODEL FOR ESTIMATING THE MODIFICATION OF HEAT BUDGET INTRODUCED BY HEDGES

JEAN-PIERRE CHIAPALE

*INRA Station de Bioclimatologie d'Avignon—Montfavet,
84140 Montfavet (France)*

ABSTRACT

The proposed numerical model allows to compute the surface temperature of a plant canopy for each point of the protected zone, taking into account modification of direct solar and infrared radiation, latent and sensible heat fluxes. An increase of surface temperature is found during the day for the protected zone compared to the unprotected, while a diminution of the same order of magnitude is observed during the night. This computed accentuation of continental climatological characters is in good agreement with results of other studies concerning land roughness influence on regional climatology.

LIST OF SYMBOLS

T_s, T_H surface temperature of the plant canopy and of the hedgerow (°C)
T_a, T_m air and soil temperatures (°C)
ϕ_s, ϕ_L, ϕ_{ss} sensible heat, latent heat and soil heat fluxes (W.m^{-2})
G_1, G_2 direct solar radiation on a vertical and horizontal surface (W.m^{-2})
D_f, R_a diffuse sky and long-wave atmospheric radiations (W.m^{-2})
R_n, $G\downarrow$, $G\uparrow$ net radiation, incident global and reflected global radiations (W.m^{-2})
H, D height of hedgerow, alley width (m)
a_1, a_2 albedos of the hedge and of the plant canopy
F_i different angle factor
σ Stefan-Boltzmann constant (W.m^{-2} °K^{-4})
k Von Karmann constant (= 0,4)
u_*, z_0, z friction velocity, roughness length, height above the ground
ρ, C_p density and specific heat of air at constant pressure
h, I_0 sun elevation and solar constant (= 1 400 W.m^{-2})
L latent heat of vaporization of water
K heat capacity of the soil
$F(z)$ specific humidity at height z
A amplitude of air temperature

INTRODUCTION

Climatic factors such as solar radiation, temperature, humidity and wind-speed interact with the landscape structure (particularly hedgerow systems or fruit orchards) and the crop inside hedgerows, thus producing a distinctive climate. It is necessary to understand these interactions and the effects of hedgerow systems in order to provide optimum conditions for crop growth and development. An energy balance approach is developed here on a general basis which allows prediction of the climate of plant canopy between two hedges from the knowledge of external climatic factors and given properties of the structure and of the crop.

During the night, hedges act as radiative shelters : emitted infrared radiation is substituted to a fraction of atmospheric radiation in a varying proportion according to the distance to obstacles. During the day, this effect is coupled with the modification of direct solar radiation. One part, corresponding to the shade, is suppressed, while an other part is reflected by hedges.

Variations in turbulent exchanges coefficients are superposed to these modifications of radiative components. The creation of isolated cells near the ground results in global reduction of heat and mass transfers between the surface and the atmosphere.

PRINCIPLES OF NUMERICAL MODEL

The numerical model of hedgerow system is constituted by network of parallel and infinite elements, each being continuous in the direction of the hedge. The hedgerows are considered as non-transmitting solid object, with a reflectance coefficient a_1, and are separated from adjacent hedgerows by a vegetative cover with a reflectance coefficient a_2 (cf. figure 1).

Direct light calculations

Computer evaluation of direct light is made from algorithm of USHER [1], according to day of the year and hour during the day. The choosen latitude and longitude in this example are respectively 43°42 N and 4°35 E.

The direct solar radiation on the soil or on the vertical surface of hedgerow is attenuated by the atmospheric absorption. The atmospheric transmission is taken into account for the calculation : (JACKSON and PALMER [2])

$$G_2 = I_0 \, (0,796)^m \sin h$$

where m is air mass. A general relationship for m is given by ROBINSON [3], with allowance for the curvature of the earth.

One part of the direct solar radiation is reflected by the hedgerow. The following assumptions are made : the reflected radiation is diffuse and independant of the wavelength and the Lambert's cosine law is supposed valid.

JAKOB [4] gives the angle factor for one surface of differential size and the other of arbitrary profile, whose generating lines are parallel to the first surface. The schematic diagrams in figure 1 present the nomenclature used in the different calculations.

A Numerical Model for Estimating the Modification of Heat Budget

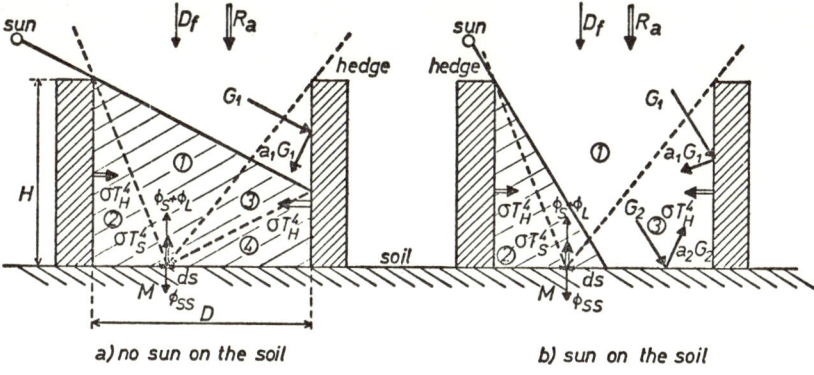

a) no sun on the soil   b) sun on the soil

PROTECTED ZONE

Figure 1 - Model hedgerow system used in study.

a) If the sun does not reach the soil between the hedgerows, the direct solar radiation reflected, at the point M, is equal to : $F_3 \cdot a_1 \cdot G_1$
F_3 being the angle factor for ds and the sunny surface of hedgerow

b) If the sun reaches the soil, we have :

$$\delta (1-a_2) \cdot G_2 + F_3 \cdot a_1 \cdot G_1$$

where δ is equal to 0 or 1 if this point is shaded by the hedge or not. F_3 is the angle factor for the surface ds and the right hedgerow.

Diffuse sky radiation and long-wave atmospheric radiation

The whole diffuse sky radiation and atmospheric radiation do not reach the soil between two hedges. If the sky is supposed to be an uniform luminous source of diffuse light, we have, for the sum of these two radiations at the point M :

$$F_1 \cdot (Df + Ra)$$

F_1 being the angle factor for ds and the top.

For this procedure we have supposed that the ratio of diffuse sky radiation on direct solar radiation is constant and equal to 0,16, and that atmospheric radiation Ra is equal to σT^4, where T is the apparent temperature of atmospheric emission equal to 243° K.

Infrared radiation of hedgerow

Radiation of hedge can be considered as the total black-body radiation of the surface of hedgerow at T_H surface temperature.

Emitted infrared radiation is substituted to a fraction of atmospheric radiation. At the point M, we can write, if we assume that this surface temperature is constant :

a) $\sigma T_H^4 \cdot (F_2+F_3+F_4)$ for no sun on the soil
b) $\sigma T_H^4 \cdot (F_2+F_3)$ for sun on the soil.

We suppose also that the surface temperature of hedgerow is given by $T_H = T_a + 5°$. This hypothesis seems to be valid in most cases.

Net Radiation

It will be assumed that all surfaces are perfectly black radiators. The net radiation equation for the protected zone is :

a) $Rn = F_3 a_1 G_1 + F_1(Df+Ra) + \sigma T_H^4 (F_2+F_3+F_4) - \sigma T_s^4$

b) $Rn = \delta(1-a_2)G_2 + F_3 a_1 G_1 + F_1(Df+Ra) + \sigma T_H^4 (F_2+F_3) - \sigma T_s^4$

For the unprotected zone, we have (cf. figure 2) for net radiation :

$$Rn = (1 - a_2)G_2 + Df + Ra - \sigma T_s^4$$

This net radiation Rn is divided in three parts : evapotranspiration, sensible heat flux and heat storage in the soil. Assuming that only non divergent vertical fluxes occur over the plant canopy, and taking the fluxes towards the surface of canopy as positive and those away from it as negative, the energy balance equation is :

$$R_n + \phi_S + \phi_L + \phi_{ss} = 0$$

Sensible heat and latent heat fluxes

In order to calculate the latent heat and sensible heat fluxes using the aerodynamical method, data on the vertical profiles of air temperature, specific humidity and horizontal component of wind speed are needed. Under neutral conditions, if vertical wind velocity, temperature and humidity profiles are supposed to be logarithmic, latent and sensible heat fluxes can be computed by using corresponding exchange coefficients.

The sensible heat flux is equal to :

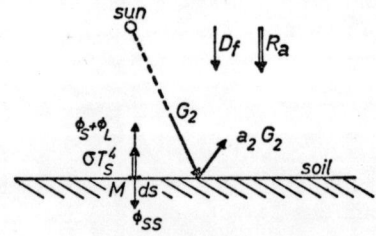

Figure 2 - Energy balance for unprotected zone

$$\phi_s = k \rho C_p u_* \frac{(T_a - T_s)}{\text{Log } z/z_0}$$

and the latent heat fluxes to :

$$\phi_L = L \rho k u_* \frac{F(z) - F_0}{\text{Log } z/z_0}$$

We can assume that near the surface of plant canopy the air is at the saturation vapour pressure, i.e. the plant is at the potential evapotranspiration.

The velocity profile in semi-logarithmic representation shows two different straight lines corresponding at two values of friction velocity u_* /5_/ /6_/. In the procedure, different values of u_* are introduced for the unprotected and the protected zones, deduced from experimental results. This friction velocity is supposed to be constant for various distance from the hedgerow. This last hypothesis is a rough one, because u_* depends on the rate of development of the boundary layer between the hedgerows.

The mean air temperature T_a is not constant along the day. So a daily variation of this temperature was deduced from the results of HALLAIRE /7_/ by mean of an approximation with Tchebycheff and sine-cosine polynomials. An annual variation of mean temperature is also introduced. Then :

$$\overline{T_a} = 14 + 8 \cos 2\pi \frac{(J - 195)}{365}$$ where J is the day of the year ; and

for the amplitude of air temperature

$$A = 8,5 + 2,5 \cos 2\pi \frac{(J - 195)}{365}$$

For air temperature, we have : $T_a = \overline{T_a} - \frac{A}{2} + \rho A$, where ρ is a number between zero and one (see /7_/.

Heat flux on the soil

For the heat flux into the soil, we have the relation :

$$\phi_{ss} = K \frac{(T_m - T_s)}{z}$$

i.e. the heat flux is supposed a stationnary phenomenon, where K is the heat capacity of the soil and z the depth where soil temperature T_m is taken. Heat flux into the soil is a small part of the energy budget.

The procedure consists in solving the energy balance equation for protected and unprotected zones, where surface temperature T_s of plant canopy is the unknown variable. This equation which can be written

$$a T_s^4 + b T_s + c = 0 \qquad \text{(a, b, c are constant coefficients)}$$

is solved by Newton method of bi-partition. The calculations are made by mean of a FOCAL numerical program on a PDP 8/E minicomputer. Copies of this procedure can be obtained from the author.

RESULTS

In the following numerical example, we have choosen for height of hedgerow H = 10 m and for width D = 80 m (ratio $\frac{D}{H}$ = 8). This ratio is currently observed in the design of hedgerows in Brittany (West France).

Distribution of surface temperature

In figure 3, we present the distribution of surface temperature of plant canopy between hedgerows for different values of local time. Calculations are made for east-west orientation of hedgerows, and for the 150th day of the year. The straight line represents the surface temperature of the unprotected zone.

During the night and after the sunrise, the surface temperature at the center of alley is lower than that of the unprotected zone (temperature difference - 0,8° C). We can also see in figure 3 that near the hedgerow, the surface temperature is higher than that near the center (difference of about 2° C), because emitted infrared radiation of hedgerow is important.

During the day, we have the same result in temperature distribution except for the part shaded by the hedge. For the choosen day, the shadow during first

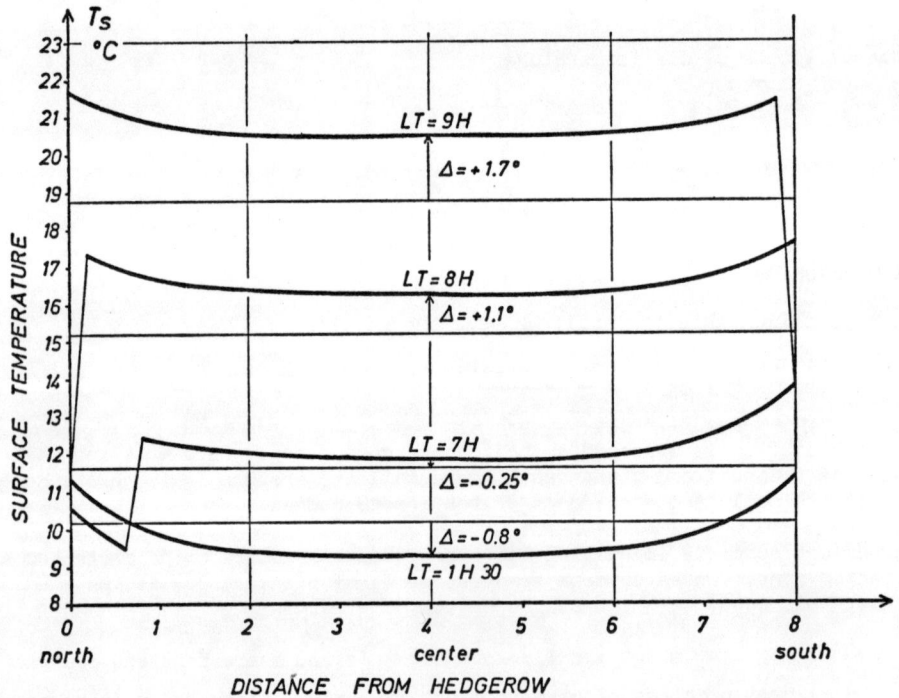

Figure 3. Surface temperature vs. distance to hedgerow
LT = local time   Δ = temperature difference

A Numerical Model for Estimating the Modification of Heat Budget

hours is on the south side of hedgerow and after, on the north side (azimuth sun at sunrise is greater than 90° from south).

The difference temperature between the center of alley and the unprotected zone increases during the day. At local time of 9h00 this difference is equal to + 1,7° C.

Daily variation of surface temperature difference between protected and unprotected zones

Figure 4 shows the difference of surface temperature versus local time, between the center of alley and the unprotected zone for different days of the year. We can see that during the night the center of protected zone is colder than unprotected zone. During the whole year this phenomenon is observed and the temperature difference between the two zones is nearly constant (magnitude of - 0,8° C).

The diurnal variation of the temperature difference is very sharp after the sunrise and reaches a maximum value for the maximum sun elevation. The highest value of difference depends on the period of the year. For example :

day number	temperature difference
1	+ 1,2
50	+ 2,1
100	+ 2,6
150	+ 2,7

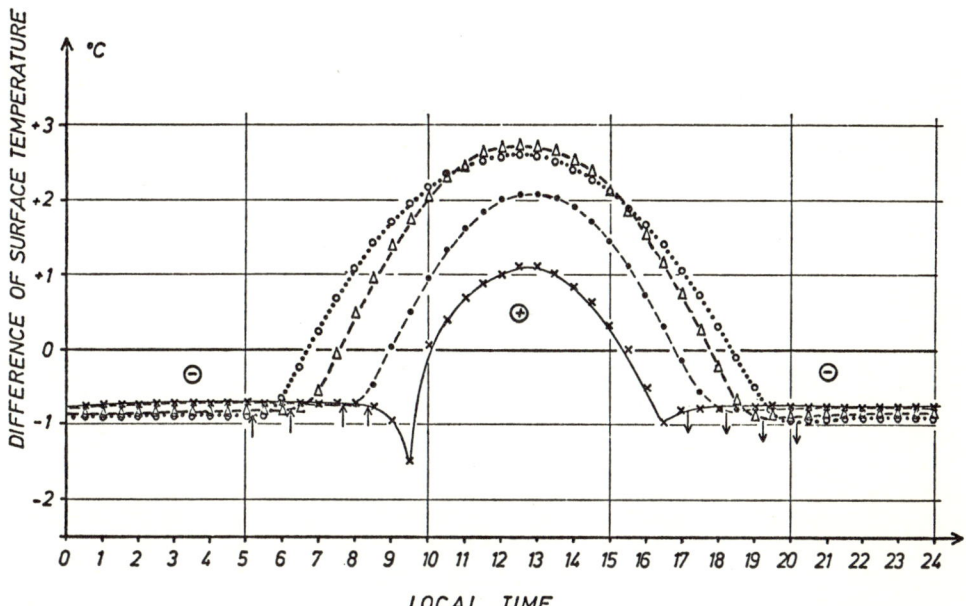

Figure 4. Daily variation in difference of surface temperature between protected and unprotected zones. ↑ sunrise ↓ sunset
Day : x 1, . 50, Δ 100, o 150.

In a given country, hedges then produce a shift toward accentuated continental climatological characters.

Variations of surface temperature for different spacing of hedgerows

In the example, with a ratio $\frac{D}{H}$ equal to 8, we have choosen for the friction velocity u* a value of 0,2 m.s^{-1}, and for the friction velocity of unprotected zone a value of 0,3. Figure 5 shows the difference of surface temperature between the center of alley of protected zone and the unprotected zone. An empirical variation of u_* with $\frac{D}{H}$ is assumed.

When this ratio is very high, the friction velocity between hedgerows is nearly equal to that of the unprotected zone. Temperature difference during the day decreases very quickly with the $\frac{D}{H}$ ratio.(4° C to 1° C when $\frac{D}{H}$ ratio changes from 4 to 32). During the night, this difference increases with $\frac{D}{H}$ ratio, because surface temperature between hedgerows is smaller than temperature of unprotected zone.

Components of energy balance

The different components of energy balance, calculated for the unprotected zone and for the center of alley between hedgerows, are presented in table 1. Two

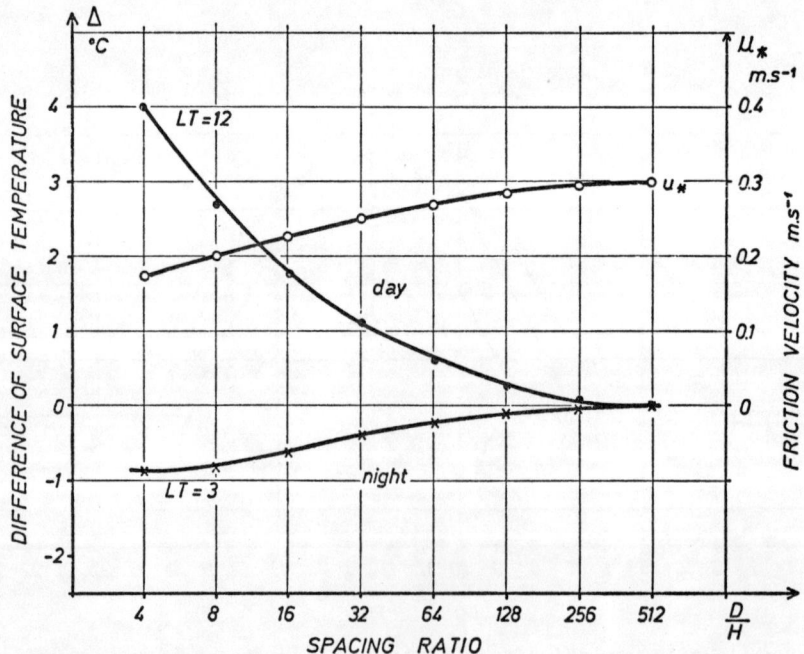

Figure 5. Variation of difference of surface temperature , between protected and unprotected zone versus $\frac{D}{H}$ ratio. o u_* local time : . 12, x 3.

Local time		T_s	$G\downarrow$	$G\uparrow$	$G\downarrow - G\uparrow$	R_n	ϕ_s	ϕ_L	ϕ_{ss}
7	∞	5,04	51	10	41	-100	162	-68	6
	E-W	4,49	51	10	41	-92	120	-35	7
	N-S	3,78	8	2	7	-123	136	-21	7
8	∞	8,62	267	53	214	54	92	-149	3
	E-W	9,09	266	53	213	57	51	-111	3
	N-S	9,13	269	54	215	59	50	-112	3
9	∞	12,42	507	102	406	227	27	-253	0
	E-W	13,81	506	101	405	225	-12	-212	-2
	N-S	13,83	508	102	407	227	-12	-213	-2
10	∞	15,05	719	144	575	382	-28	-352	-3
	E-W	17,11	718	144	574	376	-62	-310	-5
	N-S	17,12	719	144	575	377	-62	-310	-5
11	∞	17,04	878	176	702	498	-62	-432	-5
	E-W	19,53	877	175	701	490	-93	-390	-7
	N-S	19,52	876	175	701	490	-93	-390	-7
12	∞	18,13	969	194	776	566	-80	-480	-6
	E-W	20,84	968	194	775	556	-110	-438	-8
	N-S	20,82	966	193	773	555	-110	-437	-8

Table 1. Components of energy balance. Day 100 - D/H ratio = 8
Sunrise 6,22 H, sunset 19,22 H. ∞ : unprotected zone, E-W east-west orientation, N-S north-south orientation.

orientations have been choosen for the hedgerows : east-west and north-south. For the 100th day of the year, global radiation, net radiation and heat flux into the soil are nearly the same in the two zones. The latent heat flux in the protected zone is always lower than the flux in the unprotected zone (the difference magnitude is about 10 %). This result is in good agreement with experimental results /8_7 and study concerning land roughness influence on regional climatology (SEGUIN /9_7). The sensible heat flux for protected zone is greater or not than flux for unprotected zone, following that surface temperature is greater or not than air temperature.

The variations of energy balance components are more important if we consider the variations with the distance to the hedgerow. The net radiation in the zone corresponding to a distance twice the height of hedgerow varies from 3,7 times to 1,01 times the net radiation of unprotected zone with north-south orientation and from 2,6 times to 0,98 time with east-west orientation. The difference between incident global radiation and reflected global radiation varies from 1,45 times to once the difference in the unprotected zone for north-south orientation and from 1,02 to once from east-west orientation.
The shadow of the thedge, according to the day of the year and the orientation of hedgerow can vary from 2,4 times to 0,2 times the height of hedges.

DISCUSSION AND LIMITATIONS OF THE MODEL

The limitations of the model are constituted by some rough hypothesis introduced in the calculations. Among them, the assumption of a constant friction velocity u_* for various distances to the hedgerow, and the use of aerodynamical method to calculate the fluxes are not valid, especially near the hedges, because these vertical fluxes are not constant with height.

On the other hand, at the center of alley theses hypothesis are not too far from the reality, and the results obtained for the temperature difference should fit with experimental results.

The other hypothesis seem more valid (constant surface temperature of hedgerow, constant ratio between diffuse sky radiation and global atmospheric radiation, heat flux into the soil). Improvement of this numerical model is presently undertaken.

CONCLUSION

The model used in this paper, although subject to some limitations as previously described, can be an useful approach for initiating or supporting experimental studies. Results of surface temperature difference and latent heat flux modification, seem to demonstrate its suitability for modeling and predicting the climate of plant canopy between hedgerows from the knowledge of external climatic factors.

BIBLIOGRAPHY

/ 1 / USHER M.B. (1970). An algorithm for estimating the length and direction of shadows with reference to the shadows of shelterbelts.
J. Appl. Ecol., 7, pp. 141-145

/ 2 / JACKSON J.E., PALMER J.W. (1972). Interception of light by model hedgerow orchards in relation to latitude, time of year and hedgerow configuration and orientation. J. Appl. Ecol., 9, pp. 341-357

/ 3 / ROBINSON N. (1966). Solar Radiation. Elsevier Publishing Company, pp.50-51

/ 4 / JAKOB M. (1957). Heat Transfer. Volume II. John Wiley and Sons, Inc.
pp. 19-21

/ 5 / BOUCHET R.J., GUYOT G. (1972). Aménagement du territoire et microclimat. Problème du bocage. Ac. Agr. France, pp. 1224-1240

/ 6 / SEGUIN B., GIGNOUX N. (1974). Etude expérimentale de l'influence d'un réseau de brise-vent sur le profil vertical de vitesse du vent.
To be published in Agric. Meteorol.

/ 7 / HALLAIRE M. (1950). Les températures moyennes nocturnes, diurnes et nycthémérales exprimées en fonction du minimum et du maximum journaliers de température. C.R. Acad. Sci., 231, n° 25,pp.1533-1535

/ 8 / GUYOT G., SEGUIN B., VERBRUGGHE M. (1974). Modification of land roughness and resulting microclimatic effects : a field study in Brittany (France) Communication to International Seminar Heat and Mass Transfer in the Environment of Vegetation, Dubrovnik, August 26-30

/ 9 / SEGUIN B. (1973). Rugosité du paysage et évapotranspiration potentielle à l'échelle régionale. Agric. Meteorol., 11, pp. 79-98.

32

MODIFICATION OF LAND ROUGHNESS AND RESULTING MICROCLIMATIC EFFECTS: A FIELD STUDY IN BRITTANY

G. GUYOT and B. SEGUIN

INRA—Station de Bioclimatologie—84140 Avignon/Montfavet (France)

SUMMARY

Large scale modifications of the landscape in regions protected by hedges, lead to land roughness modifications which are accompanied by microclimatic and climatic variations. Together with theoretical studies elsewhere described, a comprehensive field study was undertaken in Brittany (France) in order to determine their extent. Climatological observations were continuously performed during a three-year period, while intermittent more extensive micrometeorological experiments allowed to understand the mechanism of physical processes. Hedge network effects upon turbulent transfers and radiative exchange, and resulting influence of the various microclimate components, were studied both at local and regional scales. Results, expressed in terms of differences between adjacent protected and open areas, are analyzed and discussed.

LIST OF SYMBOLS

E_p	Piche evaporimeter evaporation
H	height of hedgerow (m)
PET	potential evapotranspiration (P.A. protected area
	(O.A. open area
R_n	net radiation (W.m^{-2})
$\overline{U}$	mean wind velocity (m.s^{-1})
T	air temperature (° C)
X	distance of hedgerow (m)
z	height above the ground
z_{oL}	local roughness length (m)
z_{oR}	regional roughness length (m)
Δf	water saturation deficit in air (mb)
γ	psychrometric constant
Δ	slope of the water vapour saturation curve (mb.° K^{-1})

1. INTRODUCTION

The traditional landscape of the western part of France is called the "bocage". Small fields, frequently less than half an hectare, are

surrounded by tree hedges planted on clay banks, which constitue a very
dense network of windbreaks. Such a land structure is not convenient
for modern agriculture mechanization.

When the services of Agriculture Ministry promoted 15 years ago, a
large operation of field regrouping, the hedges were cut more or less
completly. These workings introduced a large scale modification of the
land-structures, which, consequently, may affect both microclimate and
climate in concerned areas. To determine the extent of these effects
a comprehensive study was undertaken in Brittany. Field experiments
were based upon simultaneous theoretical studies /̄1̄/ /̄2̄/ concerning
both hedge influences on turbulent and radiative transfers.

2. EXPERIMENTAL PROCEDURE

The experiments were made during a three years period with permanent
climatological measurements and intermittent short-time micrometeorological studies /̄3̄/.

2.1. Experiment Locations

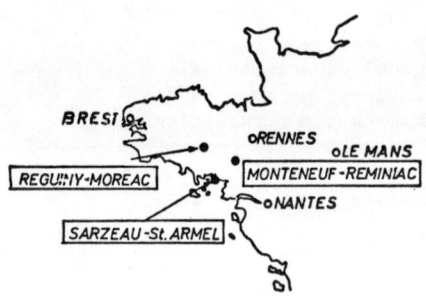

Three sites have been choosen in the
region of Morbihan (Fig. 1) One is
located on the seashore, while two
others are situated in the inland
(about 60 km from the sea). Each site
consists of two adjacent areas : one
where the original land structure
has been kept, and the other where
hedges have been cut. These two areas
are as identical as possible for
external factors like topography,
distance to the sea, type of soil,...
In each of these areas a reference field is choosen where meteorological screens are implanted and micrometeorological experiments performed. These fields correspond to grasslands, in order to get easier
comparisons. Distances between reference fields for each site vary
between 1,5 to 6,7 km.

2.2. Type of Study

Two types of studies were led simultaneously :
- global studies, in order to determine global effects of hedges on
  local microclimate and regional climate

- analytical experiments, performed with mobile apparatus, to establish
  analytical effects of hedges on local or regional energy balance.

2.3. Technical Apparatus

For global studies, classical meteorological screens were used, with
following measurements : air temperature and humidity and Piche eva-

porimeter at 2 m-height, soil temperature at 10 cm-depth. In the continental site (Reguiny-Moreac), potential evapotranspiration was measured with lysimeters, together with precipitations.

For analytical experiments, simultaneous measurements were performed in open area and protected field. 6 m-height mobile masts were used to determine local hedge effects, according to various distances from hedge. These masts were equipped with cup anemometers and aspirated psychrometers.

At the same time, the spatial variation of radiative components inside the protected field was determined by 3 pyranometers ("Kipp and Zonen" and 8 net radiometers (CEA-INRA type /̄ 4 /̄).

In order to determine regional effects /̄ 1 /̄, vertical profiles were also measured on 4 towers (30 or 20 m-height) : one in open area, as reference, and the others in the centers of different fields of the protected area. Since the typical height of hedges varied between 6 m and 10 m, this height was thought to be sufficient to avoid hedge local wakes and determine regional influences of the upper part.

Temperature and humidity measurements were performed with platinium resistances. Corresponding profiles were recorded with two data loggers, together with radiation measurements. Wind profiles were recorded with separate counting units /̄ 5 /̄. For all parameters, mean-values were obtained for 30 min-periods.

3. MICROMETEOROLOGICAL EXPERIMENT RESULTS [6]

In opposition to climatological measurements, micrometeorogical experiments give detailed instantaneous informations about the structure of microclimate parameters. These informations have a less general value, but they allow a better understanding of overall effects

3.1. Aerodynamic Effects of Hedge Network

3.1.1. Local scale

Inside of one protected field, aerodynamic effects of hedges are similar to those of shelterbelts /̄ 7 /̄ /̄ 8 /̄. The reduction of wind velocity depends upon constitutive characteristics of hedges. One example of measured reduction is given in Fig. 2. It appears to be somewhat larger than that generally observed for isolated windbreaks, which may be due to the surrounding roughness effect.

3.1.2. Regional scale

Vertical wind profiles are in good agreement with previous measurements in regions protected by dense shelterbelt networks /̄ 9 /̄. While they are semi-logarithmic in open areas, they exhibit "kinks" in protected fields, leading to two semi-logarithmic straight lines (Fig. 3) The local roughness parameter z_{oL} determined from the lower part is characteristic of the grass surface. From the upper part, it is possi-

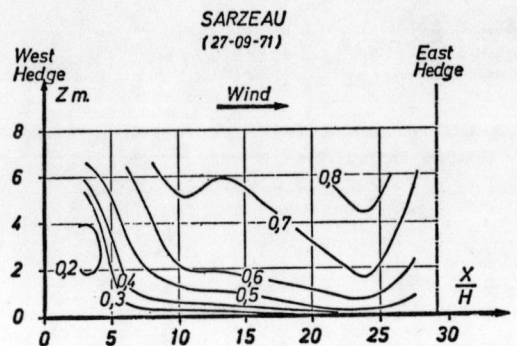

Fig. 2 - Reduction of wind velocity by natural hedges : curves of equal velocity expressed in function of the reference velocity at 2 m-height in the open area.
The abscissa is expressed in multiples of the hedge height (about 6 m in that case)

ble to determine a regional roughness parameter z_{OR}, whose order of magnitude is near 1 m. For a given location, various z_{OR} may be determined according to different wind directions. Their mean-values from repeated experiments are shown on Fig. 4 for the maritime site.

They exhibit wind directions corresponding to a relatively high roughness, while others show a significantly lower roughness, which is in good agreement with land structure around measurement location. It then appears some correlation between the regional roughness parameter

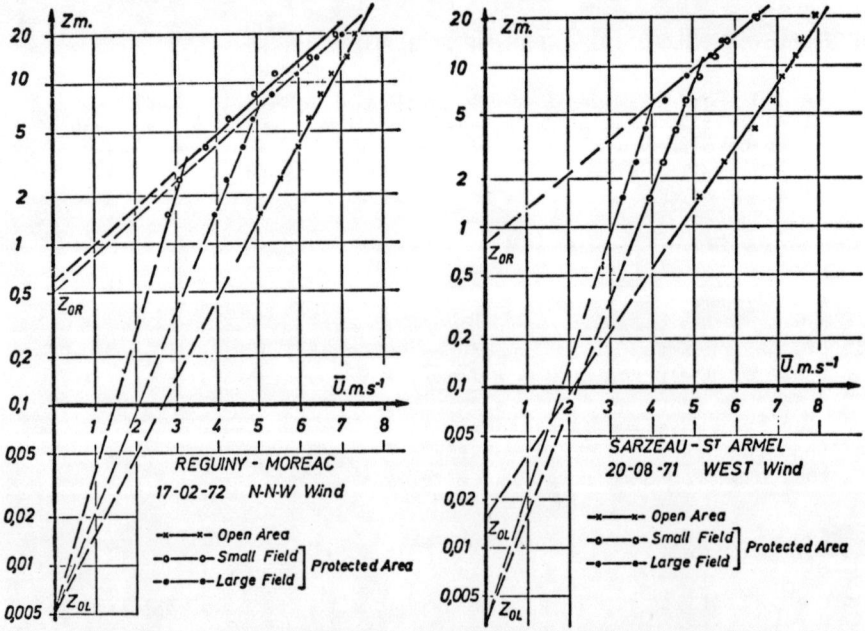

Fig. 3 - Measured wind velocity profiles on a 30 min-period, simultaneously in protected and open areas in two sites

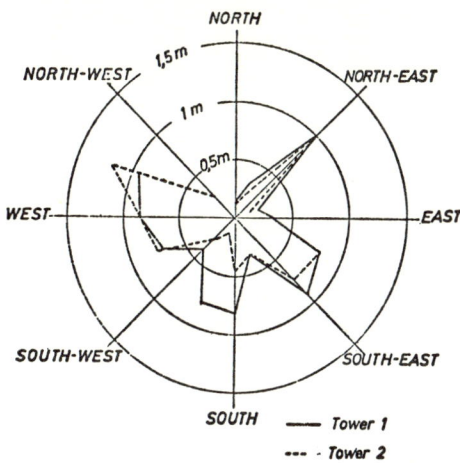

Fig. 4

Variation of the mean regional roughness parameter z_{oR}, according to various wind directions for the maritime site (Sarzeau)

and the density of hedges. Further studies will be necessary to improve this qualitative agreement.

3.2. Radiative Effects of Hedges

For radiative components, only local studies have been performed. Theoretical considerations about the modification of regional albedo /¯10¯/ could not be tested, because of the difficulty of realizing adapted measurements.

Simultaneous measurements of radiative components between the open area and the center of protected field gave quasi-identical values of global and net radiation. Only a slight increasing of net infrared radiation loss of the order of 10 $W.m^{-2}$ seems to be observed in the protected field, which may result from higher surface temperatures of about 2° in the daytime. On the other hand, horizontal profiles inside the protected field exhibit local influences of hedges.

For global radiation, measurements near hedges (at a distance of one time the height of the hedge) give negligible differences with open area when the sky is clear, while a reduction of about 10 % is observed for overcast days. This deviation could be attributed to the effect of hedges on diffuse solar radiation, whose the relative importance in global radiation is higher for overcast weather.

In spite of technical problems, which make some results doubtful, some effects on net radiation seem to appear from measurements such as those shown on Fig. 5

Higher levels of net radiation are obtained in the vicinity of hedges, which is in good agreement with theoretical predictions about the modification of atmospheric long-wave radiation resulting from sky obstruction by hedges /¯2¯/.

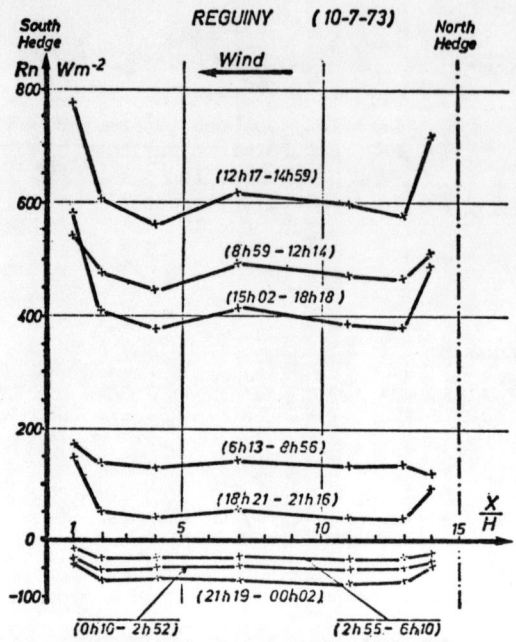

Fig. 5

Diurnal variation of net radiation intensity inside of a protected field for a clear day (10.7.73 in Reguiny)

3.3. Effects on Air Temperature

3.3.1. Compared temperature profiles

Compared temperature profiles measured simultaneously on observation towers in open area and protected field give different results according to the occurrence of wind. When wind is blowing, typical profiles like that shown on Fig. 6 occur during the daytime.

According to theoretical predictions /̲11̲/, they exhibit a same figure as for wind profiles. In the protected field, two straight lines appear. The upper part, which corresponds to regional scale, shows reduced gradients, while the reduced turbulent transfer in the lower part leads to significatively increased temperature gradients in the lower part. It results from this an increased air temperature near the ground in the protected area. It is to be remarked that the resulting difference, which amounts to about 1,5° C at 0,2 m height, is considerably damped at 2 m-height, where it is reduced to 0,1° C. The measurements in the meteorological screen are then far from being representative of surface modifications.

When the wind is blowing, the process may occur during nighttime, leading to lower air temperature near the ground. For still nights, the same fact may occur, but contradictory results may be found. Other factors like topography, occurrence of fog,... seem to interfere in theses cases.

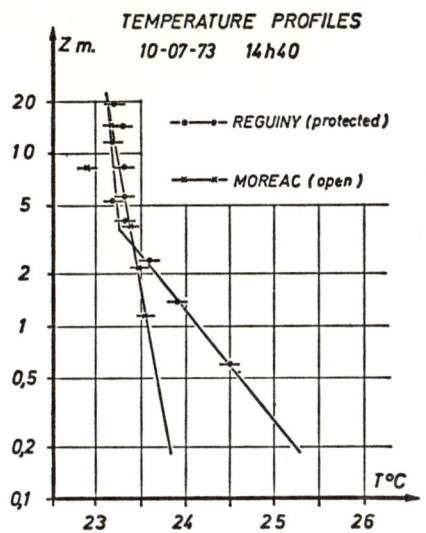

Fig. 6

Typical simultaneous temperature profiles in protected and open areas

3.3.2. Variation of air temperature inside protected field

Wind velocity and radiative balance are known to vary with the distance to hedge. A similar trend must then occur for air temperature, which is confirmed by experiment results, such as shown by Fig. 7

For daytime, air temperature appears to decrease downwind of the hedge, while for nighttime more irregular variations are observed.

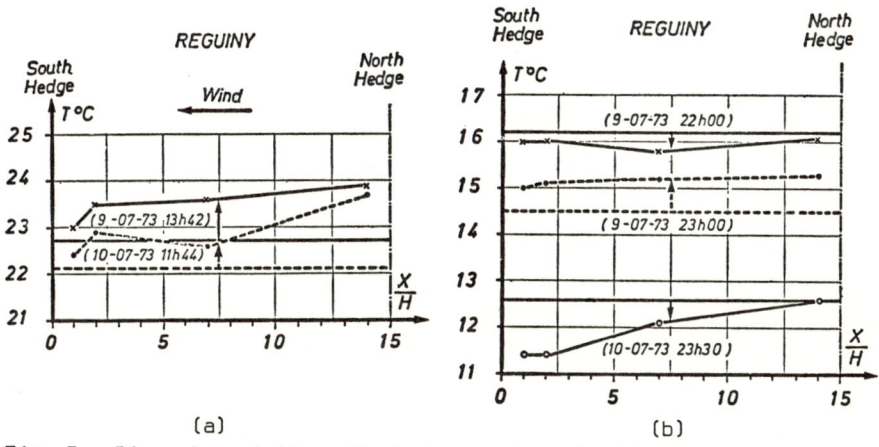

Fig. 7 - Diurnal variation of air temperature inside protected field
(Straight lines correspond to simultaneous value in open area)
    a - daytime values   b - nighttime values

4. CLIMATOLOGICAL RESULTS

These results are obtained from classical meteorogical observations. They correspond to larger time and space scales than micrometeorogical results as previously described. Observed differences are then generally less significative in magnitude, but they apply to much longer periods, so that they may be considered as expressing the global effects.

4.1. Air and Soil Temperatures

Air temperature measurements are in good agreement with micrometeorological findings. In the protected field, maximum temperatures are generally higher by about 1 - 1,5° C, while minimum temperatures may be equally reduced or increased.

A more detailed examination of results displays the fact that they are generally reduced during periods with clear sky, while they may be increased for overcast periods or when fog formation is frequent.

The general trend is illustrated by Fig. 8, where are reported frequencies of deviations of maximum and minimum temperature in protected field compared to open area, for one experiment site (Reguiny-Moreac). It results from this a significative increase of daily thermal amplitude, as shown by Fig. 9.

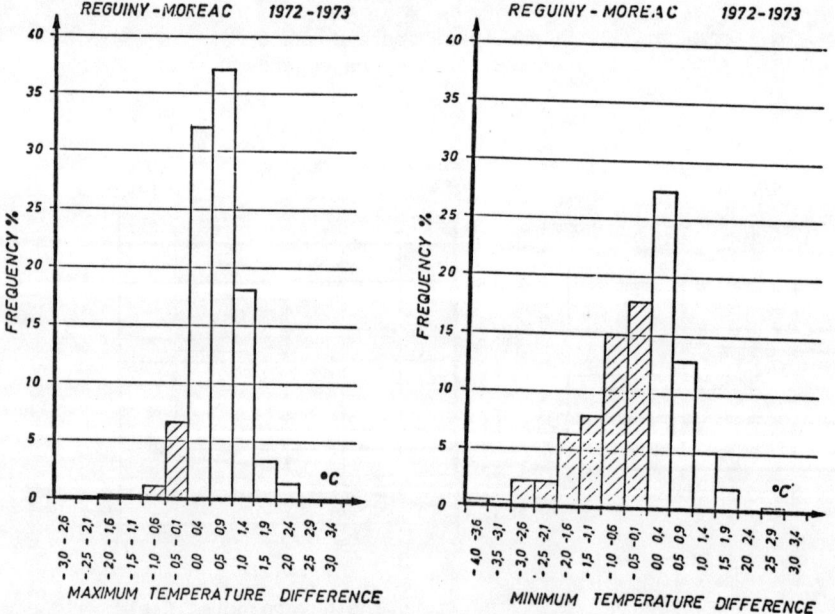

Fig. 8 - Percent frequency of temperature difference occurrence between protected and open areas for 1972-1973 year in Reguiny-Moreac.

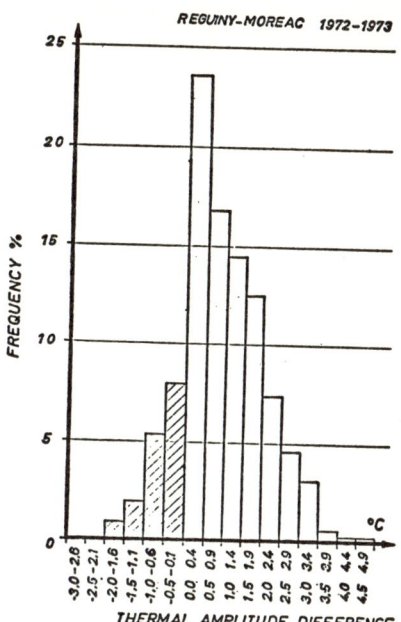

Fig. 9

Percent frequency of thermal amplitude difference occurrence between protected and open areas for 1972-1973 year in Reguiny-Moreac

The comparison between the three measurements sites shows that the modification of thermal amplitude is more noticeable for inland sites than on the seashore. It appears also to be larger for the smallest field (0,5 ha in Monteneuf-Reminiac).

The results for soil temperature appear to be somewhat dispersed. Elevation of maximum daytime temperature and reduction of minimum in nighttime regularly occur in the site of Reguiny-Moreac, which is in accordance with observed tendencies for air temperature. Less systematic effects appear for the other sites, with sometimes contradictory results. This dispersion may result from factors like type of soil, groundwater level, exact depth of measurement, etc...

4.2. Air Humidity

Effects of hedge network on air humidity may be expressed in terms of different parameters such as relative humidity or water vapour pressure.

For expressing the real effect upon air dryness, it seems better to use the saturation deficit Δf, which equals the difference between saturation vapour pressure at air temperature and the effective vapour pressure.

At the time of minimum temperature, saturation deficits are equally null in protected field and open area, so that vapour pressure is generally slightly reduced in the protected field, due to lower minimum temperature. During the daytime, saturation deficits are increased in the shelter, as shown by Fig. 10.

This overall effect varies according to periods and regions. It is however possible to observe that hedge networks generally increase the air dryness, which may be surprising if the reduction of water vapour exchange is considered. But two opposed factors may explain this result : first, air temperature is increased - and second, hedges act as obstacles opposed to maritime air penetration.

4.3. Evaporation and Potential Evapotranspiration

Measurements of evaporation were made in meteorological screens with Piche evaporimeters, while compared potential evapotranspiration measurements were performed for 1972-1973 year only on the site of Reguiny-Moreac by using two drainage lysimeters planted with short grass. Results are summarized in the following table $/\overline{\ }12\overline{\ }/$.

Hedges significantly reduce Piche evaporation by about 20 %, which may result essentially from wind velocity reduction, since air saturation deficit is slightly increased in protected field.

On the contrary, no noticeable effect can be seen for potential evapotranspiration, taking into account a relative error of 5 % for this measurement. This finding is, in fact, quite logical since windbreaks only reduce the advective contribution to potential evapotranspiration. As climate of Brittany is essentially a maritime one, the relative influence of this advective term is low. As the ratio $\frac{Ep}{PET}$ is near 0,5 and the advective term of Penman's formula may be approximated by the relation $\frac{\gamma Ea}{\Delta+\gamma} = 0,34$ Ep for the corresponding latitude $/\overline{\ }13\overline{\ }/$ its relative contribution to potential evapotranspiration is lower than 20 %. The measured reduction of about 20 % for Piche evaporimeter

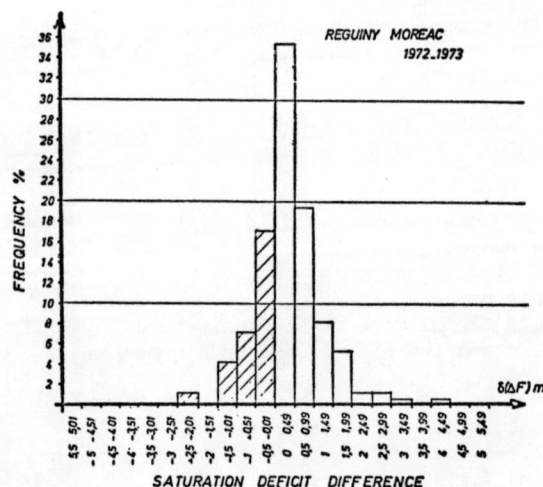

Fig. 10

Percent frequency of saturation deficit difference between protected and open areas for 1972-1973 year in Reguiny-Moreac

Table I

	REGUINY protected area		MOREAC open area		PET(PA)/PET(OA)	Ep(PA)/Ep(OA)	MOREAC Ep/PET	PET reduction %
	PET	Ep	PET	Ep				
From 4.13 to 5. 3	46,3	15,5	48,8	22,4	0,95	0,69	0,46	5
From 5. 4 to 5.30	85,0	40,5	83,4	47,9	1,02	0,85	0,57	3
From 5.31 to 6.27	107,5	44,5	102,7	59,8	1,05	0,74	0,58	5
From 6.28 to 8. 2	105,5	42,4	105,6	58,0	1,00	0,73	0,55	5
From 8. 3 to 8.30	79,5	36,3	85,7	45,1	0,93	0,80	0,53	4
From 8.31 to 9.30	78,8	26,4	76,7	34,2	1,03	0,77	-	4
From 10.1 to 10.31	32,5	-	29,5	-	1,10	-	-	-
From 11.1 to 11.28	16,7	-	18,2	-	0,92	-	-	-
Total	551,8	-	550,6	-	1,00	-	-	-

leads to a calculated reduction near 5 %, as shown in the last column of table I. Since, on the other hand, hedge effects on radiative budget and air temperature lead to a similar increase of the radiative term $\frac{\Delta Rn}{\Delta \gamma}$, these contradictory tendencies may explain the observed equality of PET between protected area and open area.

This finding is in good agreement with recent observations [14], where windbreak influence upon PET is shown to be noticeable only for important advective conditions.

5. CONCLUSION

The above described comprehensive studies display that the effects of hedges on turbulent exchanges and radiative transfers equally interfer in the overall effect upon local microclimate and regional climate. Some results are difficult to interprete because of the interposition of external factors such as topography, type of soil and ground cover, etc... The global effect may be sumarized in a tendency to increase the continental character of the local climate, by increasing amplitudes.

These modifications will, in turn, affect the conditions of plant growth and development, which are the final results for the agronomic point of view. Together with these climatic studies, parallel biological measurements have been made, the detail of which is given elsewhere [15]. The resulting ensemble allows to get a valuable idea of the various aspects of hedge network effects on the environmental conditions.

ACKNOWLEDGEMENT

The authors wish to exprime grateful thanks to MM. VERBRUGGHE and BAUTRAIS, who were in charge of realizing the field experiments.

BIBLIOGRAPHY

/1_7 SEGUIN .B (1973). Rugosité du paysage et évapotranspiration potentielle à l'échelle régionale. Agric. Met. Vol 11, p. 79-98

/2_7 CHIAPALE .J.P (1974). A numerical model for estimating the modification of heat budget introduced by hedges. Paper presented at the Dubrovnik Symposium on "Heat and mass transfer in the environment of vegetation". August 26-30

/3_7 GUYOT .G (1974). Présentation des dispositifs expérimentaux destinés aux études des effets du bocage sur les facteurs climatiques. ANDAFAR Seminar, Paris, October 3th 1974

/4_7 DENIS .P, GUICHERD .P, DAMAGNEZ .J et MERMIER .M (1973). Mesure du rayonnement net : mise au point d'un pyrradiomètre différentiel à circulation d'eau. Presented to Congress "The sun in the service of mankind". Paris 2nd-6th July 1973

/5_7 GUYOT .G (1970). Ensembles de comptage anémométriques destinés à des études écologiques. In : Techniques d'études des facteurs physiques de la biosphère, INRA, Paris, p. 397-408

/6_7 GUYOT .G (1974). Synthèse des études de climatologie. Application locale. ANDAFAR Seminar, Paris, October 3th 1974

/7_7 VAN EIMERN .J (1964). Windbreaks and shelterbelts. OMM. WMO. Techn. note 59, 188 pp.

/8_7 GUYOT .G (1972). Etude de l'écoulement de l'air au voisinage d'un obstacle poreux en couche-limite turbulente (Aérodynamique des brise-vent). Thesis. University of Paris VI. 162 pp.

/9_7 SEGUIN .B et GIGNOUX .N (1974). Etude expérimentale de l'influence d'un réseau de brise-vent sur le profil vertical de vitesse du vent. Accepted for publication in "Agric. Met".

/10_7 CHIAPALE .J.P. (1974). Etude théorique des modifications d'albédo liées à l'aménagement du territoire. Station de Bioclimatologie INRA Avignon-Montfavet. Note technique M 74-4, 51 pp.

/11_7 SEGUIN .B (1974). Modifications climatiques et microclimatiques liées a la rugosité du paysage : aspects théoriques. ANDAFAR Seminar. Paris. October 3th 1974

/12_7 GUYOT .G (1974). Les effets microclimatiques des brise-vent. Conséquences sur les composantes du bilan hydrologique. XIIIe Journées de l'Hydraulique. Paris September 16-17-18 th 1974

/13_7 BROCHET .P et GERBIER .N (1972). Une méthode pratique du calcul de l'évapotranspiration potentielle. Ann. Agron. Vol 23, n° 1, p. 31-49

/14_7 LOMAS .J et SCHLESIGNER .E (1971). The influence of a windbreak on evaporation. Agric. Met. Vol 8, n° 2, p. 107-115

/15_7 MALET .Ph (1974). Méthodes possibles pour l'étude de l'effet climatique original d'un bocage sur la production végétale. ANDAFAR Seminar. Paris. October 3th 1974

REFLECTANT INDUCED MODIFICATION OF THE RADIATION BALANCE FOR INCREASED CROP WATER USE EFFICIENCY

RAOUL LEMEUR and NORMAN J. ROSENBERG

*Laboratory of Plant Ecology, University of Ghent,
Coupure Links 553 9000 Ghent, Belgium*

ABSTRACT

Soybeans (Glycine max L., var. Amsoy) were treated with Celite and kaolinite reflectants in order to increase reflection and reduce net radiation. Such treatment should decrease evapotranspiration and improve water use efficiency.

The effect of the treatment on near infrared reflection was small, but reflection of visible radiation was doubled. Reduction of net radiation was 5 to 10%. Reflection down into the canopy was also intensified.

Treatments at the beginning of the growing season were more effective than at crop maturity. Factorial applications were not cumulative.

Celite is a better material than kaolinite, showing the desired degree of reflection with half as much material as was required with kaolinite.

INTRODUCTION

Reflectant materials applied to plant canopies reduce net radiation at the crop surface. Water consumption should be decreased by such treatment since the net radiation supplies energy for evapotranspiration.

Co- author : Dept. of Horticulture and Forestry, University of Nebraska, Lincoln, Nebraska 68503, U.S.A.

The potential water-saving effects of reflectants may be negated if the treated crop suffers a reduction in photosynthetic activity because a significant amount of photosynthetically active radiation (PAR : 400 - 700 nm) has been reflected away and does not reach the grana. Such an effect is less likely to be consequential in crops which are light-saturated at levels of solar radiation intensity lower than that occurring in the field. Evidence has been reported (1) that soybean is such a light-saturated crop, at least under growth chamber conditions and in the field when leaf area index (LAI) is less than about 4 to 6.

Our reflectant research has invoved the use of two types of reflectant materials : kaolinite and Celite^x. Kaolinite is a white natural clay, and Celite is a commercially available diatomaceous earth which is frequently used in filtration processes. Both reflectants are relatively cheap. Other new materials such as fire-dry fumed silica or micro-crystalline cellulose may also have useful properties. There is, however, good reason to resist the careless introduction of chemicals in nature as certain ecological cycles could be disturbed. The materials cited above are inert and, as such, offer no danger of plant or soil contamination.

The study reported here was undertaken to establish the specific reflection patterns of a Celite and kaolinite reflectant. Their effectiveness with respect to the modification of the radiation balance and the influence of factorial applications at two stages of crop development is also evaluated.

MATERIALS AND METHODS

Soybeans (Glycine max L., var. Amsoy) were planted at the University of Nebraska Micrometeorology Research Laboratory at Mead (41° 09' N, 96° 30' W) on June 2, 1972. Rows were 46 cm apart and oriented east-west. Seeds in the row were 5 cm apart. LAI of the growing crop was determined periodically. Three areas of 21 x 21 m,

---

(x) Celite is a registered trade mark for diatomaceous silica products manufactered by Johns-Manville Corp., Lompoc, California 93436, U.S.A.

herafter called plots A, B and C, were used for radiation experiments at different times of the growing season. The two reflectants and their applications were arranged in the following way : A1 : control subplot, A2 : kaolinite treatment on soil, A3 : kaolinite treatment on leaves, and A4 : Celite treatment on leaves. A preliminary screening of kaolinite and Celite soil treatments indicated that Celite was almost completely absorbed by the soil matrix.

The first treatment was applied on main plots A, B and C on July 12,1972. At that moment, the soybeans were developed up to LAI = 2.5 and soil cover was approximately 50% . On August 7 and August 21 a second and third treatment was applied to the main areas B and C, and to C respectively. On those dates the crop was fully developed and LAI was 5.5 and 6.4 .

The spraying of the reflectant materials was conducted with a 7-nozzled high pressure spray device in two successive operations. This resulted in two superimposed coatings which were weather resistant. The reflectants were applied as slurry mixtures, including a surfactant and a non-toxic gum.

Quantitative and qualitative (spectral) radiation measurements were made on July 13 for area A and on August 8 for area B. Unfortunately, rain storms occurred during the period Aug. 22 - Aug. 25 and the radiation experiments had to be cancelled in area C.

Total global shortwave radiation, diffuse sky radiation and reflected shortwave radiation were measured with temperature-compensated Eppley pyranometers. Net radiation data above the soybean crop was taken with a Swissteco net radiometer. Spectra of incoming and reflected radiation were determined with an Isco spectroradiometer. All instruments, excep29t the pyranometers for incoming solar and diffuse sky radiation, were suspended on a pivoting boom positioned in the center point of the experimental area. The boom was rotated in the horizontal plane 1 m above the crop surface, and consecutive readings were taken in each subplot 1, 2, 3 and 4. A complete sweep took 10 minutes. The boom supported a special framework so that the spectroradiometer could be turned upside down and thus be used for both incoming and reflected radiation.

The influence of reflectants on the radiation balance inside the canopy was measured in subplots B1, B3 and B4. Linear net radiometers were installed above the canopies and at 3 levels inside each crop. The instruments were self-made and provide for a spatial averaged reading over 33 cm.

Some definitions are required at this point to distinguish between the different types of results obtained in this study. The spectrum of incident radiation is characterized by its spectral intensity $i(\lambda)$. Hence $i(\lambda) d\lambda$ denotes the amount of energy contained in a small wavelength band $d\lambda$. A fration of this amount, i.e. $r(\lambda)i(\lambda)d\lambda$ will be reflected by the soybean canopy, and $r(\lambda)$ denotes the spectral reflectivity of the crop which is wavelength dependent. Using this notation, the total energy of radiation between two wavelength limits λ_1 and λ_2 is equal to :

$$I(\lambda_1, \lambda_2) = \int_{\lambda_1}^{\lambda_2} i(\lambda) d\lambda. \qquad (eq.\ 1)$$

At the same time the ratio of total reflected radiation to total incoming radiation will be :

$$\alpha(\lambda_1, \lambda_2) = \int_{\lambda_1}^{\lambda_2} r(\lambda)i(\lambda) d\lambda \int_{\lambda_1}^{\lambda_2} i(\lambda) d\lambda. \qquad (eq.\ 2)$$

The term on the left in eq. (2) is called "albedo". Frequently the term "shortwave reflection coefficient" is used as a synonym, although the latter is expressed generally on a percentage basis.

RESULTS AND DISCUSSION

The attenuation of incoming shortwave radiation as a function of decreasing solar elevation (β) at Mead is shown in Fig. 1.
The specific absorption patterns are caused by a number of processes. Molecular scattering and large particle scattering decrease the energy in the ultraviolet and blue wavelength band. Fig. 1 shows two ozone absorption peaks occurring at 430 and 560 nm. Some oxygen absorption also occurs at 760 nm. All other absorption peaks between 650 and 1500 nm are due to water absorption. A small CO_2-absorption effect is superimposed on a strong H_2O- absorption at 1400 nm.

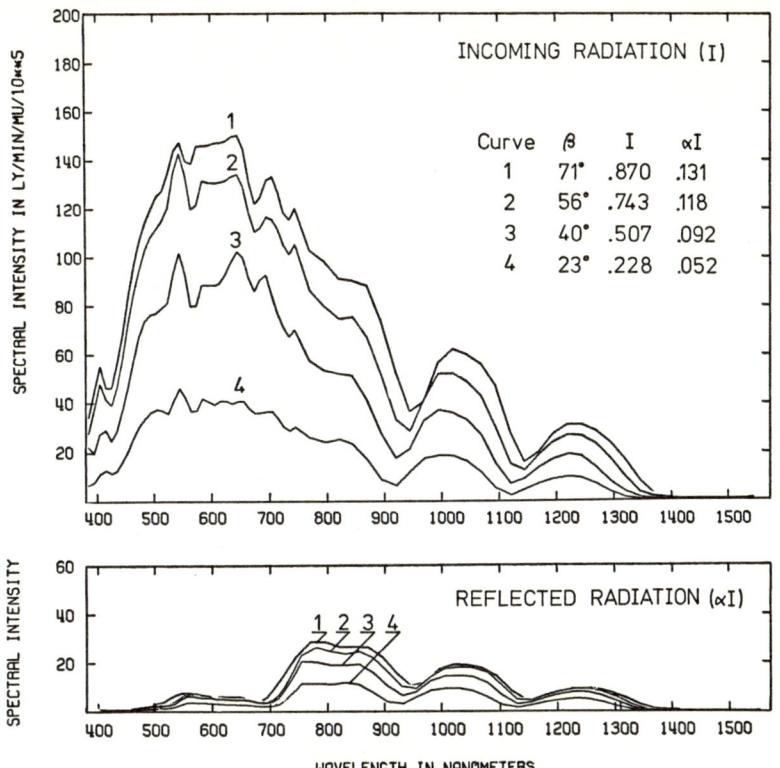

Fig. 1 Spectral intensity of incoming shortwave and reflected radiation over the green soybean crop. Integration of the area below the curves yields a measure of the total energy within the waveband 380-1550 nm. Incoming (I) and reflected (αI) energy (ly min^{-1}) within this band are listed in the figure.

Fig. 1 also depicts the spectral intensity of the radiation reflected from the green soybean crop. The wavelength distribution of reflected energy is markedly different from that of the incoming shortwave radiation. This is due to the wavelength selective absorption by the vegetative elements of the soybean canopy. Fig. 2 illustrates that the Celite and kaolinite coatings do not change the global picture of the reflection spectra. Celite reflects more energy in the PAR (400-700 nm) and NIR (750-1550 nm) wavelength bands than kaolinite. Both materials are more effective in the PAR region. The kaolinite soil treatment seems to suppress NIR reflection which is

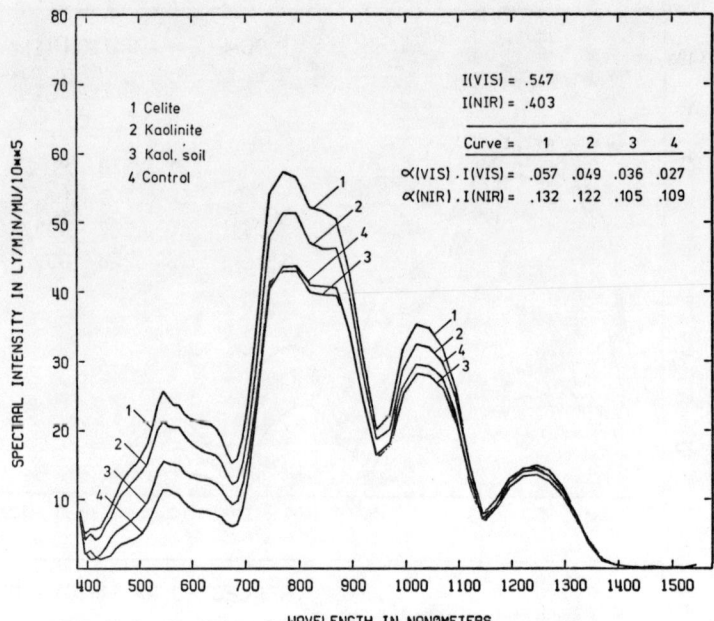

Fig. 2 Spectral intensity of reflected radiation from the 4 soybean subplots in area A ($\beta = 70°$). The energy contained in the wavelength bands of the visible (VIS : 380-750 nm) and near infrared (NIR : 750-1550 nm) is listed.

a possible indication for a higher water content of the surface soil layer.

Our findings at a number of solar elevations indicated that Celite was definitely more effective throughout the whole day. After 1 Celite treatment in area A at 50% soil cover the shortwave reflection coefficient α_s increased by 4%. The effect on α_s due to the kaolinite treatment was only half as much and the average result of the kaolinite soil treatment seemed to be inconclusive.

The treatment also reduced net radiation 1 m above the crop in both areas A and B. The reduction was, however, determined by the type of material and time of application. Kaolinite seemed to be more effective during the first stage of crop development with average R_n reductions of 10% and 6% for kaolinite and kaolinite soil treatments, respectively. The Celite coating, on the other hand, showed a smaller average decrease of 5%. This situation was inversed at full crop development with average R_n reductions of 9% and 6% for Celite and kaolinite treatments, respectively. These opposite results might indicate the influence of a temperature factor. Obviously, a check of the longwave radiation balance (3000-∞nm) would eliminate some of this ambiguity.

Fig. 3 illustrates the normalized net radiation profiles in treated and untreated soybeans. The leaf area density profile produced from discrete values of LA density within several layers, is also presented. Net radiation was increased in the upper two thirds of the treated crops, but discrimination between Celite and kaolinite is virtually impossible.

Our observations thus indicate that, in addition to the energy lost over the treated canopy, the availability of radiation was increased at the upper layers of high LA density. This secondary benefit of the reflectant treatment depends mainly on the soybean architecture itself. It was shown (2) that the soybean crop used in this experiment displayed an erectophile leaf angle distribution. Hence, the effect will be smaller in planophile varieties where the downward scattered radiation is more limited. These considerations

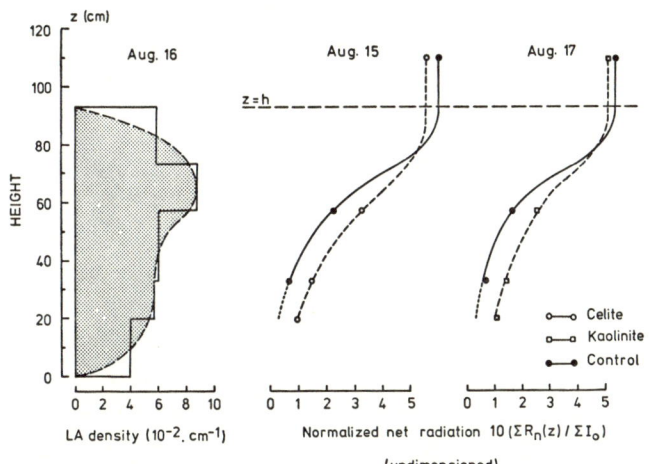

Fig. 3 Normalized net radiation profiles inside treated and untreated soybeans. The profiles are weighted averages based on the summed totals of incoming shortwave(ΣI_o) and net(ΣR_n) radiation. The leaf area (LA) density profile is presented for comparison.

lead us to the problem of the optimum soybean architecture for reflectorization. The problem is one of optimizing upward reflection, for which a planophile canopy would be best, while permitting light penetration into the depths of the canopy, for which an erectophile canopy would be best.

The effect from one or two reflectant coatings during the growing season may be evaluated from Fig. 4 for Celite treated crops. The corresponding graph for kaolinite was almost identical. The difference between the spectral reflection coefficients for the NIR waveband are due to the difference in LAI and not to the influence of the reflectant treatment itself. It has been shown (3) that the radiation spectrum from soybean has a strong energy shift towards the NIR with increasing leaf area development. Therefore, the influ-

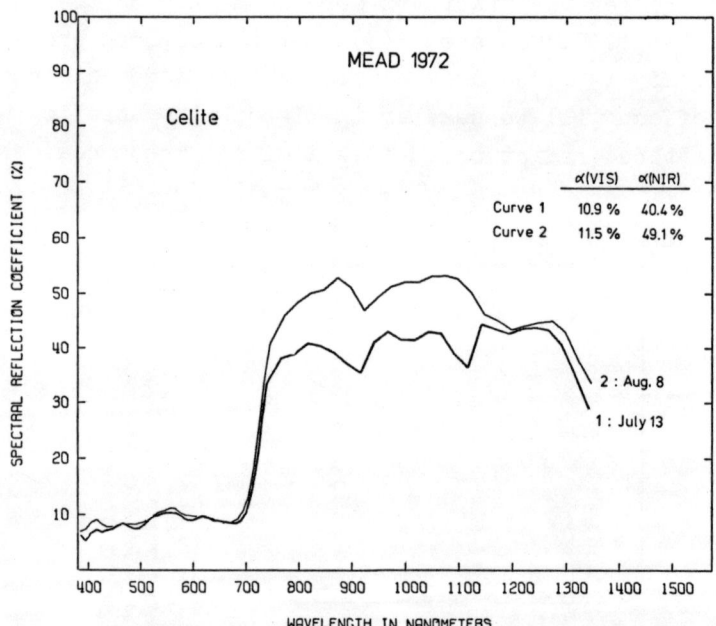

Fig. 4 Spectral reflection coefficients for Celite treated soybeans after one (July 13, LAI = 2.5) and two (Aug. 8, LAI = 5.5) applications. Integrated values for the VIS and NIR wavelength bands are listed. Incoming shortwave radiation and solar elevation for curve 1 was 1.03 ly/min and 46°, respectively. The corresponding values for curve 2 were 1.04 ly/min and 45°.

ence of increasing LAI should be eliminated to estimate the effectiveness of two applications. This is readily done by indicating the percentage differences of reflection coefficients and net radiation relative to the corresponding value of the untreated crop. The relative percentage increase of the reflection coefficientsfor treated soybeans is expressed by the ratio $[\alpha(treated) - \alpha(control)]$ / α(control). A similar expression can be used for net radiation.

Table 1 lists daily averages of such percentage increases for reflection coefficients and net radiation. We notice that the effect is not cumulative after two separate applications. Indeed, two applications in area B result in a smaller percentage increase of shortwave reflection and a smaller percentage reduction for net radiation. This might indicate that additional applications are necessary only to maintain the quality of the coating. The amount of cover is eventually decreased by rainstorms or lodging , and the appearance of new leaves.

CONCLUSIONS

Our studies of the radiation balance over a reflectorized soybean crop indicate that upward reflection is greatly increased by Celite and kaolinite coatings on the leaf surfaces. Because of their

Table 1 Effect of Celite and kaolinite treatments on the components of the radiation regime. All values are daily averages expressed as the percentage increase relative to the corresponding data of the untreated crop.

	Reflection Coeff.			Net Rad.
	Shortwave	PAR	NIR	
July 13, LAI = 2.5				
Celite	+26%	+131%	+16%	-5%
Kaolinite	+13%	+87%	+0.4%	-10%
Kaolinite soil	-3%	+25%	-9%	-6%
Aug. 8, LAI = 5.5				
Celite	+12%	+135%	-2%	-9%
Kaolinite	+10%	+118%	-4%	-6%

white colour, both materials are more effective in the PAR wave band than in the region of the near infrared. Since photosynthesis requires light in the visible wavelengths, some possibility exists that the treatment might have a detrimental effect on crop yield.

A decrease of net radiation was shown for the two reflectant types and for the soil treatment. Observations inside the soybeans indicated that a secondary benefit should be intensified reflection of radiation down into the canopy.

As the effect of different treatments is not cumulative, applications may be restricted to maintain the quality of the coating. The best time for application is probably the middle of the growing season when water stress is likely to occur. From a commercial standpoint, the stage of podfilling is most important. The preceding period of maximum growth rate could, therefore, determine a suitable interval for reflectant treatments.

In the field we found that Celite is a more effective material, producing the desired degree of reflection with half as much material as was rquired with kaolinite. Kaolinite, on the other hand, is cheaper and shows a higher resistance to weather deterioration.

LITERATURE CITED

1. Doraiswamy, P.C. and N.J. Rosenberg. 1974. Reflectant induced modification of soybean (Glycine max L.) canopy radiation balance. I. Preliminary tests with a kaolinite reflectant. Agron. J. (in press).

2. Lemeur, R. 1973. A method for simulating the direct solar radiation regime in sunflower, Jerusalem artichoke, corn and soybean canopies using actual stand structure data. Agr. Meteorol. 12 : 229-247.

3. Lemeur, R. and N.J. Rosenberg. 1974. Reflectant induced modification of soybean (Glycine max L.) canopy radiation balance. II. A quantitative and qualitative analysis of radiation reflected from a green soybean canopy. Agron. J. (in press).

34

THE USE OF ANTI-TRANSPIRANTS TO CONTROL WATER CONSUMPTION IN ECO-SYSTEMS- AN EXPERIMENTAL STUDY OF SHORT- AND LONG-TERM EFFECTIVENESS OF VARIOUS TRANSPIRATION-REDUCING CHEMICALS

FRANK KREITH and ASHOK TAORI

University of Colorado, Boulder, Colorado, USA

ABSTRACT

This paper presents experimental results on the short- and long-term effectiveness of film-forming and physiologically-active antitranspirants in reducing the rate of evapotranspiration from tobacco leaves under controlled experimental conditions. Physiologically-active antitranspirants can initially reduce water loss to less than 30 percent of that for an untreated specimen, but their effectiveness diminishes sharply after approximately 48 hours. The best currently available antitranspirants appear to be capable of reducing evaporation loss appreciably if applied approximately every 6 or 7 days.

1. INTRODUCTION

As discussed in detail by Pojakoff-Mayber and Gale [1], there are four methods for reducing transpiration from green plants:

1. To reduce the infrared radiation absorbed by the leaves by increasing the leaf reflectance with chemical sprays.
2. To decrease the mass transfer coefficient between the plant surface and the environment by setting up windbreaks.
3. To decrease the partial pressure difference of the water vapor between the interior of the plant and the surrounding air by growing the plant in an enclosure and
4. To apply chemical sprays which produce an increase in the resistance to the diffusion of water vapor from the plant to the surroundings.

Research on the use of chemical sprays as a means of reducing the transpiration from plants has developed to the point where application on a large scale is being considered. Two of the principal questions to be answered before using the chemical spray method are how long an application with a given spray remains effective and which of the available antitranspirant agents is most economical. These questions become particularly important when considering antitranspirant sprays to facilitate survival of mesomorphic plants in semi-arid regions with a large evaporation potential [1] or to increase water runoff in watersheds where phreatophites grow along the streambed and their physical destruction could cause erosion [2]. This research was undertaken to obtain a preliminary evaluation of the long- and short-term effectiveness of the two major types of chemical antitranspirant agents currently available, i.e. film-forming and metabolically active antitranspirants.

The rate of water loss or transpiration from a plant is often described by a simple equation of the form [3]

$$T = \frac{\rho_0 - \rho_\infty}{R_1 + R_2} \qquad (1)$$

where T is the rate of transpiration from the leaf,
ρ_0 is the density of the saturated water vapor in the leaf,
ρ_∞ is the density of the water vapor in the surrounding air,
R_1 is the internal resistance of the leaf to water vapor transport, and
R_2 is the resistance of the boundary layer over the surface of the leaf to the transport of water vapor.

In most plants the main pathway for water vapor diffusion is through the stomates because the cuticular resistance is very large compared to the stomatal resistance [4]. Thus, water loss can be reduced by closing the stomates or by placing a film over the leaf surface. The first mentioned method can reduce diffusion of water vapor from the plant more than diffusion of CO_2 from the surrounding atmosphere to the site of fixation within the chloroplast of the plant because in the CO_2 path there is in addition to the two resistances R_1 and R_2 in Eq. (1), a third resistance in series in the liquid phase of the mesophyll cells [5]. If this third resistance is of the same order of magnitude as $R_1 + R_2$, an increase in R_1 by closure of the stomates causes a larger reduction in transpiration than in photosynthesis.

Film-forming antitranspirants will interpose a resistance to diffusion between the entire leaf surface, including the openings of the stomates, and the surrounding atmosphere. An increase in the ratio of photosynthesis to transpiration can therefore only be achieved if the film is more permeable to CO_2 than to H_2O vapor. Otherwise, reduction in transpiration will cause a larger reduction in photosynthesis. However, if the objective is mainly in evaporation a film-forming antitranspirant of any type can be useful.

For an antitranspirant to be economical, the cost of the material and its application must be less than the value of the water saved and/or the increased yield of the crop. Since the effectiveness of antitranspirants in reducing evapotranspiration diminishes with time, the frequency with which the antitranspirant must be applied to achieve a given reduction in the water requirement is important for an economic evaluation. The goal of this investigation was to provide information on the change in the effectiveness of antitranspirants of the stomata-closing and film-forming varieties with time. Two metabolically active materials, phenyl mercuric acetate and monoglycerol ester of n-decenvl succinic acid, and two film-forming antitranspirants, one a wax base material (Mobileaf manufactured by Mobile Oil Company, New Jersey, U.S.A.), the other a polymer base material (Wilt-Pruf manufactured by Nursery Specialty Products, Connecticut, U.S.A.) were used in this study.

2. EXPERIMENTAL TECHNIQUES

Previous investigations on the effectiveness of antitranspirants have been handicapped by two basic experimental difficulties:

1. Experiments have been performed "in vivo" and it was therefore impossible to maintain sufficient control over environmental conditions to compare results between treated and control specimens.
2. Under normal conditions plants grow during an experiment and without an analytical procedure to take the growth of the plant into account, it is not possible to compare results from one experiment with results from another.

In this investigation the first of these difficulties was overcome by performing the experiments in an ecological wind-tunnel under reproducible and controlled conditions, while the second difficulty was overcome by utilizing an analytical procedure, described in Sec. III, which introduced into the evaporation Sherwood Number and the Flow Reynolds Number a significant length based on the instantaneous size and shape of the leaf.

The experiments were performed in a closed-circuit high turbulence wind tunnel which had been modified to simulate the conditions in an ecological system as described in Ref. 2. Humidity in the tunnel was controlled by passing the air over a water bath or over water vapor absorbing pellets, while the temperature was controlled by passing the air over heating or cooling coils. A window in the top of the test section provided for irradiation of the plants by a light source consisting of Grolux and incandescent lamps to simulate solar conditions. The velocity of the air in the tunnel could be controlled between 1 and 25 feet per second.

The plants used in the tests were tobacco (Nicotiana Tabacum, hybrid 1159), grown from seedlings in plastic pots in a greenhouse. When the plants were about 2 feet tall, they were sprayed with an insecticide, Vapona Carmel GH-19 (2, 2 dichlorovinyl dimethyl phosphate) to deter insects. Then the plants were allowed to grow to a height of about 4 feet before they were used in a test. Before each test, all leaves, except one, were removed from the stem. After this operation, to prevent water loss from those areas on the stem from which leaves had been severed, the plants were kept in the greenhouse for 12 to 24 hours, until a scar formed. Then the pots were wrapped in plastic bags, tied around the stem of the plant to prevent water loss from the soil. Next, a pot with a plant was placed on the balance in the tunnel and the leaf was placed between fine nets to keep its surface flat and parallel to the wind [6]. In this manner, the leaf resembled a flat plate for which data on boundary layer thickness, friction-, and heat- and mass-transfer are available. This permitted a comparison of the experimental results on a living system with results of previous boundary layer studies in the literature for the flow and mass transfer from a flat surface. Transpiration studies were conducted at velocities ranging from 3 to 32 feet per second, corresponding to Reynolds Numbers based on the length of the leaf, between 10,000 and 50,000.

The temperature of the air in the test section of the wind tunnel was measured with a thermistor, the relative humidity with a Lithium-chloride sensor, and the wind speed with a hot wire. Leaf temperatures were measured continuously by placing a glass-coated thermistor bead on the top surface of the leaf and pressing the bead into the leaf. Intravenous temperatures of the mid riff of the leaf were also measured, using a very fine (40 gauge) copper-constantan thermocouple.

Transpiration rates were measured gravimetrically, using the balance shown schematically in Fig. 1. The pot containing the plant specimen was placed on one side of the balance and counter balanced by weights on the other. The side with the weight rested on a highly sensitive pressure transducer cell whose output was processed and recorded on a precision voltmeter. However, because of environmental noise, it was necessary to average the electrical signal for each gravimetric reading over a period of 20 seconds by means of an integrator, also shown in the block diagram. Gravimetric readings were

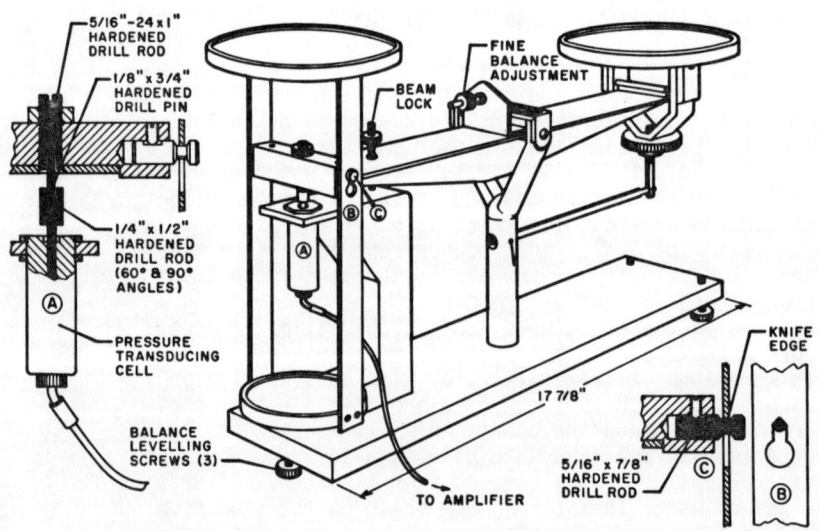

Figure 1
Schematic Diagram of Transpiration Balance

taken at 1/2 hour intervals and evapotranspiration rates calculated by dividing the change in weight by the time interval.

The experimental measurements in the wind tunnel were of much greater accuracy than comparable measurements would be under natural conditions. Nevertheless, because one is dealing with a rather complex system, the estimated accuracy of the evapotranspiration rate is only about 5%. The measured values of temperature, relative humidity and wind speed are accurate to within ± 1%.

Each series of tests commenced with a test on the unsprayed plant over the entire Reynolds Number range. After the controlled data had been obtained the plant leaf was sprayed with the antitranspirant, using a spray gun which provided a uniform spray over the upper and lower surfaces. All the antitranspirant materials studied were diluted in pure water to which 0.1 percent by volume of surfactant (Triton X-100) was added. The spraying distance was kept at 25 cm and the spraying time was 2 seconds. Before being placed again in the wind tunnel for further study the leaf surface was allowed to dry under Grolux lamps for about 30 minutes. After the test specimen had again been placed in the wind tunnel, the series of tests which had previously been conducted with the untreated leaf, was repeated over the same Reynolds Number range. Then, test data were obtained periodically over a period of 5 to 6 days at a mean Reynolds Number of 27,500, until the evapotranspiration rate climbed to 90% of that for the control. In view of the fact that the leaf was growing during that period, appropriate adjustments had to be made for the change in surface area. This was done by measuring the leaf area before and after a test and assuming a linear growth rate during the time interval of the test to determine the area at intermediate time periods. The leaf was periodically inspected visually to determine whether or not the antitranspirant had damaged the leaf.

3. METHOD OF ANALYSIS

In order to compare the rate of evaporation from one leaf with the rate of evaporation from another, a common basis for comparing mass transfer rates from irregular shaped surfaces must be established. In conventional engineering systems experimental data for heat- and mass-transfer from geometrically similar systems are usually correlated by means of dimensionless parameters, some of which contain a length scale to characterize the size of the system. For flow, as well as heat- and mass-transfer over a flat plate, it can be shown [7] that the appropriate length is the distance between the leading and the trailing edge. For flow over irregular shaped bodies, the choice of the length scale in dimensionless parameters such as the Reynolds, Nusselt and Sherwood Numbers is somewhat arbitrary and must be based on an idealization of the real system. Thus, in order to use dimensionless parameters such as the Reynolds and Sherwood Numbers to correlate experimental data of mass transfer experiments from irregular shaped surfaces, it is necessary to define a characteristic length dimension for the system. An approach, which has been used successfully to correlate convection heat transfer data from broad leaves of plants [8], is to define an equivalent length in the flow direction equal to the length L of a rectangle of width D which would yield the same average flux as a leaf with a maximum width D, but with the actual leaf shape as shown in Fig. 2. In other words, the leaf is treated as though it consists of lamina of width dy and length w(y) in the direction of the flow, with no interactions between the fluid flowing over adjacent laminae. With this assumption it is possible to apply the results of two-dimensional boundary layer flow to flow over a leaf.

As shown in detail in Ref. 9, the local Sherwood Number at a distance x from the leading edge, Sh_x, in flow over a flat plate is given by a relation of the type

$$Sh_x = 0.332 \, Re_x^{1/2} Sc^{1/3} \tag{2}$$

where $Sh_x = K_T \, x / \mathcal{D}$
$Re_x = U_\infty \rho_\infty x / \mu_\infty$
$Sc = \mu / \rho \mathcal{D}$

If the concentration (or density) of the water vapor is uniform over the surface, the local molar flux J_x is given by

$$J_x = K_T(x)(\rho_0 - \rho_\infty) \tag{3}$$

Integrating Eq. 3 over the leaf surface area yields the total flux J_T as

$$J_T = \bar{C} \int_0^D \int_0^{w(y)} x^{-1/2} dx \, dy \tag{4}$$

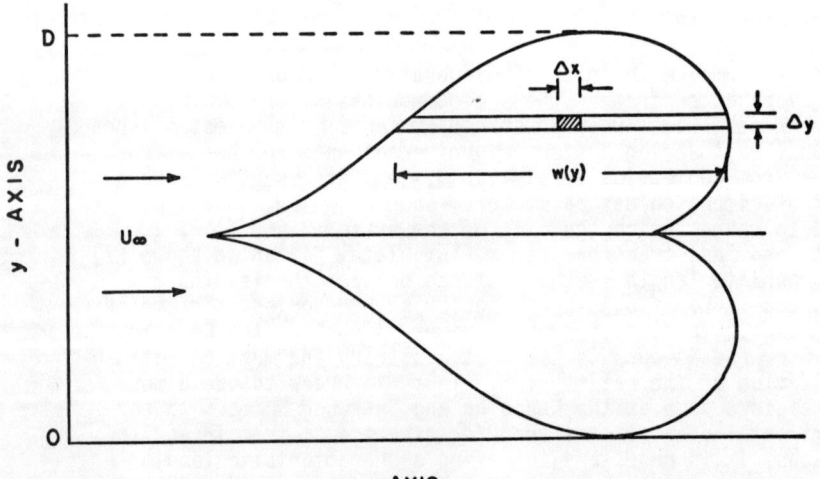

Figure 2
Nomenclature Defining the Effective Length Dimension of a Leaf

where $\bar{C} = 0.332 \, Sc^{1/3}(\rho_0 - \rho_\infty)(U_\infty \rho_\infty/\mu_\infty)^{1/2}$

The average molar flux $\bar{J}_T$ is then

$$\bar{J}_T = J_T \int_0^D \int_0^{w(y)} dx \, dy \tag{5}$$

which after integration gives

$$\bar{J}_T = 2C \, Sc^{1/2}(\rho_0 - \rho_\infty)(U_\infty \rho_\infty L/\mu_\infty)/L \tag{6}$$

where the equivalent length L is defined by

$$L = \left[\int_0^D w(y) dy \Big/ \int_0^D w(y)^{1/2} dy \right]^2 \tag{7}$$

Thus, L is the length of a rectangle of width D which would yield the same average evaporation rate as the leaf. It can be obtained by numerical integration once the leaf shape and size is known. This procedure was used to evaluate L for all the leaves used in this investigation.

4. EXPERIMENTAL RESULTS

Transpiration from a plant leaf is a mass transfer process which differs from mass transfer in conventional engineering systems principally by the discrete nature of the sources from which diffusion occurs. In conventional engineering systems the transfer interface is continuous and the surface area across which mass is transferred is the same as the area over which frictional forces act. For a leaf, however, mass transfer occurs through a finite number of stomatal openings, while skin friction and heat transfer take place over the entire leaf area. As shown in more detail by Cannon [2], to relate experimental results for mass transfer from leaves with transport phenomena in systems for which friction and heat transfer data are also available, it is convenient to define an overall mass transfer coefficient K_T in terms of the entire leaf surface A (upper + lower) according to

$$\bar{K}_T = T/A(\rho_0 - \rho_\infty) \tag{8}$$

The dimensionless parameter used to correlate engineering mass transfer data is the Sherwood Number, Sh, defined by the relation

$$\overline{Sh} = \bar{K}_T L/\mathcal{D} \tag{9}$$

where L is the effective length of the leaf [8] defined by Eq. 7 and $\mathcal{D}$ is the diffusivity of water vapor in air at the temperature of the surrounding. It is generally assumed [3] that the water vapor in the substomatal cavities of the leaf is saturated so that ρ_0 was taken as the density of water vapor at the leaf temperature while ρ_∞, the density of the water vapor in the vicinity of the leaf, was based on the partial pressure at the temperature of the air and water vapor in the test section. The independent control parameter for each test was the velocity of the air stream in the test section, U_∞, expressed in terms of the Reynolds Number, Re_L ($= \rho_\infty U_\infty L/\mu_\infty$), based on the effective length of the leaf in the flow direction L given by Eq. 7.

The effectiveness of an antitranspirant E is defined as the ratio of the Sherwood Number for the leaf treated with the antitranspirant Sh_T to the Sherwood Number of the untreated control leaf, Sh_C, or

$$E = \overline{Sh_T}/\overline{Sh_C} \tag{10}$$

Thus, when the driving potential $(\rho_0 - \rho_\infty)$ is the same for the treated leaf and the untreated leaf, E is equal to the ratio of the transpiration flux from the leaf sprayed with the antitranspirant agent to the flux from an untreated leaf of the same size and shape.

Using the above definitions, the experimental results are summarized in Figs. 3 and 4. Fig. 3 shows the Sherwood Numbers for the untreated control leaf and for leaves treated with two kinds of film forming antitranspirants (Mobileaf 1:5 by volume and Wilt-Pruf 1:4 by volume) and two stomata closing antitranspirant materials (mono-glycerol ester of n-decenyl succinic acid and phenylmercuric acetate) as a function of the Reynolds Number, approximately three hours after application of the antitranspirants. The succinic acid was sprayed in a 0.15 millimolal solution whereas the PMA was used in a 1.0 millimolal and a 0.3 millimolal solution. An inspection of Fig. 3 shows that the short term effect of PMA was to reduce the transpiration rate over the entire range of Reynolds Numbers to Between 25 and 35% of the value for an untreated plant, whereas the monoglycerol ester spray achieved reductions between 55 and 80% over the Reynolds Number range of the tests. Film forming type antitranspirants, on the other hand, only achieved reductions in evapotranspiration rates to between 50 and 65% of the values for an untreated plant. Similar results have been reported by other investigators [10].

To determine the feasibility of using antitranspirants, it is necessary to know how their effectiveness changes with time. Fig. 4 presents the ratio of the Sherwood Number for leaves treated with antitranspirants to the Sherwood Number for similar untreated leaves as a function of time after application at a Reynolds Number of 27,500. An inspection of the results in Fig. 4 shows how the effectiveness of the antitranspirants varies with time. Except for the leaves treated with a 1 millimolal PMA spray, the transpiration rate of leaves treated with all other antitranspirants increased after application, but the effectiveness of both of the film forming type anti-transpirants tested diminished less rapidly than the effectiveness of the other two metabolically active agents. After three days only the 1 millimolal PMA antitranspirants had an effectiveness below the 70 percent level. However, the plants sprayed with 1 millimolal PMA began to wilt after three days and the results can therefore not be considered as a valid test of antitranspirant treatment. On the other hand, none of the other plants showed any visible ill

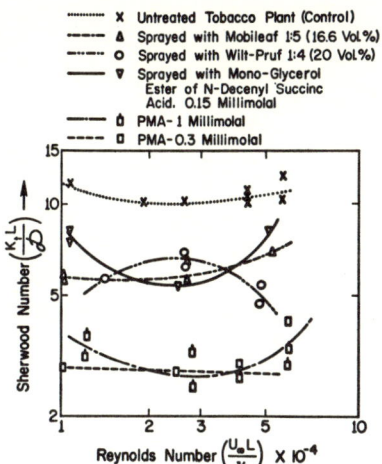

Figure 3
Short-Term Effectiveness of Various Antitranspirants - Sherwood Number vs. Reynolds Number for Control Specimen and Tobacco Plants Treated with Antitranspirants

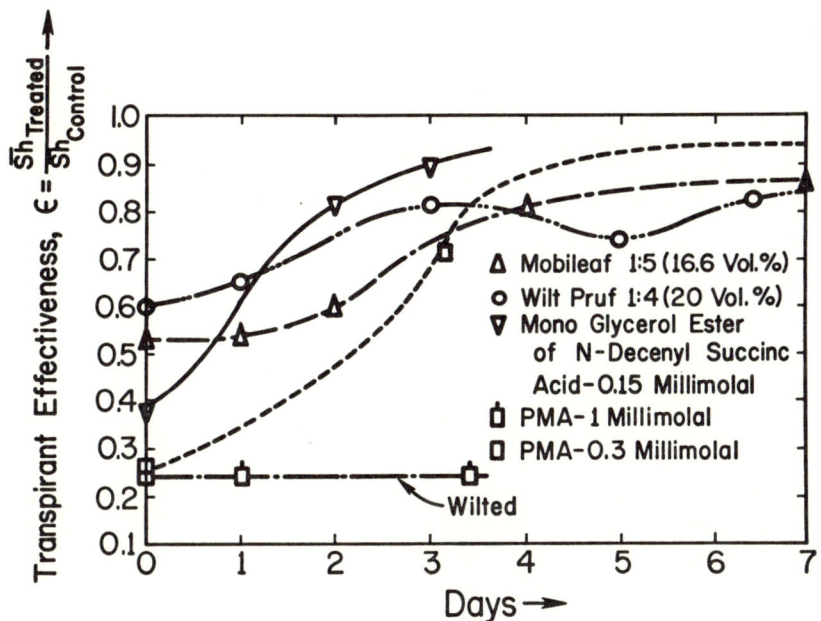

Figure 4
Long-Term Effectiveness of Various Antitranspirants - Effectiveness vs. Time

effects as a result of their antitranspirant treatment and the data shown in Fig. 4 can therefore be considered a valid indication of the longer term effectiveness that can be expected from antitranspirant sprays.

5. SUMMARY AND CONCLUSIONS

1. An experimental technique for measuring the rate of water consumption of living plants under controlled and reproducible conditions has been developed.
2. For the short term antitranspirants can be very effective, some reducing water consumption to 30 percent of the consumption by an untreated specimen.
3. The average effectiveness of antitranspirants during the first week after application is between 70 and 80 percent.
4. Since the cost of water varies considerably in different parts of the world, the economic viability of using antitranspirants cannot be ascertained by a technical study alone. The results of this investigation suggest, however, that in arid regions and/or under conditions where water is expensive or not available in sufficient quantities, the use of antitranspirants may be economical provided no adverse ecological effects occur. However, under normal conditions commercially available antitranspirants do not have a sufficient long range effectiveness to be economical today.
5. An effort should be made to develop ecologically safe antitranspirants with high long-term effectiveness and the practical engineering aspects of their application should be studied.

ACKNOWLEDGMENT

The authors gratefully acknowledge the financial support of the U.S. National Science Foundation under Grant No. GK 17184, as well as assistance provided by the University of the Negev, Israel, in the course of this work.

NOMENCLATURE

A - area of leaf

$\bar{C}$ - constant (see Eq. 4)

D - maximum width of leaf

$\mathcal{D}$ - diffusivity of water vapor in air

E - effectiveness of antitranspirant

J_T - mass transfer flux

K_T - total mass transfer coefficient for evaporation from leaf

L - equivalent length of leaf (see Eq. 7)

R - resistance to water vapor transport

Re_L - Effective Leaf Reynolds Number

Re_x - Reynolds Number at x

Sc - Schmidt Number

Sh_x - Sherwood Number at x

$\overline{Sh}$ - Average Sherwood Number

T - rate of mass transfer from leaf

U_∞ - free stream velocity

w(y) - width of leaf

x - distance from leading edge

y - leaf coordinate perpendicular to x

μ - viscosity

ρ - density

REFERENCES

1. Poljakoff-Mayber, A., and J. Gale, "Physiological Basis and Practical Problems of Reducing Transpiration," pp. 277-306 in Vol. III, <u>Water Deficit and Plant Growth</u>, Acad. Press, Inc., New York, 1972.

2. Cannon, J. N., <u>A Model Study of Transpiration from Broad Leaves</u>, Ph.D. Thesis, University of Colorado, 1971.

3. Gaastra, P., "Photosynthesis of Crop Plants as Influenced by Light Carbon-Dioxide, Temperature, and Stomatal Diffusion Resistance," Mede. Lanbouwhogesch Wageningen, 1959, pp. 1-68.

4. Meidner, H., and T. A. Mansfield, <u>Physiology of Stomata</u>, McGraw-Hill Book Company, New York, 1968.

5. Cowan, I. R. and C. A. Troughton, "The Relative Role of Stomata in Transpiration and Assimilation," <u>Planta</u>, 97: 325-336, 1971.

6. White, H. and F. Kreith, "Thermal Analysis of Temperature Fluctuations in Plant Leaves," <u>Heat Transfer 1970</u>, Elsevier Pub. Co., Amsterdam, Netherlands, Cu 3.8: 1-9, 1970.

7. Kreith, F., <u>Principles of Heat Transfer</u>, 3rd ed., International Text Book Co., Scranton, Pa., 1973.

8. Parkhurst, D., P. R. Duncan, D. M. Gates, and F. Kreith, "Wind Tunneling Modelling of Convection of Heat Between Air and Broad Leaves of Plants," Agr. Meteorol., 5: 33-47, 1968.

9. Kays, W. M., Convective Heat- and Mass-Transfer, McGraw-Hill Book Company, New York, 1968.

10. Mishra, D. and G. C. Pradhren, "Effect of Transpiration Reducing Chemicals on Growth, Flowering, and Stomatal Opening of Tomato Plants," Plant Physiol., 50: 271-274, 1972.

WATER TRANSFER TO GERMINATING SEEDS AS AFFECTED BY SOIL HYDRAULIC PROPERTIES AND SEED-WATER CONTACT IMPEDANCE

AMOS HADAS

Institute of Soils and Water, Agricultural Research Organization,
The Volcani Center, Bet Dagan, Israel

ABSTRACT

Seed germination as affected by soil water hydraulic properties (e.g. matric potential, hydraulic conductivity) and seed-soil water contact, was studied experimentally. The concept of contact impedance combining seed-soil water contact area contact zone-hydraulic properties was introduced and determined experimentally. It was found that reducing soil matric potential or the conductivity to water postpones germination as expected, but to a lesser extent than contact impedance does. Applications to field conditions are discussed.

LIST OF SYMBOLS

$\bar{a}$ — mean seed radius [m]
$\bar{D}$ — mean diffusivity to water [m^2/s]
h — defined by $L/\bar{a}$ [1/m]
k — defined by $\theta^i_{soil}/\theta^g_{seed}$ [dimensionless]
L — defined by $\bar{a}\bar{\alpha}/\bar{D}_{seed}$ [dimensionless]
M — water gain by seed [kg/m^3]
n — integers 1,2,3,....n
r — radial distance from seed's center [m]
t — time [s]
$\bar{\alpha}$ — factor related to seed-soil mean contact impedance [m/sec]
β_n — roots of $\beta_n \cot\beta_n + L - 1 = 0$

θ — volumetric water content [m^3/m^3]

Π — 3.1416 ....

Subscripts

The letters t, i, g are used as sub or top script and signify time t, initial state and state at germination respectively. The letter s or seed signify soil and seed respectively.

INTRODUCTION

Experimental data concerning seed germination under water stress relate the affected germination rate and final germination to various soil factors. These factors are decreasing soil water potential [1,2,3], soil water hydraulic conductivity [2,4], and seed-soil water contact area [3,5]. Even though all agree that these factors do affect germination, there are conflicting opinions regarding the relative importance of these factors and the ranges of magnitudes which affect germination.

In comparing the results reported in the literature listed above, one can not avoid the feeling that at present the knowledge is based on incomplete and imperfect experimental evidence, since the combined effects of the soil water potential, hydraulic conductivity and seed-soil water contact area were not completely separated.

In the reported work, special attention is paid to separate the effects the various factors have on germination.

EXPERIMENTAL METHODS

Chickpea (<u>Cicer arietinum</u> L. local var.) seeds were germinated under three different sets of conditions: (a) in aerated pure water or solutions of polyethylene-glycol (m.w. 20,000) having different osmotic values (ranging from -160 to -10^4 Joules/kg); (b) buried in pure sand fractions of different diameters (0.6 to 4 mm), through which a flow of pure water was maintained at a rate equaling the appropriate hydraulic conductivity (ranging 1 x 10^{-3} to 10^{-12} m/sec) for a given water content; and (c) buried in soil aggregates equilibrated at different matric water potentials on pressure plates (ranging from -30 to -1500 Joules/kg). All experiments were carried out at a constant temperature of 22 $\pm$ 1°C.

The hydraulic conductivity of the sand and aggregates fractions was determined by the methods of Youngs [6] and Collis-George and Rosenthal [7], respectively. Seed-soil water contact area was determined according to Collis-George and Hector [3].

THEORETICAL CONSIDERATIONS AND DATA ANALYSIS

By assuming a spherical seed placed into wet soil the water flow from the soil into the seed can be described by equation

$$\frac{\partial \theta_{seed}}{\partial t} = \frac{\bar{D}_{seed}}{r} \left[\frac{\partial^2 (r\theta_{seed})}{\partial r^2} \right] \quad (1)$$

for the following initial and boundary conditions

$$0 < r \leq \bar{a} \quad t < 0 \quad \theta_{seed} = \theta^i_{seed}$$

$$r = \bar{a} \quad t \geq 0 \quad -\bar{D}_{seed} \frac{\partial \theta_{seed}}{\partial r} \bigg|_{r=\bar{a}} = \bar{\alpha}(\theta^D_{seed} - \theta^t_{seed})$$

A solution to equation (1) for $t \to \infty$ is given by Crank [8,p.91], but since seed germination occurs at finite time period, his equation was changed accordingly and is given in equation (2),

$$\frac{M_t/M_\infty}{M_g/M_\infty} = \frac{M_t}{M_g} = \frac{1 - \sum_{n=1}^{\infty} 6L^2 \exp\{-\beta_n^2 \bar{D}_{seed} t/\bar{a}^{-2}\}/\beta_n^2\{\beta_n^2 + L(L-1)\}}{1 - \sum_{n=1}^{\infty} 6L^2 \exp\{-\beta_n^2 \bar{D}_{seed} t_g/\bar{a}^{-2}\}/\beta_n^2\{\beta_n^2 + L(L-1)\}} \quad (2)$$

Knowing $\bar{D}_{seed}$, $\bar{a}$, and M_t/M_g (from seeds weighed at different times), one can estimate L by matching the experimental M_t/M_g values with precalculated M_t/M_g curves for different L, $\bar{a}$, t and $\bar{D}_{seed}$ values.

A solution of equation (1) for the soil, $r > \bar{a}$, will yield an estimate of M_t/M_g assuming the seed to be a perfect sink. Such a solution is given in equation (3) [ref. 9, p. 350],

$$\frac{M_t}{M_g} = 1 - \left\{ \frac{\bar{a}^{-3}}{2\pi^{1/2} k (\bar{D}_s t)^{3/2}} + \frac{3\bar{a}^{-4}[2+\bar{a}h(2-k)]}{4\pi^{1/2} k^2 h (\bar{D}_s t)^{5/2}} + \ldots \right\} \quad (3)$$

Knowing $\bar{D}_s$, θ^i_{soil}, θ^f_{seed}, L, and $\bar{a}$, one can estimate the time needed to reach a given M_t/M_g.

From relative water uptake – time sequence for the seeds $\bar{D}_{seed}$ was determined according to [10]. A family of curves of M_t/M_g as functions of various L, $\bar{D}_{seed}$ and t values, was constructed.

Experimental $M_t \cdot M_g$, when matched to computed values for given $\bar{D}_{seed}$ and t, yielded the L values sought for the various sand fractions at different water contents.

These L values were then related as functions of the relative wetted seed surface area. These values were used to predict relative water uptake by seeds from soil aggregates.

RESULTS AND DISCUSSION

a. Effects of soil water potential

The germination course of chickpea seeds as a function of time and external water potential is given in Figures 1a and b. The data show that complete germination is attained even at external water potentials of -1×10^3 to -2×10^3 Joules/kg. There is excellent agreement between the germination percentage obtained in osmotic solutions (Figure 1a) and in soil aggregates (Figire 1b). These water potentials are very low compared with those reported by Collis-George and Hector [3] to reduce germination. However, the time for germination increases as the external water potentials decrease. According to the data, for chickpea seeds the shortest germination time can still be maintained at water potentials down to $-3.8 \times 10^{+2}$ Joules/kg.

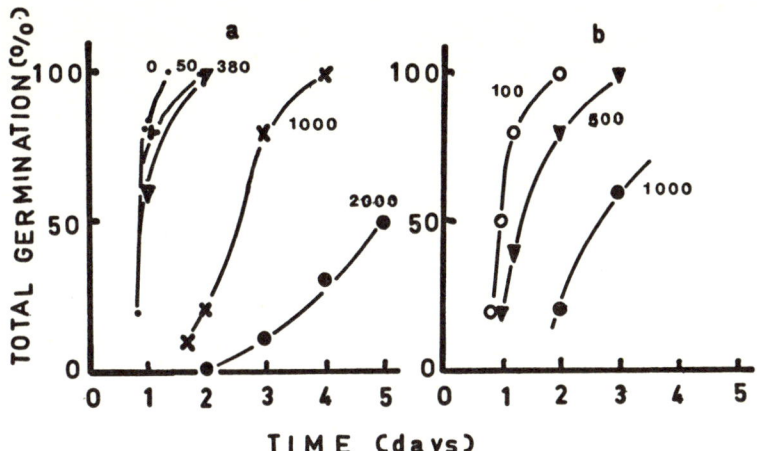

Fig. 1: Total germination of chickpea seeds as functions of time and water potential; a) seeds in osmotic solutions and b) buried seeds in soil aggregates of 0.25 - 0.5 mm. Numbers indicate water potential in [Joule/kg].

In previous work [11] it was found that 80% relative water gain (on an air-dry seed weight basis) ensures germination, yet there is a time delay between the attainment of this level of seed hydration and its radicle emergence. The time lag found was 4-6 and 24-48 h for seed germinated at water potentials of 0 to $-3.8 \times 10^{+2}$ and 2×10^3 Joules/kg, respectively.

b. Effects of seed-soil water contact area and the soil capillary conductivity

Experimentally, it was extremely difficult to differentiate between the effects the contact area and the soil's bulk capillary conductivity have on germination. In Figure 2 germination percentage is given as a function of time and relative contact area. The data show, as expected, that germination is delayed as the contact area decreases. The data presented in Figure 3 show that germination is delayed as the contact area and the capillary conductivity decrease. It should be noted that five orders of magnitude change in the capillary conductivity did not change the course of germination where the contact area is quite appreciable (35 to 12% of the seed surface), although the soil water potential is very high.

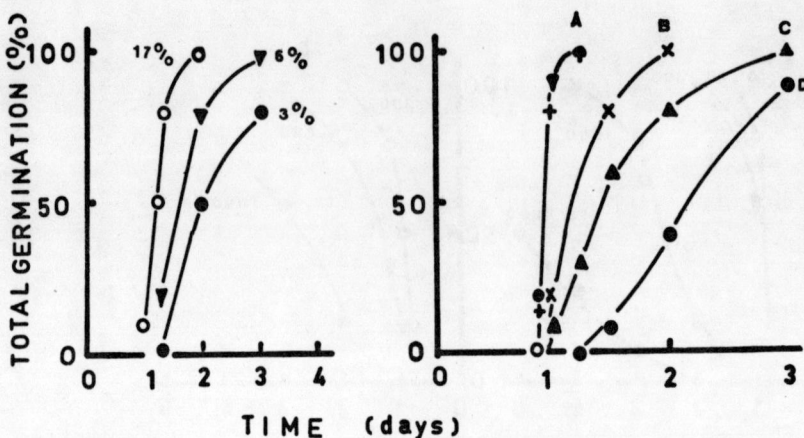

Fig.2: Total germination of chickpea seeds as functions of relative wetted area. Hydraulic conductivity K = $10-5 \times 10^{-10}$ [m/s] range. Numbers indicate relative wetted area.

Fig.3: Total germination of chickpea seeds as a function of relative wetted area, hydraulic conductivity and time. a) Hydraulic conductivity range 1×10^{-3} - 3×10^{-8} [m/s], wetted area 35-12%. b) Hydraulic conductivity 1×10^{-9} [m/s] wetted area 8%. c) Hydraulic conductivity 1×10^{-10} [m/s] wetted area 6%. d) Hydraulic conductivity 1×10^{-12} [m/s] wetted area 17%.

c. Effect of contact impedance

During germination, water can move from the soil into the seed only across the seed-soil water contact areas, of which the capillary conductivity and the actual areas are not known. Because of the special geometrical configuration of the contact areas, neither the soil bulk's capillary conductivity nor the estimated contact areas alone can characterize the water transfer across the seed-soil water contact zone, but a combination of both - such as contact impedance - can.

Contact impedance factor L values are given as a function of the computed relative contact area in Figure 4. In Figure 5 computed values of relative contact area as functions of aggregate sizes and water matric potentials are given. These data show that the larger the seed is, compared with the aggregate size, for a given matric potential value, or the larger the matric potential is for a given seed-soil aggregate size, the larger the contact area is.

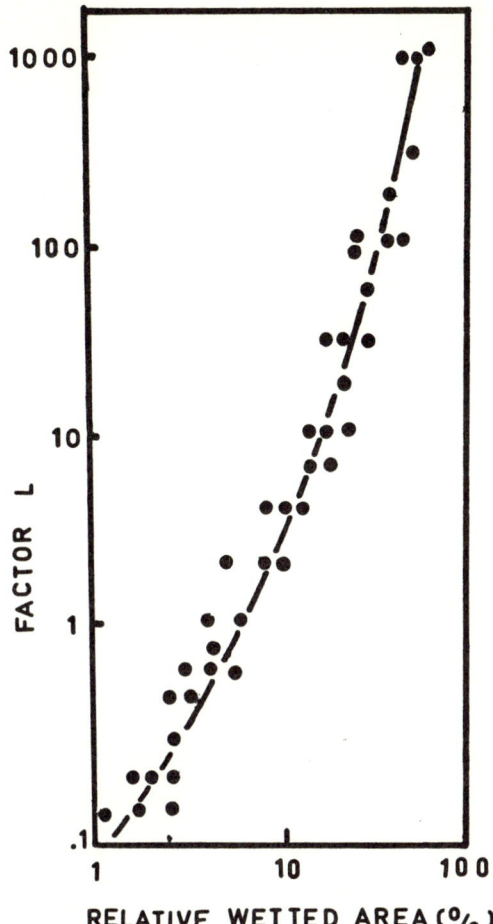

Fig. 4: Dependence of L factor on the relative wetted area.

In Figures 6 and 7, predicted relative water gains by vetch and chickpea seeds and experimentally determined values are compared and agree fairly well. It should be stressed here that the soil water diffusivity values used for the computations were lower than those reported in the literature [12] or computed from data of [13]. Using higher soil water diffusivity values would have yielded higher computed relative water gains and thus spoil the fairly good agreement shown in Figures 6 and 7. The predicted data were computed by assum-

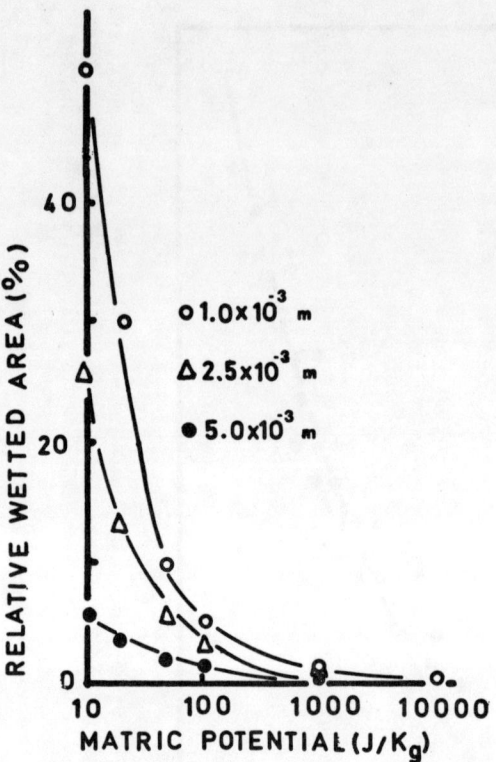

Fig. 5: Relative wetted area as a function of aggregate diameter and soil water matric potentials for seeds of 10×10^{-3} m diameter.

ing the seed to be a perfect sink, which it is not, nor is its surface evenly permeable to water. Higher soil water diffusivity values would have shortened the computed time to reach a given relative water gain by the seeds. This observation suggests that the soil can furnish more water than the seed actually imbibes, under adverse conditions of low soil water diffusivity and high contact impedance. This phenomenon can be attributed to the following factors: scale effect, the non-uniform and non-symmetric arrangement of the seed organs.

As long as the seed is much larger than the soil aggregates or particles surrounding it, the contact points are numerous and probably evenly distributed around the seed: therefore, the contact impedance concept - which assumes these conditions - holds. However,

Water Transfer to Germinating Seeds

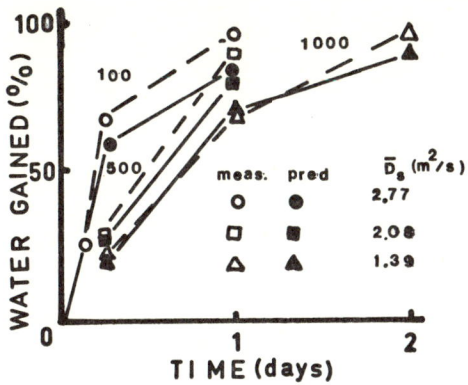

Fig. 6: Comparison between measured and predicted values of relative water gain (% of dry seed weight) as a function of time for Vetch seeds in soil.
(Numbers indicate matric water potential [Joule/kg]).

as the seed size approaches the soil aggregates' size or is smaller, the number of contact points decreases toward the limit of one contact point, which makes the contact impedance concept inapplicable. The seed behaviour will depend then on where the contact point is located on the seed surface. Since most seed coats are not evenly permeable to water [14] or even sometimes totally impermeable except to a special area (e.g. the micropile zone in pea seeds, ref. [15]), the contact points location becomes crucial to germination. Still, it can not simply be taken into consideration by the model used here. In order to analyse and predict more accurately the seed's behaviour, one should analyse the water pathway in the soil and into and inside the seed, a formidable task with the meager knowledge we have at hand. However, some field recommendations can be drawn up.

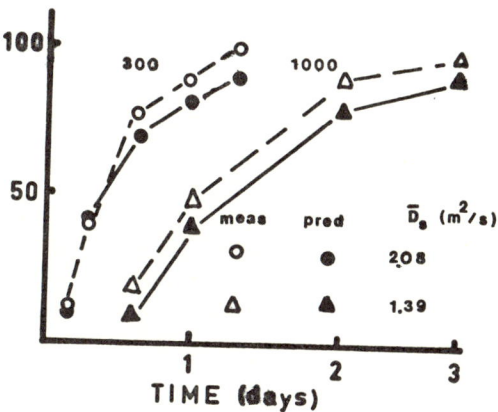

Fig. 7: Comparison between measured and predicted values of relative water gain (% of dry seed weight) as a function of time for chickpea seeds in soil.
(Numbers indicate matric potential [Joule/kg]).

d. Field applications

Since the soil can furnish more water than the seed can imbibe, it seems that the limiting factors are the contact impedance and the location of the contact zone on the seed surface; hence, means of improving the seed soil contact consequently improve seed germination. Data given in Figure 5 suggest that seed-soil water contact area increases as the aggregates' size decreases with respect to the seed size, yet if the clod-to-seed-size ratio is larger than 5 to 1 there will be minute increase in the contact area at moderately low and very low matric water potentials values. Hence, one can recommend that seeds should be sown in aggregated beds where the aggregates are one-fifth smaller, or less than the seed, so as to ensure good seed-soil water contact for a given water contact. This is true for dry farming seed-bed preparations, where large seeds are concerned, but for irrigated areas or humid regions the aggregates can be larger than one-fifth of the seed size so as to allow fast relief from excessive water and improved aeration without impairing water uptake, since contact impedance will be low due to higher water contents or matric potentials even for large aggregates. If small seeds are to be sown, this prerequisite of the seed-aggregate size can not be met without pulverizing the soil. For dry land farming the seeds should then be either punched into the soil, or sown and the soil then compacted around the seeds, to ensure a better seed soil contact and germination [16].

REFERENCES

1. Hunter, J.R., and Erickson, A.E. (1952) Relation of seed germination to soil moisture tension. Agron. J. 44: 107-110.

2. Collis-George, N., and Sands, J.E. (1959) The control of seed germination by moisture as a physical property. Aust. J. agric. Res. 10: 628-636.

3. Collis-George, N. and Hector, J.B. (1966) Germination of seeds as influenced by matric potential and by area of contact between seed and soil water. Aust. J. Soil Res. 4: 145-164.

4. Shaykewich, C.F. and Williams, J. (1971) Influence of hydraulic properties of soil on pre-germination water absorption by rape seed (Brassica napus L.). Agron. J. 63: 454-457.

5. Sedgley, R.M. (1963) The importance of liquid-seed contact during germination of Medicago tribuloides. Aust. J. agric. Res. 14: 646-653.

6. Youngs, E.G. (1964) An infiltration method of measuring the hydraulic conductivity of unsaturated porous materials. Soil Sci. 97: 307-311.

7. Collis George, N. and Rosenthal, M.J. (1966) Proposed outflow method for the determination of the hydraulic conductivity of unsaturated porous materials. Aust. J. Soil Res. 4: 165-180.

8. Crank, N. (1956) Mathematics of Diffusion. Oxford Press, Oxford.

9. Carslaw, H.S. and Jaeger, J.C. (1959) Conduction of Heat in Solids. Clarendon Press, Oxford.

10. Philips, R.E. (1968) Water diffusivity of germinating soybean, corn and cotton seed. Agron. J. 60: 568-571.

11. Hadas, A. and Stibbe, E. (1973) An analysis of soil water movement toward seedlings prior to emergence. In:"Physical Aspects of Soil Water and Salts in Ecosystems." Hadas, A. et al., Eds. Ecological Studies Vol. 4: 97-106, Springer Verlag, Heidelberg.

12. Doering, E.J. (1965) Soil-water diffusivity by the one-step method. Soil Sci. 99: 322-326.

13. Amemiya, M. (1965) The influence of aggregates size on moisture content-capillary conductivity relations. Proc. Soil Sci. Soc. Amer. 29: 744-748.

14. Mayer, A.M. and Poljakoff-Mayber, A. (1963) The Germination of Seeds. Pergamon Press, London.

15. Manohar, M.S. and Heydeker, W. (1964) Effects of water potential on germination of pea seeds. Nature 202: 22-24.

16. Dasberg, S., Hillel, D. and Arnon, I. (1966) Response of grain sorghum to seed bed compaction. Agron. J. 58: 199-201.

ENERGY AND AGRICULTURE: A NATIONAL CASE STUDY

G. STANHILL

Volcani Center, Agricultural Research Organization
Bet Dagan, Israel

Agriculture can be described as man's effort to increase the rate at which plants fix solar energy into forms convenient for his use. The large improvements achieved in the efficiency of this process during the last third of a century illustrated in Fig. 1, can be attributed largely to inputs which, either directly or indirectly, require large amounts of non-solar energy. These requirements are now so large that the overall energy balance in many branches of intensive agriculture is negative. In such circumstances, modern agriculture can be described as the process of converting concentrated fossil fuel into an edible form.

Thus, the food and energy crises are linked organically. On the one hand there is an urgent need to increase world food production; on the other hand, to implement this increase will further add to the demand on reserves of fossil fuels whose current global rate of depletion has been estimated to exceed the rate of deposition by a thousandfold (Johnson, 1970).

The relationship between energy input and food output in agriculture is therefore a matter of considerable importance. There is, however, limited information available on the subject, especially with regard to alternate

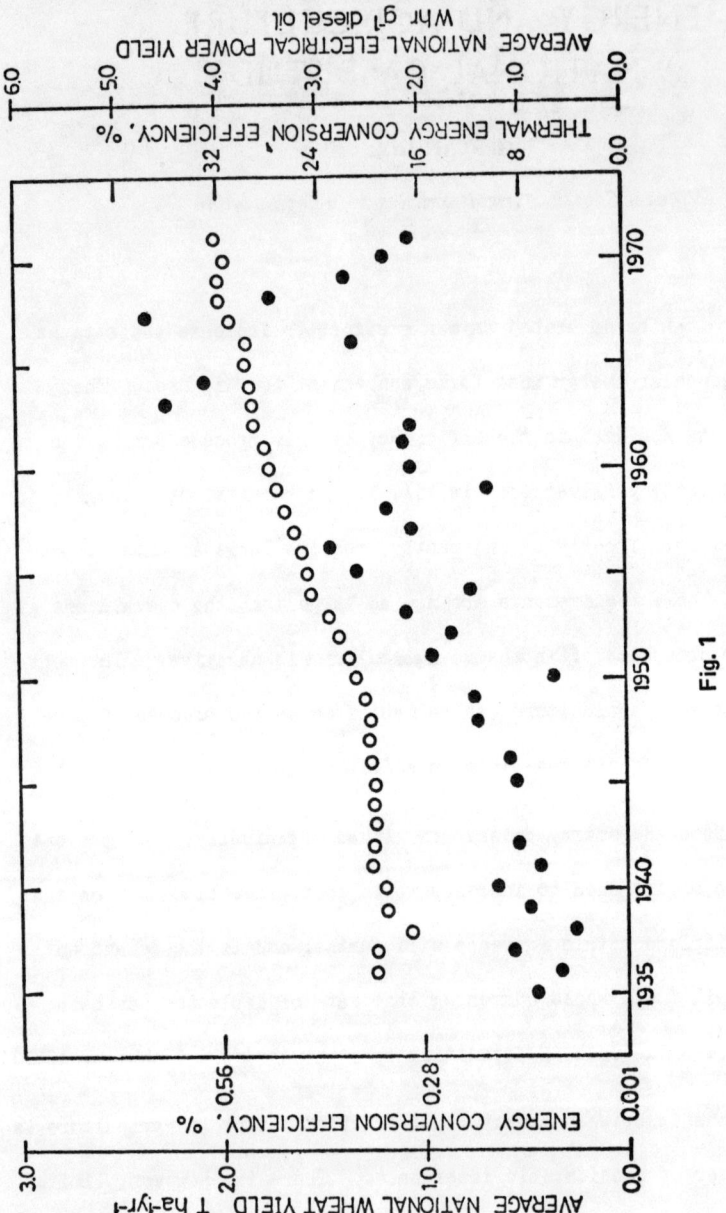

Fig. 1

Fig. 1. Energy conversion in wheat (●) and electrical power (○) production in Israel, 1935-1971. The annual, average wheat yields, obtained from government statistics, were doubled to allow for straw production; a calorific value of 4.0 Kcal g^{-1} was allowed for both grain and straw. The energy equivalent was divided by the estimated solar radiation within the photosynthetically active waveband absorbed by the crop. This was estimated from local measurements of global radiation (Stanhill, 1970), of which the photosynthetically active fraction above the crop was estimated to be 9.5, with a crop reflectivity and transmissivity of 0.10 and 0.15, respectively. The statistics on efficiency of electrical power production in Israel were kindly supplied by M. Nelken of the Israel Electrical Corporation.

---

and complex production systems. To date, the published information includes preliminary energy flow charts for a number of widely different food production systems (Odum, 1967) a brief summary of four systems currently found in Israel (Stanhill, in press) a very detailed study of one productive monoculture, U.S. corn production (Pinmental et al. 1973) and of U. S. agriculture as a whole (Steinhart and Steinhart, 1974) and a brief summary of the British agricultural situation (Blaxter, 1974).

This article examines the energy input and production, internal and external transfers and overall balance in the food production and consumption, of Israel, a country almost entirely dependant on inports for its energy requirements. Although the values obtained may be expected to differ for each country examined, it is thought that the results presented may be representative of a wide range of developing agricultural systems in energy-poor regions.

ENERGY INPUT INTO AGRICULTURE

The non-solar energy input in Israel's agriculture falls into three categories. The first is that fraction used on the farm, e.g. the fuel needed to operate tractors or heat greenhouses. The second is the off-farm energy input into agriculture e.g. that needed to generate the electricity for distributing irrigation water or that needed to manufacture fertilizers. The third fraction is that energy input required to maintain and expand the agricultural infra-structure - including the manufacture of farm equipment, processing of feeds and the transport of agricultural inputs and products.

The various energy inputs for the agricultural year 1969/70 are detailed in Table 1. They total 7.83×10^{12} Kcal, equivalent to a flux density of 1903 Kcal m^{-2} yr^{-1} when averaged over the 4105 Km^2 of cultivated land; some three orders of magnitude below that for the incident solar radiation which is 1805×10^3 Kcal m^{-2} yr^{-1} (Stanhill, 1970).

Fifty eight percent of the non-solar input was in Israel, 42% was abroad in the manufacture of agricultural inputs imported into Israel. The local energy input during 1969/70 is detailed in Table 2, according to energy source.

The changing energy requirements in agriculture have been determined from the linearized rates of change of the major inputs (Table 3). Summing and weighting the major changes in energy input detailed in Table 3 give an estimated increase in annual input of 290×10^9 Kcal or 71 Kcal m^{-2} yr^{-1}.

TABLE 1

ANNUAL ENERGY AND MASS INPUT INTO AGRICULTURE
ISRAEL 1969/70

INPUT	MASS Tons$\times 10^3$	ENERGY Kcal$\times 10^9$	FLUX DENSITY[1] Kcal m^{-2} yr^{-1}
ON FARM			
Fuel, lubricants and grease[2]		764	186
Human labor[3]		23	6
OFF FARM IN ISRAEL			
Water distribution[4]	1,340,000	2919	709
Fertilizer manufacture			
Nitrogenous[5]	113	432	105
Phosphatic[6]	60	22	5
Potassic[7]	15	15	4
Total	200	504	123
Pesticide manufacture[9]			
Insecticides	6.1	135	33
Fungicides	1.9	46	11
Herbicides	1.1	25	6
Total	11.8	285	69
Animal feed processing[10]	1108	65	15
OFF FARM, ABROAD			
Animal feed stuffs[11]	721	3040	740
Machinery manufacture[12]	5.5	103	25
INFRASTRUCTURE			
Transport – of produce[13]	1,600	76	18
– of inputs[14]	1,043	50	12
TOTAL		7829	1903

1. Total cultivated area in 1969/70 was 4105 Km^2 (13). 2. Calculated from number of tractors and average fuel use, 220 liters ha^{-1} yr^{-1} (13). 3. Assuming a 42-hour week for the 84000 labor force (13). 4. 1340 x $10^6 m^3$ water applied, average electrical use 0.81 KWH m^{-3} (13), generated at 32% thermal efficiency (Fig.1). 5. Energy requirements for manufacture average 15.145 x 10^6 Kcal per ton N. 6. Energy requirements, including rock processing, average 1.743 x 10^6 Kcal per ton P_2O_5. 7. Energy requirements average 1.637 x 10^6 Kcal per ton K_2O. 9. Using U.S. values of 24.2 x 10^6 Kcal per ton of plant-proctection chemicals (9). 10. Local data average 58x10^3 Kcal per ton of processed feeding stuff. 11. Average imported feeding stuff 3.05 x 10^6 Kcal per ton (13, 15). 12. Estimated from increase and replacement of 5500 tons of machinery per year using U.S. value of 18.8 x 10^6 Kcal per ton of machinery manufactured (2) 13. Each ton of agricultural produce transported requires 47.7x10^3 Kcal (13). 14. Rate per ton as in 13.

TABLE 2

ENERGY SOURCES AND SINKS FOR LOCAL AGRICULTURAL USE
ISRAEL 1969/70

SOURCE	SINK	AMOUNT
FUEL OIL tons$\times 10^3$	Cultivation, etc.	77,020
	Fertilizer manufacture	35,757
	Pesticide manufacture	19,586
	Transport of produce and inputs	6,993
	Animal feed processing	0.400
ELECTRICITY KWH$\times 10^6$	Water distribution	1,085
	Fertilizer manufacture	50
	Animal feed processing	14
	Pesticide manufacture	29
MANPOWER Man-years	Management etc.	84,800

A comparison of the total energy input into Israel agriculture with average values for the input into agricultural production in the U.S.A., (Steinhart and Steinhart, 1974) shows one very significant difference. In Israel, a very large proportion, (39%) of the energy input into crop production represents irrigation, in the United States the proportion is 7%. The cause for the large local value is the energy required for lifting water some 200 m from Lake Kinneret, the main reservoir of the National Water Carrier, into the distribution system.

Even larger energy inputs will be needed if substantial quantities of irrigation water are to be supplied by desalination, even with the most energetically favorable processes. For example the calculated energy requirement for desalination in giant, dual-purpose nuclear installations (Agro-Industrial Group, 1968) producing 3.78×10^6 m^{-3} day^{-1}, is 2700 Kcal m^{-3} for a conventional light water nuclear reactor linked to a multistage, flash evaporator and 1600 Kcal m^{-3} for an advanced breeder reactor with vertical-tube evaporators. With an average depth of water application in Israel of 78 cm yr^{-1} to the irrigated area, as in 1969/70, and allowing 0.3 KWh m^{-3} for distributing the desalinated water over a mean distance of 30 km, the total energy requirement for irrigating with desalinated water would be 2748 Kcal m^{-2} yr^{-1} for the near-term nuclear plant and 1864 Kcal m^{-2} yr^{-1} for a far-term, breeder reactor.

Greater environmental control - as in heated greenhouses - is another rapidly expanding trend in intensive agriculture. The energy

TABLE 3

ANNUAL CHANGES IN ENERGY AND MASS INPUTS

INPUT	PERIOD LINEARIZED	ANNUAL MASS INPUT INCREASE	ANNUAL ENERGY INPUT INCREASE Kcal×10^9
Fuel	1967 - 1972	0.8×10^6 liters	7.6
Human labor	1960 - 1970	- 4500 man years	- 1.2
Irrigation water distribution	1960 - 1972	30×10^6 m^3	51.8
Fertilizer manufacture N	1960 - 1972	1.3×10^3 T	19.7
P$_2$O$_5$	1965 - 1972	0.6×10^3 T	1.1
K$_2$O	1965 - 1972	1.3×10^3 T	2.2
Pesticide manufacture	1966 - 1970	0.5×10^3 T	12.2
Animal feed processing	1960 - 1970	-	0.2
Animal feed import	1960 - 1970	31×10^3 T	94.5
Machinery manufacture	1962 - 1970	5.5×10^3 T	103.0

requirement is prodigious; 265 tons of fuel oil being needed annually to heat each hectare of greenhouses. Together with the additional manpower, electricity and fertilizer requirements (Frumkin, 1971), the total energy input is estimated at 278×10^3 Kcal m^{-2} yr^{-1}. Enrichment of the greenhouse atmosphere with CO_2 requires some 48 ton Propane ha^{-1} yr^{-1} in Israel (Enoch, 1974); an additional energy requirement equivalent to 53×10^3 Kcal m^{-2} yr^{-1}.

ENERGY PRODUCTION BY AGRICULTURE

The energy equivalent of Israel's agricultural production in 1969/70 for both primary plant and secondary animal products, is given in Table 4. The values listed refer to marketed produce and disregard the energy content of non-utilized materials.

In this year the total primary production by crop plants was equivalent to 4.98×10^{12} Kcal, or 1210 Kcal m^{-2} yr^{-1} averaged over the whole cultivated area, representing a solar energy fixation efficiency of less than 0.1%. Table 4 shows that the productivity of different crop types differed considerably; the figures in this table can be compared with those reported for even more intensive systems of agriculture. Thus, the average U.S.A. corn crop yielded an equivalent of 2040 Kcal m^{-2} yr^{-1} during 1969/70 (Pinmental et al., 1973) whilst the estimated energy fixation by intensive, year-round double-cropping in the Middle-East with crops specially selected for their high calorific yield has been estimated at 4710 Kcal m^{-2} yr^{-1} (Agro-Industrial Group, 1968). The energy equivalent of the total dry matter yield obtained in heated greenhouses in Israel is even higher: 8400 Kcal m^{-2} yr^{-1} (Stanhill et al., 1973).

TABLE 4

ANNUAL ENERGY AND MASS YIELD IN AGRICULTURE
ISRAEL 1969/70

CROP PRODUCTION CROP TYPE	MASS YIELD tons×10³	ENERGY YIELD[1] Kcal×10⁹	ENERGY FLUX Kcal m^{-2} yr^{-1}	ANIMAL PRODUCTION PRODUCT		MASS YIELD tons×10³	ENERGY YIELD[1] Kcal×10⁹
CEREALS incl. straw	291	1085	68	MEAT	Poultry	102	112
					Beef	36	85
INDUSTRIAL AND OIL CROPS	361	679	1284		Sheep, etc.	16	27
FRUIT incl. citrus	1553	551	645	EGGS		14	99
VEGETABLES incl. potatoes	741	255	983	FISH ponds and lakes only			16
ROUGHAGE excl. natural pasture	1722	2422	4142	MILK		502	332
				HONEY		2	5
T O T A L	4668	4982	1214	T O T A L		658	676

[1] Calculated from data on calorific values of food products (Watt and Merrill, 1963).

The potential dry matter yield in Israel of crops grown in the open, whose yield is limited only by solar radiation and the biochemical efficiency of the photosynthetic process, has been estimated at 8.03 kg m^{-2} yr^{-1}, equivalent to 33,750 Kcal m^{-2} yr^{-1} (Stanhill, in press).

Seventy percent, of the total energy fixed by plants in Israel's agriculture during 1969/70 was used for feeding animals. This represented 45% of the total energy of 7.01×10^{12} Kcal (including labor and fuel), invested in animal production in that year. The total animal products obtained from this input had an energy equivalent of 0.676×10^{12} Kcal, illustrating the order of magnitude loss of energy commonly found between primary and secondary trophic levels in the food chain.

The energetic implications of increasing agricultural production were calculated and the annual increases for the major products are listed in Table 5. The annual increase in the energy equivalent of primary plant productivity totaled 210×10^9 Kcal or 51 Kcal m^{-2} yr^{-1} while that of secondary, animal products totaled 34×10^9 Kcal. This annual increase of 5% is similar to the 4% increase in plant productivity.

These figures, derived from linarized time trends, ignore the large year to year, random variations in agricultural output which are much larger than occur with the inputs. In a semi-arid area such as Israel crop yields are very dependent on climate as illustrated in Fig. 1 with data for the wheat yield per unit area for the last 35 years. The data also illustrate that variability of yield increases with advancing technology. In addition to yield variability the size of the area cropped and of the animal population vary considerably due to both economic and climatic factors outside the farmers' control.

TABLE 5

ANNUAL CHANGES IN ENERGY AND MASS YIELDS

PRODUCTION	PERIOD LINEARIZED	ANNUAL MASS INCREASE $10^3 \times T$	ANNUAL ENERGY INPUT INCREASE[1] $Kcal \times 10^9$
PRIMARY - PLANT			
CEREALS including straw	1955 - 1972	27	98.8
INDUSTRIAL AND OIL CROPS	1949 - 1972	33	62.0
FRUIT including citrus	1959 - 1972	95	33.6
VEGETABLES including potatoes	1966 - 1972	46	15.8
ROUGHAGE excluding natural pasture	1959 - 1972	0	0
SECONDARY - ANIMAL			
MEAT	1948 - 1972	6.7	16.4
EGGS	1967 - 1972	50×10^6 units	3.8
FISH ponds and lake only	1966 - 1972	0.9	1.0
MILK	1948 - 1972	19.6	12.9

[1] Calculated from data on calorific values of food products (Watt and Merrill, 1963).

THE ENERGY BALANCE OF AGRICULTURE

The energy balance for primary crop production in Israel's agriculture in 1969/70 was approximately zero. In that year, the total energy input for crop production was 258×10^9 Kcal less than that harvested. Per unit cultivated land, the net energy gain was 66 Kcal m^{-2} yr^{-1}. The data in Tables 3 and 5 show that this balance is increasing very slightly at a rate of 22 Kcal m^{-2} yr^{-1}.

The balance for the various branches of Israel agriculture varied considerably. Thus the energy input, production and balance for irrigated farming is much larger than for dry land farming. Besides the 1690 Kcal $m^{-2} yr^{-1}$ average energy requirement for water application in 1969/70, the extra fertilizer and plant protection chemical application, as well as fuel, labor and machinery used in irrigation farming represent an additional energy input. Assuming that the size of these extra input terms was proportional to that of the extra yield, the total energy input in irrigated crop production in 1969/70 was calculated to be 2568 Kcal $m^{-2} yr^{-1}$. Despite the fact that the energy yield in primary production per unit irrigated area was 2425 Kcal $m^{-2} yr^{-1}$, double that for agriculture as a whole, the net energy balance per unit irrigated area was negative compared with the small positive balance for the entire agricultural area. Normalized to energy input, crop production as a whole produced 1.06 Kcal per 1 Kcal input compared with 0.94 Kcal produced by irrigated crops. If all the water for irrigation had been supplied from a near term nuclear desalination plant, the energy deficit in irrigated agriculture would rise to 1200 Kcal $m^{-2} yr^{-1}$.

Other advanced systems of agriculture show even larger net energy deficits. That for heated greenhouse has been calculated as 269,400 Kcal $m^{-2} yr^{-1}$ even when accounting for the energy equivalent of total, aboveground dry matter production, rather than the marketed produce.

The above examples, as well as previous comparisons with less advanced West Bank agriculture and primitive Sinaitic shepherding systems (Stanhill, in press), suggest that as the intensity of agricultural production and the degree of environmental control increase, so does the net energy deficit. This conclusion is strongly supported by data showing the increasing energy subsidy to the food production of U.S. agriculture (Steinhart and Steinhart, 1974).

The energy balance in Israel's agriculture, is even more serious if the final food products, both plant and animal, rather than the primary plant production hitherto considered. As already indicated, only 30% of the energy fixed in crop plant products is directly utilized by man, the rest being used to feed livestock. If the portion of crops directly used in Israel for human consumption is added to the total animal products, a total energy yield of 2.04×10^{12} Kcal or 780 Kcal m^{-2} yr^{-1} is obtained, giving a net deficit in local food production of 1123 Kcal m^{-2} yr^{-1}. If animal husbandry were reduced to the level which could be supported by local crop wastes, e.g. bran, straw, cotton seed, sugar beet pulp and tops, which supply some 325×10^9 Kcal, and the natural pastures which supply some 550×10^9 Kcal a year (Hunea, 1968), the agricultural energy deficit could be reduced to 710 Kcal m^{-2} yr^{-1} but animal production would be reduced by some 80 per cent.

The energy balance of agriculture could be made positive, even with intensive production systems such as desalinated irrigation agriculture, if crops were selected solely for their energy conversion efficiency. The 'high-calorie' system envisaged in the Oak Ridge study (Agro-Industrial Group, 1968), using high yielding crops of high calorific content in an intensive double cropping system without secondary animal production, is estimated to yield a surplus of 780 Kcal m^{-2} yr^{-1}.

THE HUMAN AND FOOD ENERGY BALANCE

In 1969/70 the three main sinks for the primary plant products of Israel's agriculture were as follows: 70% of its energy equivalent was used to support animal husbandry, 20% was directly consumed by man within Israel, and 10% was exported for human consumption outside Israel. A very small amount of the total energy production, approximately 2%, consisted of non-food products, e.g. cotton lint and tobacco. In 1969/70, the energy equivalent of exported agricultural produce was 0.58×10^{12} Kcal, over 90% in citrus fruit. For the same period, imports of wheat and sugar had an energy content of 1.30×10^{12} Kcal and 0.33×10^{12} Kcal, respectively. The net energy balance of imported over exported primary plant products was 1.04×10^{12} Kcal. Exports of secondary animal products in the same period were negligible; the net import amounted to 0.172×10^{12} Kcal, and consisted mainly of beef (0.12×10^{12} Kcal) and dried milk (0.04×10^{12} Kcal).

Thus the total net import of food during the year amounted to an energy equivalent of 1.214×10^{12} Kcal which together with the 2.04×10^{12} Kcal of local food production, gave a total annual food energy sum of 3.25×10^{12} Kcal.

This figure agrees to within 2% with that calculated from statistics of the national diet (Central Bureau of Statistics, 1972). During 1969/70 the per capita calorific intake averaged 2988 Kcal per day, 20% of which was of animal origin. The calorie intake shows a small annual increase, averaging 20 Kcal per capita per day, less than 1% per year. However two-thirds of this increase (14 Kcal per day), represents an increasing consumption of animal foods.

During 1969/70 the energy fixed in primary crop production could have fed 1.69×10^6 people at the then prevailing dietary levels. This figure was calculated with the assumptions that the entire primary production was used for local food consumption, and that the fraction of the calorific intake derived from animal products (currently 20%) was produced at the observed 10:1 plant-to-animal energy conversion ration.

The number of people that could be fed from local agriculture by 1979/80 was calculated to be 1.96×10^6, allowing for the observed 4.2% annual increase in primary production (Table 5) and the previously indicated dietary trends.

The observed population increase in Israel during the last 15 years was 81.5×10^3 per year. The extra food requirement entailed by this increase indicates that the proportion of the population whose food could be supplied by local agriculture will drop from 58% in 1969/70 to 51% by 1979/80.

Thus in Israel, as in the world as a whole, the demand for human food energy is increasing rather more rapidly than is agricultural productivity.

SUMMARY AND CONCLUSIONS

The energy flows in Israel's agricultural and food balance have been summarized graphically (Fig. 2) in an energy network diagram using Odum's conventions and symbols (Odum, 1967). In the energy flow from sun to man the greatest drops in energy level, and hence the lowest energy conversion ratios, are found in the first and the final links. The first link, in which solar radiation is absorbed by plant tissue and photosynthetically fixed, shows a three orders of magnitude drop in energy flux. The desirability and possibility of increasing the efficiency of energy conversion at this stage are generally accepted and a major goal of agricultural research. The final link, the conversion of food energy into new human biomass shows, in

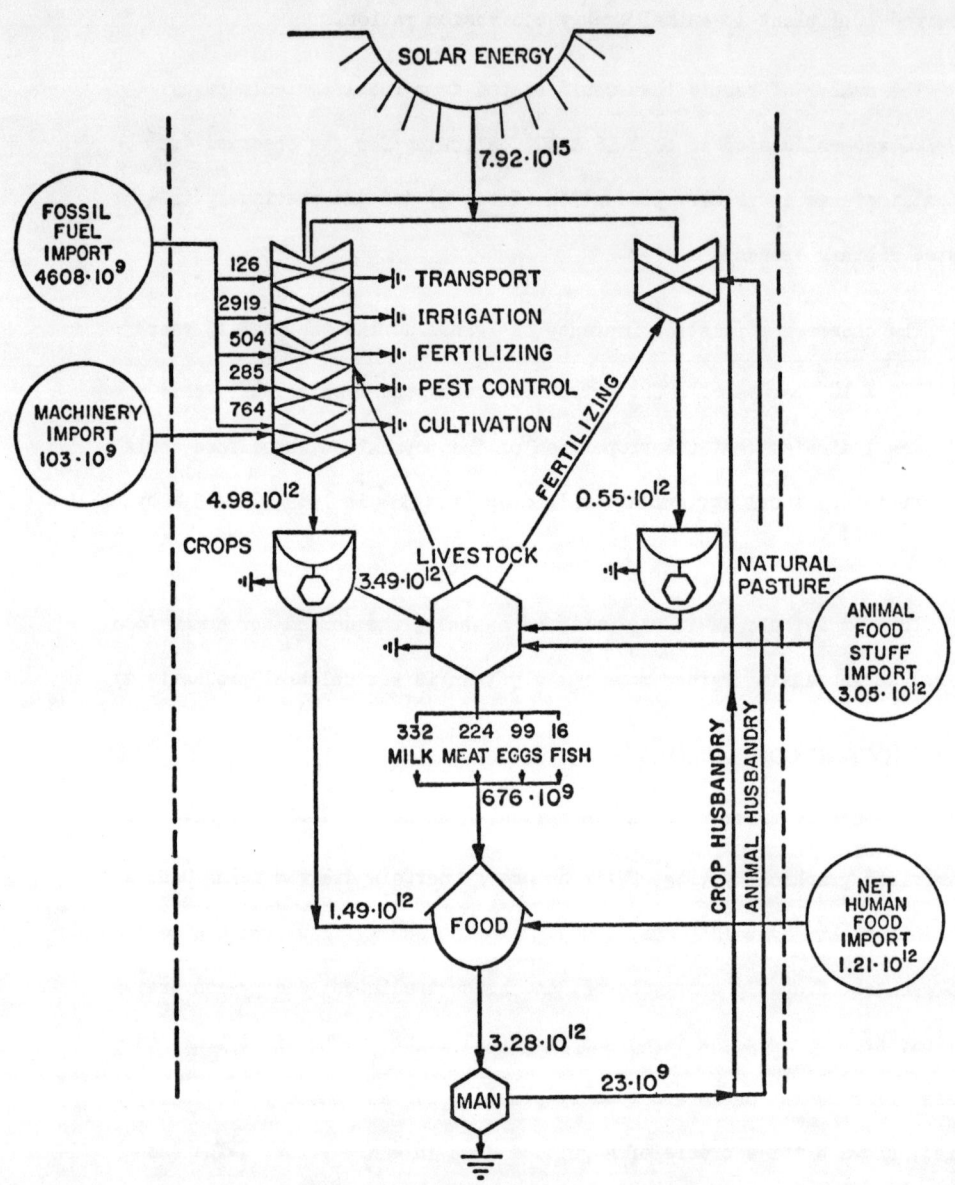

Fig. 2

Fig. 2.   Energy flow in Israel agriculture and food balance, 1969/70.

The broken lines represent the pre-1967 borders containing 4105 km^2 of cultivated land. (Central Bureau of Statistics, 1972). All energy fluxes are in Kcal yr^{-1}. Symbols used are as below, a full explanation can be found in Odum (1967).

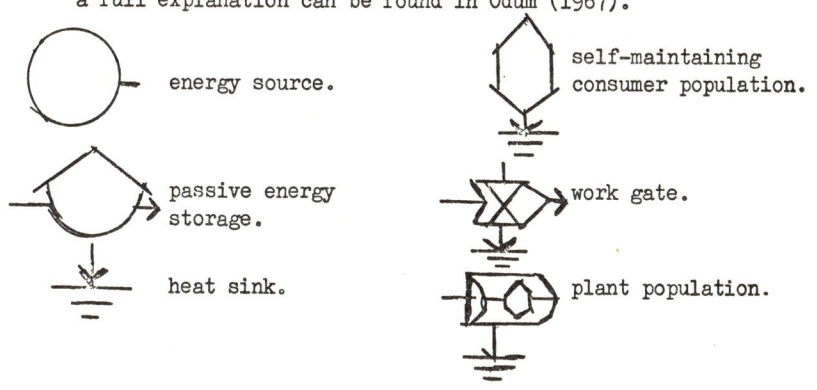

Israel, a two orders of magnitude frop in energy flux. The desirability and possibility of further reducing this energy flow, via population control, are outside the scope of this article, but the importance of population increase as the forcing function for agricultural development should not be ignored.

To revert to the subject of the agricultural energy balance, a powerful case has been made (Blaxter, 1974) for the proposition that the replacement of human and animal energy by that of fossil fuel is the factor that has been largely responsible for the increase in agricultural productivity, approximately 5% per year in the case of Israel. This increase however, involves an increasingly negative energy balance. For agriculture as a whole in Israel, an input of 2.44 Kcal fossil fuel is used to produce 1 Kcal of plant and animal food products.

Israel agriculture currently requires an annual, internal per capita energy input of 0.05 tons of fuel oil, 386 KWH of electrical power and 10 days of labor to produce slightly over half the human food energy consumed. The remaining fraction is supplied by importing approximately one-third of a ton of grain per capita per year, one-third of which is used directly as human food, and two-thirds of which is fed to livestock.

This import of fossil fuel and grain, low in monetary but high in energy content, has in the past been largely paid for by exporting fruit, flowers and vegetables - crops high in monetary but low in energy content. This international energy - money exchange, follows the steep solar energy gradient between Israel and Western Europe.

Unfortunately, this exchange system seems unstable for a number of reasons: firstly, increases in the cost of fossil fuel will lead to increases in both the cost of grain imports and of local food production; second, the world's population and its food needs are outstripping its food production, which will, presumably, lead to food shortages and hence further price increases; and third, agricultural pollution, which is closely linked to the energy input level, is being increasingly recognized - its control will lead to greater production costs and to restrictions in production.

The inescapable conclusion is that serious attention will have to be paid to the more significant elements in the energy aspects of agricultural and food policies, especially if national aims of an increasing population and rising dietary levels are to be maintained.

Some possibilities for energy economies are obvious, but many of these are trivial. Here, only the two most important will be mentioned. The need to economize in livestock production and consumption is primary; two obvious corollaries are the greater exploitation of natural pasture and of crop and food wastes. The second, locally critical area, is in irrigation farming. The trend toward a closed water distribution system, where water is transported considerable distances under pressure and often lifted to considerable heights, involves a large expenditure of energy; so does desalination even under the most favorable conditions. Under such circumstances the energetic aspects of any new large-scale irrigation scheme should be examined very carefully, especially if desalination is involved.

ACKNOWLEDGEMENTS

I wish to thank the many persons who supplied me with the information used in this paper. In particular Mr. Nelken of the Israel Electric Corporation for data on the energetics of electrical power generation, Mr. E. Kali of Water Planning for Israel, Ltd. for the power requirements of water movement, Drs. Dahlia Greidinger (Chemicals and Phosphates, Ltd.) and T. Zisner (Dead Sea Works, Ltd.) for information on energy use in fertilizer manufacture and Mr. Z. Rosenberg for data on animal feed processing.

I also wish to thank Pauline Cohen and J. Levitt of the Volcani Center, Bet Dagan for their many helpful comments on an earlier draft of this manuscript.

REFERENCES

1. Agro-Industrial Study Group (1968) Nuclear Energy Center: Industrial and Agro-Industrial Complexes. Oak Ridge National Laboratory Report 4290. Oak Ridge.

2. Berry, R. S., and Fels, M. F. (1973) The energy costs of automobiles. Science and Public Affairs 29: 11-17, 58-60.

3. Blaxter, K. (1974) Power and agricultural revolution. New Scientist 61: 400-403.

4. Enoch, Z. (1974) Personal communication.

5. Frumkin, D. (1971) Economic calculations in the Rose export sector. (in Hebrew) Ministry of Agriculture, Ha Qirya, Tel Aviv.

6. Hunea, L. (1968) Feeding stuffs balances in the Israel livestock farming. (in Hebrew with English summary) Ministry of Agriculture, Ha Qirya, Tel Aviv.

7. Johnson, F. S. (1970) The oxygen and carbon dioxide balance in the earth's atmosphere. in Global Effects of Environmental Pollution. pp. 5-11 Reidel, Dortrecht.

8. Odum, H. T. (1967) Energetics of world food production. Chapter 3. in The World Food Problem. A report of the President's Science Advisory Committee. Vol. III. pp. 59-94. The White House, Washington, D.C.

9. Pinmental, D., Hurd, L. E., Bellotti, A. C., Forster, M. J., Oka, I. N., Scholes, O. D., and Whitman, R. J. (1973) Food production and the energy crisis. Science 182: 443-449.

10. Stanhill, G. (1970) Measurements of global solar radiation in Israel. Israel J. Earth Sci. 19: 91-96.

11. Stanhill, G. (in press) Solar radiation and crop production in Recent Advances in Biometeorology. Ed. L.P.Smith.

12. Stanhill, G., Fuchs, M., Bakker, J. and Moreshet, S. (1973) The radiation balance of a glasshouse rose crop. Agric. Meteorol. 11: 385-404.

13. Statistical Abstract of Israel, 1971 (1972) No. 22. Central Bureau of Statistics, Herusalem.

14. Steinhart, J. S., and Steinhart, C. E. (1974) Energy use in the U.S. food system. Science 184: 307-316.

15. Watt, Bernice, K. and Merrill, Anabel, L. (1963) Composition of Foods. Agriculture Handbook No. 8, Agricultural Research Service, U.S.D.A., Washington, D. C.

SECTION 2

POLLUTION IN THE PLANT ENVIRONMENT

PROBLEMS OF CHEMICAL REACTION AND BIOLOGICAL PROCESSES IN SOILS

D. E. ELRICK, P. H. GROENEVELT and T. J. M. BLOM

*Department of Land Resource Science, University of Guelph,
Guelph, Ontario, Canada*

ABSTRACT

A macroscopic approach, based on soil as a continuum, is used to incorporate the chemical reactions and biological processes that may take place within soils. The equation of continuity for a component in soil is presented and extended to include the effects of coupled processes. Some of the important concepts of adsorption and production within soils is reviewed, with particular reference to pesticides and nitrogen compounds in soils.

LIST OF SYMBOLS

A = amount of substrate oxidized per unit mass of biomass synthesized, dimensionless
B = constant, L^3/Mt
C_i = the amount of component i per cm^3 liquid, M/L^3
D^* = dispersion coefficient, L^2/t
g = gravitational acceleration, L/t^2
j^D = volumetric diffusion flux, L/t
j^V = volume flux, L/t
K = hydraulic conductivity, L/t
K_1 = constant, dimensionless
K_2 = constant, L^3/M
K_3 = constant, $(L^3/M)^{1/n}$
K_4 = constant, L^3/M
k_m = saturation constant, M/L^3
k'' = specific rate constant, t^{-1}
k^* = constant, $M/L^3 t$
k_i = rate constant, t^{-1}
L_D, L_{DV}, L_V, L_{VD} = transport coefficients, $L^3 t/M$

m = biomass concentration M/L^3
m_B = mass of dry soil, M
m_o = amount of biomass at $t = 0$, M/L^3
n = constant, dimensionless
n_i = flux of component i in grams per cm^2 bulk per sec, $M/L^2 t$
p = hydraulic pressure, M/Lt^2
R = gas constant, $L^2/t^2 T$
s = space variable, L
S_i = adsorbed amount of component i in grams per cm^3 bulk volume, M/L^3
T = absolute temperature, T
t = time, t
v_i = velocity of component i, L/t
x = mass of adsorbate, M
α = constant, t^{-1}
α^* = proportionality factor, t^{-1}
α^*_∞ = maximum proportionality factor for a large substrate concentration, t^{-1}
β = amount of enzyme per unit biomass involved in waste metabolism, dimensionless
θ = volumetric moisture content, dimensionless
λ = proportionality coefficient, $M/L^3 t$
ρ_i = bulk density of component i, M/L^3
ρ_B = bulk density of dry soil, M/L^3
σ = reflection coefficient, dimensionless
Φ_i = production term of component i, $M/L^3 t$
ϕ_i = volume fraction of component i in cm^3 per cm^3 bulk, dimensionless
ω = solute permeability, t
∇ = the "del" or "nabla" operator

INTRODUCTION

To some extent the study of soil science has suffered from compartmentalization. There are soil specialists in the fundamental areas of physics, chemistry, genesis (geology) and microbiology, but there has been little attempt to integrate many of these studies. Nowhere is this more evident than when attempting to utilize the available information on chemical reactions and biological processes to evaluate their effects on mass transport in soils.

By their very nature, soils are highly reactive materials. Although extremely variable, soils may possess some or all of the

following properties; a large surface area available for chemical
reactions, a diverse microbial population capable of carrying out a
number of microbial transformations, a porous matrix which is capable
of physical and chemical (osmotic) sieving as well as exhibiting
swelling effects and lastly a "home" for plant roots, earthworms and
many soil animals. As W.R. Gardner (1972) so aptly put it: this area
"is concerned with those processes which the soil physicist frequently
attempts to eliminate as factors in his experiment". Simple models of
the soil system have been necessary in order to make any progress at all
in understanding the mechanisms involved in mass transport. The need
to understand many of the previously neglected effects is now apparent.

Recent studies applying some of the basic concepts of chromatography and reaction bed dynamics are presented by Reiniger and Bolt(1972),
Golubev and Garibyants (1971) and Goring and Hamaker (1972). In this
article we do not attempt a comprehensive review of the subject of
chemical reactions and biological processes in soils. The review is
restricted to certain aspects of coupled processes, adsorption effects
and production processes.

THE EQUATION OF CONTINUITY

For a macroscopic volume element of a soil the following equation
of mass balance can be written:

$$\frac{\partial (S_i + \theta C_i)}{\partial t} = - \nabla \cdot n_i + \Phi_i , \qquad (1)$$

where S_i is the adsorbed amount of component i in grams per cm^3 bulk volume;

C_i is the amount of component i in the solution phase (grams per cm^3 liquid);

θ is the liquid content of the soil in cm^3 liquid per cm^3 bulk;

$n_i = \theta C_i v_i$ is the flux of component i in grams per cm^2 bulk per sec.;

Φ_i is a production term of component i in grams per cm^3 bulk per sec.

It must be noted that the above variables are all macroscopic in
nature and refer to averages based on a representative volume of the
soil which is large enough to give a representative value but yet
small enough to be considered a point in the macroscopically continuous
system. For example, the representative volume would be of the order
of 100 pores.

Both S_i and Φ_i are, in the most general sense, functions of the
concentrations of all the ions present in solution as well as space
and time. The resulting equations of (1) give a set of n coupled
differential equations containing the n variables C_i. Provided that
suitable relationships can be obtained for S_i and Φ_i, the set of
equations can in principle be solved either analytically if amenable
to such analysis or numerically with the aid of a computer. Fortunately,
many of the problems of interest in soils can be reduced to relatively
simple cases.

For non-volatile components the flux n_i is limited to the liquid phase and this discussion will be limited as well to transport taking place only in the liquid phase.

Introducing $C_i(v_i - j^V/\theta)$, the flux of component i relative to the mean pore water velocity j^V/θ, gives

$$\frac{\partial(S_i + \theta C_i)}{\partial t} + \nabla \cdot C_i j^V = -\nabla \cdot [\theta C_i (v_i - j^V/\theta)] + \Phi_i \quad . \tag{2}$$

where the Darcy velocity (volume flux) j^V is defined as

$$j^V \equiv \sum_m \phi_m v_m \quad \text{(summation over the liquid components only)}$$

expressed in cm^3 liquid per cm^2 bulk per sec.

The volume fraction ϕ_m is a macroscopic variable. The sum of the volume fractions of the components present in the liquid phase is equal to the liquid filled fraction θ. It should be noted that the liquid in this macroscopic sense cannot be thought of as incompressible if the volumetric liquid content is allowed to change, which occurs in general in inhomogeneous soils and under unsaturated conditions; i.e. field soils. Thus in general

$$\nabla \cdot j^V \neq 0.$$

It is only in the case of steady state liquid flow that one may use $\nabla \cdot j^V = 0$.

From here on we will restrict ourselves to a liquid phase consisting of the components water and a single solute indicated by the subscripts w and s respectively. It should be noted here that one often encounters in the literature the relative solute velocity $(v_s - v_\ell)$ which is the difference between the velocity of the solute and the mass average velocity of the liquid. The latter is defined by

$$v_\ell \equiv (\rho_w v_w + \rho_s v_s)/\rho_\ell \quad . \tag{3}$$

where ρ_k is the bulk density of k (grams per cm^3 bulk) and

$$\rho_\ell = \rho_w + \rho_s. \tag{4}$$

The relation between the relative velocity $(v_s - v_\ell)$, i.e. relative to the mass average liquid velocity and the relative velocity $(v_s - j^V/\theta)$, i.e. relative to the volume average liquid velocity is

$$(\theta v_s - j^V) = \frac{\phi_w \rho_\ell}{\rho_w} (v_s - v_\ell). \tag{5}$$

COUPLED PROCESSES

In the theory of irreversible thermodynamics, each flux is assumed to be a linear function of all the driving forces and may be generalized as

$$J_i = \sum_k L_{ik} X_k$$

where in soils the driving forces X_k are generally the gradients of the hydrostatic pressure, the osmotic pressure, the electrostatic potential and the temperature. For simplicity we assume that only the gradient ∇p and the osmotic pressure gradient $RT\nabla C_s$ are operative.

For the above system we may write (Groenevelt and Bolt, 1969)

$$j^V \equiv \phi_w v_w + \phi_s v_s = -L_V \nabla p - L_{VD} RT\nabla C_s \quad (6)$$

$$j^D \equiv \phi_w (v_s - v_w) = -L_{DV} \nabla p - L_D RT\nabla C_s. \quad (7)$$

With $\theta = \phi_w + \phi_s$ one finds that

$$j^D = \theta v_s - j^V. \quad (8)$$

Setting $i = s$ in equation (2) and substituting equations (6), (7) and (8) gives

$$\frac{\partial (S_s + \theta C_s)}{\partial t} + \nabla \cdot C_s [-L_V \nabla p - L_{VD} RT\nabla C_s] =$$

$$- \nabla \cdot C_s [-L_{DV} \nabla p - L_D RT\nabla C_s] + \Phi_s . \quad (9)$$

In a soil that exhibits semi-selectivity, the effects of osmosis on the net transport and of solute sieving on the relative transport are thus directly incorporated into the above equation. L_V and L_D are proportional to the hydraulic conductivity and the dispersion coefficient respectively, viz

$$K = \rho_\ell g L_V \quad (\text{cm/sec}) \quad (10)$$

and

$$D^* = RT C_s L_D \quad (\text{cm}^2/\text{sec}) , \quad (11)$$

where g is the acceleration due to gravity. The dispersion coefficient D^* may be a second-rank tensor and can be considered a "lumped" coefficient in that it includes the effects of macroscopization. It therefore includes the effects of microscopic velocity variations, the geometry of the void space and molecular diffusion.

The coupling effects can be demonstrated more clearly by using the reflection coefficient σ and the solute permeability ω. These coefficients are defined by Katchalsky and Curran (1965) as follows:

$$\sigma \equiv - \frac{L_{VD}}{L_V} \quad (12)$$

and

$$\omega \equiv C_s (L_D - L_{VD}^2 / L_V). \quad (13)$$

Substitution gives

$$\frac{\partial (S_s + \theta C_s)}{\partial t} - \nabla \cdot [\frac{C_s}{\rho_\ell g}(1-\sigma)K\nabla p] =$$

$$\nabla \cdot [\omega - \frac{C_s \sigma}{\rho_\ell g}(1-\sigma)K]RT\nabla C_s + \Phi_s . \quad (14)$$

Use has been made of the equality $L_{DV} = L_{VD}$.
For a non-selective porous medium (e.g. glass beads or coarse sand) the reflection coefficient $\sigma = 0$ and then the above equation reduces to the familiar form

$$\frac{\partial (S_s + \theta C_s)}{\partial t} + \nabla \cdot j^V C_s = \nabla \cdot D^* \nabla C_s + \Phi_s . \quad (15)$$

It must be remembered that only in case of stationary liquid flow may one use $\nabla \cdot j^V = 0$ and thus replace the term $\nabla \cdot j^V C_s$ by $j^V \cdot \nabla C_s$.
For perfect selectivity ($\sigma = 1$) eq. (14) degenerates into

$$\frac{\partial (S_s + \theta C_s)}{\partial t} = \Phi_s, \quad (16)$$

as $L_{VD} = L_V = L_D$ and thus $\omega = 0$. The porous medium is then impermeable for the solute and the only places where solute is allowed to be present is in the larger voids between the narrow (salt excluding) passage ways.

ADSORPTION

In equation (1) the relationship between S_i and C_i is given by the adsorption equation. In a general form the relationship may be written as:

$$S_i = S_i(C_1, C_2, \ldots C_i, \ldots C_n, s, t), \quad (17)$$

where s is a space variable. Thus the amount of ion i which is adsorbed is in general a function of all the ions, the position s and the time t.

Adsorption in soils is the process of concentrating a substance at the surface of the solid particles. The causes of adsorption are detailed by Hamaker and Thompson (1972) and Golubev and Garibyants (1971). The possible types of interaction listed by Hamaker and Thompson are as follows: (1) Van der Waals - London; (2) hydrophobic bonding; (3) charge transfer and hydrogen bonding; (4) ligand exchange; (5) ion exchange; (6) direct and induced ion-dipole and dipole-dipole forces; (7) chemisorption; and (8) magnetic bonding.

Adsorption is highly dependent on the chemical and physical nature of both the adsorbate and the adsorbent. Osgerby (1970) has reviewed the literature on sorption of non-ionic pesticides and Knight et al. (1970) have summarized the experimental information on ionic pesticides. Although there is still controversy concerning the kind of adsorption of many pesticides on soils, both reviews indicate that the adsorption of any pesticide is generally well described by the Freundlich equation and less well described by the Langmuir equation.

Langmuir and Freundlich

The adsorption equation of Langmuir has a physical basis and has been developed for the adsorption of gases onto solids. However, the equation is often applied to the adsorption of molecules on a substrate at a liquid-solid interface. The equation can be written as:

$$\frac{x}{m_B} = \frac{S_i}{\rho_B} = \frac{K_1 K_2 C_i}{1 + K_2 C_i} \quad , \tag{18}$$

where ρ_B is the bulk density in grams of (oven-dry) soil per cm^3 bulk; x/m_B is the amount adsorbed in grams per gram of soil; K_1 and K_2 are constants for the system.

In soil, the Langmuir equation is not very successful as it assumes that the energy of adsorption is constant and independent of surface coverage. In a heterogeneous mixture such as soils one would not expect this assumption to be very appropriate. Miscible displacement experiments by Youngson et al. (1967), Kay and Elrick (1967) and Davidson et al. (1968) confirm this statement.

The Freundlich equation is strictly empirical and may be written as:

$$\frac{S_i}{\rho_B} = K_3 C_i^{1/n} \quad , \tag{19}$$

where K_3 and n are constants for the system. Deviations of $1/n$ from 1.0 indicate the degree of non-linearity between the concentration of ions (or molecules) in solution and in the adsorbed phase. Davidson and McDougal (1973) and Hamaker and Thompson (1972) have presented data for a number of pesticides in soils which show that the value of $1/n$ is close to unity.

It should be noted that both the Langmuir and Freundlich isotherms reduce to the linear relationship:

$$\frac{x}{m_B} = \frac{S_i}{\rho_B} = K_4 C_i \tag{20}$$

under the appropriate conditions. The distribution coefficient K_4 is related to the adsorption - and desorption-rate constants.

Values of K_4 for several pesticides have been determined by shaking a certain concentration of pesticide with a soil and measuring the concentration of the pesticide in the supernatant liquid after a certain period of time. The difference between the initial and final concentration is assumed to be due to the amount adsorbed. Retention curves determined by Kay and Elrick (1967), Davidson et al. (1968) showed that K_4 was constant over a wide range of values of C_i.

Both the Langmuir and the Freundlich equations assume instantaneous equilibrium as well as complete reversibility between the adsorbed and solution concentrations. These models therefore oversimplify many of the adsorption processes taking place in soils.

Non-Instantaneous and Irreversible Adsorption

Miscible displacement experiments with low pore velocities gave reasonable results according to the Freundlich equation. However, similar experiments with high pore velocities showed a time-dependence of the distribution coefficient K_4.

Rate dependent adsorption models have been developed by Lapidus and Amundson (1952), Oddsen et al. (1970), Lindström and Boersma (1971) and Davidson and McDougal (1973). Assuming that the rate of adsorption can be expressed by:

$$\frac{\partial S_i}{\partial t} = \alpha (\rho_B K_4 C_i - S_i) \; , \tag{21}$$

where α and K_4 are constants which depend upon the soil and the chemical in solution. In this equation the rate of adsorption is assumed to be proportional to the difference between the amount which has been adsorbed and the equilibrium value. Experimental results with lindane showed that α did not seem to be constant over the whole period (Huggenberger et al., 1972). Letey and Oddson (1972) also reported values of α that were higher initially, indicating an initial rapid rate of adsorption followed by a slower rate. Values of α ranged from 0.03 to 0.07 [hour^{-1}] for a muck soil and from 0.6 to 5.1 [hour^{-1}] for a mineral soil.

Kahn (1973) carried out equilibrium and kinetic studies on the adsorption of 2,4-D and picloram on humic acid. The equilibrium data followed the Freundlich-type isotherm. The rate-limiting step for the initial period was shown to be the diffusion of the herbicide molecules to the surface of the humic acid particles. At longer times the rate-limiting step was interpreted to be intraparticle diffusion of the herbicide molecules into the interior of the humic acid particles.

The adsorption on soils has been shown to be irreversible for many pesticides. Geissbühler et al. (1963), Swoboda and Thomas (1968), Valoras et al. (1969) and Davidson and McDougal (1973) reported hysteresis in the adsorption isotherms.

For an extensive analysis of adsorption phenomena during flow in porous media, readers are referred to the recent review by Reiniger and Bolt (1972).

PRODUCTION

The production term Φ_i given in equation (1) in a general form is given by

$$\Phi_i = \Phi_i (C_1, C_2, \ldots C_i \ldots C_n, s, t) \; ,$$

where, as before, s is a space variable. In soils this production term is not easily expressed. For some components it may represent processes like dissolution or precipitation whereas for others it may represent a chemical transformation due to chemical or microbial reactions.

Recent interest has focused on the nitrogen cycle in soils and in particular on the nitrification and denitrification processes. Studies on soils have used incubation, perfusion and miscible displacement techniques. In the incubation method, nitrogen compounds are added to soil and the mixture is then generally maintained at a constant temperature for a given period of time; no flow of solution is involved. The perfusion technique, on the other hand, permits a direct and continuous investigation of nitrification by recycling a given solution through a soil column. Changes in the physical and chemical properties of both the perfusate and the soil may affect the nitrification process. Direct leaching techniques described by Macura and Kunc (1965) and modified by Erh et al. (1967) permit a continuous leaching of the soil in which the products of nitrification are removed from the system.

The nitrification process is described by the following equation:

$$NH_4^+ \xrightarrow{i} NO_2^- \xrightarrow{ii} NO_3^- .$$

The above reactions are generally microbial in nature. The two reaction rates are strongly dependent on the activity of the bacteria, which in turn is influenced by the nature of the substrate, the oxygen and carbon dioxide exchange within the soil, the temperature, the available carbon, the pH of the soil, the presence of inhibitors, etc.

McLaren (1970) has given a mathematical equation for the rate of microbial oxidation for both reactions (i) and (ii) under controlled conditions:

$$\Phi_i = -A \frac{dm}{dt} - \alpha^* m - \frac{k''\beta m C_i}{k_m + C_i} , \qquad (22)$$

where
  C_i = substrate concentration (g/cm^3 soil solution);
  t = time (sec);
  A = amount substrate oxidized per unit mass of biomass synthesized;
  m = biomass concentration (g/cm^3 bulk);
  α^* = substrate oxidized per unit mass biomass per unit time for maintenance (g/g biomass / sec);
  β = amount of enzyme per unit biomass involved in waste metabolism (g enzyme/g biomass);
  k'' = specific rate constant (g/g enzyme / sec);
  k_m = saturation constant (g/cm^3 solution);

Different simplifications can be applied to equation (22). McLaren (1971) and Cho (1971) have assumed a steady state situation in which the mass of bacteria is constant; i.e. $dm/dt = 0$. If we now set: $\alpha^* = \alpha_\infty^* C_i / (k_m + C_i)$ we then obtain:

$$\Phi_i = - \frac{k^* C_i}{k_m + C_i} , \qquad (23)$$

which has been used by Ardakani et al. (1973).

A further simplification can be made if the amount of substrate C_i is small in comparison with k_m. Under these conditions,

$$\Phi_i = -k_i C_i . \tag{24}$$

If we let C_1, C_2 and C_3 represent the concentrations of ammonium, nitrite and nitrate in solution, the coupled production terms may be written as follows:

$$\Phi_1 = -k_1 C_1$$
$$\Phi_2 = k_1 C_1 - k_2 C_2$$
$$\Phi_3 = k_2 C_2 - k_3 C_3,$$

where k_i are the associated rate constants. Cho (1971) also assumed a linear adsorption relationship for ammonium and solved the resulting coupled differential equations (equation 1) for a stepwise change in concentration of C_i at the surface of a semi-infinite column.

Cho (1971) used values of 0.01 hr^{-1} and 0.1 hr^{-1} for k_1 and k_2 respectively. This is in good agreement with values given by Knowles et al. (1965) of 0.025 hr^{-1} and 0.04 hr^{-1} respectively and more recent values of 0.01 hr^{-1} reported by Misra et al. (1974) for the entire ammonium to nitrate transformation.

The analytical solutions reported by Cho (1971) and by Misra et al. (1974) assume a steady state production model. The enrichment phase of bacterial growth is assumed to have taken place. Experiments such as those reported by Elrick and Maclean (1966) on the transport and degradation of the herbicide 2,4-D in soil indicate that the biomass increased greatly during the experiment. As a first approximation one may assume a linear growth of the biomass:

$$m = m_o + \lambda t , \tag{25}$$

where m_o is the amount of biomass present at the start of the experiment and λ is a proportionality coefficient. Equation (22), using the assumptions which led to equation (24), may now be written as:

$$\Phi_i = -A \frac{dm}{dt} - BmC_i , \tag{26}$$

where the coefficient k_i as in equation (24) has been replaced by Bm. Combining equations (25) and (26) gives

$$\Phi_i = -A\lambda - k_i C_i - B\lambda C_i t . \tag{27}$$

Substitution into equation (15) then gives for a non-selective medium with constant values of D^* and j^V:

$$\frac{\partial(S_s + \theta C_s)}{\partial t} = D^* \frac{\partial^2 C_s}{\partial x^2} - j^V \frac{\partial C_s}{\partial x} - kC_s - B\lambda C_s t - A\lambda . \tag{28}$$

This equation may provide a proper mathematical model to describe the experimental results of Elrick and Maclean (1966).

These two simple examples of nitrogen and 2,4-D reactions in soils illustrate the complexity of microbial effects on transport through soils. There is a pressing need for a better understanding of these reactions and for a quantification of the various component reaction rates.

REFERENCES

1. Ardakani, M.S., J.T. Rehbock, and A.D. McLaren. 1973. Oxidation of nitrite to nitrate in a soil column. SSSAP 37: 53-56.
2. Cho, C.M. 1971. Convective transport of ammonium with nitrification in soil. Can. J. Soil Sci. 51: 339-350.
3. Davidson, J.M. and J.R. McDougal. 1973. Experimental and predicted movement of three herbicides in a water-saturated soil. J. of Env. Qual. 2: 428-433.
4. Davidson, J.M., C.E. Rieck, and P.W. Santelmann. 1968. Influence of water flux and porous material on the movement of selected herbicides. Soil Sci. Soc. Amer. Proc. 32: 629-633.
5. Elrick, D.E., and A.H. MacLean. 1966. Movement, adsorption and degradation of 2,4-Dichlorophenoxyacetic acid in soil. Nature 212: 102-104.
6. Erh, K.T., D.E. Elrick, R.L. Thomas, and C.T. Corke. 1967. Dynamics of nitrification in soils using a miscible displacement technique. Soil Sci. Soc. Amer. Proc. 31: 585-591.
7. Gardner, W.R. 1972. Physico-Chemical and Microbial reaction effects on transport in porous media. Proceedings second Symposium of IAHR-ISSS, University of Guelph, Vol. 2: 667-682.
8. Geissbühler, H., C. Haselbach, and H. Aabi. 1963. The fate of N'-(4-chlorophenoxyphenyl) - NN - dinethylurea (C-1983) in soils and plants. I. Adsorption and leaching in different soils. Weed Research 3: 140-153.
9. Golubev, V.S. and A.A. Garibyants. 1971. Heterogeneous Processes of geochemical migration. Consultants Bureau. New York-London.
10. Goring, C.A.I., and J.W. Hamaker (eds.). 1972. Organic chemicals in the soil environment. Marcel Dekker, Inc., New York. Vol. 1.
11. Groenevelt, P.H., and G.H. Bolt. 1969. Non-equilibrium thermodynamics of the soil water system. J. of Hydrology 7: 358-388.
12. Hamaker, J.W. and J.M. Thompson. 1972. Adsorption. In: Goring and Hamaker cf. ref. 10. pp 49-143.
13. Huggenberger, R., J. Letey, and W.J. Farmer. 1972. Observed and calculated distribution of Lidane in soil columns as influenced by water movement. Soil Sci. Soc. Amer. Proc. 36: 544-548.
14. Katchalsky, A., and P.F. Curran. 1965. Non-equilibrium thermodynamics in biophysics. Harvard University Press, Cambridge.
15. Kay, B.D. and D.E. Elrick. 1967. Adsorption and movement of lindane in soils. Soil Sci. 104: 314-322.

16. Khan, S.U. 1973. Equilibrium and kinetic studies of the adsorption of 2,4-D and picloram on humic acid. Can. J. Soil Sci. 53: 429-434.
17. Knight, B.A.G., J. Coutts, and T.E. Tomlinson. 1970. Sorption of ionised pesticides by soil. S.C.I. monograph No. 37. Soc. of Chemical Industry.
18. Knowles, G., A.L. Downing, and M.J. Barrett. 1965. Determination of kinetic constants for nitrifying bacteria in mixed culture with an electronic computer. J. Gen. Microbiol. 38: 263-278.
19. Lapidus, L. and N.R. Amundson. 1952. Mathematics of adsorption in beds. VI. The effect of longitudinal diffusion in exchange and chromatographic columns. J. Phys. Chem. 56: 984-988.
20. Letey, J. and J.K. Oddsen. 1972. Mass transfer. In: Goring and Hamaker cf. ref. 10. pp 399-440.
21. Lindström, F.T., and L. Boersma. 1971. The theory on the mass transport of previously distributed chemicals in a water saturated sorbing porous medium. Soil Sci. 111: 192-199.
22. Macura, J., and F. Kunc. 1965. Continuous flow method in microbiology: V. Nitrification. Folia Microbiol. 10: 125-135.
23. McLaren, A.D. 1970. Temporal and vectorial reactions of nitrogen in soil. Can. J. Soil Sci. 50: 97:109.
24. McLaren, A.D. 1971. Kinetics of nitrification in soil: Growth of the nitrifiers. Soil Sci. Soc. Amer. Proc. 35: 91-95.
25. Misra, C., D.R. Nielsen and J.W. Biggar. 1974. Nitrogen transformation in soil during leaching. I. Theoretical considerations. Soil Sci. Soc. Amer. Proc. (in press).
26. Oddson, J.K., J. Letey, and L.V. Weeks. 1970. Predicted distribution of organic chemicals in solution and adsorbed as a function of position and time for various chemicals. Soil Sci. Soc. Amer. Proc. 34: 412-417.
27. Osgerby, J.M. 1970. Sorption and transport processes in soils. S.C.I. monograph No. 37. Soc. of Chemical Industry.
28. Reiniger, P. and G.H. Bolt. 1972. Theory of chromatography and its application to cation exchange in soils. Neth. J. Agric. Sci. 20: 301-313.
29. Swoboda, Allen R., and Grant W. Thomas. 1968. Movement of parathion in soil columns. J. Agr. Food Chem. 16:923-927.
30. Valoras, N., J. Letey, and J.F. Osborn. 1969. Adsorption of nonionic surfactants by soil materials. Soil Sci. Soc. Amer. Proc. 33: 345-348.
31. Youngson, C.R., C.A.I. Goring, and R.L. Noveroske. 1967. Laboratory and greenhouse studies on the application of fumazone in water to soil for control nematodes. Down Earth 23: 27-32.

PREDICTION OF SOIL- AND GROUNDWATER POLLUTION

DR IR L. WARTENA

Possibilities of Developing Forecasting Techniques

ABSTRACT

The title contains two key words: forecast and pollution.
Forecasting has always been a precarious buseniss. The difference between forecasting and fortunetelling is that the former is based on a model of the processes involved.
The definition of pollution gives a rise to a lot of problems, among them the connection between pollution and the environment and the assesment of the harmfulness of polluting matter.
For a forecasting system it is necessary to classify the sources and the released matters. This classification must be based on relevant characteristics.
An example makes it clear that the choice between harmless and dangerous can be conditional.
A third section asks attention for chemical and physical processes, which change the nature of the substance or the amount.
Transport processes play an important role. Some complicating features are pointed out.
The different steps are condensated to a flowdiagram.
In the second half of the paper a number of case studies are described, to illustrate the various ways of analysis. A straight forward use of the developed scheme is only possible in exceptional cases, but the diagrams can serve as a guide.

STATEMENT AND ANALYSIS OF THE PROBLEM

Mankind always wished to have a look into the future. Fortunetelling is a very old profession. The Greek had their oracles. In the Bible it is mentioned as early in the books of Moses, e.g. Lev. 19:26. In modern times fotune-telling has become less popular and has given way to forecasting and planning. The future develops from to-day, because processes now start which change the present situation. The difference between fortune-telling and forecasting is that the latter is based on a quantitative or at least a qualitative model of the processes involved, while planning aimes at influencing the course of events by intervention in the processes.

Studying forecasting possibilities means studying the relevant processes. Discussing a problem is only useful if the subject is well defined. Consequently, we have to deal with the meaning of pollution first. Only after doing this we are able to distinghuish all the steps relevant to forecasting systems.

The word pollution usually has bad connotations. A usefull definition for our purpose is the following: Pollution of soil or water is any addition of matter or properties which impedes its use for mankind. The relation to the use that mankind wishes to make of water and soil is an essential one. It means that it is inaccurate to conclude that e.g. fertilizers pollute the environment.

If they are used on arable land, destined for crop production, the possibilities of obtaining the required results are served. But if an area of poor heathland is destined to keep it in its present state, for instance for recreational purposes, fertilizer is a polluting matter. It is in this connection that a warning is justified. Pollution prevention is not sufficient to guarantee a good environment. Mankind can make bad use of soil and groundwater in an environmental sense, even without any pollution occuring.

The reverse, i.e. the case where a wrong use is partly corrected by pollution is so exceptional that in general it can be stated that prevention of pollution is favourable to the environment.

However, in this lecture I cannot go into the important problem of the task of human beings in the management of the earth and in the care of nature.

The extent to which the environment is defiled depends, of course, on the nature of the polluting matter, and also on its concentration as a function of space and time. Any approach to predict the variation of concentration must be based on a knowledge of all the transport processes which play a role. Moreover, the distribution of sources and the initial conditions of the discharge process must be known. For soil and groundwater pollution the most important transport is effectuated by water as a carrier. Groundwater flow calculations are therefore of major importance. At least changes in composition of the polluting matter during the time interval between discharge and the moment its effect is considered must be born in mind.

A review of the important steps in a pollution process is given in the following flow-diagram.

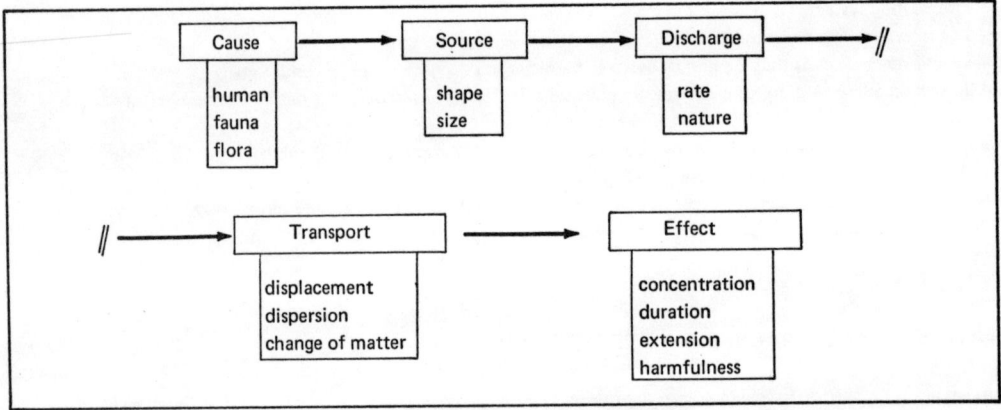

Fig. 1

One of the most difficult problems is to assess the harmfulness of the pollution. The effects of DDT and CO_2 may serve as an illustration of this point. Twenty-five years ago, DDT in a concentration as is found to-day in the Northern Atlantic, looked absolutely harmless, while to-day its harmfull effects are known. CO_2 was looked upon as absolutely without danger. But according to the climate model of Budyko an increase of some tenth's of percents will change the climate to such an extent that human life on our planet is seriously affected. If a given substance would be harmless than the whole procedure of predicting its occurrence seems to be senseless. But we only have a very limited knowledge of all the processes in nature in which natural components are involved. From the behaviour of artificial components we have hardly any notion. This implies that we cannot be too careful in handling materials and that predicting of their concentrations and persistence of occurrence is a worthwile task.

Prediction of Soil- and Groundwater Pollution

CHARACTERISTICS OF SOURCES AND DISCHARGES

In the foregoing the difficulties which arise when we try to recognize harmful matter was already mentioned. This even can lead to uncertainties about its real source. I shall explain this on the basis of an example, pertaining to irrigation.

Practically no destilled water exists in nature and therefore irrigation water contains certain quantities of different ions. Among these Na^+, Mg^{++}, Ca^{++}, Cl^- and SO_4^{--} are very common. The Cl^- ion is already toxic for most plants in low concentrations. Na^+ and Mg^{++} are necessary nutrients but can easily impede the growth due to antagonistic effects. The reaction to SO_4^{--} ion is in general indifferent. The Ca^{++} ion is a nutrient which is seldom antagonistic and which can have a good influence on soil structure. In many irrigated areas the latter is not important because the soils in those areas possess Ca^{++} ions in abundant quantities.

But all the ions contribute to the osmotic value of the soil solution. Too high an osmotic value of the irrigation water can hamper the water intake by the roots and consequently be harmfull to the crop. The osmotic value in the root-zone depends not only on the water quality and the irrigation practice but also on rainfall and because of leaching, on drainage. The crop only uses a small part of the ions applied through irrigation. If the surplus is not removed by a groundwater flow, the concentration of ions in the root-zone increases. This increase can lead to crop damage, but only a thorough hydrological study allows a conclusion as to the effects.

It is seen that irrigation water containing the ions mentioned above can pollute the soil. Grounwater thus depends on a number of conditions which have not even been summed up completely.

Examples of salt pollution are numerous and they have contributed to the end of mighty empires as that of Babylon and that of Ninive. Until some decades ago such polluted agricultural areas were abandoned or gave extremely poor harvests. The natural groundwater outflow was so small or non-excistent, that the accumulation of salts was catastrophic. Installation of a drainage system and leaching could make such areas usable for agriculture again. The drainage water was up to 2.5 times as saline as the ocean at first. The irrigation could start again, but now the transport processes had been changed. Drainage, together with a number of other measures on irrigation practice, now gives the possibility to prevent pollution.

This example shows that the circumstances under which an effect is realised are of enormous importance. The ancient people could irrigate for decades or even centuries without suspecting either the irrigation water or the irrigation system, of having catostrophic effects in the long run. Another example of self confidence originating from lack of knowledge, just as in the DDT case.

Improvement of predictions can be obtained by the classification of processes and matters. An example is a classification made in the Netherlands of pesticides for use in and along ditches and rivers and on public vegetated areas such as sports fields, parks and so on.

A list for this purpose was drawn up by a Dutch Firm, so no subtances forbidden in the Netherlands occur in the list. A White, Grey and Black Class are distinguished. Three criteria are used: the persistency in the soil, the acute and the chronic toxicity. As a measure of persistency the half life time is used here.

Toxicity is defined in a quite arbitrary way; the toxicity for human beings, and for flora and fauna as far as interpreted at the time and place of application. This restriction must be made, because pesticides are only used with the purpose of profiting from their poisonous properties.

	Persistency in soil	Toxicity acute	Toxicity chronic
white	< 2 months	low	not allowed
grey	< 3 months	medium	medium
black	> 3 months	high	high

Figure 2 — Classification of pesticides

Application of the white class only confirms with the general directions being in force. The grey pesicides may be applied only if no other means are available and there is an absolute damage imminent. The use of the black ones is dissuated. A well-known example of a substance falling in the black class is DDT: it has a low toxicity but a persistency of 4 to 40 years.

Quite another problem present sources with unknown place and time of operation: the calamities. It is not possible to predict at which time and place a traffic accident will be the cause that a transported dangerous liquid penetrates into the soil. The same holds for a leak occurring in a pipeline, etc. This problem can only be approached in a statistical sense, based on the probability that a given source arises. This can be coupled on the chance that a certain mount of substance is exceeded. The conclusion can be that the nature of the discharge can vary from one of high toxicity to one that is only harmful under certain conditions. The source can be classified as (1) well defined, e.g. in case of seepage water originating from a rubbish dump; (2) difficult to predict as in the case of traffic accidents; (3) vague and difficult to analyse as in the case of irrigation.

PROCESSES THAT CHANGE THE NATURE AND AMOUNT OF MATTER BETWEEN DISCHARGE AND EFFECT

Organic matter is destroyed by chemical and biochemical processes. In most cases this implies a decrease of toxicity and in general a decrease of the potential pollution effect. But increase of toxicity is also possible. Formation of poison in anaerobe destruction of albuminoids is well known. Photochemical destruction can be of importance in surface waters and if the polluting matter remains on the surface during a sufficient time, as is the case with many pesticides. That photochemical reactions are not always favourable is known from the smog formation in the atmosphere. Finally a remark about reactions (chemical, photochemical or biochemical) between different released matters must be made. The result can again be very different according to circumstances. Precipitation sometimes restricts be dispersion, less dangerous substances can be formed but very toxic matters can arise too. An example is the Mercury-methylene, which can be formed in a solution with Hg^{++} ions and carbohydrogenes.

TRANSPORT PROCESSES

That transport is one of the main features of the pollution problem and is essential for forecasting systems is self-evident. However, that many transport processes play a role in soil and groundwater pollution and that at least rough qualitative estimations about the influence of soil water movement on the dispersion of pollutants can be made is mostly forgotten in environmental engineering. And indeed, if we realize that from the source to the effect in soil and groundwater all kinds of transport in surface waters and even in the atmosphere can occur, it becomes clear that molecular and turbulent diffusion, potential flow, buoyancy, evaporation and condensation, codestillation, solution and presipitation, two- and multi fase transport all can play a role. Often some of these processes or effects are working simultaneously or successively.

Most of these processes cause an increasing dispersion; and in cases where only acute danger with high concentration needs to be feared, dispersion is favourable, because the concentrations are lowered. Examples are the SO_2 concentration after release from a chimney and the mixing of sewage water with clean river water downstream of the outlet of a sewage system. In both cases the concentrations are lowered as a result of a turbulent diffusion process. But, if the duration of exposure is of importance for the effect, then the benefit of a lower concentration can be cancelled by the longer exposure time. This can be the case with nuclear radiation after a nuclear explosion. If dispersion is followed by accumulation, the danger is very great, because then spots with high concentration are spread over a vast area or volume. This is the case with DDT accumulation in the tissue of some animals. Precipitation and absorption in general lower the acute danger due to the decrease of mobility, but can act as a new source under other circumstances.

Prediction of Soil- and Groundwater Pollution

For groundwater pollution groundwater flow in the saturated zone is of special interest. For hydrological problems are solutions giving rates of discharge or potentials of interest. For pollution problems the speed along a streamline often is of more importance. That means that other solutions than those normally used in hydrology are required.

An example can be found in the prediction of the duration of seepage of saline water in some low lying areas in the Netherlands as is shown in fig. 3.

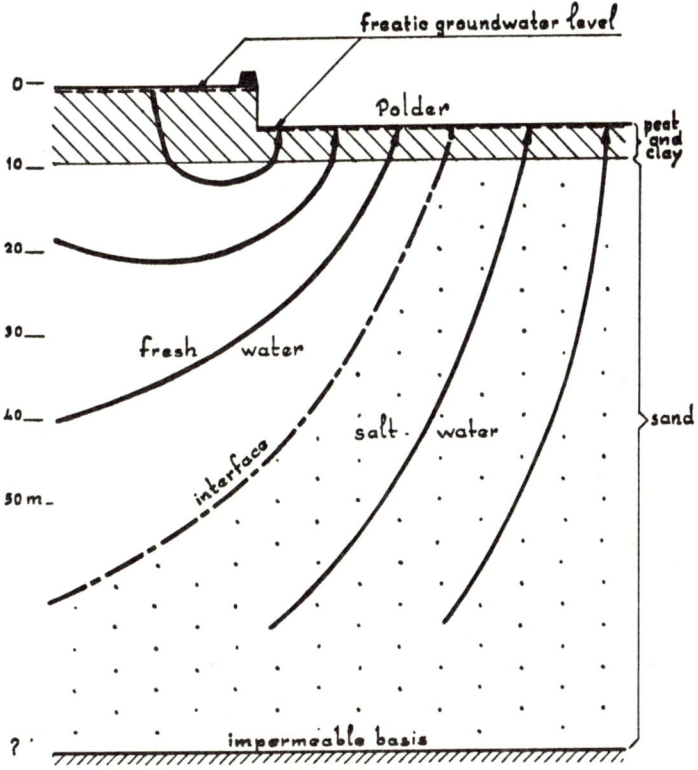

Fig. 3

The infiltrating fresh water displaces the saline water in the sand deposists. This fossile sea water now appears as seepage in the low lying polder (fig. 3). Hydrological models can, for instance, also be used to predict the displacement of polluted water which seeps out of a rubbish dump or the change in water quality of water pumped out of a well.

SCHEMATIC SUMMARY

For practical applications the scheme of fig. 1 can act as a guide for the analysis. Some blocks of the flow diagram can be elaborated somewhat by way of example. The diagram is by no means complete.

Cause	
industry	all
agriculture	direct
domestic	and
traffic	indirect
dumping	

Source
point source
areal source
volume source
surface source
deep source
For each sourve the time dependence is crucial

fig. 4

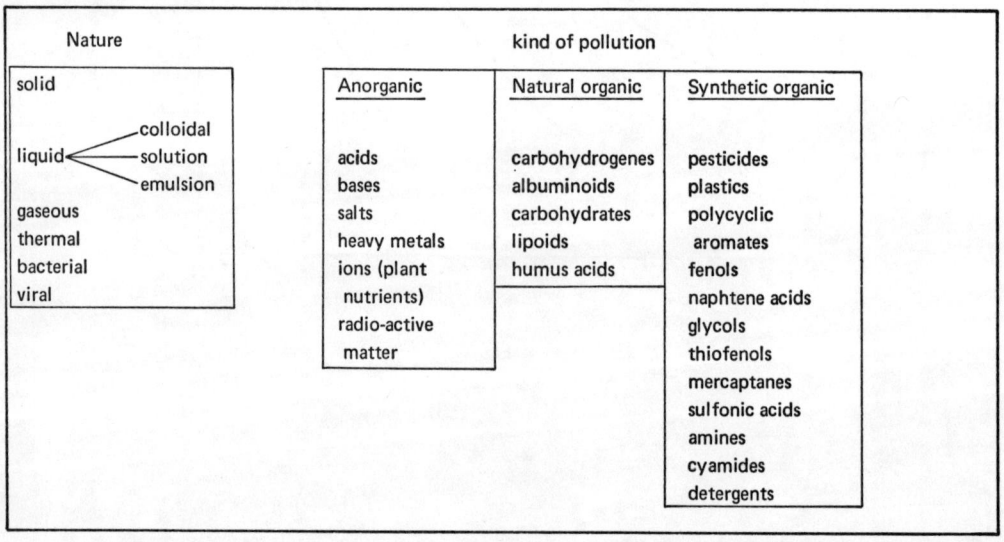

Nature	kind of pollution		
	Anorganic	Natural organic	Synthetic organic
solid			
liquid — colloidal / solution / emulsion	acids	carbohydrogenes	pesticides
gaseous	bases	albuminoids	plastics
thermal	salts	carbohydrates	polycyclic aromates
bacterial	heavy metals	lipoids	fenols
viral	ions (plant nutrients)	humus acids	naphtene acids
	radio-active matter		glycols
			thiofenols
			mercaptanes
			sulfonic acids
			amines
			cyamides
			detergents

Prediction of Soil- and Groundwater Pollution

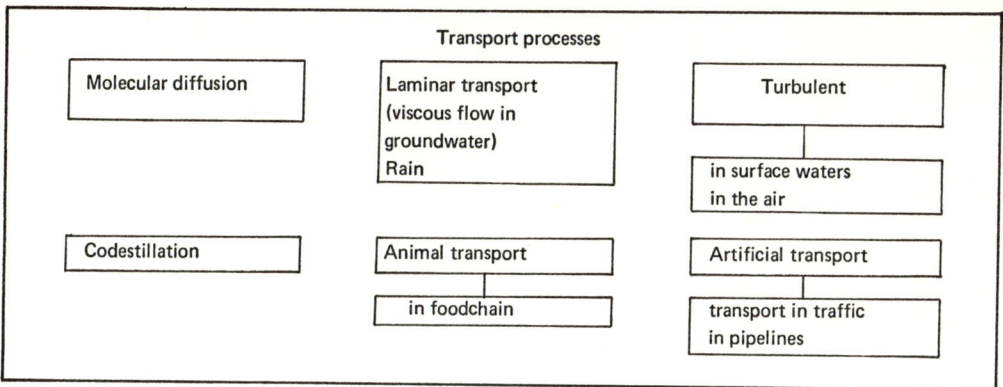

fig. 6

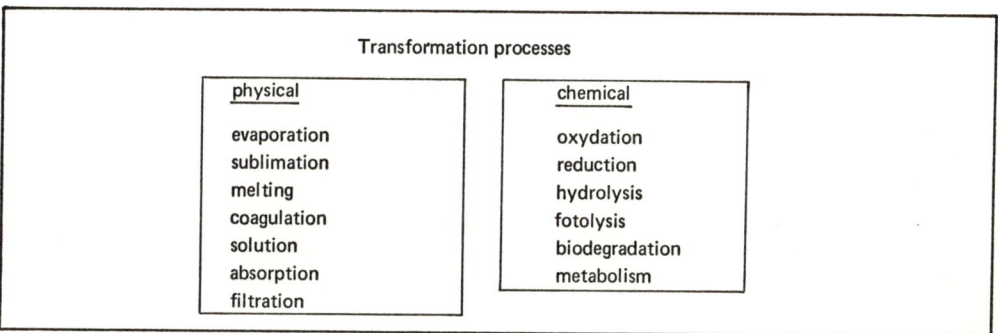

fig. 7

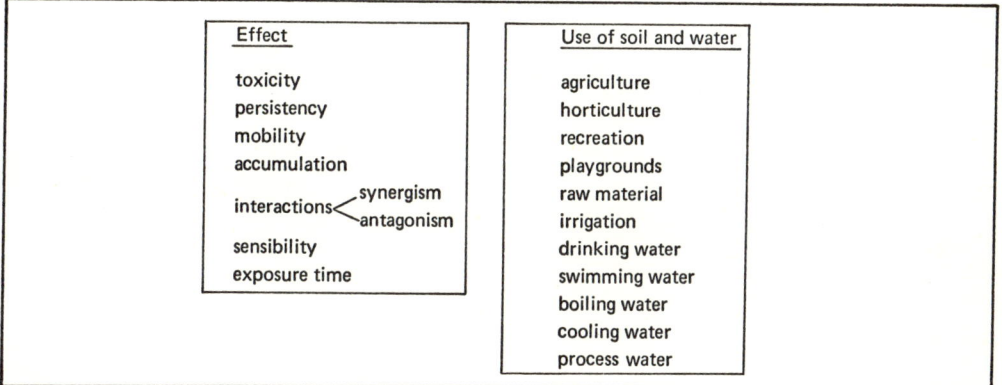

fig. 8

EXAMPLES OF ANALYSIS

In this section a number of case studies will be described to illustrate the various ways of analysis.

1. Cases where a straightforward use of the developed scheme is possible are exceptional. An example where this could be done is provided by the salt release in the river Rhine from the French potassium mines. The rate of discharge of the Cl^- is well-known. Its contribution to the Cl^- concentration at any point of the river at any time is not so easy to forecast. The mathematical difficulties which arise in calculating the rate of water discharge, taking into account all the discharge waves of the tributaries, can hardly be overcome.

   Besides the relation between precipitation and discharge is not so well-known that discharge forecasts of the different rivers on the basis of amounts of precipitation are possible. Some predictions can be given on a statistical basis. The difficulty for good forecasts of quantity and quality of surface waters is in general lack of knowledge of the transport problem.

2. The Bodensee is a bog lake in the upper Rhine between Germany and Switzerland. The water and the bottom deposits are polluted by polycyclical aromates to a higher than normal extent.
   Borneff and Kunte studied this problem in detail. Two main points were investigated:
   a. what is the source;
   b. which part is cancerogenic.
   Concerning the first problem an obvious approach is to analyse the water feeding sources of the lake; i.e. the different rivers and the groundwater flowing into it. Another source could be dust of industries or of roads, which can be expected to pollute the surface water especially during precipitation. Only a limited amount of heavily polluting industries are situated in the surroundings of the Bodensee. An analysis of dust from industrial areas, as Dortmund, revealed both much lower amount of polycyclical aromates and a lower fraction of cancerogenic ones than were found in the lake. Dust of some roads with much traffic (Ludwigshafen) showed high concentrations of both substances but the amount of dust brought by precipitation into the Bodensee seems to be negligible. The data on the dry matter in different contributing rivers show that rivers containing water from urban areas were the wrong-doers. The data disclosed a very big scatter which could not immediately be explained. Analysis of faeces and urine showed very low amounts of the dangerous matter. Finally there appeared to be a connection with rainfall. Especially after heavy rain the total amount as well as the percentage of cancerogenic pollutants increased. Separation of industrial and other urban water was quite simple. The urban water could be analysed by analyzing the sewage water just before it entered and immediately after it left the sewage treatment plant. The data shown in the next table are convincing.

	dry weather		heavy rain	
	before entering the plant	after leaving the plant	before entering the plant	after leaving the plant
polucyclic aromates in mg/m^3 sewage water	139	1243	37880	31560
% cancerogenics	16	29	43	46

fig. 9

Now the source could be identified as being the dust on freshly tarred tarmacstreets. This dust was washed into the sewage system. Part of it is kept back during the mechanical purification in the treatment plant. This part comes free at a later stage, causing the increasing effect of the plant with dry weather.

3. A consulting firm was asked to study the groundwater flow near a dump of chemical waste. In the near surroundings many trees badly suffered from groundwater pollution. The dump was situated in the eastern parts of the Netherlands. There was no chemical industry situated in the neighbourhood. When the consultants asked questions about the originig and composition of the dumped material the simple answer was that these were "secret".

During the first visit they saw a hole, caused by sand production, filled with a black liquid with a bituminous odour. It was located in a sand ridge, part of a morain, consisting of very permeable sands. These sands hardly contained clay, silt or organic matter. In a number of sand production pits in the same hill in the neighbourhood the highest groundwater table was at least 10 m and probably more than 20 m below the surface of the threatened area. Besides it was very unlikely that water movement disturbing layers existed here. At groundwater depths of more than 10 m in this soiltype hardly if any capillary rise of groundwater into the root zone of pine and birch forest can be expected. Further investigation showed clearly that the damage was caused by air pollution. Farmers told that formerly during and a short time after dumping damage on crops on distances up to 500 m were not an exception. Sometimes fire started during the dumping, but mostly it extinguished spotaneously after a couple of hours. Once there was a big fire of a 24 hours duration.

Chemical analysis of the fluid showed that the pit was filled with a clay containing sulfonized bituminous matter, mixed with sulfonicacids, H_2SO_3 and H_2SO_4. The top layer of 0.1 to 0.2 m thickness contained a diluted solution of H_2SO_3 and H_2SO_4 with a pH of approximately 1.5. On the bottom a sludge deposit of clay and sulfonized bituminous products was found. It seems to be practically impermeable for liquids. Near the shore at approximately 1.10 m below the bottom no pollution could be found even with a very careful chemical analysis.

The conclusion was that the trees suffered from air pollution. When no new dumping would take place, the danger for air pollution would be negligible. Danger for groundwater pollution was not probable for a couple of decades or more. Over a long period groundwater pollution was considered possible and moreover the pit was very dangerous for men and a deadly peril for birds.

With burners it would be possible, at least with additional fuel, to burn the sludge but a problem arises of SO_2 release in the air which could lead to the formation of H_2SO_4, which could penetrate into the soil and cause groundwater pollution. A method for neutralizing the material with marl and lime was developed which gave transformation of the pit material into a tough substance with a sufficient bearing capacity and conductivity for water. This could be covered with soil and planted with trees.

4. Soil pollution causes much damage on trees in urban areas. Languishing and suffering due to other reasons than physiological old age is a frequently observed phenomenon. Extensive studies have already been made and the cause can vary greatly. Some kinds can be established diagnostically on the ill trees, but frequently this is impossible. Direct reasons are e.g. Cl^- ion poisoning, poor respiration, shortness of minerals or shortness of water. Many times combinations occur but all these reasons are no causes in the sense of the analysis scheme given above. One of the simplest causes is Cl^- poisoning. Chlorides are often used for the prevention of slippery roads at temperatures below zero and this leads to tree damage most frequently.
If the leaching in winter is insufficient, there is a possibility of a high concentration in the groundwater and eventually also in the unsaturated zone.

In the case where saline soil water penetrates the root-zone, both from above or from below poisoning is possible. Prevention by using less salt on the street would be the best way to eliminate tree damage, but is difficult to realize. Analyzing soil and groundwater can be a base for forecast but is in general not sufficient. Much more must be known about the water supply in summer from above as rain, but also from below as groundwater.
The same information is necessary if one wants to come to curing measures. The only measure in general is water infiltration to wash out the root-zone and to prevent capillary rise from below. Washing out is but only possible of the existing natural or artificial drainage is sufficient. Sometimes the drainage has to be improved.

A more complicated cause is shortness of water supply. If a plant suffers from shortness of water it means that the roots are not able to take up enough water. The reasons for this can be varried. A cause can be mechanical damage of the root system. Excavations near the stems (e.g. for placement of pipelines or a sewage system) somestimes can reduce the total root volume by more than 25%, which diminishes the possibility of water uptake, while the demand remains unchanged. Another possibility is that the salt concentration of the soil water is too high. High osmotic values can occur without poisoning effects of any of the ions as has been discussed before. A third and very common cause is that the water content of the soil is insufficient. Many urban areas have bad top soils, as e.g. course sand or gravel, with a low water holding capacity in the root-zone. In a silvicultural sense these soils are marginal. In addition, a road construction with an inpermeable toplayer prevents the penetration of rain water. If this water flows straightly into the sewage system or in drainage ditches, the water replenishment is lower than under natural circumstances. For about the same reasons as shortness of water also shortness of nutrients can occur, because one has to deal with poor soils, poor root systems and no application of manure of fertilizers.

Soil analyses in combination with a hydrological investigation can form a basis of a good prediction of such damage and at the same time for advising measures which prevent pollution. A poor respiration gives rise to the most complex problem of all. In the root-zone the pores in the soil matrix are partly filled with water, partly with a gas, called air. The composition of soil air is different from that of the lower atmosphere. Roots of plants and trees respirate. The transport of the carbon dioxide surplus and the replenishment of the oxigen mainly is a diffusion process. The diffusion process and the respiration tend to be in equilibrium which each other. Is the CO_2 concentration so high or the O_2 concentration so low that the respiration is hampered, then the plants will suffer. This means that asphalt is bad and asphaltation of a brick-paved road can cause the death of near by big old trees! But also young trees can begin to suffer without changing the road construction. When the crowns grow the root system must do so as well and the total rate of root respiration grows. A diffusion rate which was sufficient for small, young trees can be insufficient for bigger ones. Another reason for a bad root respiration is a too low pore volume. If the soil is getting wetter, the diffusion resistance increases. The same can be said for a too high a compaction of the soil under traffic conditions. In both cases the root development is also stunted, so the damaging cause works twofold. In different countries (e.g. the U.S.A., the Netherlands, the U.K.) tehere are firms that can investigate systematically the chemical and the physical (including hydrological) behaviour of the soil where trees will be planted. They can predict difficulties and give advices how to prevent them.

A respiration difficulty of a completely different origin is that caused by leakage of natural gas from a pipeline. In the immediate surroundings of the leak, the methane displaces all other gases. Next a region exists where the diffusion takes a greater part in the transport and air and CO_2 diffuse towards the leak while CH_4, diffuses away from the leak. (fig. 10).

This is the situation near a fresh leak or when the soil has a temperature lower than say 5^oC. But at higher temperatures a microbial CH_4 oxydation can take place. This means O_2-consumption and CO_2-production occur. A situation as in fig. 11 is found. It is clear that in a large volume respiration of vegetation is hampered or has become impossible. Besides it is a difficulty that the steady state never is approximated. The microbial activity changes with temperature and the latter varies during the year.

A mathematical description by Hoeks leads to results that are in satisfactory agreement with the measurements. This means that measurements of O_2 and CO_2 content of the soil air can help to find leaks. Also the distance of a pipeline on which such measurements must done in order to be warned can be calculated.
Measures can then be taken to avoid damage on vegetation. After reparing the pipelines e.g. a forced ventilation can be applied.

Gentleman, many other examples could be given and probably there are many other illustrative exemples. The only thing I wanted to do (and that was my task) was to promote analytical thinking in the field of soil and groundwater pollution. Too often the physisist has too little knowledge of the environmental problems and the environmental

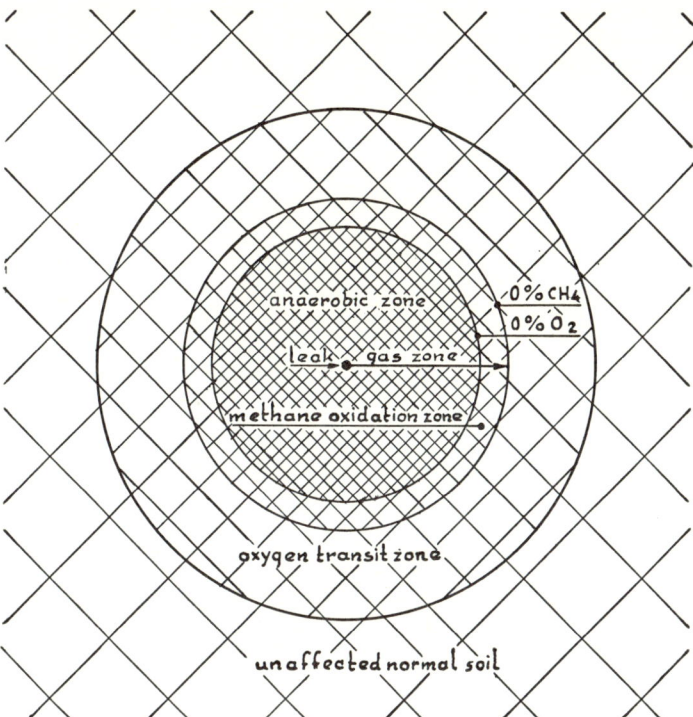

fig. 10

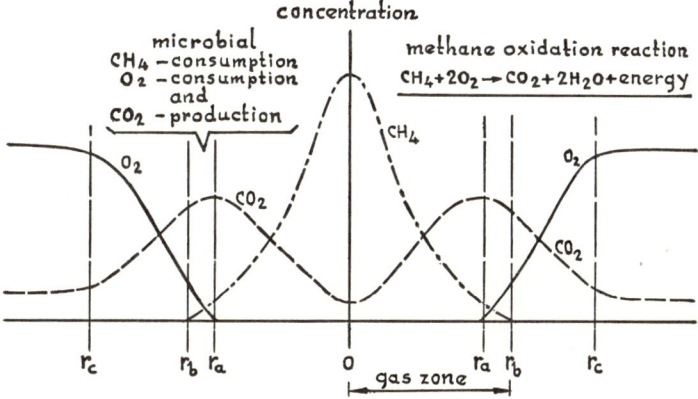

fig. 11

engineer knows little of transport problems in order to predict pollution. Putting the problem more generally, it becomes clear that workers in different disciplines have a task, in particular physisists and engineers. With their cooperation forecasts can be bettered. The use of good predicting techniques in itself does not add to obtaining a better and stable environment for human beings. Only as a tool in the hands of men who realize that they have as creatures the mission to superintent the creation can lead to improvement of the nature and the physical environment of mankind.

POLLUTION IN PLANT CANOPIES

ARTHUR CHAMBERLAIN

*Atomic Energy Research Establishment,
Harwell, United Kingdom*

ABSTRACT

Methods of determining the flux of pollutant gases to plant canopies are summarised, and results reported in the literature are analyses in terms of the components of resistance. The corresponding problems of transfer of particulate matter are briefly summarised. Some instances are given of the implications of mass transfer of pollutants to crops.

LIST OF SYMBOLS

B	dimensionless sub-layer Stanton number (Owen and Thompson 1963)	
c_d	drag coefficient = u_*^2/u_1^2	
c_D	bulk drag coefficient	
c_i	particle impaction coefficient	
c_v	mass transfer coefficient = $F/\chi u_1$	
D	molecular diffusivity	$m^2 s^{-1}$
d	zero displacement	m
d_o	chord diameter of leaf	m
F	flux of matter to surface	$gm^{-2} s^{-1}$
k	Von Karman's constant	
K_V, K_H	eddy diffusivities for vapour, heat	$m^2 s^{-1}$
Re_*	roughness Reynolds number = $u_* z_0 \nu^{-1}$	

r	resistance to mass transfer = $F^{-1}\chi$	$m^{-1}s$
r_a, r_b, r_c	aerodynamic, boundary layer, surface resistances	$m^{-1}s$
u	wind speed	ms^{-1}
u_1	wind speed at reference height	ms^{-1}
u_*	friction velocity	ms^{-1}
v_g	velocity of deposition = F/χ	ms^{-1}
v_s	sedimentation velocity of particle	ms^{-1}
v_T	turbulent transport velocity of particle	ms^{-1}
z	height	m
z_o	roughness length	m
α, β	constants	
ϕ_H	dimensionless shear of heat	
ν	molecular viscosity	$m^2 s^{-1}$
σ	Schmidt number = ν/D	
χ	mass concentration in air	gm^{-3}

1. INTRODUCTION

There are at least four reasons for studying the mass transfer of pollutants to and from vegetation

(a) Effects of the pollutants on the vegetation

(b) Effects on herbivores eating the vegetation and on humans eating flesh or drinking milk from the herbivores

(c) Loss of pollutants at the earth's surface as a factor in the life cycle of the pollutants

(d) Use of pollutants as a tracer for studying mass transfer processes.

Pollutants may be gaseous or particulate. The analogies between mass transfer and heat or momentum transfer apply most closely to gaseous species, but the transport of particles to and from surfaces is of long-standing agricultural and horticultural interest, and offers a number of unsolved problems to the aerodynamicist and particle physicist.

2. METHODS OF MEASURING TRANSPORT OF POLLUTANTS TO PLANT CANOPIES

2.1. Mass balance in crop

The uptake from the atmosphere is deduced from the net accumulation of pollutant in the crop, allowing for uptake from and translocation to the soil and re-emission to the atmosphere. The sulphur balance studies of Thomas et al (1943) and Cowling, Jones and Lockyer (1973) working with growth chambers, and of Johansson (1959) with pot culture in the open air, can be used to derive the velocity of deposition or canopy resistance for uptake of SO_2 by crops. The method is analogous to the use of lysimeters to measure transpiration. In growth chambers, good sulphur balances have been obtained, but, with uncertain air movement, the aerodynamic resistance is unknown. The work of Johansson (1959) is of historical interest, but his method probably suffered from the 'oasis' effect − adrection of SO_2 to crops with small horizontal extent.

Two variants of this method using radioactive tracers may be mentioned. A radioactive gas ($^{35}SO_2$, $^{131}I_2$) or radioactively tagged particles may be released in a wind tunnel or the open air, and the uptake by crops measured. This overcomes the difficulty of accounting for non-aerial uptake, but care is needed that the fetch is adequate in relation to the height of sampling so that the constant flux hypothesis is reasonable. More subtly, ^{35}S has been added to the nutrient media of crops (Olsen 1957, Bromfield 1972), to label the sulphur taken up from the roots, thus allowing the foliar absorption to be distinguished.

2.2. Mass balance in gas phase

The rate of uptake of SO_2, O_3 or other pollutants may be estimated from the rate of loss from the atmosphere over the crop. The atmosphere may be contained in a real box (Aldaz 1969, Aldaz and Regener 1969), or wind tunnel (Hill 1971), or it may be contained in an imaginary box, or part of the atmosphere defined in some way.

Experiments in real boxes suffer from the usual disadvantages of growth chamber experiments. An early example of attempts to strike mass balances in the free atmosphere was that of Meetham (1950) who was probably the first to appreciate that uptake of SO_2 by the earth's surface by gaseous absorption is an important aspect of its life cycle. Raynor and his co-workers (1970, 1972) have attempted to apply this method to assess deposition of pollen grains in the field, by erecting sampling arrays downwind of natural sources of pollen and measuring the total horizontal mass flux at various distances of travel. The main difficulty with this type of experiment is assessment of the 'leakage' through the top and sides of the imaginary box.

2.3. Gradient method

The measurement of flux from the vertical gradient above a crop, well established for water vapour and CO_2, is now being used extensively to measure transport of pollutants such as SO_2 and O_3 to crops (Fig. 1). For this method to work the following conditions must apply, namely

(a) accurate measurement of concentration in air

(b) uniform terrain

(c) absence of local sources of pollution

(d) sufficiently rapid uptake to generate a measurable gradient.

To deduce the vertical flux F from the gradient $\partial\chi/\partial z$ the vertical eddy

Pollution in Plant Canopies

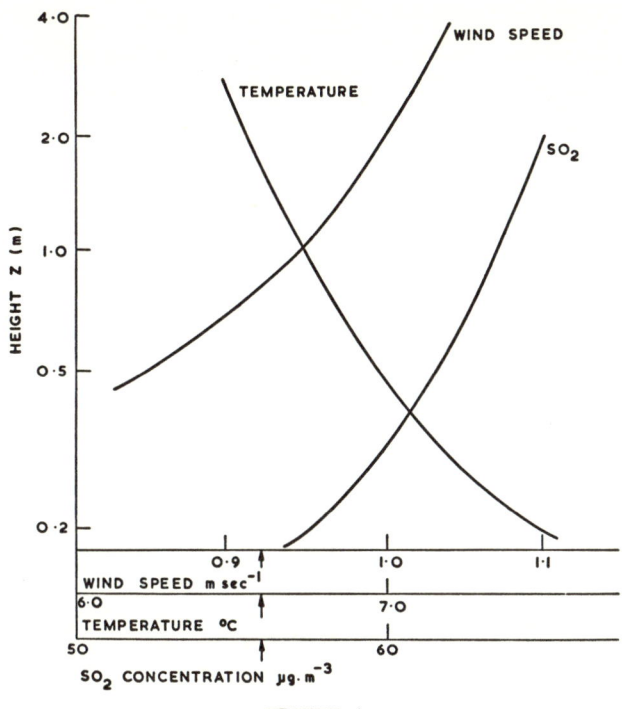

FIGURE 1

Profiles of SO_2, wind speed and temperature over grass

diffusivity $K_V(z)$ must be known. This is usually equated to the eddy diffusivity for heat, given by

$$K_H = ku_* (z-d)/\phi_H \qquad (1)$$

where ϕ_H is the dimensionless shear of heat related empirically to the Richardson number by several workers including Businger et al (1971) and Dyer and Hicks (1970).

3. RESISTANCE TO UPTAKE OF REACTIVE GASES

Published measurements of the resistances to uptake for SO_2 are given in Table 1 and for I_2 and O_3 in Table 2. Strictly the resistance should be quoted with respect to a reference level where the air concentration χ is measured. Where possible the results have been corrected to reference height $z = 1$ m, but in some instances the reference height was unstated. The range in values

TABLE 1

Resistances of canopies to uptake of SO_2

Method	No. of Expts.	Crop	u (1 m) m/s (mean)	u_* m/s (mean)	r m^{-1} s	$r_b + r_c$ m^{-1} s	Notes	Ref.
Mass balance in crop		Alfalfa	–	–	100	–	Growth chamber	1
		Rye-grass	–	–	160	–	Growth chamber	2
		Mustard	–	–	100	–	Glass-house (soil tagged with ^{35}S)	3
Mass balance in air	7	Alfalfa	3	0.4	40	–	Wind tunnel, limited fetch over crop	4
Fumigation with $^{35}SO_2$	6	Grass	4.0	0.41	120	90	Field	5
Fumigation with $^{35}SO_2$	3	Grass	2.8	0.25	130	70	Field	6
Gradient	9	Grass	2.45	0.22	200	130	Field	5
Gradient	14	Short grass	3.0	0.18	360	–	Field	7
Gradient	19	Grass	3.0	0.29	220	180	Field	8
Gradient	13	Bare chalky soil	3.3	0.25	115	34	Field	9

References: (1) Thomas et al 1943, as interpreted by present author; (2) Cowling 1974; (3) Bromfield 1972; (4) Hill 1971; (5) Garland, Clough and Fowler 1973; (6) Owers and Powell 1974; (7) Garland, Atkins, Readings and Caughey 1974; (8) Shepherd 1974; (9) Garland, unpublished.

TABLE 2

Resistances to uptake of O_3 and I_2

Gas	Method	Number of Experiments	Crop	u (1 m) m/s	u_* (m/s)	r m^{-1}s	$r_b + r_c$ m^{-1}s	Notes	Reference
O_3	Gradient	14	Grass	5.2	0.46	160	135	Mean of 4 sets of experiments at different sites	1
O_3	Gradient	12	Bare soil	2.8	0.18	260	170	–	2
I_2	^{131}I	8	Grass	3.8	0.40	70	40	–	3
I_2	Mass balance in crop	7	Grass	2.4	0.28	170	130	–	4

References: (1) Galbally (1971)

(2) Turner, Rich and Waggoner (1973)

(3) Chamberlain and Chadwick (1966)

(4) Vogt et al (1973)

of r for uptake of SO_2 is less than a factor 10, from 40 to 360 $m^{-1}s$. By subtracting $r_a = u_*^2/u_1$ from r, the sum $r_b + r_c$ of the boundary layer and canopy surface resistances is obtained. Of these the former takes into account that the resistances across the inner boundary layer are not the same for gas transport and for momentum transport. There are three reasons for this. Diffusion across the viscous part of the layer depends on molecular diffusivity. Form drag has no analogue in mass transfer. Momentum is a vector quantity and only components parallel to the flow contribute to skin friction, whereas heat and mass are scalar quantities. For surfaces with bluff roughness elements, the following relations apply (Owen and Thomson 1963, Chamberlain 1966, 1968, Garratt and Hicks 1973).

$$r_b = (Bu_*)^{-1} \qquad (2)$$

$$B^{-1} = \alpha (Re_*)^{0.45} \sigma^{0.8} \qquad (3)$$

where $Re_* = u_* z_0 \nu^{-1}$ is the roughness Reynolds number, $\sigma = \nu/D$ is the Schmidt number and α is a constant depending on the shape and spacing of the roughness elements. Owen and Thompson obtained an average value of $\alpha = 2.4$ from wind tunnel experiments but in a review of field and wind tunnel data Garratt and Hicks suggest $\alpha = 1.7$.

For comparison with the resistance to uptake of SO_2 by bare soil in the bottom row of Table 1, the following values of the parameters: $u_* = 0.25$ ms^{-1}, $z_0 = 0.003$ m, $Re_* = 50$, $\sigma = 1.3$, $\alpha = 1.7$, when inserted in equation (3) give $B^{-1} = 12$ and $r_b = 48$ ms^{-1}. This more than accounts for the value of $r_b + r_c$ measured by Garland, indicating that r_c for uptake of SO_2 by a moist calcareous soil is small.

For surfaces with fibrous roughness elements including probably most plant canopies, B^{-1} does not increase so rapidly with Re_* as equation (3) implies.

Thom (1972) has deduced theoretical evidence that B^{-1} for foliage is proportional to $u_*^{\frac{1}{3}}$. From wind tunnel work (Chamberlain 1966) it appears that $B^{-1} = 5$ for water vapour, and $B^{-1} = 7$ for SO_2 ($D = 1.15 \times 10^{-5}$ $m^2 s^{-1}$) may be reasonable estimates. With $B^{-1} = 7$ and $u_* = 0.30$ ms^{-1}, (a typical value for air flow over crops) $r_b = 20$ $m^{-1}s$, which is a fairly small component of the total resistance.

The remaining part of the resistance, r_c, is also complex. For transpiration from leaves with open stomata, r_c can be equated approximately to the stomatal resistance, since cuticular resistance is usually high in comparison. For uptake of SO_2, O_3 and I_2 by leaves, the degree of stomatal opening affects r_c (Spedding 1969, Rich, Waggoner and Tomlinson 1970, Thorne and Hanson 1972, Adams and Voilleque, 1971) but the relative importance of uptake through stomata and cuticular uptake are not yet known. There is evidence that the uptake of all three gases on both leaf surfaces and inert surfaces is more rapid when humidity is high than when it is low (Spedding 1969, Barry and Chamberlain 1963, Cox and Penkett 1972), probably because of surface reactions.

Nevertheless, the order of magnitude of $r_b + r_c$ for all three gases in crop canopies (Tables 1 and 2) is not greatly different from the bulk surface resistance of crops to transpiration (minimum values 30 to 50 sec m^{-1} in good growing conditions, Monteith 1963) when allowance is made for the molecular diffusivity ($D = 1.15 \times 10^{-5}$ $m^2 s^{-1}$, 1.4×10^{-5} $m^2 s^{-1}$, 8×10^{-6} $m^2 s^{-1}$ for SO_2, O_3 and I_2 respectively).

4. TRANSPORT OF PARTICULATES TO SURFACES

Particles in the sub-micrometre size range have appreciable Brownian motion, which can be considered in the same way as molecular diffusivity as a process for transport across viscous sub-layers. Thom (1968) and Chamberlain (1974) have shown that the resistance for transport to and from leaf-like laminas can be described by equations of the Pohlhausen type

$$r = \beta \left(\frac{d_o}{u}\right)^{\frac{1}{2}} \sigma^{\frac{2}{3}} \qquad (4)$$

where d_o is the chord diameter of the leaf, u the wind speed, $\sigma = \nu/D$ is the Schmidt number and β depends on the orientation of the leaf. A spherical unit density particle of diameter 0.1 µm has a diffusivity $D = 7 \times 10^{-6}$ cm^2/s, giving $\sigma = 2 \times 10^4$ and $\sigma^{\frac{2}{3}} = 760$. If equation (4) was in fact valid for so large a value of σ, it would suggest that the resistances to uptake of 0.1 µm particles by leaves and plant canopies would be about 3 orders of magnitude higher than the resistances to water vapour and heat transfer.

Sehmel (1971) has calculated the rate of transport of particles to smooth surfaces and wind tunnel experiments (Clough 1973) have verified the theory. For particles of diameter 0.1 µm and a friction velocity $u_* = 0.24$ ms^{-1} the resistance to transport to a smooth upwards facing copper surface was 5×10^4 m^{-1}s ($v_g = 2 \times 10^{-3}$ cm/s), compared with an aerodynamic resistance of 100 m^{-1}s, so the extrapolation according to $\sigma^{\frac{2}{3}}$ does appear to be approximately valid.

It seems, however, that a small degree of fibrous-type roughness, for example a flat filter paper surface, considerably reduces the resistance to particle transport, even though the surface may be aerodynamically smooth (Wells and Chamberlain 1967). The reason for this is that hair-like protuberances, surrounded by very small viscous sub-layers, are relatively more important in promoting transport to the surface when σ is very large. Experiments in a wind tunnel (Chamberlain 1966, and Fig. 2) show that the resistance for uptake of 0.1 µm particles is an order of magnitude less for a short grass canopy than for a smooth surface.

When the particle size exceeds about 1 µm, sedimentation, impaction and interception, and not Brownian diffusion are the mechanisms of transport

Pollution in Plant Canopies

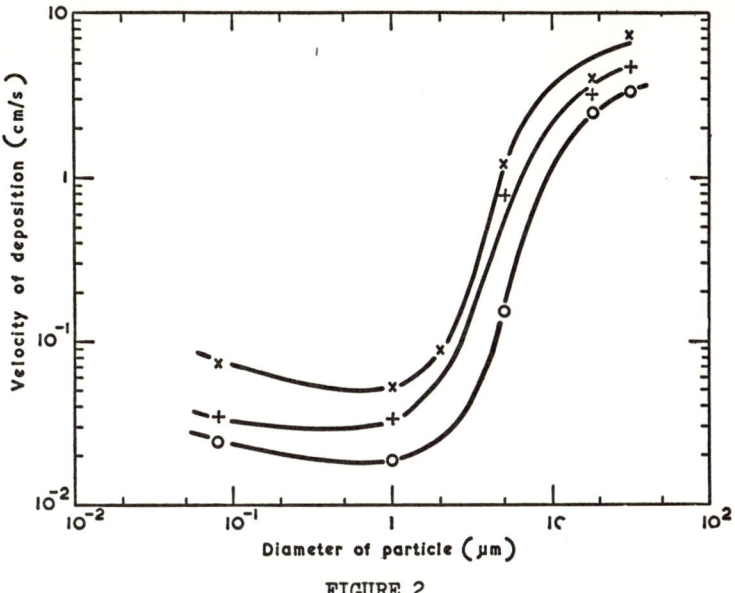

FIGURE 2

<u>Deposition of particles to short grass in wind tunnel</u>

$(X) u_* = 1.4$ ms^{-1}, $(+) u_* = 0.7$ ms^{-1}, $(\odot) u_* = 0.36$ ms^{-1}

across the boundary layer. The impaction velocity may be generated by turbulence in the air stream, so this mechanism will operate to transfer particles to a smooth surface, but in plant canopies the impaction takes place on leaves stems and other roughness elements. Gregory (1961) has given an account of the impaction processes in relation to the deposition of spores and pollen on plants.

In discussing the transport of particles it is more convenient to use the velocity of deposition v_g ($= r^{-1}$), or the ratio of velocity of deposition to wind speed, rather than the resistance. This is because the concept of additivity of resistances, so useful in gaseous transfer, is no longer applicable in the same way. Also, to a first approximation, for upwards facing surfaces

$$v_g = v_s + v_T$$

where v_s is the sedimentation velocity and v_T is the transport velocity due to impaction.

It is instructive to compare the transport coefficient for particles with the analogous drag coefficient and vapour transport coefficients. This can be done theoretically for a very simple type of roughness element, a cylinder transverse to the flow. The drag coefficients C_d and vapour transport coefficients C_v of cylinders are known[1]. The particle impaction coefficient C_i is numerically equal to the impaction efficiency (the ratio of the number of particles impacted on the surface to the number which would have passed through the space occupied by it had it not been there), for which both theoretical and experimental estimates are available (May and Clifford 1967). A comparison of C_d, C_v and C_i is shown in Fig. 3 for three sizes of cylinders, 10, 3 and 1 mm in diameter, and for two sizes of unit density particles, 5 and 20 μm diameter. The impaction process is potentially most effective for small obstacles and high wind speeds, and it is theoretically possible for C_i to exceed C_d for particles of 20 μm diameter and cylinders of diameter similar to crop stems.

A high rate of transport of such particles to crops has been considered by plant pathologists (Gregory 1961) as a possible explanation of observed 'infection gradients' - the rate of fall off with distance from the source of plant lesions caused by spore forming organisms.

This question is not fully resolved, but field work (Chamberlain and Chadwick 1972, Raynor et al 1970, 1972) with radioactively tagged and natural

---

[1] The fluxes are calculated per unit presentation area of the cylinders.

spores and pollen suggest that the coefficient v_T/u does not exceed C_D. There are two reasons for this. If v_T/u was greater than C_D, the boundary layer near the crop would be depleted of particles by impaction at a faster rate than they could be replenished by eddy diffusion from above. Secondly, the

Pollution in Plant Canopies

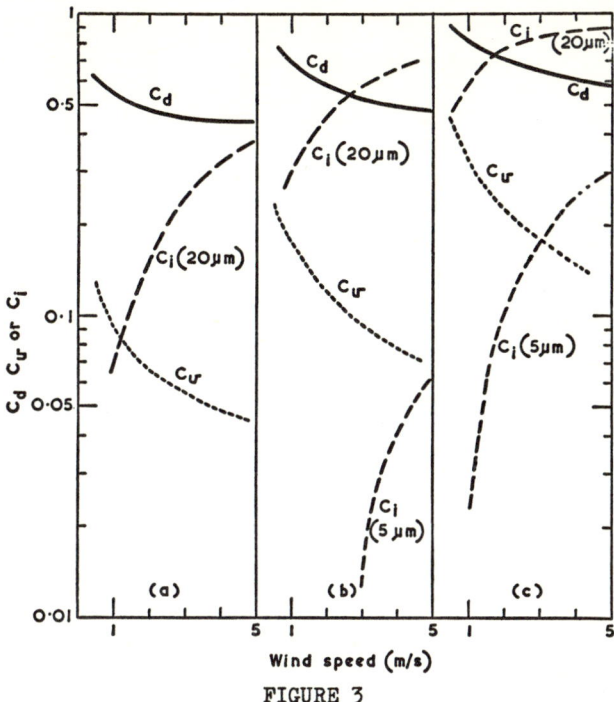

FIGURE 3

Coefficients of drag (C_d), water vapour transport (C_v) and impaction of particles (C_i) for cylinders of diameter (a) 10 mm, (b) 3 mm and (c) 1 mm

high theoretical impaction efficiencies for particles of 20 μm and greater on small obstacles are often not achieved because the particles bounce off the surface unless either the particles or the surface have a moist or sticky coating (Chamberlain 1966, Chamberlain and Chadwick 1972).

This limitation does not apply to the transport of droplets to surfaces, because droplets nearly always fragment on impact. A very effective mechanism for the transport of acid pollutants such as SO_2 and HF to plant canopies is available in a driving mist. The pollutants are absorbed in the mist droplets (typically 10 to 30 μm in diameter) and the impaction of the droplets speeds up the transfer rate to the vegetation. The phytotoxic effect may be most marked if the amount of precipitation is small, for example a driving mist may contain enough droplets to absorb the pollutants and transfer them to the vegetation, while the rate of precipitation is inadequate to wash the pollutants off the leaves.

5. IMPLICATIONS OF FOLIAR UPTAKE OF POLLUTANTS

A few examples may be given of the implications of transfer by diffusion, impaction or sedimentation of pollutants from atmosphere to crops. This is sometimes referred to as dry deposition to distinguish it from washout of pollutants in rain.

5.1. Wet and dry deposition of oxides of sulphur

This subject, discussed by Meetham (1950) in relation to the sulphur balance of the UK, and by Eriksson (1960) in relation to Europe, is still important in the assessment of the mean life of S in the atmosphere, acidification of soil, and S nutrition of plants. The deposition in rain is mainly as SO_4^{--}, derived partly from particulate sulphate incorporated in cloud droplets and partly from solution and oxidation of SO_2 in cloud and rain drops. It follows that the relative importance of 'wet' and 'dry' deposition depends on the vertical distribution of SO_2 and SO_4^{--} in the atmosphere, as well as on the velocity of deposition of SO_2 (the velocity of deposition of SO_4^{--} particles is small and no gradient of SO_4^{--} can be detected in the lowest few metres of the atmosphere). For this reason, washout is relatively more important at long range from the sources of SO_2, when the pollution has diffused up to the cloud layer and where also the SO_4^{--}/SO_2 ratio tends to be enhanced. In Table 3 some examples of the wet and dry deposition (expressed as kg S/ha y) are given.

TABLE 3

Deposition of S
possible contributions of wet and dry deposition assuming $v_g = 0.5$ cm/s

	Heavy pollution	Moderate pollution	Light pollution
In rain			
mg S/litre	3.9[a]	2.1[b]	
Rainfall mm/y	700	700	
Deposition kg S/ha y	27	15	9[c]
Dry deposition			
SO_2 concentration µg S/m^3	120[d]	35[b]	4[c]
Deposition kg S/ha y	190	55	6
Total deposition			
kg S/ha y	217	70	15
% dry deposition	87	79	40

(a) results of Stevenson (1968) for Leeds
(b) results of Stevenson for Rothamsted
(c) rural station in Sweden
(d) central Liverpool, 1966-71.

A mean velocity of deposition of 0.005 m/s ($r = 200$ ms^{-1}) has been assumed for SO_2. It appears that gaseous transfer of SO_2 is the major contribution to total deposition of S in polluted areas, and a substantial contributor in remote areas.

Under what conditions the uptake of SO_2 is beneficial to plants, by redressing sulphur deficiency, and under what conditions it is detrimental, is still very much under discussion. There is no doubt that SO_2 in concentrations in the range 50 - 100 $\mu g/m^3$ (annual average) is toxic to lichens (Ferry et al, 1973) and bryophytes (Gilbert 1968). A depression of yield has been found in experiments in which rye grass was grown in cabinets in which a concentration of 190 $\mu g/m^3$ of SO_2 was maintained for 26 weeks (Bell and Clough, 1973). These experiments were done in winter, when growth was in any case restricted, and from work in Manchester, Bleasdale (1952) had previously concluded that pollutants are particularly toxic to overwintering plants. In good growth conditions, with ample light and nitrogenous soil nutrients, but with moderate sulphur deficiency in the soil, Thomas et al (1943) and Cowling, Jones and Lockyer (1973) found increased yields with SO_2 concentrations of 270 and 130 $\mu g/m^3$ respectively compared with controls exposed to air from which SO_2 had been scrubbed.

It may be supposed that plants require sulphate for good growth, but an accumulation of SO_4^{--} ions beyond a certain point is harmful. If the resistances for uptake of SO_2 and transpiration of water are similar, the uptake of SO_2 may be more nearly proportional to the transpiration of water by a crop over a period than to the production of dry matter. In restricted growth conditions, water transpired per unit production of dry matter is likely to be greater than in good growth conditions (this was so in Thomas's experiments). It follows that the risk of reaching toxic levels of SO_4^{--} per unit dry matter, is greater when growth is slow.

5.2. Effect of deposition in reducing air concentrations

Transfer to vegetation of gaseous and particulate pollutants helps to clean the atmosphere. This operates in two ways. The lateral spread is limited. From the measured values of the velocity of deposition of SO_2, and the mean levels in country areas of the UK, it can be calculated that even in dry weather only about 50% of the SO_2 emitted crosses the coast (and much of this is absorbed in the UK). Secondly, deposition reduces the concentration of gases such as SO_2 and O_3 immediately above the plant canopy relative to that at a height of, say, 100 m by a factor which may be considerable, particularly at times of low winds and stable stratification, when the aerodynamic resistance above the canopy is a larger proportion of the total resistance than in turbulent conditions (Regener and Aldaz 1969). Fig. 4 shows the fall of SO_2 (and increase in radon which is emitted from the soil) during a calm clear night. The concentrations of SO_4^{--}, which was present as particulate, was relatively stable. Within the canopy, the reduction is much greater, and Gilbert (1968) considered that the survival of bryophyte colonies in woodland and other sheltered locations in areas which were otherwise 'bryophyte deserts' was related to the reduction of SO_2 levels by competition of other vegetation.

5.3. Effects on animals

Herbivores eat foliage from a large area of ground each day. A cow will eat grass from an area of 160 m^2 in normal grazing conditions, and will inhale about 1.5 litres of air per second (Comar 1966). Suppose a pollutant is present in the air for a time T secs at a concentration

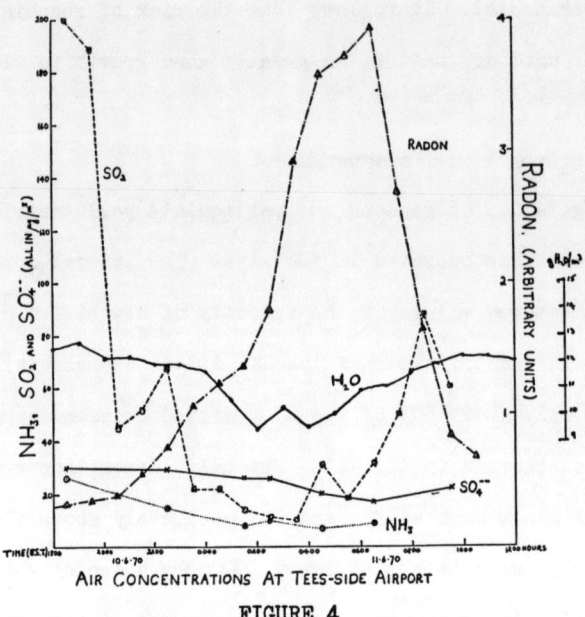

FIGURE 4

Variations of concentrations of
SO_2, H_2O, NH_3 and radon gas, and SO_4^{--}
<u>particulate in air near the ground on a clear night</u>

χ $\mu g/m^3$, the bulk resistance to uptake by the foliage is 500 ms^{-1}, 50% of the uptake takes place on the part of the foliage which is eaten, and that the pollutant persists on the foliage for a mean time of 10 days before being washed off by rain or lost in other ways. Then the amounts ingested and inhaled by the cow are

Ingested dose = $\frac{\chi T}{500}$ x 0.5 x 10 x 160 = 1.6 χ T

Inhaled dose = 1.5 x 10^{-3} χ T

It follows that the ingestion route is about 1000 times as effective as the inhalation route for uptake by cattle. This has been found by experience to be true in particular for the uptake of radioiodine released to atmosphere in nuclear bomb tests and in other ways (Comar 1966, Chamberlain 1970). It also explains why horses and cattle are often poisoned by lead particles emitted over the countryside from smelters, whereas people living in the same area may show little enhanced uptake.

REFERENCES

Adams, D.R. and Voilleque, P.G. (1971) Effect of stomatal opening on the transfer of $^{131}I_2$ from air to grass, Health Phys. 21, 771-5.

Aldaz, L. (1969) Flux measurements of atmospheric ozone over land and water, J. Geophys. Res. 74, 6943-5.

Barry, P.J. and Chamberlain, A.C. (1963) Deposition of iodine onto plant leaves from air, Health Phys. 9, 1149-57.

Bleasdale, J.K.A. Thesis, University of Manchester, 1952, also Environ. Pollut. 5, 275-285 (1973).

Bromfield, A.R. (1972) Absorption of atmospheric sulphur by mustard (Sinapsis alba) grown in a glass-house, J. Agric. Sci. Camb. 78, 343-4.

Businger, J.A., Wyngard, J.G., Izumi, Y. and Bradley, E.F. (1971) Flux profile relationships in the atmospheric surface layer, J. Atmos. Sci. 28, 181-9.

Chamberlain, A.C. (1966) Transport of gases to and from grass and grass-like surfaces, Proc. Roy. Soc. A. 290, 236-65.

Chamberlain, A.C. and Chadwick, R.C. (1966) Transport of iodine from atmosphere to ground, Tellus, XVIII, 226-37.

Chamberlain, A.C. (1966) Transport of Lycopodium spores and other small particles to rough surfaces, Proc. Roy. Soc. A 296, 45-70.

Chamberlain, A.C. (1968) Transport of gases to and from surfaces with bluff and wave-like roughness elements, Q. J. Roy. Met. Soc. 94, 318-22.

Chamberlain, A.C. (1970) Deposition and uptake by cattle of airborne particles, Nature 225, 99-100.

Chamberlain, A.C. and Chadwick, R.C. (1972) Deposition of spores and other particles on vegetation and soil, Ann. Appl. Biol. 71, 141-58.

Chamberlain, A.C. (1974) Mass transfer to bean leaves, Boundary-layer Meteorology (in the press).

Clough, W.S. (1973) Transport of particles to surfaces, Aerosol Sci. 4, 227-34.

Comar, C.L. (1966) Radioactive materials in animals - entry and metabolism, in Radioactivity and human diet, ed. R. Scott Russell, Pergamon, Oxford.

Cowling, D.W., Jones, L.A.P. and Lockyer, D.R. (1973) Increased yield through correction of sulphur deficiency in rye-grass exposed to sulphur dioxide, Nature 243, 479-80.

Cox, R.A. and Penkett, S.A. (1972) Effect of relative humidity on the disappearance of ozone and sulphur dioxide in contained systems, Atmos. Environ. 6, 365-8.

Dochinger, L.S., Bender, F.W., Fox, F.L. and Heck, W.W. (1970) Chlorotic dwarf of Eastern White Pine caused by an ozone and sulphur dioxide interaction, Nature 225, 476.

Dyer, A.J. and Hicks, B.B. (1970) Flux gradient relationships in the constant flux layer, Q. J. Roy. Met. Soc. 96, 715-21.

Eriksson, E. (1959, 1960) The yearly circulation of chloride and sulphur in nature, Tellus XI 375-403 and XII 63-109.

Ferry, B.W., Baddeley, M.S. and Hawksworth, D.L. (1973) Air pollution and lichens, University of London, Athlone Press.

Galbally, I.E. (1971) Ozone profiles and ozone fluxes in the atmospheric surface layer, Q. J. Roy. Met. Soc. 97, 18-29.

Garland, J.A., Clough, W.S. and Fowler, D. (1973) Deposition of sulphur dioxide on grass, Nature 242, 256-7.

Garland, J.A., Atkins, D.H.F., Readings, C.J. and Caughey, S.J. (1974) Deposition of gaseous sulphur dioxide to the ground, Atmos. Environ. 8, 75-9.

Garratt, J.R. and Hicks, B.B. (1973) Momentum heat and water vapour transfer to and from natural and artificial surfaces, Q. J. Roy. Met. Soc. 99, 680-7.

Gilbert, O.L. (1968) Bryophytes as indicators of air pollution in the Tyne Valley, New Phytol. 67, 15-30.

Gregory, P.H. (1961) The microbiology of the atmosphere, Leonard Hill, London.

Hill, A.C. (1971) Vegetation: a sink for atmospheric pollutants, J. Air Poll. Control Ass. 21, 341-6.

Johansson, O. (1959) On sulphur problems in Swedish agriculture, Ann. Roy. Agr. Coll. Sweden 25, 57-169.

May, K.R. and Clifford, R. (1967) The impaction of aerosol particles on cylinders, spheres, ribbons and discs, Ann. Occupational Hyg. 10, 83-95.

Meetham, A.R. (1950) Natural removal of pollution from the atmosphere, Q. J. Roy. Met. Soc. 76, 359-71.

Monteith, J.L. (1965) Evaporation and environment, in The state and movement of water in living organisms, ed. G.E. Fogg, Cambridge University Press.

Olsen, R.A. (1957) Absorption of sulphur dioxide from the atmosphere by cotton plants, Soil Sci. 84, 107-11.

Owen, P.R. and Thompson, W.R. (1963) Heat transfer across rough surfaces, J. Fluid Mech. 15, 321-4.

Owers, M.J. and Powell, A.W. (1974) Deposition velocity of sulphur dioxide on land and water surfaces using a ^{35}S tracer method, Atmos. Environ. 8, 63-7.

Raynor, G.S., Ogden, E.C. and Hayes, J.V. (1970) Dispersion and deposition of ragweed pollen from experimental sources, J. Appl. Meteorol. 9, 885-95.
(1972) Dispersion and deposition of corn pollen from experimental sources, Agron. J. 64, 420-7.
(1972) Dispersion and deposition of timothy pollen from experimental sources, Agric. Meteorol. 9, 347-66.

Regener, V.H. and Aldaz, L. (1969) Turbulent transport near the ground as determined from measurements of the ozone flux and the ozone gradient, J. Geophys. Res. 74, 6935-42.

Rich, S., Waggoner, P.E. and Tomlinson, H. (1970) Ozone uptake by bean leaves, Science 169, 79-80.

Schmel, G.A. (1971) Particle diffusivities and deposition velocities over a horizontal smooth surface, J. Colloid. and Interface Sci. 37, 891-906.

Shepherd, J.G. (1974) Measurements of the direct deposition of sulphur dioxide onto grass and water by the profile method, Atmos. Environ. 8, 69-74.

Spedding, D.J. (1969) Uptake of sulphur dioxide by barley leaves at low sulphur dioxide concentrations, Nature 224, 1229-30.

Stevenson, C.M. (1968) An analysis of the chemical composition of rain water and air over the British Isles and Eire for the years 1959-64, Q. J. Roy. Met. Soc. 94, 56-70.

Thom, A.S. (1968) The exchange of momentum, mass and heat between an artificial leaf and the air flow in a wind tunnel, Q. J. Roy. Met. Soc. 94, 44-55.

Thom, A.S. (1972) Momentum, mass and heat exchange of vegetation, Q. J. Roy. Met. Soc. 98, 124-34.

Thomas, M.D., Hendricks, R.H., Collier, T.R. and Hill, G.R. (1943) The utilisation of sulphate and sulphur dioxide for the sulphur nutrition of alfalfa, Plant Physiol. 18, 345-71.

Thorne, L. and Hanson, G.P. (1972) Species differences in rates of vegetal ozone absorption, Environ. Pollut. 3, 303-12.

Turner, N.C., Rich, S. and Waggoner, P.E. (1973) Removal of ozone by soil, J. Environ. Quality 2, 259-63.

Wells, A.C. and Chamberlain, A.C. (1967) Transport of small particles to vertical surfaces, Brit. J. Appl. Phys. $\underline{18}$, 1793-9.

Vogt, K.J., Heinemann, K., Matthes, W., Polster, G., Stoeppler, M. and Angeletti, L. (1973) Untursuchungen zur Ablagerung von elementavem und organisch gebundenem Jod auf Gras, Kernforschungs anlage Julich.

TRANSPORT OF MICRONIC PARTICLES FROM ATMOSPHERE TO FOLIAR SURFACES

Y. BELOT and D. GAUTHIER

Département de Protection, Commissariat à L'Energie Atomique, B.P.n° 6 — 92260 Fontenay aux Roses

Measurements have been made in a wind tunnel of the transport of particles, in the 1-10 μm subrange, to shoots of pine and oak trees. The monodispersed particles were made of fluorescent dye to enable the deposition of small amounts to be detected. The effect of wind speed and particle size was investigated and the microstructure of deposits examined. The results are of interest for the problem of particles deposition on canopies.

INTRODUCTION

The analysis of the transport of particles on canopies of complex structure requires the knowledge of the transfer rates to isolated parts of the constitutive plants. Early works on isolated plant parts are related by ROSINSKI /1/ and LANGER /2/, who emphasized the influence of the reentrainment of particles, once a sufficient area coverage of the leaves has been attained. In the present work, the collection of particles is studied with as low coverage as possible, so as to minimize the reentrainment effect and evaluate the maximum rate of capture. The particles considered here are in the 1-10 μm subrange and are usually referred to as atmospheric dusts. The isolated plant parts exposed to particles were taken from pinus sylvestris and quercus sessiliflora as representative of evergreens and deciduous trees.

In the present experiments pine and oak shoots were mounted in the working section of a closed wind tunnel fed with a monodisperse aerosol of micronic particles. Once the vegetal sample had been exposed, the amount of tracer deposited on the sample was measured and a velocity of capture v calculated. The velocity of capture, v, was taken as the ratio F/χ where F was the flux per square centimetre of foliar surface and χ the mean concentration in air of the working section. This parameter was relevant to plant shoots, and distinct from the velocity of deposition employed for plant canopies by CHAMBERLAIN /3/ and others. From the data thus obtained, relationships were investigated between velocities of capture, size of particles and wind speed; these relationships being needed for a realistic analysis of capture processes within canopies.

EXPERIMENTAL METHODS

The main body of the work was done in a low-speed, closed-return wind tunnel. The working section had a cross-section of 0.30 x 0.30m and a length of 1.30m; the air speed was between $1m\ s^{-1}$ and $10m\ s^{-1}$.

Particles were made to circulate in the tunnel from a spinning-top atomizer /4/, fed with a mixture of uranine and methylene blue in methanol as recommended by WHITBY /5/. The particles thus produced were perfectly spherical, homogeneous and quasi-monodispersed Careful size distribution measurements by light microscopy revealed that the particles were log- normally distributed and that the geometrical standard deviation of particle diameters was about 1.1. As these micronic particles were highly charged, they were neutralized by mixing the charged aerosol with positive and negative small ions generated in the gas phase by ionizing radiation of a Thallium-204 source. There was on the particles a residual equilibrium charge in agreement with the theory /5/.

An experiment of deposition of particles in the wind tunnel went as follows. The plant shoot was mounted in the working section and exposed to the fluorescent particles during 5 to 60 minutes. The shoot and the tunnel were grounded to avoid any undesirable electrical effect. Meanwhile the aerosol flowing past the shoot was continuously sampled by aspiration through a membrane filter under careful isokinetic conditions. The fluorescent tracer deposited on the plant and filter was then determined, by submerging each sample in a buffered ammoniacal solution at pH 9.4, taking an aliquot and measuring its fluorescence. From those measurements, the velocity of capture was then calculated.

EXPERIMENTAL RESULTS AND DISCUSSION

Behaviour of plant samples in the wind tunnel

The selected pine shoots were constituted by rectilinear 20cm twigs and 200-300 sharp needlelike leaves of about 6 x 0.2cm, with total area of 600cm^2. No change was observed with wind speed, in the spatial configuration of the sample. The cross-sectionnal area of the shoot measured from photographs taken in the direction of airflow was found mearly constant, and amounted to about 1/5 of the total area of the leaves; the shoot porosity or void fraction in a plane transverse to the flow was about 1/2. Measurements of the drag force on a typical pine shoot were made with a simple moment balance. From the results obtained, a constant value of C_d=0.7 was found for the drag coefficient defined in the usual way.

The oak shoots, were constituted by a 10cm twig, and five or six flat leaves of about 8 x 5cm, with total area of 400 cm^2. The spatial configuration of the shoot changed with wind speed. At wind speeds less than 150cm s^{-1} the leaves were normal to the flow; for larger velocities the leaves fluttered about an horizontal mean position and the shoot cross-sectionnal area was about 1/6 of the total area of the leaves.

Velocity of capture as a function of wind speed and particle diameter

Measurements of the velocity of capture, v, were made in the wind tunnel at five wind speeds of 50, 100, 200, 500 and 1000 cm s^{-1}, and the experiments were replicated. In each run two shoots were mounted transverse to the flow and arranged side by side in the middle of the tunnel so as not to interfere. v to both samples was measured for particles of 2, 5 and 10 μm diameter.

As already stated, the concentration of particles in the wind tunnel and the duration of experiments were taken so as to minimize the area coverage of the leaves. This coverage C is defined as the ratio of the total cross-sectionnal area of the deposited particles to the cross-sectionnal area of the leaves. This parameter was kept at a low level and its value was under 1 per cent.

The results are shown in figure 1 as a plot of v against u. Each value of v is the mean of two or more values from replicated experiments. The segments drawn at each point give an idea of the data dispersion : the variation between replicated experiments is not considerable. The wind dependency of the velocity of capture can be approximated by the relation :

$$v = a u^3 \qquad (1)$$

This relation is valid for both plant species, at low and moderate wind speeds, which covers most of the situations actually encoun-

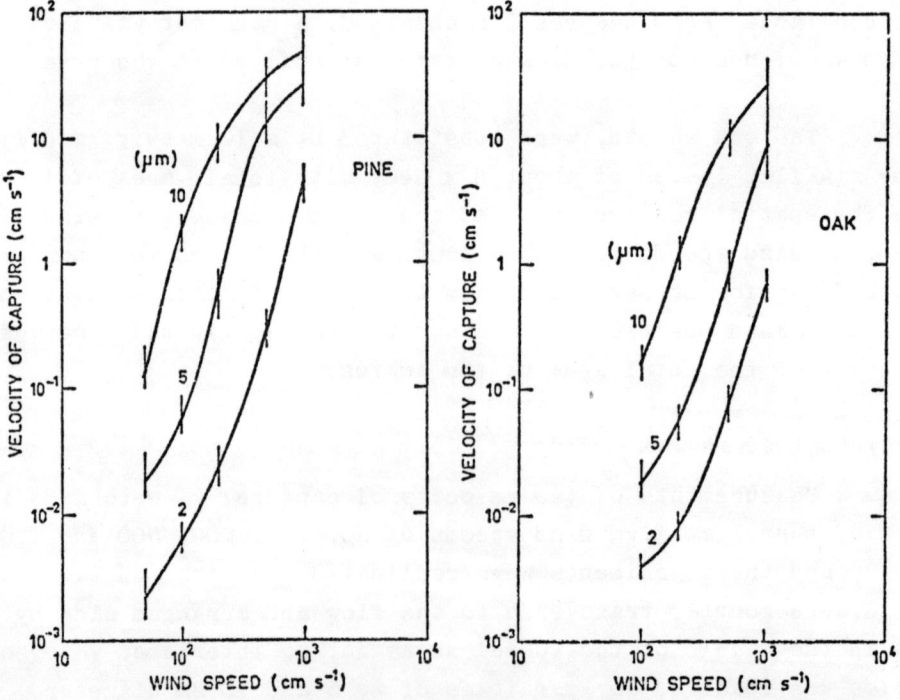

Figure 1. Velocities of capture for micronic particles on pine and oak shoots, as a function of air speed in the wind tunnel. The mass median diameters of particles were exactly 2.0, 4.8 and 9.3 μm.

tered in practice. For high wind speeds (> 5 m s^{-1}), the velocity of deposition on pine shoots is smaller then predicted by (1).

It should be noted that sedimentation has no effect on v, for wind speeds over 100cm s^{-1}. For lower velocities the relation is no longer valid, and v should be replaced by (v-kv$_s$) where v$_s$ is the sedimentation velocity and k a coefficient specific of the plant shoot.

The dependence of v on particle size can similarly be approximated by :

$$v = \alpha \, d^4 \qquad \qquad (2)$$

This relation being valid at low and moderate wind speeds. Such a steep variation of v with d, in the micronic range has abready been reported by DAVIES /6/ /7/ for the deposition of particles to smooth vertical surfaces and by CHAMBERLAIN /3/ for the deposition of particles to grass.

For both shoots, the deposition on the twig itself never exceeds 10% of the total deposition. A similarity car be noticed between the behaviour of the two vegetal species as concerns the particle deposition. But quantitatively the values of v are 1.5 to 10 times as great in the case of pine shoots. Possible reasons for the difference can be looked for in the size and shape of the leaves.

Collection efficiency of particles on plant shoots

The leaves can be compared to ribbons or plates, and the twig to a cylindrical rod. On such obstacles the particle capture is attributed to inertia and interception effects. In the first mechanism the trajectory of the particle centre intercepts the obstacle; in the second mechanism the trajectory does not meet the obstacle and the particle deposition is governed by the particle size and the structure of the boundary layer as described by DAVIES /8/.

The collection efficiency of an obstacle is usually defined as the ratio of the number of particles deposited on the obstacle to the number of particles passed through the cross-section of the obstacle. For a plant shoot, this parameter can be defined as $E=(v/u)(a/\sigma)$, where v is the velocity of capture, u the wind speed, a the total area of leaf and σ the cross-sectionnal area of the shoot.

The inertia parameter for the capture of a particle is defined by $S = (\rho' u d^2) / (18 \rho \nu L)$ where ρ et ρ' are the specific mass or air and particle, d the particle diameter. ν the kinematic viscosity and L a characteristic length taken as the smallest dimension of a single leaf, when projected on a plane normal to the flow.

In figure 2 the collection efficiency, E, has been given against the inertia parameter, S, and the experimental data agree with the dimensionless relation.

$$E = A S^2 \quad \ldots\ldots\ldots\ldots\ldots\ldots\ldots\ldots\ldots\ldots \quad (3)$$

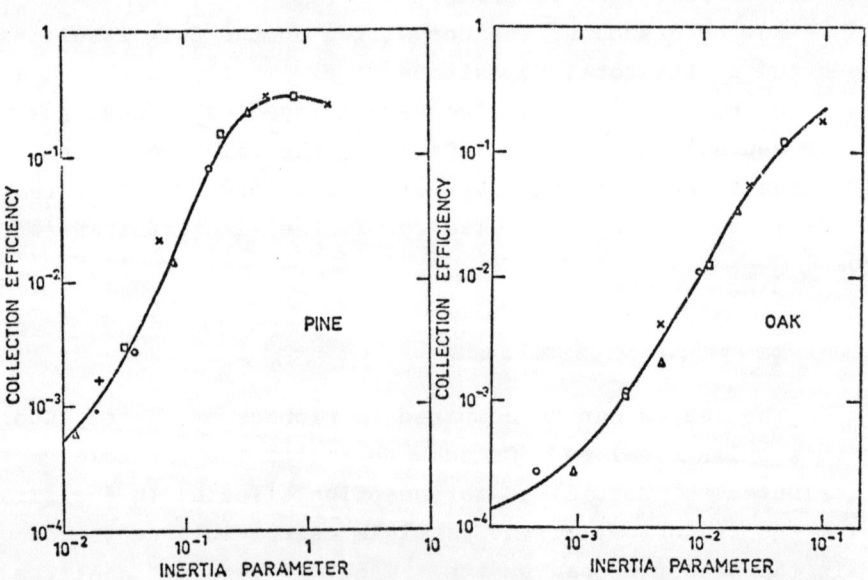

Figure 2. Collection efficiency of pine and oak shoots. The characteristic length is 0.2cm for the pine shoot, and 3cm for the oak shoot.

Valid at low inertia parameter for both vegetal species.
In the case of pine shoots a saturation effect is observed, when the flux of particles captured by the shoot is equal to the flux passing through the cross-section of the obstacle.
E does not seem to be dependent on the ratio d/L, as claimed by the theory of interception /8/. The exact reason of this discrepancy is not yet very clear.

Microstructure of deposits

The deposits of particle on the leafs were examined by light and scanning electron microscopy. There was a tendency to the development of small flocklike agglomerates of particles, as already reported by ROSINSKI /1/ and LANGER /2/. Once a particle aggregate was formed, it become large enough to act as an independent collector and enlarged further. These aggregates seemed to be located by preference on the roughness microelements of the leaf surface. As the area leaf coverage was quite low, the largest aggregates were of 100 µm diameter, and most of the time they were much smaller and consisted of 3 ou 4 particles. The explanation of this aspect is likely to be found in the symmetrical residual charges on the particles of the aerosol, and on the subsequent mutual attraction.

In the present experiments the aggregates, of moderate size, were not removed by the airflow itself. But is was necessary to manipulate the shoots with great care, so as not to dislodge aggregates. If adequate manipulation was not observed, small fragments of the deposit were lost. This observation throws light on the mechanism of particle reentrainment from plants. Individual particles of less than about 50 µm diameter are not readily removed from surfaces because they are imbedded within the viscous boundary layer of air flow /3//9/. On the contrary aggregates are easily removed by friction of leaves, vibration etc..., as a result of their size; and they are transported to the ground.

When considering the cleansing of the atmosphere by stands of plants it must be taken account of the particles deposited on the plants and also of the particles dislodged to the ground. A meager deposit on the leaves does not mean that the vegetation is inefficient in the cleansing of atmosphere. In fact most of the particles may be dislodged to the ground as aggregates. This assumption could be verified by measuring the ground dry sedimentation under a cover and within a proximate open area.

CONCLUSIONS

An advance has been made in the following points :

(a) For particles between 1 and 10 μm, the velocity of capture on plant shoots, rapidly increases with the wind speed and the particle diameter.

(b) Plant shoots behave like an array of elemental obstacles of a simple geometrical shape. Elements of a fibrous shape are likely to be more efficient that those of a bluff shape.

(c) The deposits on leaves have a tendency to form aggregates, and these aggregates are easily dislodged by friction of leaves or vibration and are likely transported to the ground by sedimentation.

An attempt is now being made to apply the results of the present laboratory experiments to the estimation of the velocity of deposition on canopies, from velocities of capture on corresponding shoots. This work based on a layer-by-layer diffusion model is under completion.

REFERENCES

/1/ ROSINSKI J. and NAGAMOTO C.T.
Kolloïd Zeitschrift 1965, 204, 111-119.

/2/ LANGER G.
Kolloïd Zeitschrift 1965, 204, 119-124.

/3/ CHAMBERLAIN A.C.
Proc. Roy. Soc. 1967, 296 A, 45-70.

/4/ MAY K.R.
   J. Sci. Instrum. 1966, 43, 841-842.

/5/ WHITBY K.T., LUNDGREN D.A. and PETERSON C.M.
   Air and water Pollution 1965, 9, 263-277.

/6/ DAVIES C.N.
   Ann. Occup. Hyg. 1965, 8, 239.

/7/ DAVIES C.N.
   Proc. Roy. Soc. 1966, 289 A, 235-246.

/8/ DAVIES C.N. and PEETZ C.V.
   Proc. Roy. Soc. 1956, 234 A, 269-295.

/9/ BAGNOLD R.A.
   Air and water Pollution 1960, 2, 357.

INDEX

Aerodynamics of vegetated surfaces, 139
Anemometry, use in determination of transfer, 337
Anti-transpirants, use of, 489
Atmosphere exchange processes, 299

Canopies, flow within, 157
Canopies, global equation for, 235
Canopies, heat and mass transfer within, 167
Canopies, theories for leaf positioning, 191
Canopy parameters, 237
Capillary surfaces, 128
Computer modeling of soils, 97
Condensation, in thin capillaries, 125

Desert, microclimatic modeling of, 275
Diffusion, in thin capillaries, 133
Diffusion porometry, 403
Displacement height, 150
Drainage, 449

Eddy-correlation method, 337
Electrochemical analog of transfer in vegetation, 265
Energy and agriculture, a study, 513
Energy and mass transfer, in plant communities, 427
Energy and mass transfer, in vegetation, 265
Energy and mass transfer, of native grassland, 311
Evapotranspiration 207, 450
Evaporation, in thin capillaries, 125
Evaporation, natural, general principles of, 207

Flow structure, 140
Fokker-Planck equations, solution of, 38
Forest canopies, models of, 254
Forest canopies, simulation of flow, above, 251
Frost, penetration in soils, 17

Grassland, energy and mass exchange in, 311
Ground-water pollution, prediction of, 549

Heat and mass transfer, from real and model leaves, 413
Heat and mass transfer, in lower atmosphere, 229
Heat and mass transfer, in plant canopies, 229
Heat and mass transfer, in plant communities, 232
Heat and mass transfer, simultaneous, 87
Heat budget, introduced by hedges, 457
Heat budget, model of, 457
Heat flux, in soils, 69

Hedges, effect on heat budget, 457
Hydraulic conductivity, of unsaturated and nonswelling soils, 32
Hydrologic cycle, 31

Infrared radiant flux, measurement and estimation of, 345

Land roughness, 467
Land roughness, modification of, 467
Laser-Doppler anemometry, use in determination of flow, 353
Leaf parameters, 237
Leaves, epidermal resistance of, 403
Leaves, heat and mass transfer from, 413
Leaves, models of, 413
Microclimatic effects of land roughness, 467
Microclimatic modeling, 275
Micrometeorological models, 277
Microscopic and macroscopic approaches, 8
Moisture capacity, 111
Moisture conductivity, 111
Momentum flow, analysis of, 287
Momentum, heat and mass transfer, determination by hot-wire anemometry, 337

Particle transport to foliar surfaces, 583
Phyto-engineering, 425
Pine stand, 287
Plant canopies, pollution in, 561
Plant canopies, radiation exchange in, 327
Plant communities, energy and mass transfer in, 427
Plant environment, pollution in, 535
Plant physiology, 200
Plants, water transfer in, 369
Pollution, in plant canopies, 561
Porous pipes, heated 87

Radiation, above canopy, 190
Radiation, interaction with leaves, 188
Radiation balance, modification of, 479
Radiative transfer in vegetation, 187
Reflectants, for modification of radiation balance, 479

Seeds, water transfer to germinating, 501
Seed-water contact impedance, 501
Shelter belts, 150

Soil environment, natural, leat and mass transfer in, 65
Soil-plant-atmosphere continuum, 444
Soil pollution, prediction of, 549
Soils, biological processes in, 537
Soils, chemical reactions in, 537
Soils, combined heat and moisture transfer in, 18, 21
Soils, diffusion in, 38
Soils, flow, in unsaturated nonswelling, 36
Soils, heat conduction in, 4
Soils, heat flux in, 13
Soils, heat and moisture transfer in, 5
Soils, heat transfer in, 5
Soils, heat transfer in, mechanism, 16
Soils, hydraulic properties of, 501
Soils, moist, computer modeling of, 97
Soils, moisture potential of water, in nonswelling, 34
Soils, temperature variation in, 13
Soils, thermal behavior of, simulation, 109
Soils, thermal conductivity of, 10
Soils, thermal diffusivity in, 15
Soils, thermal properties of, 8
Soils, total potential of water in nonswelling, 34
Soils, volumetric heat capacity of, 9
Soils, volumetric water content in, 11
Soils, water movement in, 29

Soils, water movement in, one-dimensional, 42
Soil water, 31
Soil water, disjoining pressure of, 51
Soil water flux, 69
Soil water, osmotic pressure of, 51
Soil water, rheological properties of, 49
Soil water, thermodynamic properties of, 49
Stomatal physiology, 214
Surface phenomena, in thin capillaries, 125
Swelling soil, infiltration into, 77

Transfer, irreversible, in soils, 61

Vapor movement in soils, 19
Vegetated surfaces, aerodynamics of, 139
Vegetation, radiative transfer in, 187
Vegetation, water uptake by, 443
Vegetative canopy, exchange processes within, 299

Water, consumption of in eco-systems, 489
Water transfer, 214
Water transfer, in plants, 369, 395
Water transfer, in wheat, 395
Water uptake, by vegetation, 443
Water use efficiency, 479
Wind profile, in canopies, 235

This book is due on the last date stamped below. Fines will be charged on all overdue books.

NOV 22 '77

FEB 25 '85

INTERLIBRARY LOAN

Howard-Tilton Memorial Library